Aerospace Sensor Systems and Applications

Springer
New York
Berlin
Heidelberg
Barcelona
Budapest
Hong Kong
London
Milan
Paris
Santa Clara
Singapore
Tokyo

Shmuel Merhav

Aerospace Sensor Systems and Applications

With 193 Figures

Springer

Shmuel Merhav
Department of Aerospace Engineering
Technion
Israel Institute of Technology
Haifa, 32000 Israel

Library of Congress Cataloging-in-Publication Data
Merhav, Shmuel.
Aerospace sensor systems and applications / Shmuel Merhav.
p. cm.
Includes bibliographical references and index.
ISBN 0-387-94605-5 (hardcover)
1. Avionics. 2. Detectors. 3. Aeronautical instruments.
4. Space vehicles – Guidance systems. I. Title.
TL695.M47 1996
629.135 – dc20 95-51161

Printed on acid-free paper.

Production managed by Hal Henglein; manufacturing supervised by Joe Quatela.
Camera-ready copy supplied by the author.
Printed and bound by Braun-Brumfield, Inc., Ann Arbor, MI.
Printed in the United States of America.

9 8 7 6 5 4 3 2 1

ISBN 0-387-94605-5 Springer-Verlag New York Berlin Heidelberg SPIN 10515995

To Naomi

Preface

This book is about aerospace sensors, their principles of operation, and their typical advantages, shortcomings, and vulnerabilities. They are described in the framework of the subsystems where they function and in accordance with the flight mission they are designed to serve. The book is intended for students at the advanced undergraduate or graduate level and for research engineers who need to acquire this kind of knowledge. An effort has been made to explain, within a uniform framework of mathematical modeling, the physics upon which a certain sensor concept is based, its construction, its dynamics, and its error sources and their corresponding mathematical models. Equipped with such knowledge and understanding, the student or research engineer should be able to get involved in research and development activities of guidance, control, and navigation systems and to contribute to the initiation of novel ideas in the aerospace sensor field. As a designer and systems engineer, he should be able to correctly interpret the various items in a technical data list and thus to interact intelligently with manufacturers' representatives and other members of an R&D team.

Much of the text has evolved from undergraduate and graduate courses given by the author during the past seventeen years at the Department of Aerospace Engineering at the Technion– Israel Institute of Technology and from his earlier research and development experience in flight control, guidance, navigation, and avionics at the Ministry of Defense Central Research Institute.

An effort has been made to create a text that is as self-contained as possible. The professional background of potential students for the sensor systems area may be aerospace, mechanical, or electrical engineering. Accordingly, the different levels of students' prior knowledge may require different approaches in introducing this interdisciplinary material. In order to minimize this potential difficulty, Chapters 1 and 2 are introductory. Using elementary linear systems theory and random processes, Chapter 1 introduces the principal types of aerospace sensors in generic form. Their common properties are explained in regard to their dynamic response to "legitimate" inputs, environmental interferences, and noise, which in Chapter 1 is not yet statistically defined. Chapter 2 is an introduction to random processes that provides the basis for statistical sensor error modeling and for the operations of filtering and estimation. In many cases, these have become integral functions of the sensor system. The material covered in Chapter 2 also paves the way to the operations of smoothing and prediction and to the mutual aiding of sensors by means of complementary and Kalman filtering.

The present time is a period of transition in the field of sensor technology. In addition to classical sensors like the vertical, directional, and rate gyros or analog accelerometers, all of which date back to the beginning of the century, new developments in the past thirty years have yielded optical "gyros" and quartz or silicon vibrating beam accelerometers. These are gradually gaining more ground in modern sensor systems. Sooner or later, these technologies are bound to replace the traditional ones. It appears, however, that they will coexist side by side for quite a few years to come. The reason for this is that apart from the continuing stringent sensor requirements for high-performance aircraft and spacecraft, where cost is relatively immaterial, there is a growing need for low-cost sensors in remotely piloted vehicles (RPVs) and tactical missiles. In these, the traditional sensors often "do the job" at a reasonable cost while the modern technologies, such as quartz or silicon have not yet sufficiently matured or are still too costly. For this reason, Chapter 3 is devoted in part to traditional analog accelerometers and Chapter 4 to electromechanical gyros. Although a vast literature on these topics is available, they are presented here in a manner that is particularly useful for the system engineering level. Combining skeleton 3-D drawings of all the functional parts and components along with the corresponding notations of variables and parameters facilitates the formulation of the mathematical models that provide the responses to the "legitimate" inputs and to noise and environmental interferences. An important class of errors due to imperfections in electromechanical sensors, such as anisoelasticity, vibro-pendulosity, coning, sculling, and anisoinertia, is due to the interaction of vehicular vibrations with the mechanical imperfections. These have been a major issue, especially since the advent of strapdown technology where the sensors are fully exposed to external vibrations. These errors are not treated in this book for the following reasons: (i) They have been amply treated in numerous texts on inertial navigation, such as G.R. Macomber and M. Fernandez (*Inertial Guidance Engineering*, 1962), and more recently P. Savage (*Strapdown Inertial Navigation Notes*, 1990). (ii) With the recent maturing of laser gyros and their accelerating application in inertial navigation and flight control, these effects are rapidly becoming less relevant. (iii) In low-cost flight control and guidance applications, their effect is small in comparison with other error sources.

Only a modest selection of the most important rotation and force sensors is discussed. Among them are the traditional vertical (VG), directional (DG), and rate gyros (RG), the floated rate integrating gyro (RIG), and the dry tuned rotor gyro (DTG). The gas bearing and electrostatically supported gyro (ESG) are also described as outstanding engineering achievements. Chapter 5 presents a number of important applications of gyroscopic technology. Among them are the two-axis gyro stabilized platform, which is a common payload in remotely piloted vehicles (RPVs) and guided missiles, the spinning wheel seeker heads for infrared homing missiles, ground-based optical seeker heads used in beam riding missile guidance, and the gyro-stabilized three-axis platform for INS and models for its error propagation for different sensor noise statistics. Chapter 6 introduces Coriolis angular rate sensors that use rotating or vibrating accelerometers to provide measurement of inertial angular rate, sometimes along with specific force sensing implemented by quartz or silicon technology. Due to their miniature dimensions, Coriolis inertial sensors are already applied in low-cost flight control systems and are regarded as strong contenders in inertial navigation, at least for tactical applications. Chapters 7 and 8 are devoted to laser gyros. In Chapter 7, the passive interferometric fiber-optic gyro (IFOG) is presented. Its mathematical model is developed on the basis of the Sagnac effect. Performance limits due to shot noise and other effects are also discussed. The need for its implementation as a force balance sensor is explained. Recent results demonstrating INS grade performance achieved by the leading industries in the field are presented in evidence of their establishment as principal flight control and high-grade navigation sensor technologies. Chapter 8 is devoted to the active ring laser gyros (RLG). Their principle of operation and analytical model as an extremely linear and precise open-loop sensor are presented as an almost ideal incremental rotation sensor, which perfectly and naturally interfaces with digital signal processors and navigation computers. The most recent successful achievements in zero lock RLGs are also presented. Chapter 9 is devoted to mutual aiding of sensors which is also known as *measurement fusion*. The methodology is a special form of Kalman filtering basically configured by the complementary filter constraint. The application is almost exclusively to linear systems, although its effectiveness in nonlinear problems is also demonstrated. In this Chapter, examples are shown that range from applications of simple linear time invariant systems to nonlinear time-varying systems requiring extended Kalman filtering. The power of the concept is demonstrated by the exceptionally good performance which can be achieved by quite crude low-cost sensors if they are aided by other sensors that have complementary properties.

Throughout the text, the emphasis in the presentation has been to provide physical insight, not necessarily formalism and rigor.

The material is enhanced by examples, problems, and numerous illustrations. In them, label fonts are in *italics* if they refer to variables or parameters and in **Helvetica** if they refer to components or parts. Sections that are either highly specialized or are not essential to the principal topics are marked by the symbol ⊕. Their omission should not hamper the understanding and digestion of the main material presented.

Acknowledgments

It is a pleasure to express my appreciation and thanks for the support of colleagues, friends, and organizations throughout the preparation of this manuscript.

I thank Professor Dan DeBra of Stanford University, Professor Bernard Friedland of New Jersey Institute of Technology, and Professor W.M. Hollister of MIT for advocating the prospectus, and for their encouragement and enlightening discussions. My thanks also go to Dr. P.K. Menon, who read part of an early version of the manuscript, for his constructive comments and suggestions while, during a sabbatical leave, we shared an office at NASA Ames Research Center in California.

I have been fortunate to be associated with Sundstrand Data Control in Redmond, WA (now AlliedSignal) in the research and development effort of novel Coriolis multi-sensor concepts and to get to know the Instrument Division Chief, Norman Klein, to whom I am indebted for his support for my cooperation with such gifted research scientists as Rex Peters, Rand Hulsing, and Brian Norling, who provided valuable technical information and enlightened me with numerous insights which are usually unavailable in the open literature.

I am indebted to Mr. Gershon Engel of the Israel Aircraft Precision Instrument Industries for providing me with detailed and valuable material on dynamically tuned gyros, and Professor Shaoul Ezekiel of MIT who brought me up to date on interferometric optical gyros. I am also indebted to Professor I. Bar-Itzhack of the Technion Aerospace Department, who acquainted me with some of the subtleties and intricacies in analytical platform algorithms, and to Professor M. Guelman and Dr. J. Oshman, who read parts of the manuscript and whose comments are greatly appreciated. Last, but not least, I am indebted to my former graduate students, particularly to Dr. Jacob Reiner, Dr. M. Velger, and Mr. M. Koifman, whose work is reflected in some specialized sections of this book.

Shmuel Merhav

Haifa, Israel

Contents

Introduction and historical background

The purpose of aerospace sensors is to provide all the necessary information on the *variables and parameters* of the flight system for the execution of the mission at hand. A *mission* is generally defined in three dimensional space. Sometimes, on adding the fourth 'dimension', namely, time, a four-dimensional space is defined. In contrast to the three spatial coordinates which can increase or decrease, time always grows. The recent concept of 4-D guidance has special significance in coordinating flight paths such as in air traffic control or in space rendezvous operations, where the notions of 'here-now' or 'there-then' assume paramount importance.

As compared to the first generation aircraft at the turn of the century, which were equipped with a compass, a barometric altimeter, and a relative air speed sensor, a modern military aircraft or airliner is equipped with hundreds of sensors at all levels of the flight system. As such, they have become an integral part of it, not only in regular operation but also in preflight procedures for maintenance and safety. It is paradoxical that by introducing components and subsystems with more sophisticated technologies and a myriad of sensors for monitoring their performance and safety, a situation of diminishing returns may be created. This may be so, because in view of their large numbers, a failure in any of them easily causes a mission abort even if an imminent catastrophe is not necessarily indicated. It is therefore not surprising that along with the ever-increasing demands for accuracy and precision, research and development efforts continue to seek improved technologies for better reliability and maintainability. Currently, one already occasionally encounters the term 'unlimited' for the item MTBF (mean time between failures) in the data sheet of an aerospace sensor.

Modern aerospace sensors are direct descendants of the discoveries of physical laws in theoretical and experimental physics in the seventeenth and eighteenth centuries followed by the avalanche of revolutionary inventions in the nineteenth and early twentieth centuries.

At this point, it is enough to recall only a few of the most salient milestones in order to appreciate the direct link between the late twentieth century technology of aerospace sensors and systems and the discoveries in physics and the inventions that followed. Thus, for example, Torricelli's discovery of atmospheric pressure in 1643 was quickly followed by Pascal's invention of the barometric altimeter in 1647 and consequently by his discovery during 1647–1654 of the properties of our gaseous atmosphere. Newton's discovery of the law of gravitation in 1687, his first law on the conservation of momentum, and his second law, that force equals the rate of change of momentum, laid the foundation for the much later development of the accelerometer at the turn of this century following the invention of the seismometer by Milner in 1883. Other principal examples are the early invention of a steam-powered vertical gyroscope by Hiram Stevens Maxim in 1891, later perfected as an electromechanical device by Sperry in 1913. The development of the theory of electromagnetic wave propagation by Clarke Maxwell in 1864 and the invention of the spark transmitter by Marconi in 1903, followed by the invention of the thermionic vacuum tube by Lee de Forest in 1907, paved the way to radio transmission and reception. Faraday's discovery of the laws of electromagnetism very rapidly led to the invention of the electromagnet, the electric generator and the motor, the foremost transducers in modern technology. The discovery of semiconductors in the late 1930s led to the invention of the transistor by Shockley in 1947, revolutionizing electronic technology in speed of performance, size, ruggedness, low power, and voltage and paving the way to the wide spread introduction of airborne electronic communications, control and navigation systems in aircraft, spacecraft and missiles. With the advent of *very large scale integration* (VLSI), the barriers to miniature packaging of computers, communication, flight control and navigation systems have been and still are being pushed back, while spurring research and development of new concepts and technologies of sophisticated sensor systems. These often require intelligent filtering, estimation, and signal processing algorithms for guidance, navigation, avionics, and airborne computer vision.

By and large, aerospace sensors can be categorized into two major classes: The first class is *geophysical*, namely, sensors which use physical properties of the Earth, its dimensions, its known rate of rotation around its axis and around the sun, its gravitational pull, its atmosphere, and its magnetic and electrostatic fields. Examples are the magnetic compass, the gyro-compass, and terrestrial inertial navigation systems. The second class, particularly in the category of remote sensing, is based on the utilization of radiation. Its source may be either natural, such as the sun, the stars, or infrared dispersed radiation of objects on the surface of the Earth. Examples are, sun-sensors for the determination of the angular position of the sun, star trackers, infrared cameras or man-made radiation such as radio and radar transmission from Earth-based or satellite-based transmitters. In many cases, the basic physical limitations due to noise, limited dynamic range, or insufficient frequency response render a particular type of sensor deficient in a particular aspect, so that on its own it would not be suitable for the task at hand. However, another type of sensor may have complementary properties, i. e., it is superior in the aspect in which the first is inferior. The blending of the measurements of two such sensors creates a *multisensor system*, or an *aided sensor system*. This blending of measurements is accomplished by appropriate filtering algorithms, providing a combined measurement that can be by far superior to those of the individual sensors.

Measurands in systems fall into two categories, namely variables and parameters. *Variables* relate to the state of the systems, described for example, by position, velocity, or orientation, while *parameters* typically relate to vehicular dimensions or the environment, such as pressure or temperature. Variables are often vectors that may vary quite rapidly, while parameters are scalars, which typically undergo slow changes. However, it is not always easy to draw a definite line distinguishing between a variable and a parameter. For example, atmospheric pressure definitely plays the role of an environmental parameter, but it is also directly linked to barometric altitude which can be viewed as a system variable.

Any single sensor cannot distinguish between a scalar and a vector because, at any given time, its output is represented by a number. Since vectors imply direction in three dimensional space, a spatial coordinate system must be set up in which at least three suitably oriented sensors are aligned with its axes. Thus, only the combined readings of such sensor *triads* establish the measurement of a vector. A typical example is a magnetometer, which consists of an orthogonal triad of magnetic sensors for establishing the magnitude and direction of the local magnetic field with respect to the chosen coordinate system. The readings of each of the three sensors, which in this example are, the components of the magnetic field vector, are often required in a coordinate system that differs in orientation from the first one. Thus, a suitable *coordinate transformation* is called for. For its implementation, another triad of angular measurements is required. Thus, in general, the determination of vector quantities in three-dimensional space requires at least six measurements.

As a rule, three principal categories of precision in measurement requirements in terms of percentage of full-scale values are distinguished according to the task definition. These relate to the autonomous functions of vehicle *attitude*, vehicle *guidance* and vehicle *navigation*. Attitude measurements, at least, in atmospheric flight applications are mostly in the range of 10^{-3} to 10^{-2}. Guidance measurements often require accuracies on the order of 10^{-4}, while long-term navigation may require accuracies as high as 10^{-8} in strapdown mechanizations. These precision categories were attained in first generation (1910–1940), second generation (1940–1960), and third-generation (1960–1990) inertial sensors, respectively. An excellent historical sociological account on the enormous impact of high-precision inertial grade sensor technology on national military and global policy is given in a recent book, titled *Inventing Accuracy* by MacKenzie (MIT Press ,1990). In this book the author describes modern sensor technology as a key issue in the race of the superpowers toward worldwide supremacy. The far-reaching and profound intellectual and scientific effort involved led him to label the relevant sensor technology the *Gyro Culture*.

Throughout the first half of this century and earlier, measurements were exclusively within the analog world, that is, all physical quantities were represented by displayable media such as voltages, currents, or sometimes by mechanical displacements. The extreme precision required in ballistic missile guidance, autonomous navigation of aircraft, and the associated sophisticated computations and with the advent of high-speed digital computers capable of real-time processing, brought about the gradual conversion from analog technology to all-out digital technology. Since physical quantities are analog by nature, this conversion process initially involved the development of analog-to-digital converters. Being themselves a source of errors, these converters are being superseded by sensors, such as laser gyros and vibrating beam accelerometers, which represent the measured quantities by frequency variations. These naturally interface with the required digital-processing functions. The common observation that a modern aircraft or spacecraft is, in effect, an airborne instrumentation and computation center, to a large extent expresses this reality.

It is the purpose of this book to explain the functions of aerospace sensors in the framework of the system and its mission and to study their performance and imperfections not just as isolated devices, but rather in the framework of their dynamical and environmental interactions.

1

Principles and Elements of Measurement Systems

1.0 Introduction

This Chapter introduces the basic principles that underlie measurement devices and systems and the essential functions of their elements. The devices and their interconnections are described in generic form. Thus, a common conceptual framework is provided for a variety of existing aerospace sensors. The underlying requirements for standards and reference values are introduced, and the principal classes of measurement devices are defined, including passive, active, contact, remote, deflection, and balance. Characteristics, imperfections, and principal error sources are also defined.

The purpose of a device or instrument is to provide an output that represents, in a useful way, the magnitude of a physical quantity. The term useful implies that the quality of the representation is compatible with system requirements and constraints such as accuracy, precision, response time, repeatability, reliability, cost, volume, and weight. It is the knowledge and understanding of the parameters and properties underlying performance and tradeoffs that provide the designer with the skills to make the most rational choice of sensors or, if necessary, to specify new designs within the state of the art or to initiate new development.

1.1 Elements in open–loop instruments

Instruments, sensors, and systems

By *open–loop* devices, we mean that their output has no effect on their input. A number of elements and components purposefully interconnected constitute a *measurement device* or *instrument*, is often referred to as a *sensor*. These terms are essentially synonyms, and their usage depends on the context. For example, an altimeter is sometimes referred to as a sensor and sometimes as an instrument. The preferred term is related to the nature of its usage. If the altimeter functions in direct association with the pilot, it will, as a rule, be viewed as an *instrument* and as such it might be one of many on the instrument panel in the aircraft cockpit. However, if it is part of an automatic *altitude–hold* system, it will be referred to as a *sensor*. In general, any measurement device commonly intended for manipulation by a human operator, however simple or complex, will be referred to as an instrument.

Thus, a simple bubble level in the hands of a mason is an instrument just as a signal analyzer is an instrument in the hands of a researcher in the laboratory. Scale of complexity and sophistication also determine proper use of terminology. For example, a Radar, although just a range measurement device, is referred to as a *system* or *subsystem* in a broader context but not as an instrument or sensor. It is probably appropriate to view a system on the scale of a radar, as an organized system of numerous sensors and devices that in their own right and in a different context of application might be just instruments, such as a radar receiver, power amplifier, or oscillator.

Basic sensor elements

A sensor consists of three principal elements, which constitute its core. These are:

1. the sensitive element,
2. the source of the balancing force, and
3. the output device or display.

The sensitive element is chosen so that it responds to the physical medium to be measured. It translates the response into a measurable or displayable quantity and it generally manifests itself as a force, torque, current or voltage. In the thermometer, this force is the thermal expansion force. The balancing force is provided by capillary tension, and the display is the scaled thin glass tube. In the case of very simple sensors, some of these elements coalesce into one element that performs more than one of the three basic functions above. Thus, in the simple example of the thermometer, the sensitive element is the mercury, the source of the balancing force is the thin capillary tube, and the output device, or display, is the very same tube with its temperature scale marked on it. In general, many elements in addition to the three fundamental ones are required in a measurement device.

Auxiliary functions and elements

Often, the sensitive element does not provide sufficient output for the display, so that appropriate amplification is required. In order to accomplish this, it is often required to attach a *detector* to the sensitive element. The role of the detector is to convert the small response of the sensitive element into an electrically amplifiable signal. Thus, two common additional elements are a *detector* and an *amplifier*. The amplification, however, is not necessarily electrical. Turning once more to the mercury thermometer, the minute expansion of the mercury is made displayable by the very small inner diameter of the glass tube. In order to accomplish the display function, or, in other cases, the *balancing* function, it is required to convert the amplified signal into a force or torque. This function is typically performed by an electromechanical *transducer*. The output of the measurement device may be a *display* designed for observation by a human operator and it may be any of a large variety of formats and techniques.

In many cases, especially if the device output is a mechanical rotational or translatory deflection, it must be converted into a measurable, mostly electrical, signal to facilitate its interconnection with other subsystems. This conversion is accomplished by means of a pickoff, such as potentiometer, a linear, or rotational *variable differential transformer*. Typically, neither the sensitive element, nor the detector are linearly related to the input, that is, to the measured quantity. Therefore, at least in open loop measurement devices, linearization is required.

This linearization is accomplished by a fixed program that generates the mathematical inverse of the nonlinear input–output relationship. Furthermore, the output is not necessarily zero when the input is zero, thus causing a *null-point offset*. The function of eliminating this offset is known as *output compensation*. The establishment of, or correction to, a desirable constant of proportionality between input and output is known as *calibration*.

Often, the display is in a different location from the measurement device itself. Consequently, the information must be transmitted either by wires or radio signals, often found in airborne telemetry, underwater operations, or in large-scale installations such as power stations, airports, or spacecraft launching sites. A diagram describing in generic form a measurement device comprising these elements is shown in Figure 1.1.

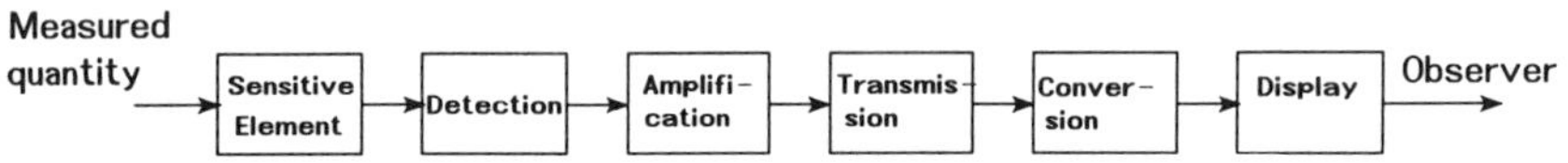

Figure 1.1. Generic form of a measurement system.

Equilibrium

A common underlying principle in all measurement devices is the *static equilibrium* of opposing forces. In a spring balance, the spring opposing the weight provides the equilibrium. In the scale balance, it is the equality of opposing torques around the suspension point, and in a freely suspended pendulous accelerometer, it is the torque due to gravity that balances the inertial torque.

Clearly, every measurement should be performed only after static equilibrium has been reached. Every measurement device is subject to inertia, which may be mechanical, thermal, electrical, or of another nature. It may take quite some time before static equilibrium is reached. For example, the plumb in building construction may swing for minutes before settling to the vertical equilibrium. In antiquity, measurements were essentially *static* because the measured quantities were practically constants, such as the vertical, the horizontal, or the north. Under these static conditions, the inertia is not a critical factor, because the measurement can almost always tolerate the necessary time lag to reach equilibrium. In modern technology, measurements are often required to determine time varying quantities rapidly. Accordingly, the inertia of sensors must be sufficiently small to prevent the measurement from lagging behind the physical variable. Devices can now be summarized by the following definitions, which provide the meanings of the terminology used in this text.

Definitions of sensor functions

1—The SENSITIVE ELEMENT has the primary role of translating a change in the measured physical variable or environmental parameters into another readily measurable quantity, such as a linear or rotational displacement, a voltage, or a current (deformation of a membrane under pressure, displacement of the proof mass in an accelerometer, change in the output voltage of a bolometer under changes of temperature).

2—The DETECTOR is a sensitive element with the secondary role of translating changes in the response of the primary sensitive element into electrical signals. It is characteristically associated with small changes provided by the primary sensitive element (a capacitive element attached to the proof mass in a force-balanced accelerometer, a piezoelectric element in a load cell, a strain gauge in a wind tunnel balance).

3—The AMPLIFIER is commonly electronic and is required for amplification of voltage or current. Its role is primarily to bring detected faint signals to a sufficient level of power or voltage to drive displays, or to be further conditioned for digitization and coding. If the output of the sensitive element is a mechanical displacement or a fluid flow, amplification is often achieved by mechanical levers or suitable tubing.

4—The TRANSDUCER is a device actuated by power from one system or energy source, supplying power in the same form or any other form to a second system, such as a loudspeaker actuated by electric power and supplying acoustic power, or a quartz crystal that produces electric power from mechanical input, such as pressure. In the specific context of sensors, the transducer primarily has the role of converting electrical power into mechanical power (for example, an electrical motor or, in the role of torquers, or forcers to be applied for angular or linear displacement or to exert force or torque).

5—The PICKOFF is an output device mostly referring to a linear and precise translation of substantially large mechanical linear or rotational displacements into electrical signals or code (potentiometers, synchros, resolvers, magnetic or optical shaft encoders in orientation gyroscopes, stabilized platforms, linear or rotational variable differential transformers (LVDT or RVDT).

6—The DISPLAY is the output device required in conjunction with the tasks of a human operator in monitoring and control. In technical jargon, the display device is often called the *monitor*. It represents the measured variables or parameters in the most meaningful form in the context of the objective of the control or supervision task. Therefore, displays take on a large variety of formats: analog scales, alphanumeric, monochrome, or color.

1.2 Measures and units

Basic measures

All measurements must be assigned units. These units are defined in terms of basic measures. There are four basic measures of an absolute nature, namely, *length*, *mass*, *time*, and *temperature*; these are denoted by l, m, t, and T respectively. All other measures in physics and engineering derive from these.

Units and standards

The common units in physics for these measures are the *centimeter*, *gram* (mass), *second*, and *degree Kelvin*, respectively. This set of units is known as the CGS system. In engineering, the units are *meter*, *kilogram* (force) or *Newton*, *second* (known as the MKS system), and *degree Celsius*. Thus, for example, acceleration has the engineering unit (m sec^{-2}). With the advent of modern physics, the absolute values of the standards, which were previously based on geophysical and astronomical constants, have been redefined in terms of basic atomic constants. In 1960 the meter was redefined as 1,650,763.73 wavelengths in vacuum of a Krypton-86 lamp. The *second* was redefined in 1964 in terms of the frequencies of atomic resonators. It is defined as the period corresponding to the atomic resonant frequency of Cesium 133. Mass is the property of an object known as inertia or the force of resistance to a change of momentum. It is measurable in terms of the reaction force when subjected to an acceleration. Thus, a body has unit mass if it reacts by 1 gr−f when accelerated by 1 cm sec^{-2}. Temperature is defined with respect to the *absolute zero* which corresponds to -273^0 C measured in units of degree Kelvin, at which all molecular thermal agitation comes to absolute rest. For further details on lower level standards, namely, National reference standards, working standards, and InterLaboratory standards, see Doebelin [1].

Reference values

All measurements are made with respect to some chosen reference quantity that sets the origin of the measurement. Thus, the direction of the local gravity vector sets the local vertical (LV) reference. The true geographical north, determined by the direction of Earth's spin axis by means of gyro compassing as shown in Chapter 4, sets the local north (LN) reference. It is important to note that LV and LN are the two orthogonal reference vectors that make possible precise angular 3D transformations from one coordinate system to another, such as vehicle body axes to Earth axes or vice versa. The LV allows measurement of pitch and roll angles θ and ϕ, while the LN allows measurement of yaw and pitch angles ψ and θ.

Fortunately, planet Earth provides these two essential and firm reference vectors, which can be measured with the great precision required in navigation. In attitude control, the local horizon sets the visual reference for the pitch and roll angles of an aircraft. Sea level sets the reference for topographic elevation, and barometric pressure at 0 degC at sea level sets the reference for barometric altitude; the direction of the local Earth magnetic field is the reference for magnetic north. None of these reference quantities is absolutely time-or position-invariant. Thus, the local gravity vector, which is the vector sum of the local *gravitation*, *centripetal*, and *Coriolis* forces, depends on the position on the surface of the Earth and whether it is observed from a moving or stationary vehicle, and on the trajectory and speed at which the vehicle is moving.

Magnetic north strongly depends on the location on the surface of the Earth and on the altitude at which the magnetic heading is being determined. Clearly, the precision of reference values is determined by two principal factors: (1) The consistency of the reference value under variations in space, time and external environmental influences; and (2) the accuracy of the measurement instruments by which the particular reference value is determined. For example, the exact direction and intensity of the gravitation vector is a function of the location on the Earth's surface. It varies in accordance with local variations of the mass density of the Earth's crust. For its measurement, a sufficiently precise pendulous or seismic instrument is required. Since no additional comparison reference exists, the reading of a pendulous device at rest at a given location is set to zero, thus establishing a nominal value for the vertical. Similar considerations hold for the local north and other reference quantities.

1.3 Passive and active instruments

Contact and remote sensing

We distinguish two principal categories of measurement devices: (1) contact instruments; and (2) noncontact, or remote sensors. Remote sensors respond to natural or man-made radiation, or to force fields such as magnetic, electric, electromagnetic, or gravitation. In the category of contact instruments, we distinguish two types: (1) passive sensors; and (2) active sensors. Every sensor requires energy for its actuation. The source of this energy may be the medium to be measured. Well known examples are the piston-type pressure gage for the determination of the gas pressure in a sealed tank, or the measurement of a battery voltage by means of a conventional galvanometric voltmeter. In an active sensor, the energy is provided by an external power supply.

Tapping of energy sources

The tapping of the measurand in passive instruments may decrease its value. In the first example, the expansion due to the displacement of the piston increases the total volume of the confined gas thus causing a pressure drop. In the second example, the galvanometer draws current from the battery causing a voltage drop across its internal resistance. Active contact instruments load the measured medium only by a negligible amount. For example, in an active manometer, the piston, which is the sensitive element, is constrained to essentially zero motion, so that no increase in volume is incurred. Constraining the piston, however, provides a force that is proportional to the pressure. Force, however, is not a displayable quantity. A detector, in the form of, for example, a piezoresistive or other force-sensitive element attached to it, provides a representative electrical signal that almost always implies amplification requiring energy from an independent electrical source that has no relation to the gas pressure.

Input impedance

In the example of the battery voltage measurement, the voltage drop can be prevented if the voltmeter has a very high input impedance so that only a minute current is drawn from the battery. But this low current may not be enough to actuate the galvanometer. Therefore an amplifier that can provide the required current from a separate supply must be used.

The common denominator of the previous examples is that the sensors are designed to have high input impedances, which implies high amplification.

In remote sensing, the measured quantity is, for all practical purposes, unaffected by the function of the sensor, which typically operates on the radiation of a high-energy remote source, such as infrared, visible light, X-rays, and nuclear. Yet the sensitive elements, such as photo resistive or photo voltaic elements, need their own power sources. However, remote sensors can essentially be passive because the measured medium almost always has infinite energy storage which for all practical purposes is not tapped by the action of the sensor. A similar situation holds for atmospheric passive sensors like the air speed sensor or the barometric altimeter, even though, in a sense, they can be viewed as contact sensors. For all practical purposes, the atmosphere is an infinite energy storage which can be simultaneously tapped by any conceivable number of pressure sensors without noticeable effect.

In summary, it is the need for high-impedance sensors requiring amplification that categorizes an *active* sensor.

1.4 Characteristics, resolution, and dynamic range

Domain and range

All the elements described so far are *input-output* devices relating an input to a desired output under conditions of *static equilibrium*. This relationship is referred to as the *characteristic* of the device, which in mathematical language maps the input X into the output Y, preferably by a bi-unique function $Y = f(X)$.

The minimum and maximum *admissible* values of the input X determine the *domain* $\{D\}$ of the device and the minimum and maximum *reachable* values of the output Y determine its *range* $\{R\}$ [6].

Linearity, resolution, and dynamic range

The ideal characteristic is a straight line with $\{D\} = \infty$ and $\{R\} = \infty$ intersecting the origin of the coordinates XOY and having infinite *resolution*, that is, there is no lower limit on an increment Δx in x that is not reflected in a corresponding increment Δy in y. Accordingly, the following definitions apply:

$$\lim_{\Delta x \to 0} \frac{\Delta y}{\Delta x} \triangleq f_x \neq 0 \; ; \; \forall \, x \tag{1.4.1}$$

If

$$\lim_{\Delta x \to 0} \frac{\Delta y}{\Delta x} = \text{const.} \; ; \; \forall \, x \tag{1.4.2}$$

the device is linear everywhere. Figure 1.2 illustrates an element with a characteristic having finite resolution denoted by x_r which is the largest increase Δx in x for which $\Delta y = 0$. Then, if x_r = const. for $\forall \, x$, the device has *uniform resolution*. This would be the case of a wire-wound potentiometer. x_r is generally randomly distributed, as, for example, in a plastic film potentiometer, in which case one can characterize the resolution by its standard deviation σ_r. Figure 1.2 also illustrates saturation of the device for large values of x. The largest admissible value x_m determines the domain $\{D\}$, while y_m determines the range $\{R\}$. The ratio between x_m and x_r is defined as the *dynamic range* D_r of the device

$$D_r \triangleq \frac{x_m}{x_r} \tag{1.4.3}$$

D_r is one of the essential features defining the precision of a measurement device. In aerospace technology, we broadly distinguish three categories of sensors, or instruments:

$D_r = 10^2 - 10^3$; medium quality, typical for attitude control.

$D_r = 10^3 - 10^4$; good quality, often required in guidance systems.

$D_r = 10^4 - 10^8$; very high precision, required in long-range inertial navigation.

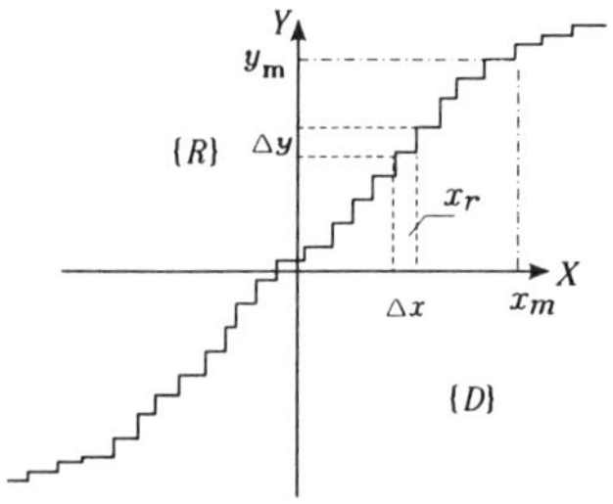

Figure 1.2. Characteristic with a random finite resolution and soft saturation.

A broad in-depth inquiry into the strategic-and technological implications of high precision inertial sensors, in particular, gyroscopes, is given in [2].

Bias, dead zone, and saturation

The value $y(0)$, which is the residual output at $x=0$ after all possible compensations have been made, is an unknown constant known as the *bias*. If $y(0)$ is a different constant for each turn on, it is called *turn-on-to-turn-on* bias. If $y(0)$ varies relatively slowly as a function of time during operation, it is called *drift* or *zero-point drift*. If it varies rapidly, it is known as *noise*. In all cases, $y(0)$ is represented as an *additive error*.

A device whose output is zero below certain threshold values of the input $x=-\Delta_1$ and $x=\Delta_2$ has a *dead zone* of $\Delta_1+\Delta_2$. If the output y is a constant y_1 for $x \geq x_{s_1}$ and is y_2 for $x \leq -x_{s_2}$, the characteristic has *hard saturation*. If y gradually assumes constant values above certain values of $|x|$, the device has *soft saturation*. A characteristic with a dead zone and saturation is shown in Figure 1.3. Saturation places upper bounds on the range $\{R\}$ measurement device. The dead zone places lower bounds on its domain $\{D\}$.

A characteristic with saturation and an infinite slope within the range R is a *switching function*. An example of a device with a dead zone is a direct current electrical motor, the characteristic of which relates $y=\omega\,(\text{sec}^{-1})$, its angular speed, with $x=u\,(\text{volt})$, its input voltage. Due to friction, the angular speed ω is zero below a certain threshold value $|u_{th}|$.

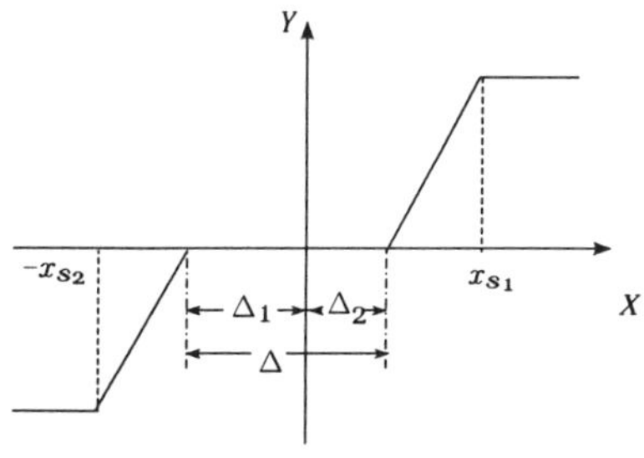

Figure 1.3. Characteristic with dead zone and hard saturation.

Hysteresis

The characteristic of a physical measurement device usually has hysteresis, accompanied by dissipation of energy resulting in a characteristic that does not have a bi-unique relationship $Y=f(X)$. A schematic description of a hysteresis type characteristic is shown in Figure 1.4. Typically, the residual value due to hysteresis, y_H, is specified by the manufacturer for many sensors in terms of percentage of full scale-value. Even a simple spring, which relates a loading force f with an expansion or contraction l, displays hysteresis resulting from internal friction in the inner structure of the material while undergoing distortions. Any mechanical linkage is accompanied by friction due to the sliding of surfaces against each other. A typical static characteristic,

$Y = f(F)$, relating the force F to displacement y under friction is shown in Figure 1.5. Clearly, the area enclosed in the parallelogram represents the dissipated energy.

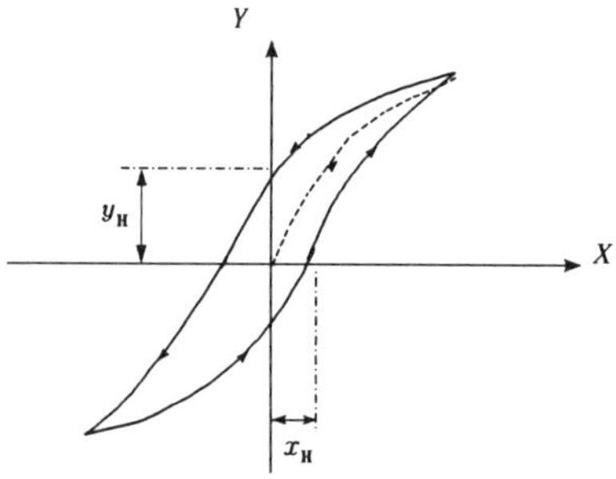

Figure 1.4. Characteristic with saturation.

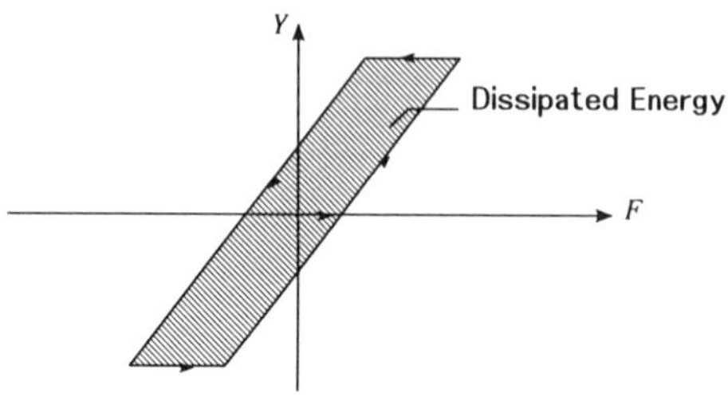

Figure 1.5. Characteristic of linkage with friction.

Figure 1.6 illustrates a mechanical linkage element with backlash. In the corresponding characteristic shown in Figure 1.5, $Y = f(X)$ relates motion y to motion x. Therefore, although this input-output relationship is not bi-unique, it has no hysteresis since no energy is dissipated in the to-and-fro cycle.

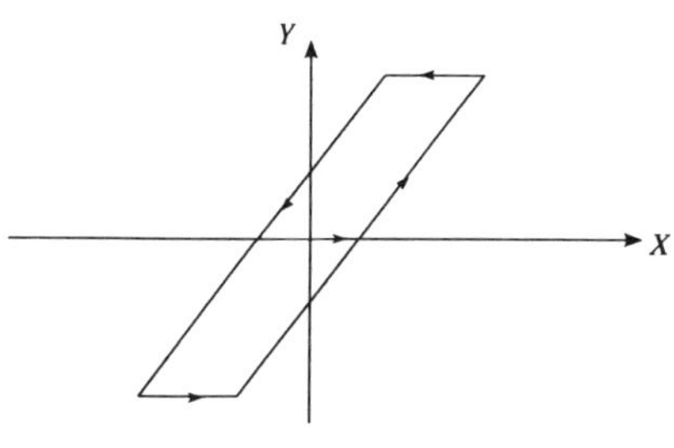

Figure 1.6. Characteristic of linkage with backlash.

An important operation in sensors is *linearization.* It provides the inverse of $Y=f(X)$, i. e., $X=f^{-1}(Y)$, which operates on the output Y. Denoting the combined output by Y^*, we have $Y^*=f^{-1}(Y)=f^{-1}[f(X)]=X$, which is ideally linearized. Clearly, only bi-unique functions and functions with finite derivatives dy/dx are invertible. Thus, for example, the switching function and the hysteresis function cannot be inverted.

In a real dynamic situation, the variables are time functions, i. e., $x(t)$, so that $y=y(t)$. As a result of the hysteresis, the output $y(t)$ will lag behind the input $x(t)$. This lag will often be the cause of instability resulting in sustained oscillations in closed-loop control systems.

Further details on nonlinear characteristics, their equivalent linear gain and their effect on control system performance can be found in [3] and [4].

1.5 Errors due to dynamics, nonlinearity, and noise

System and measurement equations

As stated in the Introduction, the sensor system is an integral part of the dynamical system, commonly described by the linear state space equations,

$$\dot{\mathbf{x}} = A\mathbf{x} + B\mathbf{u} + D\mathbf{w} \tag{1.5.1}$$

in which $\mathbf{x}$ is the $(n\times 1)$ state vector, $\mathbf{u}$ is the $(m\times 1)$ control vector, $m\leq n$, and $\mathbf{w}$ is the $(r\times 1)$ process disturbance vector, $r\leq n$. A is the $(n\times n)$ system matrix, B is the $(n\times m)$ control matrix, and D is the $(n\times r)$ disturbance matrix. To complete the system model description, (1.5.1) is augmented by the measurement equation,

$$\mathbf{z} = H\mathbf{x} + E\mathbf{u} + \mathbf{v} \tag{1.5.2}$$

where $\mathbf{z}$ is a $(q\times 1)$ measurement vector, H is a $(q\times n)$ measurement matrix consisting of ones and zeros, E is a $(m\times q)$ measurement matrix consisting of ones and zeros, and $\mathbf{v}$ is a $(q\times 1)$ measurement noise vector. (1.5.2) defines which system variables x_i, $i=1,2,..,n$, and control variables u_j, $j=1, 2,..,m$, are measured for the purpose of controlling the system. The ones in H imply that all the measurements are obtained with perfect one-to-one scaling. Often, $E\equiv 0$, i. e., $\mathbf{u}$ is not measured at all. In the special case $H=I$, where I is the identity matrix, all the elements in $\mathbf{x}$ are measured so that *full state measurement* is provided. Furthermore, (1.5.2) assumes that all the measured variables z_k, $k=1,2,..,q$, are linearly related to x_i without time lags and delays and that all the noise components v_k are additive and uncorrelated with $\mathbf{x}$ and $\mathbf{u}$. None of these assumptions are strictly valid, and the deviations from them must be taken into account by means of reliable error models. In the following control-systems-oriented framework, a central issue is the development of error models for the sensor systems.

The error-free output of the linear dynamical system, defined in (1.5.1) with respect to $\mathbf{u}$ only, can be expressed in complex frequency domain s by the transfer function matrix $F(s)$,

$$\boldsymbol{Y}(s) = F(s)\boldsymbol{U}(s) \tag{1.5.3}$$

where $\boldsymbol{Y}(s)$ is the output vector and $F(s)$ is the transfer function matrix given by

$$F(s) = H(sI - A)^{-1}B \tag{1.5.4}$$

$F(s)$ is a $(q \times m)$ consisting of the elements $f_{jk}(s)$, each of which is a single-input-single-output (SISO) transfer function relating any u_j to any $y_k(s)$; see [5].

Classification of errors

As indicated in Section 1.4, as a result of the errors in the sensors, the representations of the true outputs $y_k(s)$ are not perfect. The actual sensor outputs will be denoted by $z_k(s)$. The relation between y_k and z_k will be denoted by an operator $O(\cdot)$,

$$z_k = O(y_k) \tag{1.5.5}$$

$(\cdot)$ may assume different forms, the most important of which are:

1. Nonlinearity: $O(y) = N(y) = z$ (1.5.6)

2. Dynamics: $O(y) = G(s)y(s)$ (1.5.7)

3. Additive noise: $O(y) = 1 \times y + v = z$ (1.5.8)

The following model development assumes that it is possible to describe the dynamics $G(s)$ and the nonlinearity $N(y)$ as separate cascaded operators with $N(y)$ preceding $G(s)$. The errors due to the above three basic imperfections are defined as follows:

1. $G(s)=1;\ v=0;\ \epsilon_N = y - z = y - N(y)$: Nonlinearity (1.5.9)

2. $N(y)=y;\ v=0;\ \epsilon_D = y - z = [1 - G(s)]y(s)$: Dynamics (1.5.10)

3. $N(y)=y;\ G(s)=1;\ \epsilon_v = y - z = y - y - v = -v$: Noise (1.5.11)

The block diagram shown in Figure 1.7 summarizes the foregoing under the assumptions made. The nonlinearity is placed so that it precedes the dynamical element $G(s)$. There is no theoretical reason for this order, but it is often made on physical grounds.

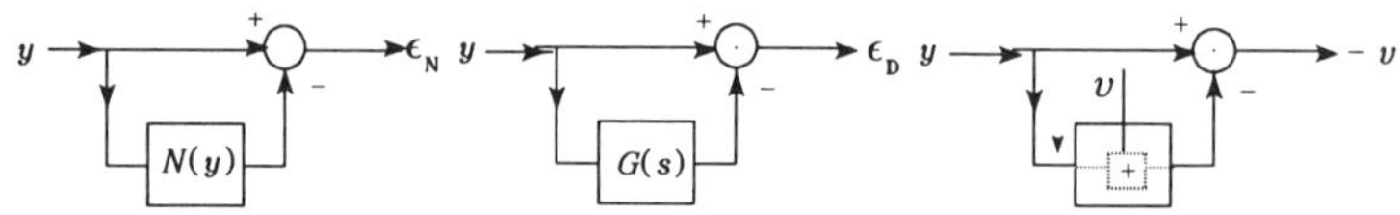

Figure 1.7. Basic error sources in measurement systems.

Sensor errors are often expressed relative to the input y. Thus,

$$\delta_N = \frac{\epsilon_N}{y} = \frac{y-N(y)}{y} = 1 - \frac{N(y)}{y} \tag{1.5.12}$$

$$\delta_D = \frac{\epsilon_D}{y} = \frac{y-G(s)y}{y} = 1 - G(s) \tag{1.5.13}$$

$$\delta_v = \frac{y-(y+v)}{y} = -\frac{v}{y} \tag{1.5.14}$$

The last expression is generally known as the noise–to–signal ratio.

A common characterization of errors is in terms of percent of full scale input y_{max}. Specifically, if v is an unknown constant, δ_v is the bias expressed as a percentage of full scale, and likewise if v is an RMS noise level. Many medium quality aerospace sensors have values of δ on the order of 1%. Note that δ_D alone is independent of the value of $y_{\max}$.

EXAMPLE 1.5–1: Error due to linearity in the characteristic

A characteristic with soft saturation can be analytically described by :

$$N(y) = \tanh y$$

$$\delta_N = 1 - \frac{\tanh y}{y} \simeq \frac{y^2}{3} - \frac{2y^4}{15} + \frac{17y^6}{315} - \dots$$

Permitting a nonlinearity error $\delta_N = 0.01$, the maximum admissible value of y is found to be $y_{\max} = 0.173$. The corresponding domain is $|D| = 0.346$. $N(y)$ and δ_N are shown in Figure 1.8–a and b respectively.

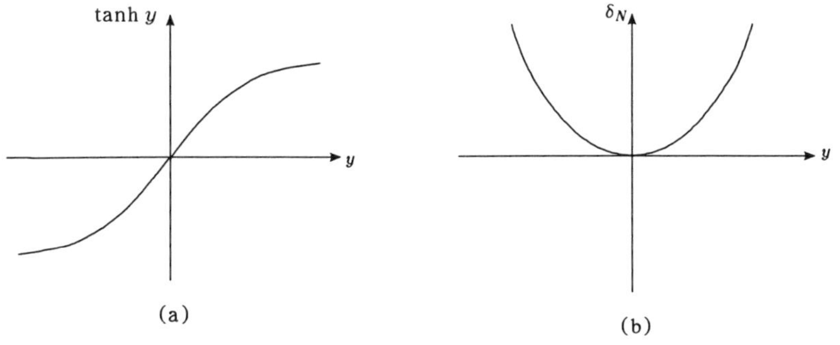

Figure 1.8. Example of soft saturation characteristic and associated static error due to the nonlinearity.

EXAMPLE 1.5–2: Rectification error due to asymmetry of characteristic

The characteristic of a sensor is not ideally symmetrical with respect to the y–axis, as indicated in Figure 1.9.

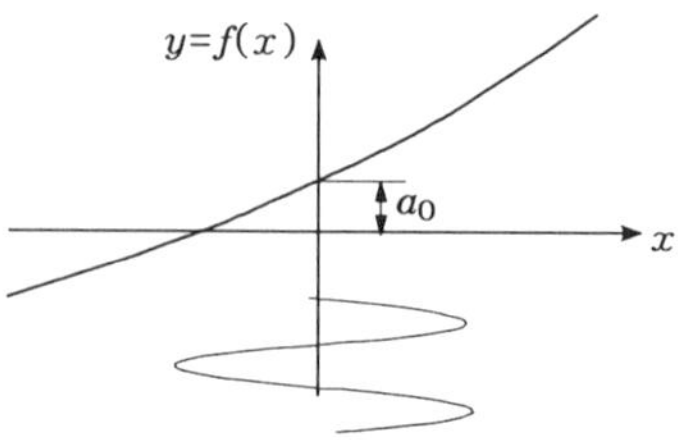

Figure 1.9. Asymmetric characteristic and its rectifying feature.

$y=f(x)$ can be approximated by the series expansion

$$y=f(x)=a_0 + a_1x + a_2x^2 + a_3x^3 + \ldots$$

A sinusoidal input, e. g., a vibration $x = v\sin\omega t$ affecting an accelerometer, as indicated in Figure 1.9, would cause the output

$$y=a_0 + a_1v\sin\omega t + a_2v^2\sin^2\omega t + a_3v^3\sin^3\omega t + \ldots$$

$$=a_0 + a_1v\sin\omega t + a_2v^2(1-\cos2\omega t)/2 + a_3v^3\sin\omega t(1-\cos2\omega t)/2 + \ldots$$

Clearly, besides the harmonic distortions caused by the odd–power terms, the even–power terms generate constants such as $a_2v^2/2$, which are in effect vibration–dependent bias terms that may be detrimental to the system performance. This phenomenon is known as *vibration rectification*. Thus, substantial bias errors can develop due to this nonlinearity.

EXAMPLE 1.5–3: Error due to dynamics of the sensitive element

The dynamics of a hot wire gage, which relates its electrical resistance R to changes in temperature θ, is described by a first order lag transfer function $G(s)$:

$$G(s) = \frac{\theta(s)}{R(s)} = \frac{1}{1+\tau s}$$

The percentage dynamical error $\delta_D(s)$ is:

$$\delta_D(s) = 1-G(s) = \frac{\tau s}{1+\tau s}$$

For a time constant $\tau = 5$ msec, $|\delta_D(j\omega)|$ is shown in Figure 1.10.

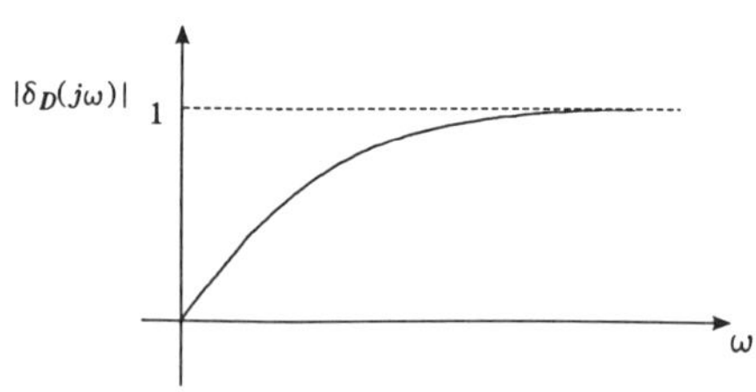

Figure 1.10. Effect of dynamical error as a function of frequency.

1.6 Environmental interference

Error model formulation

In Section 1.4, the relationship between the input X and the output Y was presented as a single-input-single-output relationship $Y=f(X)$. However, in reality, in addition to the 'legitimate' input X, there may be numerous undesirable inputs affecting the sensor output Y. These may result from environmental influences on the sensor response, such as variations in temperature, supply voltage, barometric pressure, humidity or from the influence of linear or rotational accelerations. Any one of these may affect the performance separately or in combination. The sensor, consisting of a number of cascaded elements, G_1, G_2,...G_n, each of which may be particularly sensitive to any of the environmental variations i_1, i_2,...i_n, is schematically described in Figure 1.11. The total sensor response y is the combined effect of the legitimate input x and the interferences i_k, $k = 1,...n$. We define quiescent, or nominal values for x, y, and i, and we denote these by upper case characters X, Y, and I, respectively. The deviations from these nominal values will be denoted by the corresponding lower-case letters. A block diagram relating inputs and output is shown in Figure 1.11.

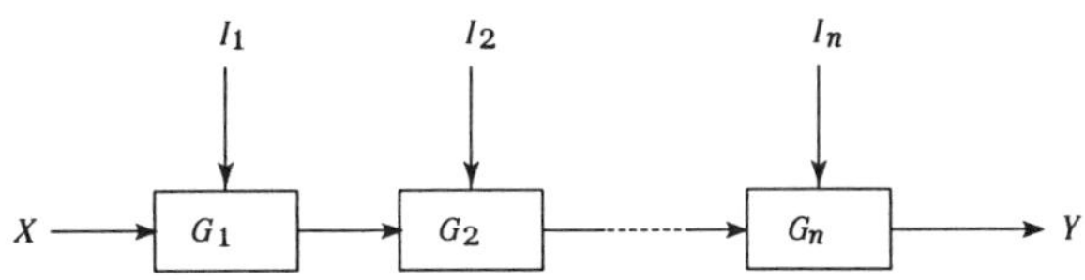

Figure 1.11. Sensor system with input and interferences.

The nominal values are, X_N, Y_N, I_{k_N} and the corresponding deviations are x, y, and i_k. Often, $X_N = 0$. Thus we have the multivariable function

$$Y = G(X, I_1,..., I_n) \tag{1.6.1}$$

Expanding (1.6.1) into a Taylor series, we have

$$y = \left.\frac{\partial Y}{\partial X}\right|_{X=X_N} x + \frac{1}{2}\left.\frac{\partial^2 Y}{\partial X^2}\right|_{X=X_N} x^2 + \left.\frac{\partial Y}{\partial I_1}\right|_{I_1=I_{1N}} i_1 + \left.\frac{\partial Y}{\partial I_2}\right|_{I_2=I_{2N}} i_2 + ...$$

$$+ \frac{1}{2}\left.\frac{\partial^2 Y}{\partial I_1^2}\right|_{I_1=I_{1N}} i_1^2 + \left.\frac{\partial^2 Y}{\partial X \partial I_1}\right|_{\substack{I_1=I_{1N}\\ X=X_N}} x i_1 + ... + \left.\frac{\partial^2 Y}{\partial I_1 I_2}\right|_{\substack{I_1=I_{1N}\\ I_2=I_{2N}}} i_1 i_2 + ... \tag{1.6.2}$$

In (1.6.1), we disregard the second term and higher, the fourth term and higher and the fifth term and higher. In this expansion we assume that the perturbations x, and i are sufficiently small and that the cross-sensitivities between the interferences i_j, and i_k are negligible. The remaining terms are formulated as follows:

$$y = G_X x + \sum_{k=1}^{n} G_{I_k} i_k + \sum_{k=1}^{n} \mu_{I_k} x i_k \tag{1.6.3}$$

where

$$G_X \triangleq \frac{\partial Y}{\partial X}; \quad G_{I_k} \triangleq \frac{\partial Y}{\partial I_k}; \quad \mu_{I_k} \triangleq \frac{\partial}{\partial X} G_{I_k} \tag{1.6.4}$$

Additive and scaling errors

Equation (1.6.3) can now be rewritten as follows:

$$y = (G_X + \sum_{k=1}^{n} \mu_{I_k} i_k) x + \sum_{k=1}^{n} G_{I_k} i_k = G_X' x + b \tag{1.6.5}$$

In (1.6.5), G_X is the slope of the characteristic at the nominal operating point $X=X_N$, and the sum in the second term in the parenthesis is the total error in the slope due to the interferences i_k, $k = 1,\ldots n$. As such it behaves like a scaling error or a multiplicative error that modifies G_X to G_X'. The last term adds to the output y and thus behaves like an additive error b, also known as the null-point-error or bias. Numerical values for these environmental sensitivity coefficients are given in the sensor data-sheets.

EXAMPLE 1.6-1: Combined effect of environmental interferences

The nominal environmental operating parameters I_k, $k=1,\ldots n$, are also given in the data sheets. For example, for a rate gyroscope the data may be:

$I_{1_N} = T_N = 300$ deg K; $I_{2_N} = G_N = 1$ g; $I_3 = V_N = 28$ volt;

The output of the gyro is measured in (deg/sec); The corresponding environmental sensitivity coefficients in the null-point offset of (1.6.5) might be:

$G_{I_1} = 0.2$ deg/sec/deg C; $G_{I_2} = -1$ deg/sec/g;
$G_{I_3} = 0.8$ deg/sec/volt;

Assume, for example: $i_1=20$ deg C; $i_2=3$ g; $i_3=-2$ volt; the resulting null-point offset would be:

$$y = 0.2 \times 20 + (-1) \times 3 + 0.8 \times (-2) = 0.6 \text{ deg/sec.}$$

In Example 1.6-1 the algebraic summation due to (1.6.5) almost cancels out the effects of environmental interference. In reality, the signs of i_k are not known, and y can only be expressed in statistical terms, usually assuming that the various environmental variables are statistically independent. This aspect will be addressed again in Section 2.3.

A graphical interpretation of (1.6.5) is given in Figure 1.12.

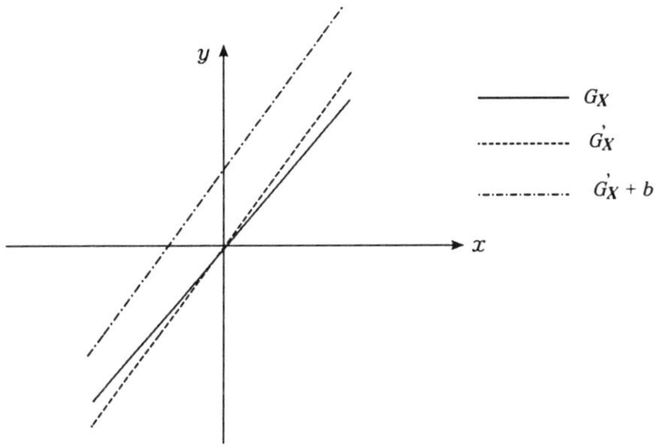

Figure 1.12. Graphical interpretation of Equation (1.6.5).

1.7 Error compensation

Isolation and protection

Within a given state-of-the-art of design and materials, the effect of environmental deviations from their nominal values, may have a significant effect on sensor precision. This is particularly true in aerospace systems in which environmental parameters vary widely. Temperature often varies by more than 180 deg C, and acceleration levels exceed hundreds of g's and in some cases, as in guided artillery ordnance, thousands of g's.

One approach to cope with this problem has been to provide protection by passive or active isolation, especially from temperature, shock, and vibration. Passive isolation from temperature variations is simply accomplished by means of special isolating jackets. This method can be effective for relatively short periods of time. Active isolation involves jackets that are heated to a predetermined temperature level that is automatically regulated to keep the protected sensor within a narrow temperature range. This had been the practice, for example, in the early high-precision inertial grade rate-integrating gyroscopes, which had to be operated within a temperature range of 0.1 deg C. Passive isolation with regard to g's, especially with regard to vibrations, is normally accomplished by *shock-mounts.* These may be quite effective in the frequency range of hundred Hz and higher, but for the lower frequencies, they become ineffective, and very bulky. It is clearly impossible to isolate sensors from sustained acceleration. Active isolation, which can be accomplished by a closed-loop electrodynamical regulator, is technically feasible but not practical. In any case, the concept of isolation, whether passive or active, always adds to the weight, volume and cost and impairs reliability. As such, it is contrary to the ever-growing need for miniature, lightweight, reliable sensors.

Output compensation

Consequently, a current trend in reducing the effects of environmental variations is by means of *active output compensation.* This method can be implemented if the sensitivity coefficients G_{I_k} are known. The environmental variations i_k must be measured by a set of auxiliary sensors S_{I_k} that provide signals proportional to i_k, which, when weighted by G_{I_k} and summed, are subtracted from the sensor output y. A block diagram describing this scheme is shown in Figure 1.13. However, since G_{I_k} are not precisely known and S_{I_k} are not perfect, the offset error in y cannot be precisely compensated, and a residual error ϵ_r will persist. This error cannot be compensated using the assumed state-of-the-art precision of sensor measurements, and it must be regarded as random bias, which can be characterized only in statistical terms.

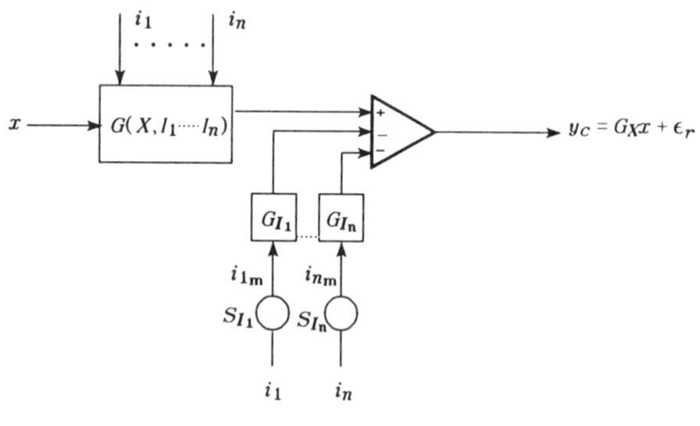

Figure 1.13. Output compensation of a sensor with interferences.

Referring to (1.6.5) and disregarding the scaling error, the compensated output is determined by

$$y_c = G_X x + \sum_{k=1}^{n} G_{I_k} i_k - \sum_{k=1}^{n} \hat{G}_{I_k} i_{k_m} = G_X x + \epsilon_r \tag{1.7.1}$$

$\hat{G}_{I_k}$ are the estimated values of G_{I_k}, and i_{k_m} are the measured values of i_k provided by the auxiliary sensors S_{I_k}. In principle, it is also possible to compensate for the scaling errors in G_x resulting from the products $\mu_{I_k} i_k$ defined in (1.6.5). However, this implementation is much more difficult because the measured values i_k must be multiplied by μ_{I_k} which are not as precisely known as G_{I_k}. The resulting product must be multiplied by x, which is not known because it is the input to be measured. It follows that in practice, output compensation can be effective for additive errors only, and not for scaling errors. In Section 1.10 it will be shown how scaling errors are reduced by implementing feedback control in the sensor.

Input compensation

Instead of compensating for the errors at the output y, it is possible to conceive of a similar concept at the input level x. This is referred to as *input compensation.* We now reformulate G_{I_k} in an alternative form:

$$G_{I_k} \triangleq \frac{\partial Y}{\partial I_k} = \frac{\partial Y \partial X}{\partial X \partial I_k} = G_X \xi_{I_k} \tag{1.7.2}$$

where the coefficients ξ_{I_k} are defined with respect to the input x by

$$\xi_{I_k} \triangleq \frac{\partial X}{\partial I_k} \tag{1.7.3}$$

ξ_{I_k} determines by what amount X would need to be changed to compensate for the effect of a variation i_k. They are, in effect, the sensitivity coefficients equivalent to G_{I_k}, but referred to the input X. The input-compensated output y_c is given by

$$y_c = G_X x + \sum_{k=1}^{n} G_{I_k} i_k - \hat{G}_X \sum_{k=1}^{n} \hat{\xi}_{I_k} i_{k_m} = G_X x + \epsilon_r \tag{1.7.4}$$

$\hat{G}_X$ and $\hat{\xi}_{I_k}$ are the corresponding estimated values. The third term in (1.7.4) represents the compensating force generated by suitable transducers in response to i_{k_m}. A block diagram of input compensation is shown in Figure 1.14.

There is a fundamental difference between output compensation and input compensation. In output compensation, the correction is carried out at the level of voltages, so that subtraction of the compensating signal is straightforward. However, compensation of x, which is a physical quantity, requires special provisions. Thus, if x represents a force, the compensation also must be a force. This implies the use of a transducer to generate a force opposing the one resulting from the input force. It is obvious that such subtractions cannot always be physically implemented. For example, if x is a temperature, subtracting temperature, would imply cooling the temperature sensor, which obviously is an awkward and costly solution. Similar complications would be encountered with humidity. In general, the notion of input compensation may be suitable for the measurement of variables but not environmental parameters.

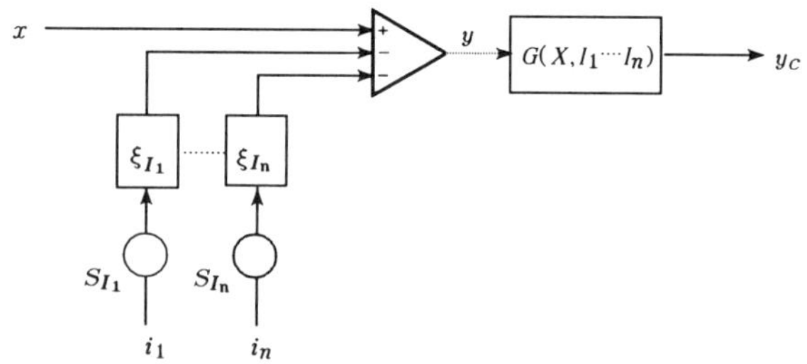

Figure 1.14. Input compensation of sensor with interferences

Figure 1.15 shows the implementation of output compensation in a membrane pressure gage. Implementation of input compensation for this pressure gauge is shown in Figure 1.16.

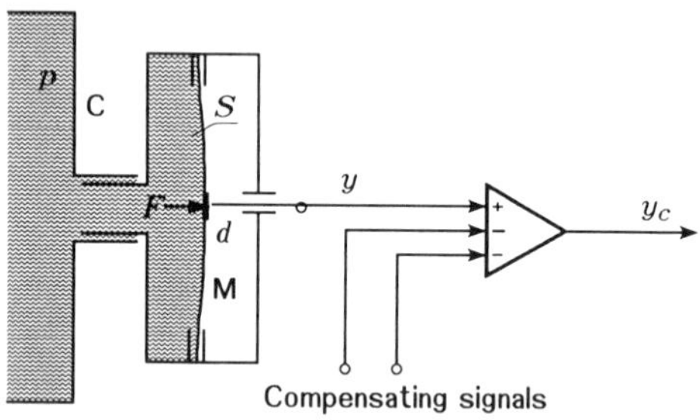

Figure 1.15. Output compensation in a menbrane pressure gauge.

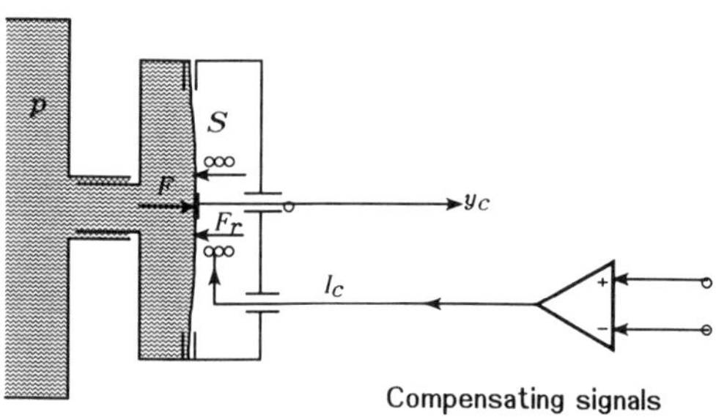

Figure 1.16. Input compensation of a membrane pressure gauge.

In this example, x is the pressure p of a fluid confined in a container C. p manifests itself as a force $F = Sp$, where S is the area of the membrane M. This force causes a deflection η of the membrane. The characteristic $\eta = h(F)$ is not linear, therefore only small values of η are allowed. This deflection is not sufficient for display purposes and therefore it must be translated into a measurable quantity. This is achieved by a suitable transducer, in this case a strain gauge or a piezovoltaic or piezoresistive detector d, which has the characteristic $v = g(\eta)$. Thus, the overall characteristic of the device is

$$y = v = g[h(F)] + b \tag{1.7.5}$$

where b is a bias error of the detector or the result of a residual deflection of the membrane. Their combined effect may be due to the influence of temperature variations. In accordance with Fig. 1.14, a compensating signal, such as $i_t = i$, provided by the temperature sensor $S_t, = S_1$ and weighted by $G_t = G_1$, subtracts b from the output y.

Input compensation, as stated earlier, must be accomplished by means of force, since x is pressure. This is described in Fig. 1.16. An electromagnetic force transducer, providing a force F_r, opposing F, is driven by a compensating current I_c that is proportional to the third term in (1.7.4), that is,

$$I_c = K\hat{G}_X \sum_{k=1}^{n} \hat{\xi}_{lk} i_{km} \tag{1.7.6}$$

where K is a constant of proportionality relating the current I_c to the force F_r.

This technique of exerting a force or a torque directly on the sensitive element is widely used in a variety of sensors, not necessarily for the purpose of input compensation but for testing, balancing and calibration. This feature will be described later in appropriate detail in conjunction with electro dynamical sensors such as accelerometers and gyroscopes.

With the advent of improved materials, in particular crystals and mineral based components, long term stability and repeatability is approaching perfection. For this reason, and because of the precision and stability of micro electronic devices, the compensation methods outlined above have become very effective.

1.8 Estimation of characteristics by regression

Error modeling

The characteristic of a sensor is determined experimentally by providing an input vector $\boldsymbol{x} \triangleq [x_1, x_2, ..., x_n]^T$ at its input and by observing its corresponding output $\boldsymbol{y} \triangleq [y_1, y_2, ..., y_n]^T$ by means of measurements. $\boldsymbol{y}$ could be plotted against $\boldsymbol{x}$, and the ensuing collection of points relating $\boldsymbol{y}$ to $\boldsymbol{x}$ would establish the required characteristic. However, since the measurement process and its instrumentation are not error-free, measurements of $\boldsymbol{y}$ would be accompanied by uncertainties $\boldsymbol{v} \triangleq [v_1, v_2, \ldots, v_n]^T$, where v_i, $i = 1, \ldots n$, is a random variable and is regarded as additive noise. Thus, the actual measurement vector denoted by $\boldsymbol{z} \triangleq [z_1, z_2, \ldots, z_n]^T$ is given by

$$\boldsymbol{z} = \boldsymbol{y} + \boldsymbol{v} \tag{1.8.1}$$

or

$$z_i = y_i + v_i \tag{1.8.2}$$

This implies that the "true" value $\boldsymbol{y}$ cannot be determined exactly. However, by applying a suitable statistical averaging process based on a well-chosen mathematical model, it is possible to estimate $\boldsymbol{y}$ with very good accuracy. The mathematical model is chosen in accordance with prior theoretical knowledge of an underlying physical relationship between $\boldsymbol{x}$ and $\boldsymbol{y}$ or a hypothesis based on experience or a choice based on expediency, such as selecting a straight line for relating $\boldsymbol{y}$ to $\boldsymbol{x}$.

In order to demonstrate this underlying principle, assume that a series of radar range measurements z_i of a target yields the readings shown in Figure 1.17.

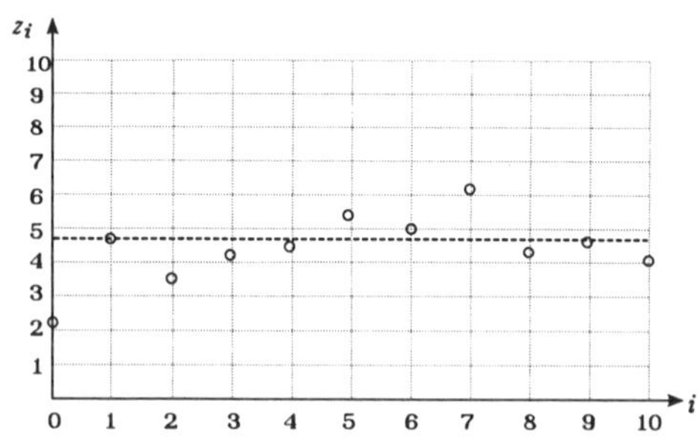

Figure 1.17. A series of noisy measurements of a constant.

From the mere observation of this collection of measurements, it is hard to decide whether the measurements z_i represent some form of quasi-periodical behavior or a constant masked by substantial measurement noise. Assume now, that from prior information we have good reason to assume that the target had moved on a circular path around the radar but at an unknown radius. Adoption of this assumption is the basis for the choice of the model, i.e., $r = b$, where b = const. In doing so, we have selected the model, i.e., $y = b$. Because of the radar noise, we cannot determine b exactly, but we can determine an estimate $\hat{b}$ of b. $\hat{b}$ is an unknown constant to be determined. Let us define an error ϵ_i as follows:

$$\epsilon_i \triangleq z_i - \hat{b} \tag{1.8.3}$$

$$J \triangleq \frac{1}{n}\sum_{i=1}^{n} \epsilon_i^2 \triangleq A(\epsilon_i^2) \tag{1.8.4}$$

where $A(\cdot)$ represents the finite sample average over n measurements. Substituting (1.8.3), we seek the optimal value of $\hat{b}$ that minimizes J. This is accomplished by setting $\partial J/\partial \hat{b}$ to zero

$$\frac{\partial}{\partial \hat{b}} A(\epsilon_i^2) = \frac{\partial}{\partial \hat{b}} A[(z_i - \hat{b})^2] = 0 \tag{1.8.5}$$

which results in

$$-2A(z_i) + 2A(\hat{b}) = 0 \tag{1.8.6}$$

Or, since $\hat{b}$ = const., $A(\hat{b}) = \hat{b}$. Thus, from (1.8.6).

$$\hat{b} = A(z_i) = \frac{1}{n}\sum_{i=1}^{n} z_i \tag{1.8.7}$$

This, rather expected and obvious result, conveys the idea that the simple average is a direct consequence of the *hypothesis* that the target maintained a constant range. However, had we conceived a different hypothesis, e.g., that the range $r = ki$, k being a constant so that r increases linearly with i, the simple average of (1.8.7) would not have been valid.

Linear regression

We now proceed to extend the modeling and estimation process to more general forms. It is common practice to model a characteristic as a straight line formulated by the equation,

$$y = mx + b \tag{1.8.8}$$

which is considered as the *truth model* for the input-output relationship. The *truth model* indicates an x,y relationship by virtue of a physical law, that is known to apply, or, by virtue of the expediency of the chosen relationship. For the input vector $x \triangleq [x_1, x_2, \dots x_n]^T$, and in the presence of measurement noise v_i, the output z_i would be,

$$z_i = y_i + v_i = mx_i + b + v_i \tag{1.8.9}$$

where m and b are constants of unknown values. The objective is to find the best estimates $\hat{m}$ for m and $\hat{b}$ for b from the measurements z_i. Accordingly, we formulate the equation

$$\hat{z} = \hat{m}x + \hat{b} \tag{1.8.10}$$

$\hat{m}$ and $\hat{b}$ are constants to be determined. (1.8.11) is the *linear regression* line. Again, we define the error vector

$$\epsilon_i = z_i - \hat{m}x_i + \hat{b} \quad i = 1, 2,...,n \tag{1.8.11}$$

The forgoing is illustrated in Figure 1.18.

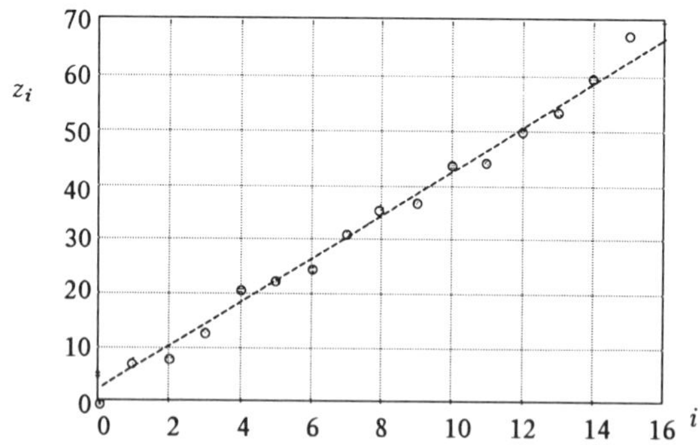

Figure 1.18. Noisy measurements of a linear relationship

Using the cost function J of (1.8.4), we have

$$A(\epsilon_i^2) = A\{[z_i - (\hat{m}x_i + \hat{b})]^2\} \tag{1.8.12}$$

By minimizing J with respect to $\hat{m}$ and $\hat{b}$,

$$\frac{\partial}{\partial \hat{m}} A(\epsilon_i^2) = 0 \tag{1.8.13}$$

$$\frac{\partial}{\partial \hat{b}} A(\epsilon_i^2) = 0 \tag{1.8.14}$$

and rearranging we have

$$\hat{b} + \hat{m}A(x_i) = A(z_i) \tag{1.8.15}$$

$$A(x_i)\hat{b} + \hat{m}A(x_i^2) = A(x_i z_i) \tag{1.8.16}$$

In matrix-vector form we have

$$\begin{bmatrix} 1 & A(x_i) \\ A(x_i) & A(x_i^2) \end{bmatrix} \begin{bmatrix} \hat{b} \\ \hat{m} \end{bmatrix} = \begin{bmatrix} A(z_i) \\ A(x_i z_i) \end{bmatrix} \tag{1.8.17}$$

leading to the solution

$$\begin{bmatrix} \hat{b} \\ \hat{m} \end{bmatrix} = \begin{bmatrix} 1 & A(x_i) \\ A(x_i) & A(x_i^2) \end{bmatrix}^{-1} \begin{bmatrix} A(z_i) \\ A(x_i z_i) \end{bmatrix} \tag{1.8.18}$$

$\hat{b}$ and $\hat{m}$ are the best least square estimates determined on the basis of the measurements of the input–output vectors $x=[x_1,z_2,\ldots,x_n]^T$ and $z=[z_1, z_2,\ldots,z_n]^T$. The explicit solutions are,

$$\hat{b} = \frac{A(z_i)A(x_i^2)-A(x_i z_i)A(x_i)}{A(x_i^2)-[A(x_i)]^2} \tag{1.8.19}$$

$$\hat{m} = \frac{A(x_i z_i) - A(x_i)A(z_i)}{A(x_i^2)-[A(x_i)]^2} \tag{1.8.20}$$

The last two expressions are known as the *linear regression* process .

Being based on a finite number of n measurements, $\hat{b}$ and $\hat{m}$ are random numbers that would differ somewhat for different choices of x and z, i. e., from experiment to experiment. The denominator is given by

$$A(x_i^2) - [A(x_i)]^2 \triangleq \hat{V}_x = \hat{\sigma}_x^2 \geq 0 \tag{1.8.21}$$

where $\hat{V}_x$ is the estimate of the variance V_x of the input x. We define

$$\hat{\mu}_x \triangleq A(x_i) \tag{1.8.22}$$

which is an estimate of the mean value of x.

Clearly, $\hat{m}$ is not quite equal to m and $\hat{b}$ is not quite equal to b. The linear regression analysis per se does not provide information about the corresponding estimation errors defined by

$$\tilde{b} \triangleq \hat{b} - b \tag{1.8.23}$$

$$m \triangleq \hat{m} - m \tag{1.8.24}$$

It will be shown later that these errors decrease as n is increased. Resubstituting

$$A(\cdot) \triangleq \frac{1}{n}\sum_{i=1}^{n}(\cdot) \tag{1.8.25}$$

into (1.8.19) and (1.8.20), we have

$$\hat{b} = \frac{\sum_{i=1}^{n} z_i \sum_{i=1}^{n} x_i^2 - \sum_{i=1}^{n} x_i z_i \sum_{i=1}^{n} x_i}{n\sum_{i=1}^{n} x_i^2 - [\sum_{i=1}^{n} x_i]^2} \tag{1.8.26}$$

$$\hat{m} = \frac{n\sum_{i=1}^{n} x_i z_i - \sum_{i=1}^{n} z_i \sum_{i=1}^{n} x_i}{n\sum_{i=1}^{n} x_i^2 - [\sum_{i=1}^{n} x_i]^2} \tag{1.8.27}$$

We now extend the method of linear regression to the more general case of polynomial regression in which the relationship between y and x is given by

$$y = \sum_{j=0}^{q} a_j x^j = \mathbf{a}.\mathbf{x} \tag{1.8.28}$$

where $\mathbf{a} \triangleq [a_0, a_1, \ldots, a_q]^T$ and $\mathbf{x} \triangleq [x_0, x_1, \ldots, x_q]^T$. The parameter vector $\mathbf{a}$ is represented by the estimate

$$\hat{\mathbf{a}} \triangleq [\hat{a}_0, \hat{a}_1, \ldots, \hat{a}_q]^T \tag{1.8.29}$$

Thus,

$$\hat{z} = \sum_{j=0}^{q} \hat{a}_j x^j \tag{1.8.30}$$

The series of measurements, as before, is

$$z_i = y_i + v_i \tag{1.8.31}$$

The estimation error is defined by

$$\epsilon_i \triangleq z_i - \hat{z}_i \tag{1.8.32}$$

We minimize $J = A(\epsilon_i^2)$

$$\frac{\partial}{\partial \hat{a}_j} A(\epsilon_i^2) = 0 \; ; \; j = 0, 1, \ldots, q \tag{1.8.33}$$

The assumption is that $n \geq q+1$. Carrying out (1.8.33) with equations (1.8.29), (1.8.30) and (1.8.33), we have

$$\begin{bmatrix} \hat{a}_0 \\ \hat{a}_1 \\ \cdot \\ \cdot \\ \hat{a}_q \end{bmatrix} = \begin{bmatrix} \Sigma 1^2 & \Sigma x_i & \ldots \Sigma x_i^q \\ \Sigma x_i & \Sigma x_i^2 & \ldots \Sigma x_i^{q+1} \\ \cdot & \cdot & \cdot \\ \Sigma x_i^q & \Sigma x_i^{q+1} & \ldots \Sigma x_i^{2q} \end{bmatrix}^{-1} \begin{bmatrix} \Sigma z_i \\ \Sigma x_i z_i \\ \cdot \\ \cdot \\ \Sigma x_i^{2q} z_i \end{bmatrix} \tag{1.8.34}$$

where

$$\Sigma \triangleq \frac{1}{n} \sum_{i=1}^{n} (\cdot) \tag{1.8.35}$$

and $\hat{\mathbf{a}}$ is the estimate defined in (1.8.29) based on n measurements. The square matrix in (1.8.35) is always symmetrical and it is known as the *information matrix* [7]. The column vector is known as the *covariance vector*. The solution (1.8.34), when substituted in (1.8.30), provides the polynomial that is the best least square fit to the polynomial of (1.8.28).

Precision of the estimate

Being based on a finite set of n measurements, the estimate $\hat{a}$ is itself a random number that would differ for different sets of measurements. The error $\tilde{a} = \hat{a} - a$ is expressed by the variances

$$\sigma_{\hat{a}}^2 = [\sigma_{\hat{a}_1}^2, \sigma_{\hat{a}_2}^2, \ldots, \sigma_{\hat{a}_q}^2]^T \tag{1.8.36}$$

which is a statistical measure of $\tilde{a}$. In order to compute the variances of (1.8.37), we first determine the variances of the measurement error estimates

$$\hat{z}_i = \hat{a}_j x_i^j \tag{1.8.37}$$

$$\sigma_z^2 = \frac{1}{n} \sum_{i=1}^{n} (\sum_{j=0}^{q} \hat{a}_j x_i^j - z_i)^2 \tag{1.8.38}$$

It can be shown that the vector of variances of (1.8.36) is determined, in accordance with [7], by

$$\begin{bmatrix} \sigma_{ao}^2 \\ \sigma_{a_1}^2 \\ \cdot \\ \sigma_{a_q}^2 \end{bmatrix} = \text{diag} \begin{bmatrix} \Sigma 1^2 & \Sigma x_i & \ldots \Sigma x_i^q \\ \Sigma x_i & \Sigma x_i^2 & \ldots \Sigma x_i^{q+1} \\ \cdot & \cdot & \cdot \\ \Sigma x_i^q & \Sigma x_i^{q+1} & \ldots \Sigma x_i^{2q} \end{bmatrix}^{-1} \sigma_z^2 \tag{1.8.39}$$

For example, we observe the result for $q = 2$, namely, the case of linear regression. The explicit solution in which $\hat{a}_0 = b$ and $\hat{a}_1 = m$ is given by

$$\sigma_{\hat{b}}^2 = \frac{\sum_{i=1}^{n} x_i^2}{n\sum_{i=1}^{n} x_i^2 - (\sum_{i=1}^{n} x_i)^2}\sigma_z^2 \tag{1.8.40}$$

$$\sigma_{\hat{m}}^2 = \frac{n}{n\sum_{i=1}^{n} x_i^2 - (\sum_{i=1}^{n} x_i)^2}\sigma_z^2 \tag{1.8.41}$$

Dividing through by n and rearranging, we have

$$\sigma_{\hat{b}}^2 = \frac{\frac{1}{n}\sum_{i=1}^{n} x_i^2}{\frac{n-1}{n}\sum_{i=1}^{n} x_i^2 + \frac{1}{n}\sum_{i=1}^{n} x_i^2 - \frac{1}{n}(\sum_{i=1}^{n} x_i)^2}\sigma_z^2$$

$$= \frac{P_x}{(n-1)P_x + \sigma_x^2}\sigma_z^2 \tag{1.8.42}$$

Assuming that $\mu_x = 0$, we have $\sigma_x^2 = P_x = \frac{1}{n}\sum_{i=1}^{n} x_i^2$ so that

$$\sigma_{\hat{b}}^2 = \frac{P_x}{nP_x}\sigma_z^2 = \frac{\sigma_z^2}{n} \tag{1.8.43}$$

Similarly, we obtain

$$\sigma_{\hat{m}}^2 = \frac{1}{\frac{n-1}{n}\sum_{i=1}^{n} x_i^2 + \frac{1}{n}\sum_{i=1}^{n} x_i^2 - \frac{1}{n}(\sum_{i=1}^{n} x_i)^2}\sigma_z^2$$

$$= \frac{1}{(n-1)P_x + \sigma_x^2}\sigma_z^2 = \frac{\sigma_z^2}{nP_x} \tag{1.8.44}$$

Equations (1.8.43) and (1.8.44) disclose that the variances of the estimates of b and m decrease with $1/n$ or that the rms values of these estimation errors decrease with $1/\sqrt{n}$.

The foregoing is now illustrated by a numerical example for the case of linear regression.

EXAMPLE 1.8–1: Estimation of characteristic by regression

The true linear characteristic, unknown to us, is

$$y = mx + b = 3x + 1$$

The number of available measurements is $n = 10$ and $x_i = 1, 2, \ldots, 10$.

Let the measurement noise v be specified as Gaussian zero-mean, that is, $\mu_x=0$ and an rms value of $\sigma_v = 0.2$. This noise intensity is not known to the observer, it is given here only for the purpose of illustrating a hypothetical experiment. The observable measurement values are $z_i = y_i + v_i$. It is necessary to provide the estimates of b and m and the variances of these estimates based on the available $n=10$ measurements.

In Table 1.1, all the required numerical values are prepared for the computation of $\hat{b}$ and $\hat{m}$ in accordance with (1.8.26), (1.8.27), (18.38), and (18.39). Their respective numerical values are:

$$\Sigma x_i^2 = 385; \qquad n\Sigma x_i^2 = 3850; \quad \Sigma x_i = 55$$

$$\Sigma x_i z_i = 1218; \qquad (\Sigma x_i)^2 = 3025; \quad \Sigma z_i = 175.86$$

Substituting these numbers into (1.8.26) and (1.8.27) results in

$$\hat{b} = \frac{175.86\times385 - 175.86\times55}{3850 - 3025} = 0.8573$$

$$\hat{m} = \frac{10\times1218 - 175.86\times55}{3850 - 3025} = 3.04$$

With the solutions for $\hat{b}$ and $\hat{m}$, we can compute σ_z^2 in accordance with (1.8.38),

$$\hat{\sigma}_z^2 = \frac{1}{10}\sum_1^{10}(3.04x_i + 0.8573 - z_i)^2 = \frac{1}{10}\sum_1^{10}\xi_i^2$$

$\hat{\sigma}_z$ is in itself only an estimate of σ_z, since it is based solely on 10 numbers. The corresponding numerical values are given in Table 1.1.

Table 1.1

x_i	$y=3x+1$	$z_i=y_i+v_i$	$x_i z_i$	x_i^2	ξ_i	ξ_i^2
1	4	3.746	3.746	1	.1592	0.0228
2	7	7.175	14.35	4	−.2372	0.0563
3	10	10.01	30.30	9	−.0297	0.0008
4	13	12.69	50.76	16	.3187	0.1016
5	16	16.11	80.55	25	−.0514	0.0026
6	19	19.17	115.0	36	−.0736	0.0053
7	22	22.05	154.3	49	.0822	0.0068
8	25	24.93	199.4	64	.2441	0.0596
9	28	27.78	250.0	81	.4354	0.1896
10	31	30.87	308.7	100	.3777	0.1427

The resulting values are

$$\hat{\sigma}_z^2 = 0.0588$$

In accordance with (1.8.38) and (1.8.39) and the preceding numerical values, we have

$$\hat{\sigma}_{\hat{b}}^2 = \frac{385}{825} \times 0.0588 = 0.0161; \qquad \hat{\sigma}_{\hat{b}} = 0.127;$$

$$\hat{\sigma}_{\hat{m}}^2 = \frac{10}{825} \times 0.0588 = 0.00073; \qquad \hat{\sigma}_{\hat{m}} = 0.0266;$$

This example demonstrates that after determining $\hat{b}$ and $\hat{m}$, it is possible to compute estimates of their variances based on the n measurements z_i. These estimates become more and more accurate as n increases. It can be shown that as $n \to \infty$, $\hat{\sigma}_{\hat{b}} \to \sigma_{\hat{b}}$ and $\hat{\sigma}_{\hat{m}} \to \sigma_{\hat{m}}$. The foregoing computation is useful, even though it is based on only 10 measurements.

1.9 Deflection instruments

Excitation, response, and display

Deflection instruments are a class of measurement devices in which the sensitive element is excited by a force or torque causing a translatory or rotational deflection. On settling at a position of equilibrium of forces or torques, the deflection represents the required measurand. Instruments in this class are therefore known as *deflection instruments.* This deflection is, in general, directly linked to a display, or a pickoff, which provides the interface to the monitoring or control functions required for the operation of the system. Deflection instruments are typically designed to provide a linear relationship between the input x and the output y. This linear relationship is usually assured by the inherent physical law underlying the response of the sensitive element. For example, in a conventional spring-mass accelerometer, the response, which in this case is the force acting on the proof mass, is proportional to the acceleration in accordance with Newton's second law. This force must be balanced and at the same time transduced in order to provide a displayable quantity. The simplest transduction is a linear spring that translates the force due to the acceleration to a measurable linear displacement. Thus, let a be the acceleration, m the mass, F the force, k the spring constant, y the deflection, and y_h the residual deflection due to hysteresis resulting in a bias. In simple designs, f is the dry friction opposing the motion of the proof mass. Disregarding the force of gravity, we have in the steady state

$$F = ma = k(y - y_h) + f \tag{1.9.1}$$

or

$$y = \frac{m}{k}a + y_h + \frac{f}{k} \tag{1.9.2}$$

In order to provide a measurable signal which represents the deflection y, a pickoff is required. In the simplest realizations the pickoff is a potentiometer; otherwise it is synchro or a variable differential transformer. The pickoff introduces the measurement uncertainty v, including noise. Thus, the output is

$$z = \frac{m}{k}a + y_h + \frac{f}{k} + v \tag{1.9.3}$$

A sketch of this deflection-type accelerometer is shown in Figure 1.19.

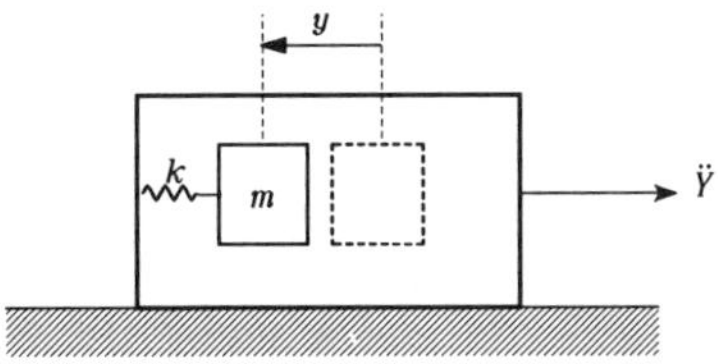

Figure 1.19. Elementary spring-loaded accelerometer.

This simple deflection instrument will be ideally linear if the spring constant k is a precise constant that is independent of the deflection y. Imperfection of k introduces nonlinearity. Imperfect knowledge of m or k causes *scale-factor* uncertainty. It should be noted that since the sensitive element in this example is excited by force, the nature of the restoring element determines what the deflection y represents. In this example, this is k^{-1}, which has the dimension (cm/gram). Thus, the spring scales the force ma into the deflection y which represents the force in cm. The example also demonstrates that even if the sensitive element ideally responds to the input, as is the case for mass and acceleration, it is the transducer or the restoring element that determines precision, linearity, and repeatability. Another critical element is the precision of the pickoff. Some important typical tradeoffs can be brought to light with the aid of the deflection accelerometer described. These will be presented in detail in Section 1.10 with the dynamics of the sensor.

Basic tradeoffs

At this point, however, some important tradeoff factors can be identified by inspection of (1.9.3). In order to reduce the friction error f/k for a given level of friction, k, the spring stiffness should be increased. But doing so would reduce the sensitivity m/k to the input a. Such a reduction decreases the deflection y. This, in turn, would result in poorer precision of the pickoff because its uncertainties would be larger in comparison with the maximum deflection y_{max}. Therefore, to preserve a desired sensitivity, m must be increased, that is, the weight of the sensor will be higher, if the required precision is higher. Conversely, a small weight implies small y so that a low -friction f is required in a high-grade pickoff.

The role of the restoring force

To reiterate; It is the *restoring force* that decides the nature of the measurement. Consider the device shown in Figure 1.20. Here, the mass m is placed in a casing C that is filled with a fluid having a viscous friction coefficient D. The device is rigidly placed in a vehicle that accelerates by $\dot{V}$. The relative velocity between the casing and the mass is v. The total acceleration with respect to inertial space is $\dot{V}-\dot{v}$, and the equation relating the forces involved is

$$m(\dot{V} - \dot{v}) = Dv \tag{1.9.4}$$

where Dv is the viscous resistance restoring force of the fluid. As long as the vehicle accelerates, that is, $\dot{V} \neq 0$, m keeps moving relative to the casing. It is easily seen that in the steady state, $\dot{v} \to 0$, and (1.9.4) reduces to

$$v = \frac{m}{D}\dot{V} \tag{1.9.5}$$

The distance y that the mass has traveled in the casing is the displayed quantity given by

$$y = \int_0^t v d\tau = \frac{m}{D}\int_0^t \dot{V} d\tau = \frac{m}{D}V \tag{1.9.6}$$

which is proportional to the velocity V. Thus, with a viscous restoring force which is proportional to velocity, we have, in essence, devised a *velocimeter*.

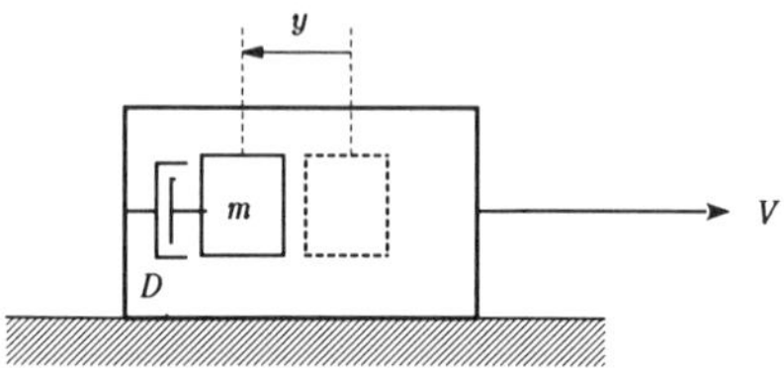

Figure 1.20. Velocimeter based on viscous balancing force

In the prceding examples the sensitive element, namely, the mass m, responded to acceleration, producing a force that had to be balanced by force. It is this *balancing by force, or torque*, that is common to all deflection instruments. However, the sensitive element can be excited by other physical variables or parameters, such as temperature, pressure or magnetic or electrostatic fields. If they manifest themselves as forces or torques, they have the effect of additive errors.

A typical example is the conventional angle-of-attack sensor; a sketch of this instrument is shown in Figure 1.21. A unit of four small winglet-shaped blades is gimbaled freely at the two-axis pivot, as shown in Figure 1.21. It is statically balanced to remove the effect of gravity. When exposed to the free air stream, an aerodynamic torque is generated and the blades deflect so as to point into the direction of the airflow. A vertical inclination with respect to the longitudinal axis of the aircraft is the angle of attack α, and a lateral inclination is the side-slip angle β. Any angular deviation δ of the blade from α causes an aerodynamical torque in the opposite sense to δ and approximately proportional to it, thus exhibiting a stable system like a simple pendulum. In this example, the sensitive element, i. e., the blade, responds to aerodynamic force but is naturally balanced by that very same force. The measurement of α is provided by a low-torque potentiometer.

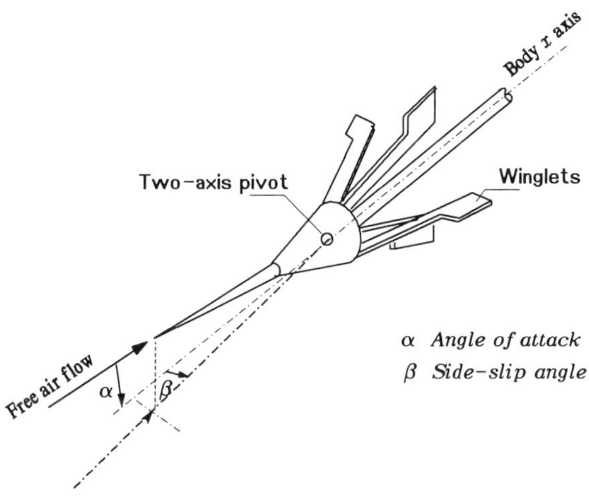

Figure 1.21. Angle-of-attack and side-slip sensor.

Another familiar example is the conventional magnetic compass. It is a deflection-type sensor responding to the force of the magnetic field, and the restoring force is again provided by the very same medium. A deviation δ of the needle from the direction of the local north-pointing magnetic field generates a restoring torque proportional to δ. In this case, only the background dial can serve as a display. Electrical output by means of a pickoff is unfeasible in its basic form. North-seeking devices with electrical readouts will be described in Chapter 4. The foregoing examples of the angle-of-attack sensor, schematically shown in Figure 1.21, the elementary compass, or any pendulous device for that matter represent deflection sensors with natural static stability. They align themselves along the direction of a force or flow field and measure its angles relative to the body axes, reporting this information to the vehicle. The measurement thus provided actually reflects the angular deflection of the vehicle from the direction of the sensor, that is, the direction of the force or flow field, in a "tail wags dog" fashion. In this sense, the vertical gyroscope, the directional gyroscope, the gyro compass and the inertial platform belong to the class of deflection devices. In the latter ones, it is the inertia that provides the physical property of maintaining the spatial orientation of the sensitive element. In this case, it is the spinning rotor of the gyroscope. The reference direction is solely determined by the initial orientation alignment of the spin axis in inertial space.

Other devices, not necessarily linear as in the foregoing examples, that transduce a physical quantity into a deflecting force are, include the *pitot tube* for determining relative air speed from the measurement of pressure difference, which actuates a deflecting element such as a manometer, or a *Bourdon tube*, or *bellows*. The common property of the spiral-like Bourdon tube and the bellows is the inherent direct amplification of deflection that their structure provides. The same techniques are used in the barometric altimeter. Detailed technical accounts on these techniques are given in [1]. Another example in which a physical quantity translates into a deflecting force is the *bi-metal* element, which bends under the influence of temperature variations.

A number of deflection measurement devices that have particular applications in aerospace systems, such as accelerometers and gyroscopes, are discussed in greater detail in Chapters 3 and 4, respectively.

1.10 Balancing instruments

The force balance principle

The principle of balancing is as ancient as measurement itself. Contrary to the previously described principle of deflection, balancing is based on nulling the deflection of the sensor. Nulling is achieved by means of well-known forces such as precise weights or by well controlled balancing forces mostly by means of electromagnetic current to force transducers. In the following elementary analysis all inherent lags and delays of the sensor elements are disregarded. Consequently, the important aspects of response time and stability are not discussed in this chapter. The principle of operation of force balance sensors is best illustrated by Figure 1.22, again using the example of the pressure diaphragm.

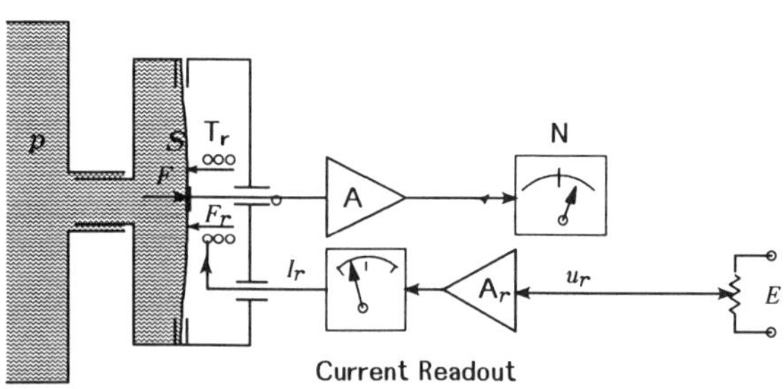

Figure 1.22. Principle of measurement by force balancing.

Linearization by force balance

The output signal y in response to pressure p may have a substantial nonlinearity due to both the nonlinear relationship between the diaphragm deflection δ and pressure p and the nonlinearity of the detector. y is amplified by the amplifier A, which drives the null-point indicator N. A balancing control command u_r feeding the amplifier A_r provides the current that drives the forcer coil T_r, which exerts the balancing force F_r, to restore the the diaphragm deflection δ to zero. The current I_r displayed by A_r is almost perfectly proportional to F_r, and thus to p. Since the diaphragm hardly deflects, its nonlinearity becomes negligibly small and the excellent linear characteristic of the force transducer T_r now governs the measurement.

Disregarding the dynamics of the various elements and effects of noise, the foregoing is best illustrated by the elementary generic description of a force balance device shown in Figure 1.23.

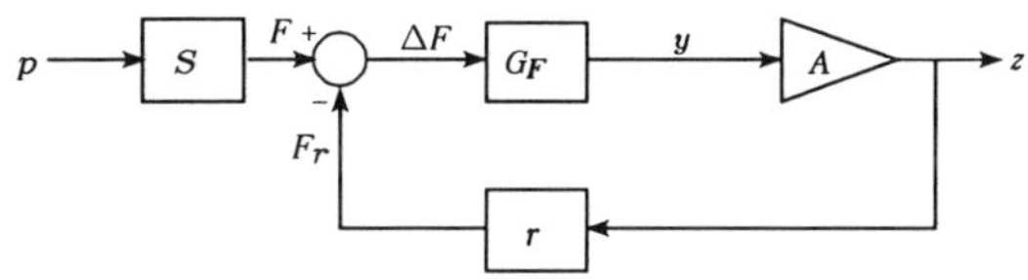

Figure 1.23. Closed loop force balance arrangement.

Let F be the force acting on the membrane **M** and let Y be the corresponding output of the detector bonded to it, obeying the characteristic

$$Y = f(F) \tag{1.10.1}$$

The small net force on the membrane is

$$\Delta F = F - F_r \tag{1.10.2}$$

Since ΔF is small, we expand (1.10.1) into a Taylor series, in accordance with

$$y = \left.\frac{\partial Y}{\partial F}\right|_{F=0} \Delta F \triangleq G_F \Delta F \tag{1.10.3}$$

where G_F is the sensitivity coefficient of the membrane-detector assembly. Define the following parameters:

S = the membrane surface in cm^2.
A = Amplification of factor of A in amp/volt
p = pressure in kg/cm^2.
p_m = measured pressure in amp
r = the force coefficient of T_r in gram/amp

We have the following relationships:

$$F_r = Ary \tag{1.10.4}$$

$$y = (F - F_r)G_F = (F - Ary)G_F \tag{1.10.5}$$

$$y = \frac{FG_F}{1 + ArG_F} \tag{1.10.6}$$

$$p_m = Ay = \frac{AG_F}{1 + ArG_F} Sp \tag{1.10.7}$$

If $ArG_F \gg 1$, we essentially have

$$p_m = \frac{S}{r} p \text{ (amp)} \tag{1.10.8}$$

Thus, the pressure p is determined by the measured current p_m and the scaling coefficient S/r, namely:

$$p = \frac{r}{S} p_m \text{ (gr-f/cm}^2\text{)} \tag{1.10.9}$$

Equation (1.10.9) shows three significant results achieved by the closed-loop force balance arrangement, namely:

1. The nonlinearity of the membrane-detector assembly is essentially eliminated as a result of the force balance principle which involves only a small net force ΔF for which the characteristic is essentially linear.

2. All possible scale factor errors due to variations in the open loop sensitivity coefficient, $G_F + \Delta G_F$, are rendered negligibly small.

3. The scaling precision is determined by the precision of an electromagnetic linear force transducer T_r which can exhibit a high degree of linearity, precision and repeatability.

4. Again, as in the example of the viscous velocimeter, it is the restoring force F_r provided by T_r that decides the nature of the measurement.

The effect of additive noise

We return to the foregoing example of the force balance pressure sensor shown in Figure 1.23, but now we include noise and use a more general notation.

The sensitive element is assumed to obey the characteristic relationship,

$$Y = f(X) \tag{1.10.10}$$

At the origin, i.e., $X = 0$, the sensitivity is

$$G_X \triangleq \left.\frac{\partial Y}{\partial X}\right|_{X=0} \tag{1.10.11}$$

If additive noise v is assumed to be at the output of the sensitive element as shown in Figure 1.24, the following relationships hold:

$$\Delta X = X - X_r \tag{1.10.12}$$

$$y = G_X \Delta X \tag{1.10.13}$$

$$z = (y + v)A \tag{1.10.14}$$

$$x_r = z_r = (y + v)Ar \tag{1.10.15}$$

Substituting and rearranging, the result for y is

$$y = \frac{G_X}{1 + G_X Ar} x - \frac{G_X Ar}{1 + G_X Ar} v \tag{1.10.16}$$

Substituting (1.10.16) into (1.10.14) and rearranging, we have

$$z = \frac{G_X A}{1 + G_X Ar} x + \frac{A}{1 + G_X Ar} v \tag{1.10.17}$$

If $G_X A r \gg 1$, the output z is approximately

$$z \cong \frac{x}{r} + \frac{1}{G_X r} v \tag{1.10.18}$$

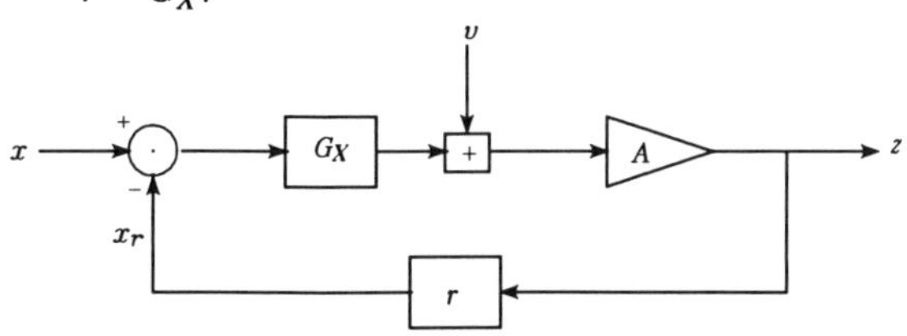

Figure 1.24. Force balance sensor with noise at the sensitive element output.

From (1.10.18), we conclude the following:

1. The input–output relationship is linearized and essentially unaffected by the nonlinearity of the sensitive element. It is precise to the precision of the

transducer T_r.

2. The additive noise v, in this simplified basic form is not attenuated by the action of the force balance configuration. The result only indicates that the noise v, defined at the level of y, is referred back to the level of the input x.

3. The scale factor is precise to the knowledge of the scale factor r of T_r.

The input-output relationship (1.10.17) which is the elementary expression for a closed-loop control system, directly discloses the very significant property of the force balance principle, namely, the reduction of its sensitivity to sensor parameter uncertainties. This is best illustrated by the following analysis.

Sensitivity to variations in parameters

For brevity, denote $G_X A \triangleq g_x$; denote a deviation in g_x by Δg_x; and denote the open-loop sensitivity by S_0, where

$$S_0 \triangleq \frac{\Delta g_x}{g_x} \tag{1.10.19}$$

Denote the closed loop transfer function by T_x, where

$$T_x = \frac{g_x}{1 + g_x r} \tag{1.10.20}$$

The sensitivity of T_x to variations in g_x is

$$\frac{\mathrm{d}T_x}{\mathrm{d}g_x} = \frac{1}{(1 + g_x r)^2} \tag{1.10.21}$$

$$\Delta T_x = \frac{\mathrm{d}T_x}{\mathrm{d}g_x} \Delta g_x = \frac{1}{(1 + g_x r)^2} \Delta g_x \tag{1.10.22}$$

Define the closed-loop sensitivity S_c by

$$S_c \triangleq \frac{\Delta T_x}{T_x} \tag{1.10.23}$$

From (1.10.20) and (1.10.22), we have

$$S_c = \frac{1}{1 + g_x r} S_0 \tag{1.10.24}$$

The conclusion therefore is that the sensitivity S_c of the closed loop sensor to sensor parameter variations is smaller by $(1 + g_x r)$ than the open loop sensitivity S_0.

The sensitivity is a measure of the *robustness* of the sensor. The smaller the sensitivity S_c, the more robust the sensor.

In high quality aerospace sensors, a loop-gain g_x on the order of 1000 or more can be implemented in the useful frequency range of 1 kHz. This implies that excellent qualities of precision and repeatability can normally be achieved by the force balance principle. The open-loop sensitivity S_0 with medium -quality technology is on the order of 1% under normal environmental variations. Thus, with a loop-gain $g_x r$ on the order of 1000, excellent robustness on the order of 10^{-5} can be achieved. The limitations on the achievable precision are discussed later.

1.11 Imperfections and limitations on precision

Sensor dynamics and parameter tradeoffs

The principal limiting factors of precision have been indicated in the foregoing sections. In this section the important aspect of the dynamics of the sensor elements is introduced and the effects of the associated time lags on the achievable precision are discussed.

The general model is given in (1.5.7), and the corresponding input-output relation is

$$z(s) = G(s)x(s) + v(s) \tag{1.11.1}$$

where $G(s)$ will be described by the rational transfer function

$$G(s) \triangleq \frac{N(s)}{D(s)} \tag{1.11.2}$$

The noise $v(s)$ is not correlated with $x(s)$, and

$$N(s) \triangleq \sum_{i=0}^{n} n_i s^i \tag{1.11.3}$$

$$D(s) \triangleq \sum_{j=0}^{m} d_j s^j \tag{1.11.4}$$

where $m \geq n + 1$. We now disregard all nonlinearities and examine the nature of $G(s)$ in the open- and closed-loop configurations. The rational transfer function form is often represented in factored form, which readily provides Bode describing functions in the real frequency domain ω and root locus characterizations in the complex frequency domain $s = \sigma + j\omega$ so that

$$N'(s) = \prod_{i=0}^{n} (z_i + s) \tag{1.11.5}$$

$$D'(s) = \prod_{j=0}^{m} (p_j + s) \tag{1.11.6}$$

The poles p_j and the zeros z_i can be real or complex. Thus, $G(s)$ is given by

$$G(s) = \frac{N'(s)}{D'(s)} \tag{1.11.7}$$

Performing a partial fraction expansion of (1.11.7), we obtain the modal form

$$G(s) = \sum_{j=0}^{m} \frac{r_j}{(s + p_j)} \tag{1.11.8}$$

where r_j are the residues of $G(s)$. The modal form is particularly useful in obtaining time responses by means of the inverse Laplace Transform, specifically, step and impulse responses of the dynamical system. These time responses take the following form

$$g(t) = \sum_{j=0}^{m} r_j e^{-p_j t} \tag{1.11.9}$$

These relationships along with stability criteria and dynamic response descriptors, are given in grater detail, among others, in [5].

With this background, we now reiterate the example of the accelerometer introduced in Section 1.9 and reformulate the force (1.9.1) to include the Newtonian force and a viscous resistance force. The nonlinearity represented by y_h and the dry friction force f are disregarded here, but the viscous resistance force f_D is introduced. A schematic of the accelerometer is shown in Figure 1.25. Under these assumptions, the force equation, in accordance with Equation (1.9.1), is replaced by

$$F = ma = ky + f_D \tag{1.11.10}$$

and a, the acceleration with respect to inertial space, is

$$a = (\ddot{Y} - \ddot{y}) \tag{1.11.11}$$

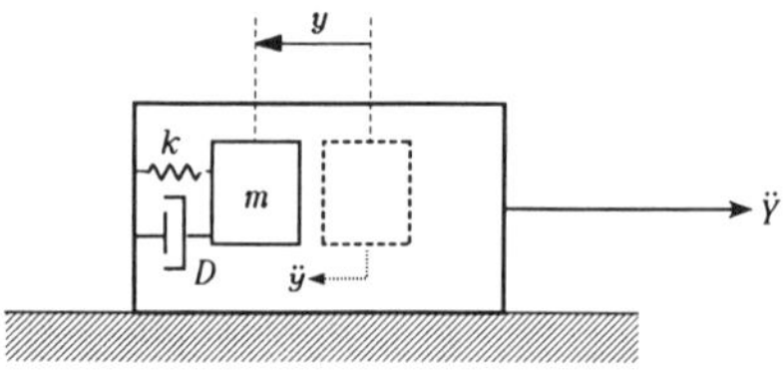

Figure 1.25. Elementary spring-loaded and damped accelerometer.

f_D is given by

$$f_D = D\dot{y} \tag{1.11.12}$$

where D is the viscous friction coefficient. Substituting the last two expressions into (1.11.10), we have

$$m(\ddot{Y} - \ddot{y}) = D\dot{y} + ky \tag{1.11.13}$$

Dividing by m and rearranging we have

$$\ddot{y} + \frac{D}{m}\dot{y} + \frac{k}{m}y = \ddot{Y} \tag{1.11.14}$$

Laplace transforming and solving for y, we have

$$y(s) = \frac{1}{s^2 + D/m\,s + k/m}\ddot{Y}(s) = G(s)\ddot{Y}(s) \tag{1.11.15}$$

This typical second-order system is characterized by its natural frequency ω_n and its damping coefficient ς_n as follows:

$$\omega_n = \sqrt{k/m} \tag{1.11.16}$$

$$\varsigma_n = \frac{D}{2\sqrt{km}} \tag{1.11.17}$$

In factored form, we have

$$y(s) = \frac{1}{(s + \omega_n\varsigma_n + j\omega_n)(s + \omega_n\varsigma_n - j\omega_n)}\ddot{Y}(s) \tag{1.11.18}$$

Performance parameters and further design tradeoffs

Additional tradeoff considerations come to light. A desirable value for ς_n is $\varsigma_n = 0.7$. This accelerometer model uses three primary independent physical parameters for the sensitive element, namely, m, k, and D, and two secondary parameters $y_{\max}$ and y_r relating to the pickoff. To analyze performance quality and provide the design basis, these physical parameters must be tied to the principal specification parameters. These parameters typically are

ω_n or $f_n = \omega_n/2\pi$, commonly set at 20–50 Hz;
ς_n normally set in the range of 0.5– 0.8;
$a_{\max}$, the maximum value of input acceleration;

a_r, the smallest resolvable a or null-point uncertainty;
y_{max}, the largest permissible deflection;
y_r, the smallest resolvable deflection;
D_r, the dynamic range, typically 100—10000 and
w_m, maximum permissible mass in gr-f.

Recall that in accordance with (1.4.3), $D_r \triangleq a_{max}/a_r$. We define the *static sensitivity* S, which from the foregoing, is

$$S \triangleq \frac{y_{max}}{a_{max}} = \frac{y_r D_r}{a_{max}} = \left.\frac{y(s)}{\ddot{Y}(s)}\right|_{s \to 0} = \frac{1}{\omega_n^2} = \frac{m}{k} \quad (1.11.19)$$

The following relationships between the physical parameters and the specification parameters result. From (1.11.19) we have

$$k = \frac{w_m}{g}\omega_n^2 \quad (1.11.20)$$

$$\omega_n^2 = \frac{a_{max}}{y_r D_r} \quad (1.11.21)$$

From (1.11.16)

$$D = 2\sqrt{km}\,\varsigma_n = 2\sqrt{m^2\omega_n^2}\,\varsigma_n = 2m\omega_n\varsigma_n \quad (1.11.22)$$

EXAMPLE 1.11-1

The following performance specifications are given:
$a_{max} = 10$ g; $\varsigma_n = 0.7$; $f_n \geq 20$ Hz; $D_r = 1000$.

The technical specifications are:

$w_m \leq 20$ gr- f; $y_r = 0.5\times10^{-5}$ meter

The natural frequency from (1.11.20) is,

$$\omega_n^2 = \frac{10\times9.81}{0.5\times6\times10^{-4}\times6\times1000} = 19620; \quad \omega_n = 140 \text{ rad/sec};$$

$$f_n = 22.3 \text{ Hz};$$

For $w_m = 20$ gr-f.,

$$k = \frac{20}{9.81\times1000}\times 19620 = 400 \text{ gr/cm}$$

From (1.11.22), the viscous coefficient is:

$$D = 2\times\frac{20}{9810}\times140\times0.7 = 0.4 \text{ gr-f/cm sec}^{-1}$$

From equation (1.11.21), one can see that a high-bandwidth accelerometer implies a low-sensitivity S or a high stiffness and, consequently, it can measure an acceleration not smaller than 0.01 g = 10 mg.

The forgeoing example reveals that in order to increase the bandwidth f_n for example by a factor of 10, it is necessary to decrease y_r by a factor of 100 or to decrease D_r by the same factor. Thus, within a given state of the spring-mass technology, it is evident that the bandwidth f_n and the dynamic range D_r are strongly conflicting design parameters. In high-performance accelerometers such as those required in inertial navigation, typical performance requirements are $f_n = 800$ Hz and $D_r = 10^4$. It is impossible to meet such requirements with the conventional deflection instruments described here. As indicated earlier, the implementation of the force balance principle resolves much of this conflict so that high-performance inertial grade precision can be achieved.

Very substantial improvements in resolving this conflict are achieved by force sensors, which have only minute deflection and translate the force directly into an electrical measurable signal. Examples are piezoresistive, resistive, or variable-capacitance accelerometers, described in Chapter 3. These devices achieve bandwidths on the order of 1000 Hz and dynamic ranges on the order of 1000 and are considered to be medium quality-sensors.

Similar considerations are pertinent in other deflection-type sensors such as rate gyroscopes or angle-of-attack sensors. Fortunately, not every aerospace variable or parameter requires a high-bandwidth or very large dynamic range. The principal reasons for requiring large bandwidth are to provide stability in the operation of closed loop-control systems, such as missile or aircraft auto pilots and gimbaled stabilized platforms, or to minimize dynamical errors whenever multiple integration of acceleration is required, such as in inertial navigation.

Enhancement of bandwidth

A large bandwidth or a small response time is often a principal requirement in sensor specification for a number of reasons. It may be necessary to measure the frequency spectrum of vibrations up to 1000 Hz or more by means of an accelerometer; in the automatic control of an aerospace vehicle, the lag of a rate gyroscope may cause instability of the controlled system or impose a severe limitation on the permissible loop gain; inertial navigation technology is based on double integration of the measured acceleration of a flight vehicle. If the accelerometer has a long response time, substantial errors in the determination of the position of the vehicle may be caused as illustrated by the following example.

EXAMPLE 1.11-2: Range error due to accelerometer lag

An accelerometer is mounted in a missile with its input axis x parallel to the direction of the anticipated flight path. A booster rocket accelerates the missile for a short time at a constant acceleration $a_x = 50\text{ g} \cong 500\text{ m sec}^{-2}$. The dynamical model of the accelerometer is given by

$$\frac{y(s)}{a_x} = G_x(s) = \frac{1}{1+\tau s}$$

where the response time constant is assumed to be $\tau = 100$ msec. The impulsive acceleration is modeled as an impulse (delta) function of magnitude a_x. The error between the measured and actual acceleration is

$$\epsilon_D = \left(1 - \frac{1}{1+\tau s}\right) a_x = \frac{\tau s}{1+\tau s} a_x$$

The error in range ϵ_x is given by

$$\epsilon_x(s) = \frac{1}{s^2}\epsilon_D(s) = \frac{\tau s}{s^2(1+\tau s)} a_x = \frac{\tau}{s(1+\tau s)} a_x$$

By inverse Laplace transforming, the range error $\epsilon_x(t)$ is

$$\epsilon_x(t) = \tau\left(1 - e^{-t/\tau}\right) a_x \to a_x\tau \qquad (\text{for } t \gg \tau)$$

Thus,

$$\epsilon_x = a_x\tau = 500 \times 0.1 = 50 \text{ m}.$$

This error may be unacceptable. Clearly, a response time of around $\tau = 1$ msec would reduce this error by a factor of 100.

The natural time constant τ is dictated by the physical limitations of the proof mass and pickoff sensitivity. Substantially reducing it is not a straightforward matter. However, it will now be demonstrated that τ can be very substantially reduced by the implementation of the principle of forced balance. We also address its effect on the dynamic range. $G(s)$, as defined in (1.11.2) and (1.11.4), is placed in the closed loop-forced balance configuration as shown in Figure 1.26. At this stage, the problems of design to a desired dynamic response and the assurance of stability are disregarded. These issues will be addressed in Chapters 3 and 4.

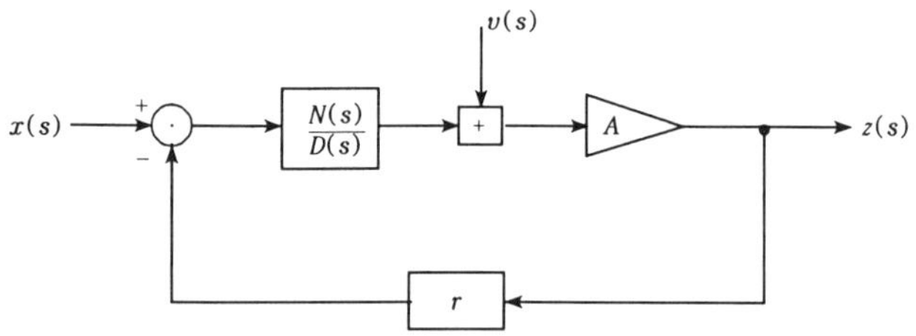

Figure 1.26. Closed loop force balance sensor with dynamical model.

The input–output relationship is

$$z(s) = \frac{\frac{AN(s)}{D(s)}}{1 + \frac{ArN(s)}{D(s)}} x(s) + \frac{A}{1 + \frac{ArN(s)}{D(s)}} v(s)$$

$$= \frac{AN(s)}{D(s) + ArN(s)} x(s) + \frac{AD(s)}{D(s) + ArN(s)} v(s) \qquad (1.11.23)$$

Denote the closed loop denominator by

$$\Delta(s) \triangleq D(s) + ArN(s) \qquad (1.11.24)$$

Setting $r = 1$ for simplicity, the characteristic (1.11.24), in explicit terms of its coefficients, takes the form

$$\Delta(s) = s^m + \sum_{i=0}^{n} (d_i + c_i A) s^i \qquad (1.11.25)$$

It is clear that the addition of the terms $c_i A$ to d_i can substantially increase the values of the poles p_i in (1.11.6), thus increasing the bandwidth of the sensor or in accordance with (1.11.9), reducing the response time. It therefore appears that the larger the gain A, the better. However, this is not quite as straightforward as it appears, for two principal reasons: 1) as will be demonstrated later in the context of stability, excessive gain tends to drive some of the closed-loop poles toward and into the right half complex frequency plane and thus drive the sensor toward uncontrolled oscillations; and 2) observation of (1.11.23) discloses that if A is made excessively large, the second term in the denominator becomes dominant so that it tends to the form

$$z(s) \Rightarrow \frac{AD(s)}{ArN(s)} v(s) = \frac{D(s)}{rN(s)} v(s) \qquad (1.11.26)$$

Since $m \geq n+1$, namely, the power of the closed-loop numerator tends to be larger than the corresponding denominator so that high frequency components in $v(s)$ are excessively amplified. Denoting the nominal scaling value of r by r_s, the scaled output z_c is given by

$$z_c(s) = r_s z(s) \qquad (1.11.27)$$

Explicitly,

$$z_c(s) = \frac{A\,N(s) r_s}{D(s) + ArN(s)} x(s) + \frac{A\,D(s) r_s}{D(s) + ArN(s)} v(s) \qquad (1.11.28)$$

$z_c(s)$ expresses the sensor output in actual physical units, such as grams, and meter/sec. Its precision is determined by the knowledge of r_s. Its uncertainty causes a *scale factor error* r_c/r as clearly seen in the first term of (1.11.28) for large values of A. The overall sensor error is now defined as

$$\epsilon(s) = x(s) - z_c(s)$$

$$= [1 - \frac{AN(s)r_s}{D(s) + AN(s)r}]x(s) - \frac{AD(s)r_s}{D(s) + AN(s)r}v(s)$$

$$= \frac{D(s) + N(s)A(r - r_s)}{D(s) + AN(s)r}x(s) - \frac{AD(s)r_s}{D(s) + AN(s)r}v(s) \qquad (1.11.29)$$

The first term in (1.11.29) is the *dynamical error* $\epsilon_d(s)$ including the *scale factor error*, and the second term is the *noise error* $\epsilon_v(s)$. Defining,

$$F_d(s) \triangleq \frac{D(s) + N(s)A(r - r_s)}{D(s) + AN(s)r} \qquad (1.11.30)$$

$$F_v(s) \triangleq \frac{AD(s)r_s}{D(s) + AN(s)r} \qquad (1.11.31)$$

We now have

$$\epsilon_d \triangleq F_d(s)x(s) \qquad (1.11.32)$$

$$\epsilon_v \triangleq F_v(s)\, v(s) \qquad (1.11.33)$$

The total error $\epsilon(s)$ can therefore be interpreted by the equivalent scheme shown in Figure 1.27.

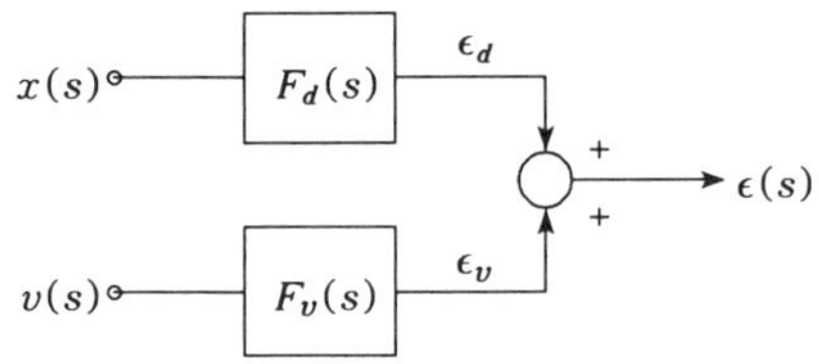

Figure 1.27. Equivalent scheme for total error.

It is of interest to formulate the asymptotic expressions $\lim F_d(s)$ and $\lim F_v(s)$ for $s \to 0$ and $s \to \infty$. With the explicit forms for $N(s)$ and $D(s)$ in (1.11.3) and (1.11.4), and in view of (1.11.31), we have for $r = r_c$,

$$\lim_{s \to 0} F_d(s) = \frac{1}{1 + A\dfrac{c_0}{d_0}}; \quad \lim_{s \to \infty} F_d(s) = \frac{A}{1 + A\dfrac{c_0}{d_0}} \tag{1.11.34}$$

$$\lim_{s \to 0} F_v(s) = \frac{A}{1 + A\dfrac{c_0}{d_0}}; \quad \lim_{s \to \infty} F_v(s) = A \tag{1.11.35}$$

These asymptotic expressions demonstrate that for large values of A, regardless of the detailed form of $G(s)$, ϵ_d approaches $1/(Ac_0/d_0)$ at low frequencies and approximately d_0/c_0 at high frequencies, while ϵ_v approaches d_0/c_0 at low frequencies and A at high frequencies. In other words, the dynamic error is effectively suppressed only at relatively low frequencies. The noise error is barely suppressed at low frequencies but is highly amplified at high frequencies.

EXAMPLE 1.11–3: Output noise of FBS accelerometer

In order to illustrate rate the effect of closed-loop force balance on bandwidth and noise, we describe the sensitive element by the simple first-order lag,

$$G(s) = \frac{N(s)}{D(s)} = \frac{1}{1 + \tau s} = \frac{a}{a + s}; \qquad a \triangleq \frac{1}{\tau}$$

In accordance with (1.11.23),

$$z(s) = \frac{aA}{a+s+aAr}x(s) + \frac{(a+s)A}{a+s+aAr}v(s)$$

$$= \frac{aA}{s+a(1+Ar)}x(s) + \frac{(a+s)A}{s+a(1+Ar)}v(s)$$

We recognize that the original bandwidth a has been increased to $a(1+Ar)$ or the original time constant τ has been reduced to the closed loop time constant τ_c,

$$\tau_c = \frac{\tau}{1+Ar}$$

Thus, implementing the closed-loop force balance principle reduces the sensor time constant by approximately a factor of A.

On the other hand, observing the noise term in $z(s)$ for large values of A is approximately determined by

$$z(s) = \frac{(a+s)A}{aAr}v(s) \cong \frac{a+s}{ar}v(s)$$

This result indeed demonstrates that as a result of the implementation of feedback, the numerator tends to have a higher degree in s than the denominator. This clearly implies that high-frequency components in $v(s)$ will be strongly amplified.

Therefore, a potential penalty for the enhancement of bandwidth as demonstrated in the foregoing can result in large unacceptable noise amplification. In order to demonstrate this effect, we assume that the detector after the sensitive element $G(s) = a/(a+s)$ is of the linear or rotary variable differential transformer, LVDT or RVDT type operating at a frequency $\omega = 2\pi f$, where f is on the order of 20 to 50 kHz. In addition to the rectified output, a residual ripple persists at the frequency $2f$ at its output. It is regarded as the noise component v, indicated in Figure 1.26. The error response (1.11.33), excited by the aforementioned sinusoidal signal $v = v_0 \sin \omega t$, describes the frequency response simply by a standard Bode plot given in the frequency range from $\omega=0$ to $\omega=1000\,\text{sec}^{-1}$. This is shown in Figure 1.28 with the amplification factor A as a parameter. The explicit expression for this Bode plot is given by setting $s=j\omega$ and formulating the norm $|F_v(j\omega)|$.

$$|F_v(j\omega)| = \frac{A\{a^4(1+A)^2+\omega^2[a^2(2(1+a)+A^2)+\omega^2]\}^{1/2}}{a^2(1+A)^2+\omega^2}$$

This example demonstrates the effect of a high gain loop closure on the noise amplification. The justification for the sinusoidal model for v is that it properly represents the actual situation where v is random wide band noise. This is so, because wide band random noise can be viewed as the sum of a large number of sine waves ranging from low frequencies to very high frequencies.

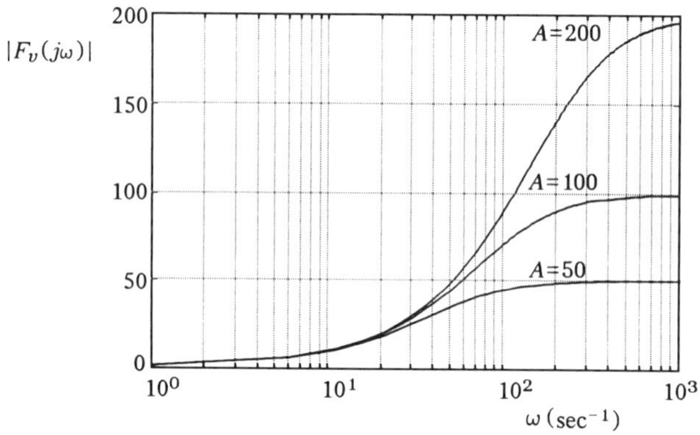

Figure 1.28. Effect of gain A on high-frequency noise.

From the foregoing it appears that for the closed loop model described in Figure 1.26, there exists no solution to overcome the conflict between bandwidth enhancement and noise amplification. The trivial remedy of placing a low-pass filter in series with the accelerometer, would of course substantially attenuate the noise power, but at the same time, it would also reduce the effective bandwidth of the accelerometer. However, this difficulty can be surmounted by providing A with a low-pass characteristic, namely, placing the low-pass filter inside the sensor loop. We shall now see that very substantial noise attenuation can be achieved while hardly reducing the closed loop sensor bandwidth. Assume again that $H(s) = a/(a+s)$ and the amplifier-filter gain be modeled by

$$A' = \frac{A}{1+s\tau_f} = \frac{A/\tau_f}{s+1/\tau_f}$$

Substituting A' into (1.11.23) and assuming for simplicity $r = 1$, we have

$$\frac{z(s)}{x(s)} = \frac{\dfrac{a}{a+s}\,\dfrac{A/\tau_f}{s+1/\tau_f}}{1+\dfrac{A/\tau_f a}{(a+s)(s+1/\tau_f)}} = \frac{A/\tau_f a}{s^2+(a+1/\tau_f)s+(a/\tau_f)(1+A)}$$

$$\frac{z(s)}{v(s)} = \frac{\dfrac{A/\tau_f}{s+1/\tau_f}}{1+\dfrac{A/\tau_f a}{(a+s)(s+1/\tau_f)}} = \frac{A/\tau_f(a+s)}{s^2+(a+1/\tau_f)s+(a/\tau_f)(1+A)}$$

The closed-loop response characteristic is now second-order. It is immediately evident that in order to prevent an oscillatory response, A cannot be made too large unless additional damping, not shown here, is introduced. Also, $z(s)/x(s) \to A/(1+A)$ as $s \to 0$, and $z(s)/x(s) \to 0$ as $s \to \infty$ while $z(s)/v(s) \to A/(1+A)$ as $s \to 0$ and $z(s)/v(s) \to 0$ as $s \to \infty$. As opposed to previous examples, the noise response tends to approach zero at high frequencies.

We now demonstrate these trends of $F_x(j\omega) \triangleq z(s)/x(s)$ and $F_v(j\omega) \triangleq z(s)/v(s)$ for the following parameters: $A = 100$, $a = 1$; and $\tau_f = 0.001$ sec, 0.005 sec, 0.01 sec. Results are demonstrated in Figures 1.29 and 1.30.

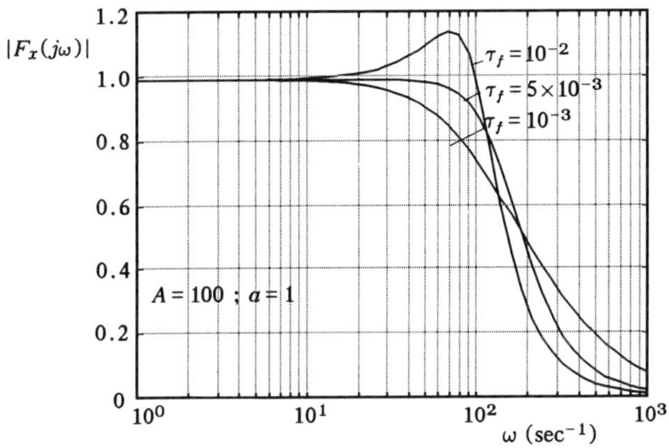

Figure 1.29. Effect of τ_f on the frquency response of the FBS.

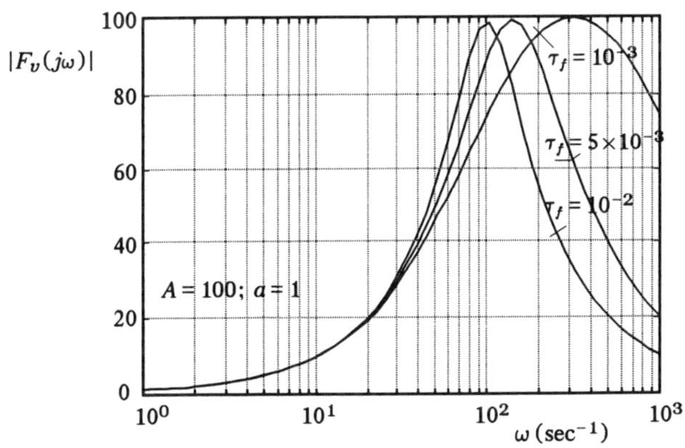

Figure 1.30. Effect of τ_f on the noise frequency characteristic.

Figure 1.29 clearly demonstrates that the closed-loop bandwidth maintains the approximate value of A and is essentially not affected by the filter time constant τ_f even though the frequency response tends to become less damped for the larger values of τ_f. However, the noise response shown in Figure 1.30 is strongly affected by τ_f. The larger τ_f, the narrower the noise bandwidth and, accordingly, the smaller the output noise power. These results should be compared to the curve corresponding to $A = 100$ given in Figure 1.28 for which $\tau_f = 0$ and for which the noise power is infinite. For example, the noise output power P_v for a unity white noise PSD corresponding to $\tau_f = 10^{-2}$, 5×10^{-3}, 10^{-3} is close to 5000, 10,000, and 50,000 respectively. The noise power is determined by spectral integration in accordance with Chapter 2. These typical noise characteristics indeed manifest themselves in such sensors as the Sundstrand Data Control Q-Flex QA-1000 series which is a highly sensitive forced balance accelerometer [8] whose performance analysis and partial data list are shown Table 3-1 in Chapter 3.

Considerations of bandwidth, dynamic range, and robustness

The foregoing discussion leads to the following conclusions:

1. With traditional open-loop electromechanical technology, the conflict between bandwidth and sensitivity imposes an unforgiving restriction on high-grade performance. Robustness to environmental variations is relatively poor.

2. Substantial improvement in terms of bandwidth and robustness is achieved by implementing force balance techniques. The penalty is increased complexity as a result of additional transducers and high-quality electronics.

3. The extremely high stability of quartz and silicon has opened new avenues of technology and design. These allow the use of open-loop devices with very good robustness to environmental variations while achieving very high bandwidths. These enhancements are due to the subminiature dimensions, which involve only minute inertias, and to the high stability of the materials, which permit precise and consistent calibration. Noise level is sufficiently low to permit high sensitivity. At present, these technologies already compete well with the force balance technique in performance, but they are by far superior in terms of reliability, volume, weight and cost. These issues will be discussed in chapters 6 and 7.

1.12 Effect of friction in instrument servomechanisms

Linear model of DC motor

In principle, the linear servomechanism is a closed-loop device in which the transducer is an electrical motor, usually direct current. It may be part of a measurement instrument, but often it operates as an actuator or signal follower in its own right. By *signal follower*, or tracker, we imply that the servomechanism copies an electrical signal into a mechanical, translatory or rotational motion.

The purpose of this section is to derive a simple model of the servomechanism that provides insight into the limiting factors that determine its achievable precision. It is assumed that the reader is familiar with the classical design methodology of linear servomechanisms with regard to stability and bandwidth. In this section, analysis focuses on the effect of bearing or brush friction in conjunction with the principal motor parameters. The torque equation is the basis of the analysis in the development of the model, namely,

$$T = K_u u - K_\omega \omega = I\dot{\omega} + T_f + D\omega \tag{1.12.1}$$

where

T — total torque on output shaft;
u — input voltage;
ω — angular rate of output shaft;
K_u — iput voltage coefficient;
K_ω — back EMF coefficient;
I — moment of inertia;
T_f — total dry friction torque;
D — viscous friction Torque.

The relationship (1.12.1) is described graphically in Figure 1.31 for the steady-state condition $\dot{\omega} = 0$ with u as a parameter.

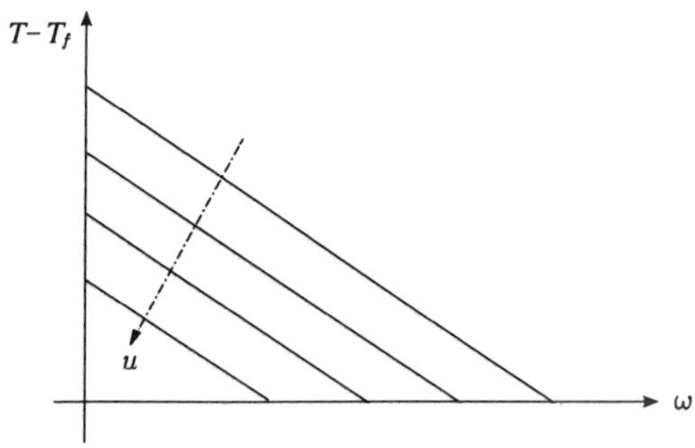

Figure 1.31. Torque-motor speed characteristic of a dc motor.

Under these conditions, we have

$$K_u u - (K_\omega + D)\omega = T_f \tag{1.12.2}$$

Setting $\omega = 0$, we have, for the threshold or starting voltage u_{th},

$$u_{th} = \frac{T_f}{K_u} \tag{1.12.3}$$

Solving ω from (1.12.2) we have

$$\omega = \frac{K_u u - T_f}{K_\omega + D} = \frac{K_u}{K_\omega + D}(u - u_{th}) \tag{1.12.4}$$

where $u_{th} \triangleq T_f/K_u$ is not necessarily a constant.

Equation (1.12.4) is graphically described in Figure 1.32. It may be a function of ambient conditions and the angular position of the shaft.

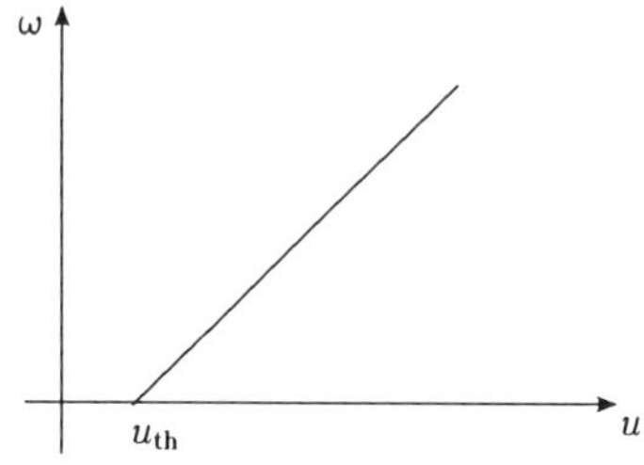

Figure 1.32. Angular rate versus input voltage of a dc motor.

In order to derive the dynamical model of the motor, we solve $\dot{\omega}$ from (1.12.1) getting

$$\dot{\omega} = \frac{K_u}{I}u - \frac{K_w + D}{I}\omega - \frac{T_f}{I} \tag{1.12.5}$$

We use the following notation:

$$\frac{K_u}{I} \triangleq \kappa \tag{1.12.6}$$

$$\frac{K_\omega + D}{I} \triangleq \alpha \tag{1.12.7}$$

$$\frac{T_f}{I} \triangleq \dot{\omega}_f \tag{1.12.8}$$

The linear motor model including friction

Substituting the preceding notation into (1.12.5), Laplace transforming and solving for ω, we have

$$\omega = \frac{\kappa}{s + \alpha}u - \frac{1}{s + \alpha}\dot{\omega}_f \tag{1.12.9}$$

Equation (1.12.9) discloses that if not for the friction term, the motor model is a linear dynamical system $F(s)=\kappa/(s+\alpha)$ for which u is the input and ω is the output. For sufficiently low frequencies, ω tracks u exactly. However, as a result of the friction term, the response ω is perturbed. $\dot{\omega}_f$ can be viewed as a noise term, but at the input of $F(s)$, not at the output as described in general earlier. This model of the motor is described in Figure 1.33.

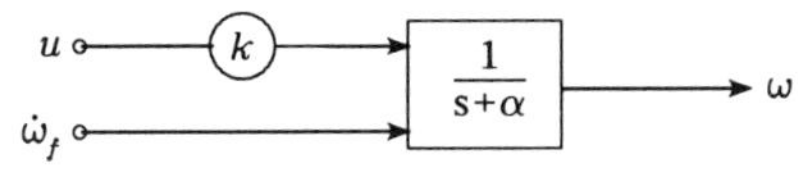

Figure 1.33. Equivalent scheme of dc motor dynamics with the friction term as an additional input.

If u is very small, its effect becomes more dominant and it manifests itself as a irregular, slow creeping motion. If u is relatively large, so that the first term is dominant, the effect of $\dot{\omega}_f$ becomes unnoticeably small, and the residual noisy motion becomes more high frequency, which is more effectively filtered by $1/(s+\alpha)$. This effect of friction actually sets the effective lower bound ω_{min} of the motor response. Together with ω_{max} of the motor, which is determined by power and heating restrictions, it determines the dynamic range.

$$D_r = \frac{\omega_{max}}{\omega_{min}} \tag{1.12.10}$$

With conventional analog amplifier and motor technology, it is difficult to achieve values for D_r in excess of 100. The ratio β of the friction term to the input term at low frequencies $s \to 0$ is equivalent to a noise-to-signal ratio

$$\beta = \frac{\dot{\omega}_f(s)}{\kappa u(s)} \tag{1.12.11}$$

which is a figure of merit of the electro mechanical drive.

Effect of loop closure on minimum speed

The value of ω_{min} can be substantially decreased by placing the motor in a closed rate loop, as shown in Figure 1.34. H represents a tachometer that is assumed to perfectly transduce ω into a proportional voltage. In reality, however, it generates some irregular ripple at its output when operating at very low angular rates.

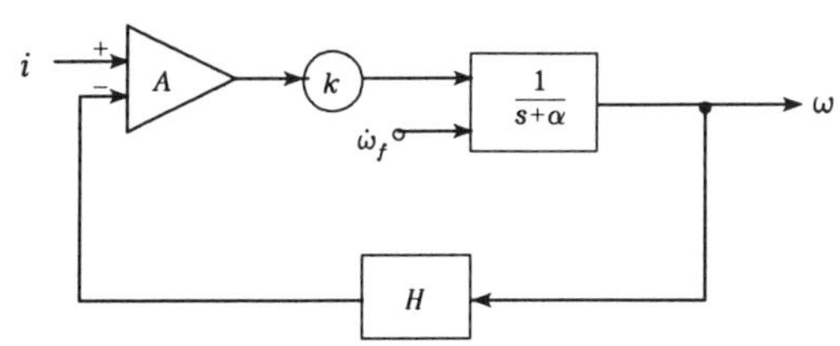

Figure 1.34. Block diagram of a dc motor in a speed controlled closed-loop servomechanism.

The input–output relationship now becomes

$$\omega = \frac{\dfrac{A\kappa}{s+\alpha}}{1+\dfrac{A\kappa H}{s+\alpha}} i(s) - \frac{\dfrac{1}{s+\alpha}}{1+\dfrac{A\kappa H}{s+\alpha}} \dot{\omega}_f(s)$$

$$= \frac{A\kappa}{s+(A\kappa H+\alpha)} i(s) - \frac{1}{s+(A\kappa H+\alpha)} \dot{\omega}_f(s) \tag{1.12.12}$$

The ratio β of the friction term to the input term at low frequencies, $s \to 0$, for $A\kappa H \gg \alpha$, is

$$\beta = \frac{\dot{\omega}_f(s)}{A\kappa i(s)} \tag{1.12.13}$$

Comparing this value of β with that in (1.12.11), we conclude that the loop-gain improves this ratio by a factor of approximately $1/A$.

From the foregoing, it follows that by the implementation of feedback, the effect of (the additive) friction noise can indeed be reduced to the factor of the inverse of the loop-gain. This result appears to contradict the conclusion of Section 1.10, which stated that the loop closure has essentially no effect on additive noise. However, recalling that friction noise is at the input and not at the output level of the controlled element, there is actually no contradiction.

Problems

Problem 1.1– Sun tracker: Implementation by force balance concept

By means of four square photovoltaic detector cells of dimensions $a \times a$ mm the sensor detects the deviation x,y of the sun's image of diameter d with respect to the center of a four quadrant array as shown. The narrow gap between the detectors is disregarded. The lens focal length determines that $d = 0.5a$. The output of each cell is proportional to the respective illuminated areas s_1, s_2, s_3, and s_4 in detectors 1, 2, 3, and 4 covered by the sun's image.

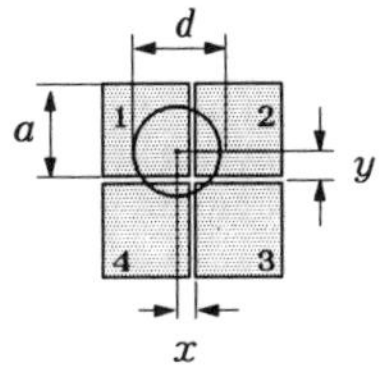

a) Compute the output characteristic $I = f(x,y)$ of the array for the deviations x and y, and determine the respective slopes

$$G_x \triangleq \frac{\partial I}{\partial x}\bigg|_{x=0} \text{ for } y \to 0 \text{ and } G_y \triangleq \frac{\partial I}{\partial x}\bigg|_{y=0} \text{ for } x \to 0$$

The optical unit is mounted on a two axis gimbaled mount driven by motors M_x and M_y, respectively. The signals $G_x x$ and $G_y y$ feed the amplifiers A_x and A_y to drive the respective motors which have known motor constants K_u and K_ω. The respective moments of inertia I_x and I_y of the complete gimbal mount are also known. The drag D of the motors is disregarded.

b) Sketch the closed-loop control system and indicate all the parameters that define it.

c) Formulate the expression for the time constant τ_c of the closed loop sun tracker.

d) Discuss the effect on τ_c of each of the parameters, a, d, A_x, A_y, I_x, I_y, K_u, and K_ω.

e) How is the closed-loop device affected by variations in the brightness of the sun's image? Explain.

Problem 1.2– Electrofluidic angular rate sensor: Force balance implementation

A fine laminar flow air jet is generated by the pumping action of a vibrating piezo electric crystal. The jet impinges on a pair of hot wire resistors R_1 and R_2 placed at the distance l from the nozzle and arranged in a Wheatstone bridge, as indicated in the drawing:

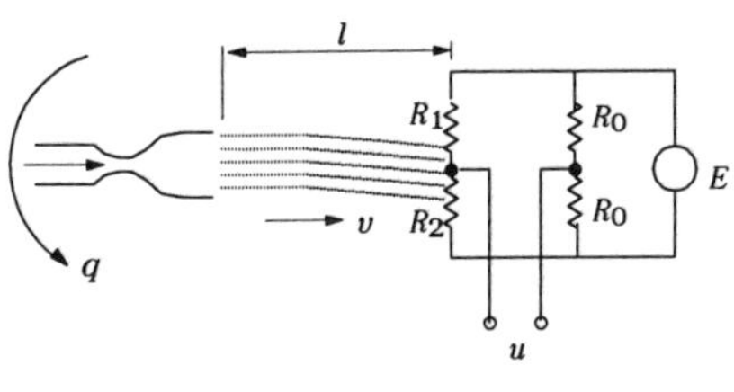

If the device rotates with respect to inertial space at, for example, an angular rate q, a Coriolis acceleration normal to the direction of the jet bends it, so that the cooling of the hot wire pair becomes asymmetrical. Thus, the resistances differentially change and the bridge becomes unbalanced. The output u of the unbalanced bridge increases as the deflection of the jet increases, which, in turn, is proportional to q. Thus, the device can perform as an inertial angular rate sensor. However, the relationship $q = f(u)$ eventually reaches saturation and cannot be expected to be strictly linear.

a) Express the transfer function $u = f(q)$ using the Coriolis acceleration $a_c = 2vq$ and the jet deflection ϵ at the resistor pair R_1, R_2. The hot wire effect is expressed by $R_1 = R_0(1+K\epsilon)$ and $R_2 = (1-K\epsilon)$, where K is a known constant. It is assumed that the wire temperatures respond without delay to ϵ.

b) Discuss the order and linearity of $q=f(u)$.

c) The device is placed in a closed-loop by means of a motor driven platform as in problem 1.1. Formulate the closed loop transfer function $q_m(s)/q(s)$ and discuss its linearity and dynamics.

d) The device operates by temperature variations of the resistors. Is it affected by environmental temperature variations? Explain.

The sensitivity of the device is $u/\epsilon = 10$ mv/mm; $l = 10$ mm; $v = 2$ m/sec; $K_u = 0.5$(rad/sec)/volt; $K_\omega/I = \alpha = 100\ \text{sec}^{-1}$. Compute the gain A that assures a closed-loop bandwidth of 30 rad/sec.

e) Is there a measurement error Δq_m in the steady state? If the answer is affirmative, how can this error be eliminated?

f) If the platform is placed in a flight vehicle that executes a pitching motion $\theta(t)$, describe the motion of the platform with respect to the vehicle and inertial space.

Problem 1.3— Force balance sensor: Effects of gain characteristics on dynamics

The following drawing describes the schematic of an FBS:

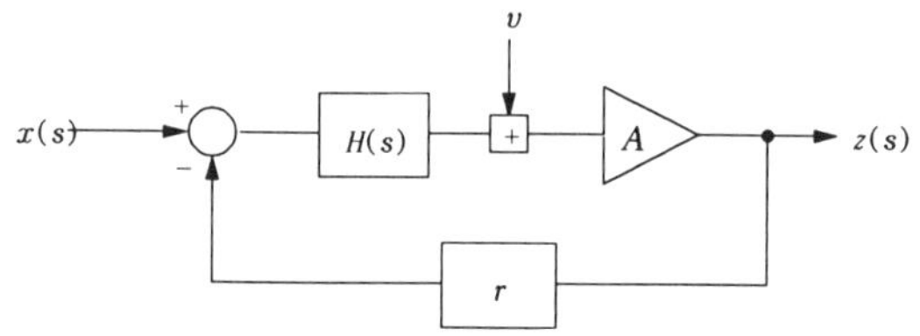

The models of the different blocks are

$$H(s) = 1/(1+s); \quad A(s) = A_0 + A_1/s; \quad r = 1$$

a) What is the order of the closed-loop system?

b) Determine the transfer functions $z(s)/x(s)$, $z(s)/v(s)$ for the following cases:

1. $A_0 = 8;\ A_1 = 0$
2. $A_0 = 8;\ A_1 = 3$
3. $A_0 = 3;\ A_1 = 8$

c) Formulate the dynamical error $\epsilon_d(t)$ in response to an x step input.

d) Compute $\epsilon_d(j\omega)$ in response to $x = x_0 \sin \omega t$ for cases 1, 2, and 3 and draw the corresponding Bode plots.

e) Repeat d) for the output $z(j\omega)$ in response to the noise $v = v_0 \sin \omega t$.

Problem 1.4— Error budget: Effects of environmental interferences

A barometric altimeter has the following percentage error coefficients G_X and is subjected to environmental variations σ_X where X represents any of the following parameters:

1. Temperature:	$G_T = 0.1\%/\text{deg C}$	$\sigma_{\Delta T} = 10 \text{ deg C}$
2. Supply voltage	$G_V = 0.2\%/\text{volt}$	$\sigma_{\Delta V} = 2 \text{ volt}$
3. Humidity	$G_H = 0.015\%/\text{H}$	$\sigma_{\Delta H} = 20 \text{ deg}$
4. Acceleration	$G_g = 0.02\%/\text{g}$	$\sigma_{\Delta g} = 0.5\text{g}$

a) δ is the overall error. Determine the *rms* error σ_δ for the follwing two cases:

1. All the interferences are statistically independent
2. A change in temperature causes a change in supply voltage ΔV in accordance with $\Delta V = 0.1 V/\text{deg C}$

b) Compare the two results.

c) In general, does statistical dependence between environmental parameters affect the overall error? Does it increase, decrease, or not generally have an effect?

Problem 1.5— Magnetic compass: By Hall effect magnetometer

The Hall effect, discovered by E. H. Hall in 1879, occurs in certain semiconductor materials such as indium antimonide. An electrical current i fed into the semiconductor chip while exposed to a magnetic field B, as shown in the figure below causes a potential difference at the output terminals that is proportional to the product of the current i and the component of B normal to the chip. The chip is bonded to a magnetic rod that concentrates the magnetic flux around it to increase its sensitivity.

$$V_m = K_H \vec{B} \times \vec{i} = K_H B i \sin\psi$$

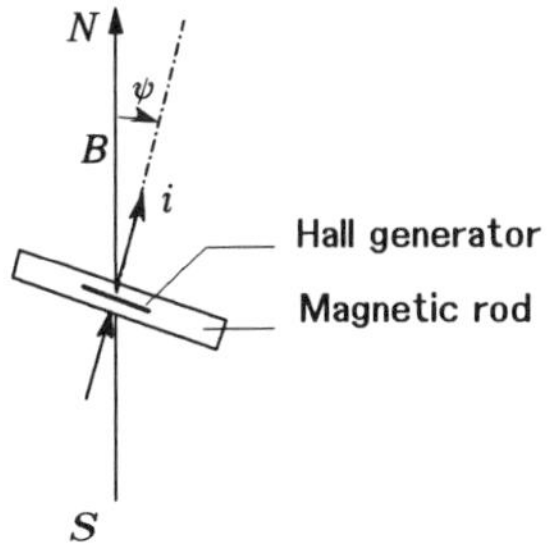

A typical sensitivity of the Hall generator is 18 mv/Gauss. The intensity of the Earth's magnetic field is ~0.5 Gauss. Therefore an output of 9 mv can be expected.

a) What are the limitations on using the Hall generator as a compass?

b) Design a closed-loop magnetometer along similar lines as in Problems 1.1 and 1.2.

c) The magnetic inclination (dip) δ influences the compass differently in different azimuth regions such as east → west or north → south. Is this effect significant in the closed-loop mechanization.

d) To the extent that this effect exists, does it cause a bias error or a scale factor error?

e) By using two perpendicular Hall magnetometers, show how you can design a solid-state compass by feeding the currents $i_1 = i_m \sin\omega t$ and $i_2 = i_m \cos\omega t$ into each of them, summing their respective outputs and phase-demodulating the sum.

f) Compare b) to e) and discuss advantages and disadvantages in regard to complexity, robustness to environmental variations, and precision.

Problem 1.6— Soft saturation: Effect of gain in FBS arrangement

A sensor has a soft saturation characteristic of the type $y = f(x) = \tanh(x)$. It is incorporated in an FBS arrangement by means of an ideal transducer T_r with $r = 1$ and an amplifier with gain A, as shown in the following diagram:

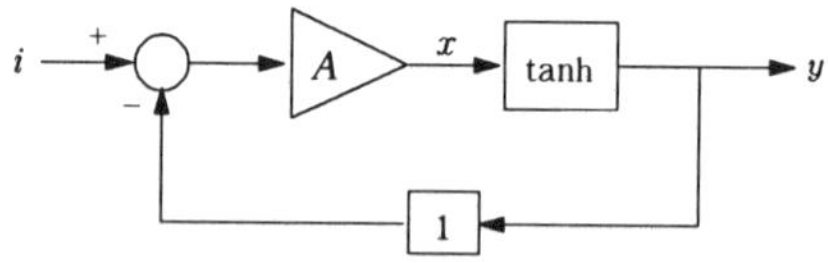

a) Express the closed-loop transfer function y/i.

b) Plot the transfer functions y/i for $A = 1$, 5, and 10.

c) Discuss qualitatively the effect of A on the linearity of y/i.

d) Indicate on the plots of y/i the points for which the error $\epsilon = 1 - y/i$ just exceeds 1%. Connect these points to obtain the 1% error contour.

e) Draw conclusions as to the effect of A on the linearization of the FBS.

References

1. Doebelin, E.O., *Measurement Systems, Applications and Design*, fourth edition, McGraw-Hill, New York, 1990.

2. McKenzie, D.A., *Inventing Accuracy—A Historical Sociology of Nuclear Missile Guidance*, The MIT Press, Cambridge, MA, 1990.

3. Gelb, A. and Vander Velde, W.E., *Multiple-Input Describing Functions and Nonlinear System Design*, McGraw-Hill, New York, 1968.

4. Merhav, S.J., "Equivalent Gain of Single Valued Nonlinearities with Multiple Inputs," *IRE Transactions on Automatic Control*, Vol. AC-7, No. 5, October 1962.

5. McRuer, D.T., Ashkenas, I., and Graham, D., *Aircraft Dynamics and Automatic Control*, Princeton University Press, Princeton, NJ, 1973.

6. Melsa, J.L., and Sage, P.S., *An Introduction to Probability and Stochastic Processes*, Prentice-Hall, Englewood Cliffs, 1973.

7. Mendel, J. M., *Discrete Techniques of Parameter Estimation*, Marcel Dekker, New York, 1973.

8. Sundstrand Data Control, "Q-Flex Servo Accelerometers," Data Sheet 2123/578, 1978.

2

Random Processes and Signals

2.0 Introduction

In this chapter the essential statistical definitions and basic analytical tools for the quantification of random processes are provided. The material, which is based on traditional text, has been selected to serve in the solution of problems arising in aerospace applications and treating measurement and process noise in filtering, smoothing, and prediction operations. These, along with sensors, have often become an integral part of a measurement system.

Physical phenomena, natural or man-induced, that occur in dynamical systems evolve as continuous time-dependent processes. Their basic mathematical representation is by ordinary or partial differential equations in which the system states are the dependent variables and time is the independent variable.

The verifiability of the correspondence between the actual physical system and its mathematical representation hinges on the condition that the process be observable by means of measurements which provide continuous signals both in time and in state-variable domain. If it is, we have a valid analog model, or, a valid analog representation of the process.

The process is often defined as a response to a deterministic forcing function, such as a sine-wave or a step-function [1]. These deterministic forms hardly ever represent reality. Yet, they are occasionally applied in aerospace and control systems for testing and system parameter identification, [2]. They are very useful as standard forcing functions in analysis, and for the definition of performance criteria such as the step response or the frequency response. In reality, processes often are random phenomena. Thus, their representation and performance measures must be formulated and defined in statistical terms. For example, while it may be interesting to observe the frequency response of an air-to-air homing missile, it is hardly of value in the evaluation of its most essential performance measure, i.e., the expected root mean square (rms) miss distance, a statistical measure that is the outcome of a complex combination of random processes.

Modern systems are controlled, or monitored, by digital computers which require interfaces that sample the measurements at discrete, usually uniformly spaced, instances in time. The resulting representation is known as a discrete event, or, *sampled data process* [3].

Moreover, computer interfacing requires analog-to-digital A/D) conversion [4] which translates the analog signals into binary numbers, e.g., from $2=1^0$ say to $2^8=256$ (8-bit conversion). This operation is known as *quantization* [4]. Thus, in many engineering applications, the physical process is represented in discrete time and quantized state-variable format. The random nature of the system pertains both to the time-dependent process of the system states and to the measurements. These are often referred to in the literature as *process noise* and *measurement noise*. However, one should not confuse randomness with noise. The latter exclusively represents the imperfections and the extraneous interferences in the sensors or the data transmission links, whereas the time history of the random roll angle of an aircraft is a description of the process itself. The mathematical model for the noise and the process may be identical, but their meaning and the roles they play are totally different.

By a *process* we mean the evolution in *real time* t of a quantitatively observable phenomenon $y=x(t)$ where t is the present. It is that instant in time that separates past from future. The observation, which must be based on measurement, is executed in the present. A deterministic process is one that enables the precise prediction of future values $y_\lambda=x(t+\lambda)$ for any arbitrary value λ on the basis of present and past observations. The mechanism that enables the prediction of future values is that the observations of the past, in conjunction with a physical law, have been abstracted as an explicit mathematical formula or algorithm that permits the exact *determination* of values of the process at any time τ in the future. Hence, the term *deterministic*. The most outstanding example is the deterministic clockwork-like process of planetary motion, which makes it possible to predict future positions of the planets millennia ahead of time. It was established on the basis of the best state-of-the-art astronomical observations by Tycho Brahe and the mathematical analysis and insight by Kepler [5] in (1609), which eventually led to his discovery of the three laws of planetary motion in (1609 and 1618) and to Newton's universal law of gravitation [6], (1687).

A random process, on the other hand, is one for which the exact prediction of future values is not possible even if it has been observed since the dawn of time. An outstanding example is the desire to predict the weather. The reason is that no explicit mathematical expression $y=x(t)$ that describes the weather as a process, is known to us. This does not imply that it will for ever be unknown. Therefore, randomness is not necessarily an objective property of nature. It may be that at any given period in time, randomness is only a manifestation of our lack of understanding of a physical phenomenon. Numerous phenomena, which in the past appeared to be random, became deterministic once their underlying physical laws were uncovered and exact explicit mathematical expressions could be formulated. In other words, the expansion of the domain of human knowledge transforms apparent Chaos into Law and Order [7].

The storage of the observations of a process, is a record. It can be in the form of a table, a graph, a magnetic tape, an optical disc, or film. In all formats, the observed variables are scaled as analog or digital quantities. Time is translated into a frozen spatial or numerical dimension. The record is of the past history and, as of any instant of the recorded time, it can be retrieved and reproduced or replayed. If the replay correctly scales the spatial or numerical registration of time back into real time by choosing the correct replay speed, an almost perfect illusion of the original real-time process is created. Its fidelity may be so perfect that it stirs emotions, as in music or motion pictures, and excites sensations, as in modern flight simulators. A human observer at the time of listening or viewing may not be aware of the difference, yet a fundamental difference prevails: Even though the record consists of many apparently random processes, it is totally predictable throughout the real-time period of the replay. In other words, the graphical representation $y=x(t)$ of a real time process never represents real time t as the separating point between past and future. There is no future in a graph, only a record of the past.

Throughout this chapter and the rest of this book, random processes of the form $y=x(t)$ will be discussed. As stated earlier, if the process is random, it is not an analytical expression as, for example, $y=\cos(\omega t+\phi)$, if ω and ϕ are given. Yet, apart from generating it in a computer using random numbers or sketching its likeness, it must be quantified and described in forms that are amenable to mathematical operations. This quantification is possible in a statistical sense only, based on probability theory [8]. In this context, $y=x(t)$ is referred to as a *random variable* or a *stochastic variable*, a concept that implies that laws and relationships apply to random processes that are amenable to analysis and design.

The purpose of this chapter is to characterize random processes and to develop the tools for mathematical operations on them, thus preparing the basis for analysis, signal processing, design, and evaluation of system performance quality.

2.1 Statistical characterization of random variables

Time averages of sample functions

A sketch of a sample of the random variable $x(t)$ is shown in Figure 2.1. The present t separates past from future. All $x(t')$, $t'<t$, can be regarded as recorded data, and all $x(t')$, $t'>t$, are unpredictable and therefore shown as a dashed line. All operations embracing an interval $\{t', t'-T\}$ are "off-line" if $t'<t$ and are "on-line" if $t'=t$.

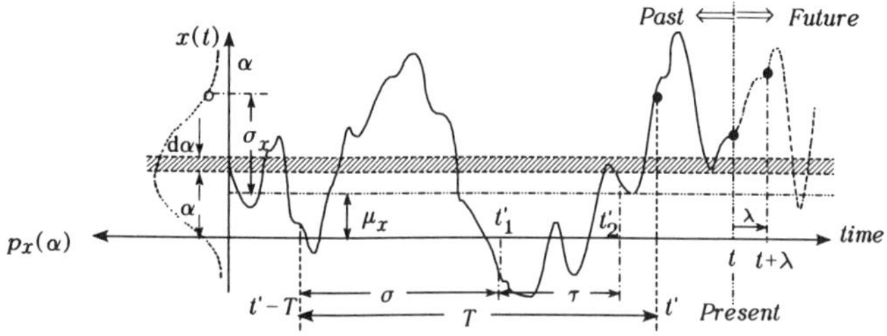

Figure 2.1. Definition of principal statistical descriptors of a sample random process.

The random variable $x(t)$ can be characterized by means of the following four basic descriptors, all of which are statistical averages. These are described here with the aid of a single sample function that under certain broad conditions to be addressed later, are representative of the entire ensemble that characterizes the process.

1. The mean value or the average — μ_x
2. The variance — $V_x = \sigma_x^2$
3. The probability density function (PDF) — $p_x(\alpha) \triangleq P\{\alpha \leq x \leq \alpha + d\alpha\}$
4. The frequency content characterized by the similarity, in a statistical sense, of $x(t'_1)$ to $x(t'_2)$ defined as the autocorrelation function — $\phi_{xx}(t'_1, t'_2)$.

These descriptors are presented here in a cursory manner and under the assumption that the relevant statistical parameters are time-invariant. They are discussed in further detail in Sections 2.2 — 2.4.

μ_x: Assuming that the sample $x(t)$ is infinitely long, μ_x is defined as

$$\mu_x \triangleq \lim_{T \to \infty} \left\{ \frac{1}{T} \int_{t'-T}^{t'} x(\tau) d\tau \right\} \tag{2.1.1}$$

Thus, μ_x is determined on the basis of the record throughout $-\infty \leq \tau \leq t'$, i.e., from the distant past $\tau = -\infty$ up to the latest recorded point in time t'. In reality, T is finite, and thus only an estimate $\hat{\mu}_x$ of μ_x can be determined, i.e.,

$$\hat{\mu}_x(t') \triangleq \frac{1}{T} \int_{t'-T}^{t'} x(\tau) d\tau \tag{2.1.2}$$

This operation averages $x(\tau)$ over a window of width T at an arbitrary point in time t'. Since $x(t)$ is a random variable, $\hat{\mu}_x$ is also random because picking a different t' implies the choice of a different time interval $\{t'-T, t'\}$ which would result in a somewhat different value of $\hat{\mu}_x$. The mathematical relationship between $\hat{\mu}_x$ and $x(t)$ is given in Section 2.2. From (2.1.1) and (2.1.2), it follows that

$$\lim_{T \to \infty} \hat{\mu}_x = \mu_x \tag{2.1.3}$$

V_x: The variance is a non-negative statistical measure of the dispersion or intensity of x around its mean μ_x. It is defined by

$$V_x = \sigma_x^2 \triangleq \lim_{T \to \infty} \left\{ \frac{1}{T} \int_{t'-T}^{t'} [x(\tau) - \mu_x]^2 d\tau \right\} \tag{2.1.4}$$

$$= \lim_{T \to \infty} \left\{ \frac{1}{T} \int_{t'-T}^{t'} x^2(\tau) d\tau - \frac{1}{T} 2\mu_x \int_{t'-T}^{t'} x(\tau) d\tau \right\} + \mu_x^2 = P_x - \mu_x^2 \tag{2.1.5}$$

$$P_x \triangleq \lim_{T\to\infty}\left\{\frac{1}{T}\int_{t'-T}^{t'} x^2(\tau)\mathrm{d}\tau\right\} \tag{2.1.6}$$

Since T is finite, only estimates of $\hat{V}_x$ and $\hat{P}_x$ be determined, namely,

$$\hat{V}_x(t') = \frac{1}{T}\int_{t'-T}^{t'} [x(\tau) - \mu_x]^2\mathrm{d}\tau = \hat{P}_x - \hat{\mu}_x^2 \tag{2.1.7}$$

where $\hat{P}_x$ is the argument of the limit in (2.1.6).

$p_x(\alpha)$: The probability density distribution determines the probability that x is in the interval $\alpha \le x \le \alpha + \mathrm{d}\alpha$. As such, it is a statistical descriptor of the wave-form of $x(t)$. This probability can be interpreted as the ratio of the time t_α to total time T, which $x(t)$ spends in the slice $\{\alpha, \alpha+\mathrm{d}\alpha\}$. For example, if $x(t)$ has the form of a symmetrical random binary signal, switching between the extremes μ_x-1 and μ_x+1, it spends half its time at these extremes. $p_x(\alpha)$ in this case is given by

$$p_x(\alpha) = \frac{1}{2}[\delta_x(\mu_x-1) + \delta_x(\mu_x+1)] \tag{2.1.8}$$

where $\delta_x(\cdot)$ is the Dirac, or delta function. Equation (2.1.8) and the binary signal are illustrated in Figure 2.2.

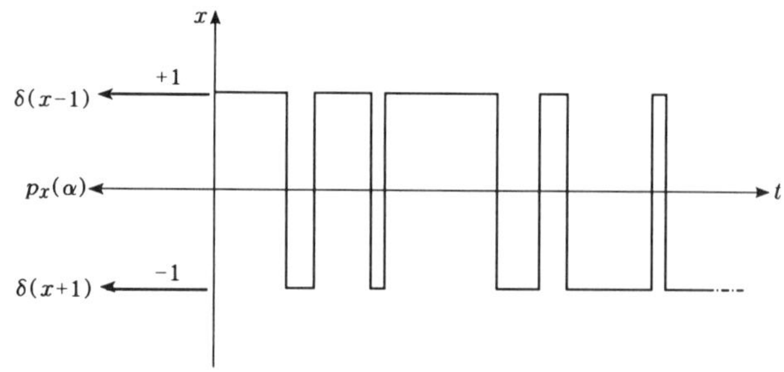

Figure 2.2. Probability density function of a binary signal.

Any PDF that exhibits high densities at distinct levels of α is characteristic of a process $x(t)$ that spends much of its time close to these distinct levels.

$\phi_{xx}(t'_1,t'_2)$: Let $x(t'_1)=x(\sigma)$ and $x(t'_2)=x(\sigma+\tau)$. Then

$$\phi_{xx}(\tau) \triangleq \lim_{T\to\infty}\left\{\frac{1}{T}\int_{t'-T}^{t'} x(\sigma)x(\sigma+\tau)\mathrm{d}\tau\right\} \tag{2.1.9}$$

The autocorrelation is a measure of the frequency content of $x(t)$. It is uniquely related to its power frequency spectrum. Again, since T is finite, only an estimate $\hat{\phi}_{xx}(\tau)$ of $\phi_{xx}(\tau)$ can be determined:

$$\hat{\phi}_{xx}(\tau,t') \triangleq \frac{1}{T}\int_{t'-T}^{t'} x(\sigma)x(\sigma+\tau)d\tau \tag{2.1.10}$$

The foregoing presentation of descriptors derives from a single sample $x(t)$ which may be just one of many possible samples. If, for example $x(t)$ represents the record of one channel, for example, of the flight test telemetered lateral acceleration of one arbitrary missile selected from a production series, the question arises whether, and under what conditions and assumptions this single record represents the performance of the entire production series. This question is pertinent because all the statistical descriptors presented earlier pertain to an entire ensemble. We therefore formulate them as ensemble statistical properties.

2.2 Ensemble averages

Ensemble of sample functions

A *random process* is defined as an entire ensemble of sample functions represented by records denoted by $\{x_k(t')\}$, $k = 1, 2, \ldots, n$, $t' < t$, where $x_k(t')$ is the kth sample. Initially, we assume $n = \infty$. For simplicity of notation, since we refer here to the record only, the symbol t' will be replaced by t. Such an ensemble is described in Figure 2.3.

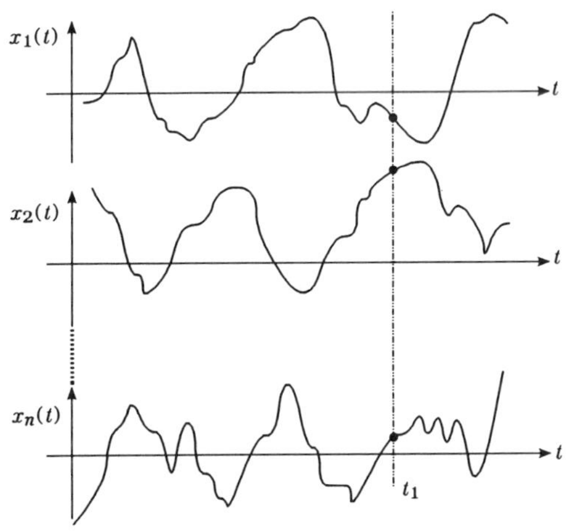

Figure 2.3. A random process as an ensemble of sample functions.

We now examine the statistical properties of the ensemble at $t = t_1$. The dash-dot line is a cut through the ensemble at t_1, which produces the collection, i.e., the ensemble of values $x_k(t_1)$, $k=1, 2, \ldots, n$, namely, $\{x_k(t_1)\}$. Its probability density function (PDF) is denoted by $p_{x_{t_1}}(\alpha)$.

The first basic axiom of probability theory [8] states that *probability is a non-negative measure.*

The probability density function

Assume that n_s samples have been taken and of those, n_α samples represent the single event $\{\alpha \leq x(t_1) \leq \alpha + d\alpha\}$. The probability of this event is

$$P_{x_{t_1}}\{\alpha \leq x \leq \alpha + d\alpha\} = \lim_{n_s \to \infty} \left\{\frac{n_\alpha}{n_s}\right\} \triangleq p_{x_{t_1}}(\alpha)d\alpha \tag{2.2.1}$$

Equation (2.2.1) represents the first-order PDF also known as the *histogram* of $\{x_k\}$ at $t = t_1$.

From the basic axiom on probability, it follows that

$$p_{x_{t_1}}(\alpha) \geq 0; \quad \forall\alpha \tag{2.2.2}$$

t_1 is arbitrary and can therefore be replaced by t. In general, the PDF of a random process will be-time dependent, i.e., $p_x(\alpha) = p_{x_t}(\alpha)$, namely, it may assume different forms for different values of t.

The probability that x is less than or equal to α,

$$P_{x_t}\{-\infty \leq x \leq \alpha\} = P_{x_t}\{x \leq \alpha\} \tag{2.2.3}$$

is the distribution function DF defined as $P_{x_t}\{x \leq \alpha\}$ as a function of α is shown as the typical S-shaped curve in Figure 2.4.

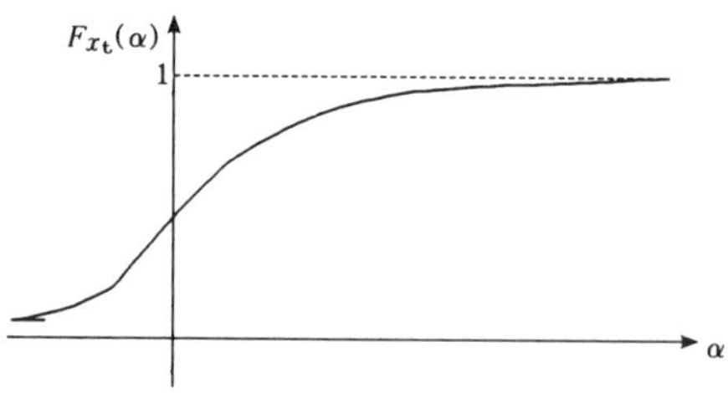

Figure 2.4. Probability distribution function of a random process.

From (2.2.3) and Figure 2.4, it follows that

$$\begin{aligned} P_{x_t}(\alpha \leq x \leq \alpha + d\alpha) &= P_{x_t}\{x \leq \alpha + d\alpha\} - P_{x_t}\{x \leq \alpha\} \\ &= dP_{x_t}(x \leq \alpha) \triangleq p_{x_t}(\alpha)d\alpha \end{aligned} \tag{2.2.4}$$

Therefore,

$$p_{x_t}(\alpha) = \frac{dP_{x_t}(x \leq \alpha)}{d\alpha} = \frac{dF_{x_t}(\alpha)}{d\alpha} \tag{2.2.5}$$

where $F_{x_t}(\alpha) \triangleq P_{x_t}(x \leq \alpha)$ is known as the *probability distribution function*, or the *distribution function* (*DF*).

The PDF therefore is the derivative of the DF with respect to α. Conversely, it follows that

$$\int_{-\infty}^{\alpha} p_{x_t}(\gamma)\,d\gamma = P_{x_t}(x \le \alpha) \triangleq F_{x_t}(\alpha) \tag{2.2.6}$$

γ is the variable of integration. Since $p_{x_t}(\alpha)$ is non-negative, $F_{x_t}(\alpha)$ is a non decreasing function of α. It follows that $P_{x_t}(x \le \infty) = 1$ since the event $\{x \le \infty\}$ is absolute certainty. Therefore,

$$\int_{-\infty}^{\alpha=\infty} p_{x_t}(\gamma)\,d\gamma = 1 \tag{2.2.7}$$

The most essential of statistical measures is the average, which in the present context is the mean value of all $x_k(t_1)$ in the infinite ensemble $\{x_k(t_1)\}$. If $p_{x_{t_1}}(\alpha)$ is not a priori known, the only way to determine the average is empirical, namely, to use the largest possible number n_s of available samples of $x_k(t_1)$ and to compute the finite average:

$$A[x_k(t_1)] = \frac{1}{n_s}\sum_{k=1}^{n_s} x_k(t_1) \tag{2.2.8}$$

Expectation

If $n_s \to \infty$, this finite average is the expected value or the expectation. Thus,

$$E(x_{t_1}) \triangleq E[\{x_k(t_1)\}] \triangleq \lim_{n_s \to \infty}\left\{\frac{1}{n_s}\sum_{k=1}^{n_s} x_k(t_1)\right\} \tag{2.2.9}$$

where $E(\cdot)$, being a limit, is a mathematical expression and $A(\cdot)$ is an estimate thereof, namely,

$$A(\cdot) = \hat{E}(\cdot) \tag{2.2.10}$$

If, however, $p_{x_{t_1}}(\alpha)$ is a priori known, $E(\cdot)$ can be analytically determined. By *a priori knowledge*, we actually imply that $p_{x_{t_1}}(\alpha)$ is either a mathematical expression that has been derived by the analysis of a physical process $x(t)$, e.g., the Poisson distribution of the process of photon bombardment [9], in a photosensitive detector, or by assuming a model based on a hypothesis such as the *uniform* distribution for the thickness errors of a mass-produced element. We thus have three classes of PDF functions: 1. empirical estimates which do not, in general, offer a theoretical explanation; 2. analytical derivations which result from rigorous analysis of a physical model, and 3. hypothetical functions, not necessarily analytic, but considered as plausible representations of reality.

It is readily shown [8] that the expectation is equal to the first moment of $p_{x_{t_1}}$ about $\alpha = 0$, namely,

$$E(x_{t_1}) = \int_{-\infty}^{\infty} \alpha p_{x_{t_1}}(\alpha)\,d\alpha \tag{2.2.11}$$

Equation (2.2.11) can be interpreted as the static *first moment* with respect to $\alpha = 0$, exerted by a flat plate cut to the shape of $p_{x_{t_1}}(\alpha)$ and positioned on the α axis. This moment is denoted by $\mu_{x_{t_1}}$ and is widely used to express the expectation $E(x_{t_1})$.

EXAMPLE 2.2–1: Demonstration of a hypothetical PDF

$p_x(\alpha)$ is a hypothetical PDF described in Figure 2.5.

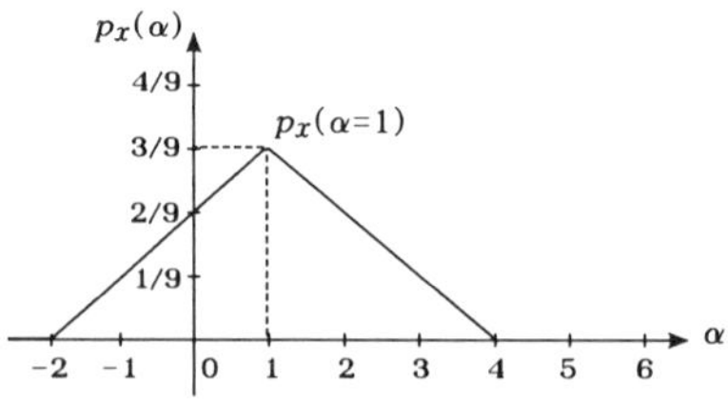

Figure 2.5. Triangular probability density function.

Determine the following:

1. $p_x(\alpha=0) = ?$
2. $E(x) = ?$
3. $P\{-1 \le x < 3\} = ?$

1. The area S under $p_x(\alpha)$, equals 1 and is

$$S = \int_{-\infty}^{\infty} p_x(\alpha)d\alpha = \frac{1}{2}[4-(-2)]p_x(\alpha=1)] = 1$$

Therefore,

$$p_x(\alpha=1) = \frac{1}{3}$$

2. $$E(x) = \int_{-2}^{1} \alpha(\frac{2}{9} + \frac{1}{9}\alpha)d\alpha + \int_{1}^{4} \alpha(\frac{4}{9} - \frac{1}{9}\alpha)\alpha = 1 \text{ (obvious from Figure 2.5)}$$

3. $$P\{-1 \le x \le 3\} = 1-(P\{x<-1\}+P\{x>3\}) = 1-\frac{1}{2}(\frac{1}{9}+\frac{1}{9}) = \frac{8}{9}$$

Generalized moments

The concept of *expectation*, or *moments* is now generalized. Consider the general bi-unique (one-to-one) time-invariant relationship $y = f(x)$ which, in the context of the ensemble discussed here, is expressed as

$$y_{t_1} = f(x_{t_1}) \tag{2.2.12}$$

Formulating $E(y_{t_1})$, we have,

$$E(y_{t_1}) = E[f(x_{t_1})] = \int_{-\infty}^{\infty} f(\alpha) p_{x_{t_1}}(\alpha) d\alpha \tag{2.2.13}$$

A special class of $f(\alpha)$ is $f(\alpha) = \alpha^m$, where $m = 1, 2, ...,$ are non-negative integers. Thus, we formulate the high-order moments,

$$E(x_{t_1}^m) = \int_{-\infty}^{\infty} \alpha^m p_{x_{t_1}}(\alpha) d\alpha \tag{2.2.14}$$

The special case of greatest practical interest is $\alpha = 2$,

$$E(x_{t_t}^2) = \int_{-\infty}^{\infty} \alpha^2 p_{x_{t_1}}(\alpha) d\alpha \tag{2.2.15}$$

For this expression the mechanical analogy is still meaningful. It is recognizable as the *moment of inertia* of $p_{x_{t_1}}(\alpha)$ with respect to $\alpha = 0$. More importantly, this second-order moment expresses the power of the process, i. e.,

$$P_{x_{t_1}} \triangleq E(x_{t_1}^2) \tag{2.2.16}$$

Stationarity

Definition 2.2–1: A process $\{x_k(t)\}$ is *stationary* in the *mth* moment if

$$E(x_{t_1}^m) = \text{const.}, \quad \forall\, t_1 \tag{2.2.17}$$

Similarly, A process is *stationary* in $f(x)$ if

$$E[f(x_{t_1})] = \int_{-\infty}^{\infty} f(\alpha) p_{x_{t_1}}(\alpha) d\alpha = \text{const.}, \qquad \forall\, t_1 \tag{2.2.18}$$

Definition 2.2–2: A process for which $p_{x_{t_1}}(\alpha) = p_{x_{t_i}}(\alpha)$, $\forall\, i$, is *strictly stationary.*

Conclusion: A strictly stationary process is stationary *in all its moments.*

Comment: The opposite is not necessarily true. A process that is stationary in any of its moments is not necessarily stationary in $p_{x_t}(\alpha)$, i.e., strictly stationary.

This follows from the recognition that the equality of definite integrals (2.2.11) does not imply the equality of the corresponding integrands $p_{x_t}(\alpha)$.

The *variance* V_{x_t} of a process is derived from P_{x_t} by means of the second central moment.,

Power and variance

$$V_{x_t} = \int_{-\infty}^{\infty} (\alpha - \mu_{x_t})^2 p_{x_t}(\alpha)\,\mathrm{d}\alpha = \int_{-\infty}^{\infty} \alpha^2 p_{x_t}(\alpha)\,\mathrm{d}\alpha$$

$$-2\mu_{x_t}\int_{-\infty}^{\infty} \alpha p_{x_t}(\alpha)\,\mathrm{d}\alpha + \mu_{x_t}^2\int_{-\infty}^{\infty} p_{x_t}(\alpha)\,\mathrm{d}\alpha$$

$$= P_{x_t} - 2\mu_{x_t}^2 + \mu_{x_t}^2 = P_{x_t} - \mu_{x_t}^2 \triangleq \sigma_x^2 \qquad (2.2.19)$$

Being a quadratic, $V_{x_t} \geq 0$. Therefore, it follows from (2.2.19) that

$$E(x_t^2) - [E(x_t)]^2 \geq 0$$

and,

$$E(x_t^2) \geq [E(x_t)]^2 \qquad (2.2.20)$$

Comment: Stationarity in the *mth* moment does not necessarily imply stationarity in the *nth* moment, $n > m$. Thus, if a process $\{x_k(t)\}$ is stationary in the mean μ_{x_t}, it is not necessarily stationary in power, P_{x_t}. This issue is addressed in the following example.

EXAMPLE 2.2—2: Stationarity of higher moments

A process $\{x_k(t)\}$ is assumed to have a *uniform* PDF. This is a hypothetical PDF, as illustrated in Figure 2.6.

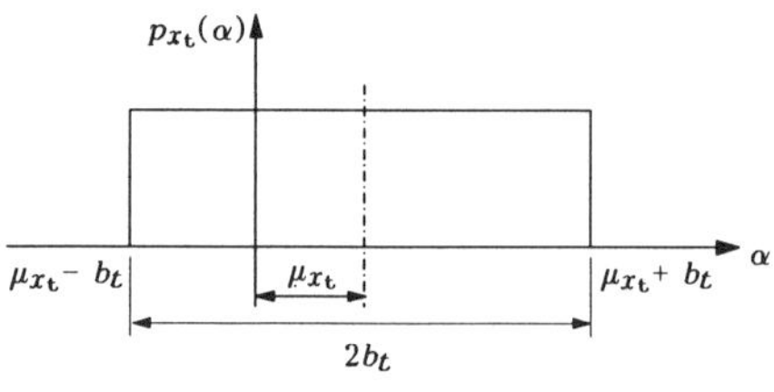

Figure 2.6. Uniform probability density distribution.

Its width is $2b_t$. $p_{x_t}(\alpha)$ *is determined by*

$$p_{x_t}(\alpha) = \frac{1}{\mu_{x_t} + b_t - (\mu_{x_t} - b_t)} = \frac{1}{2b_t}; \quad \mu_{x_t} - b_t \leq \alpha \leq \mu_{x_t} + b_t$$

$$= 0; \quad \text{elsewhere}$$

$$E(x_t^2) = P_{x_t} = \int_{\mu_{x_t}-b_t}^{\mu_{x_t}+b_t} \alpha^2 \frac{1}{2b_t}\,\mathrm{d}\alpha = \frac{1}{2b}\frac{\alpha^3}{3}\bigg|_{\mu_{x_t}-b_t}^{\mu_{x_t}+b_t} = \mu_{x_t}^2 + \frac{b_t^3}{3}$$

On the basis of (2.27),

$$V_{x_t} = \frac{b_t^3}{3}$$

It is evident that assuring the stationarity of μ_{x_t} does not imply the stationarity of the power P_{x_t}, since b_t, i.e., the variance, may be nonstationary. On the other hand, the stationarity of P_{x_t} does not imply that either μ_{x_t} or b_t is a constant. In other words, with a constant power P_{x_t}, the PDF may undergo changes according to the relationship

$$\mu_{x_t} = \sqrt{P_{x_t} - b_t^3/3}$$

The statistical averages discussed so far are dependent on the PDF defined at a given instant of time t_1. Therefore, $p_{x_t}(\alpha)$ is a function of one variable only, i.e., of α. For this reason they are named *first-order probability density distributions.* Further study of the frequency content of the process $\{x_k(t)\}$ requires the study of the second-order PDF involving joint probabilities. This is addressed in the next section.

2.3 Joint distribution, correlation

So far, only the probability $P\{\cdot\}$ of the single event $\{\alpha \leq x_{t_1} \leq \alpha + d\alpha\}$ has been addressed. We now consider two processes $\{x(t)\}$ and $\{y(t)\}$ and define the probability of the joint event $\{\alpha \leq x_{t_1} \leq \alpha + d\alpha;\ \beta \leq y_{t_1} \leq \beta + d\beta\}$, i. e.,

$$P\{\alpha \leq x_{t_1} \leq \alpha + d\alpha;\ \beta \leq y_{t_1} \leq \beta + d\beta\} \tag{2.3.1}$$

The meaning of (2.3.1) is: The probability that x at instant t_1 is within the slice $\alpha, \alpha + d\alpha$, and that y, at the same instant t_1 is within the slice $\{\beta,\ \beta + d\beta\}$.

The definition of (2.3.1) can be further generalized as follows:

$$P\{\alpha \leq x_{t_1} \leq \alpha + d\alpha;\ \beta \leq y_{t_2} \leq \beta + d\beta\} \tag{2.3.2}$$

The meaning of (2.3.2) is: The probability that x at instant t_1 is within the slice $\{\alpha,\ \alpha + d\alpha\}$, and that y at another instant t_2 is within the slice $\{\beta,\ \beta + d\beta\}$.

In analogy to the first order PDF, we define the second-order PDF or the *joint* PDF which in accordance with the basic axiom on probability is non-negative. Therefore,

$$p_{x_{t_1}, y_{t_1}}(\alpha, \beta) \geq 0 \tag{2.3.3}$$

and

$$p_{x_{t_1}, y_{t_2}}(\alpha, \beta) \geq 0 \tag{2.3.4}$$

A sketch of a symmetrical two-dimensional Gaussian-like joint PDF is shown in Figure 2.7. The mean values μ_x and μ_y in this case are indicated as the corresponding axes of symmetry. We can now proceed and define a generalized *joint moment* as follows: Let $z = g(x,y)$ be a function of the two variables x

and y, in analogy to $y = f(x)$ in the case of a single variable. In particular, consider $z = g(x_{t_1}, y_{t_2})$. In analogy to (2.2.13) and (2.2.14) the corresponding *joint expectation* can be defined as

$$E[g(x_{t_1}, y_{t_2})] \triangleq \int_{-\infty}^{\infty}\int_{-\infty}^{\infty} g(\alpha,\beta)\, p_{x_{t_1}, y_{t_2}}(\alpha,\beta)\, \mathrm{d}\alpha \mathrm{d}\beta \tag{2.3.5}$$

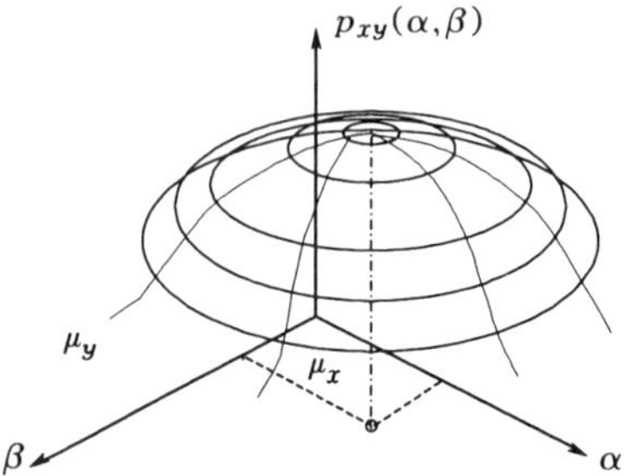

Figure 2.7. Joint probability density distribution of variables x, y.

An example of foremost importance is $g(x,y) = xy$. In this case the joint expectation is,

$$E[x_{t_1} y_{t_2}] \triangleq \int_{-\infty}^{\infty}\int_{-\infty}^{\infty} \alpha\beta\, p_{x_{t_1}, y_{t_2}}(\alpha,\beta)\, \mathrm{d}\alpha \mathrm{d}\beta \tag{2.3.6}$$

Equation (2.3.6) is a formal extension of the single variable case defined in (2.2.11). It is best interpreted by the left hand side. In Figure 2.8 the records of the two processes $\{x_k\}$, cut at t_1 and $\{y_k\}$, cut at t_2 are described side by side. The spacing is $t_2 - t_1 = \tau$.

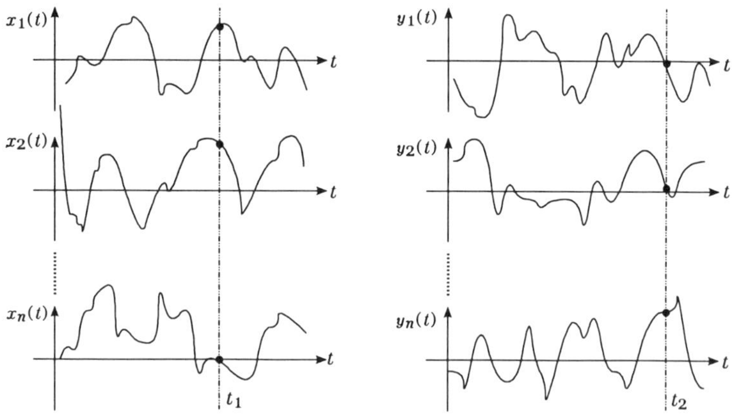

Figure 2.8. Ensembles of two random processes $\{x_k\}$ and $\{y_k\}$.

The empirical determination of the joint expectation would be performed as follows:
1. Sample the pairs $x_k(t_1)$ and $y_k(t_2)$ for $k=1, 2,..., n_s$ and multiply them.

2. Determine the arithmetic average of all these products:

$$\hat{E}(x_{t_1}y_{t_2}) = \frac{1}{n_s}\sum_{k=1}^{n_s} x_k(t_1)y_k(t_2) \tag{2.3.7}$$

Correlation

We assume that the number of the sample functions n_s is sufficiently large to yield a reliable average. If the result is zero, it means that the joint PDF $p_{x_{t_1},y_{t_2}}(\alpha,\beta)$ is zero and that there is no statistical relation or correlation between x and y at the instances t_1 and t_2. However, the result may be positive or negative. If it is positive, it means that, on the average, positive or negative values of x imply positive or negative values of y, respectively. If the result is negative, it means that, on the average, positive values of x imply negative values of y or vice versa. We may therefore find zero, positive or negative correlation between x and y. The joint expectation is therefore a statistical measure of the correlation between x and y. It may change its magnitude and sign if the location of t_1 and t_2 or the spacing τ between them is changed.

Clearly, if the joint PDF in (2.3.5) were known, the joint expectation could be analytically determined, provided the solution for the definite integral exists.

Definition 2.3–1: The processes $\{x_k(t)\}$ *and* $\{y_k(t)\}$ are *stationary in joint expectation if*, for a given $\tau=t_2-t_1$, $E(x_{t_1}y_{t_2})$=const. for $\forall$ t_1, t_2.

A sufficient condition for stationarity in joint expectation is that for a given τ, $p_{x_{t_1},y_{t_2}}(\alpha,\beta) = p_{x_{ti},y_{t_i}}(\alpha,\beta)$ for $\forall$ i,j.

Comment: Correlation has meaning only with respect to two processes or variables because it is a mutual attribute of these processes.

Definition 2.3–2: Since the joint expectation $E(x_{t_1}y_{t_2})$ expresses correlation and since it depends on t and on τ, it is a function of these variables. It is therefore termed a correlation function or cross-correlation function.

Specifically, it is denoted by

$$C_{xy}(t_1,t_2) \triangleq E(x_{t_1}y_{t_2}) \tag{2.3.8}$$

This function and its particular properties will be further addressed in Section (2.4). We now re-address (2.3.6) and consider special cases which are of great practical importance.

Case 1: Expectation of a Product:

$$g(x,y) = h(x)f(y) \tag{2.3.9}$$

$h(x)$ and $f(y)$ are both bi-unique functions.

$$E[h(x_{t_1})f(y_{t_2})]$$

$$= \int_{-\infty}^{\infty}\int_{-\infty}^{\infty} h(\alpha)f(\beta)p_{x_{t_1}y_{t_2}}(\alpha,\beta)\,d\alpha\,d\beta \tag{2.3.10}$$

If x and y are statistically independent, then it follows from fundamental probability theory [8],

$$p_{x_{t_1}y_{t_2}}(\alpha,\beta) = p_{x_{t_1}}(\alpha)p_{y_{t_2}}(\beta) \tag{2.3.11}$$

and (2.3.9) decomposes as the product

$$E[h(x_{t_1})f(y_{t_2})] = \int_{-\infty}^{\infty} h(\alpha)p_{x_{t_1}}(\alpha)d\alpha \int_{-\infty}^{\infty} f(\beta)p_{y_{t_2}}(\beta)d\beta$$

$$= E[h(x_{t_1})]E[f(y_{t_2})] \tag{2.3.12}$$

Specifically, for $h(x)=x$ and $f(y)=y$, we have

$$E(x_{t_1}y_{t_2}) = E(x_{t_1})E(y_{t_2}) = \mu_{x_{t_1}}\mu_{y_{t_2}} \tag{2.3.13}$$

Conclusion: If $\{x_k\}$ and $\{y_k\}$ are statistically independent processes, then the expectation of their product is the product of their expectations or the product of their mean values.

Orthogonality

Definition 2.3–3: *If* $E(xy)=0$, *the processes* $\{x_k\}$ *and* $\{y_k\}$ *are orthogonal.*

Comment 1: In the case of statistical independence, for these processes to be orthogonal, it is sufficient that either μ_x or μ_y be zero.

Comment 2: If $\{x_k\}$ and $\{y_k\}$ are orthogonal, they are not necessarily statistically independent.

EXAMPLE 2.3–1: Statistical independence and orthogonality

Let $\{x_k\} = \{A_k\sin(\omega t+\theta_k)\}$ and $\{y_k\} = \{B_k\cos(\omega t+\theta_k)\}$ where θ_k, the phase angle, is the random variable common to each sample pair x_k, y_k. These clearly are statistically dependent because for every x_k, y_k is uniquely determined and vice versa. Yet, the expectation $E(x_k y_k)$ is easily shown to be zero.

The foregoing example is a special case of the more general property $E\{x_k(t)\dot{x}_k(t)\}$, which will be addressed in Section 2.6.

The distribution law

Since the expectation of a constant is equal to that constant, i.e., $E(a) = a$, it follows from (2.3.12) that

$$E(ax) = aE(x) \tag{2.3.14}$$

Case 2:

$$g(x,y) = ax + by \tag{2.3.15}$$

Since the average of a sum is the sum of the averages, we have, recalling (2.3.11),

$$E[(ax_{t_1} + by_{t_2})] = E(ax_{t_1}) + E(by_{t_2})$$

$$= aE(x_{t_1}) + bE(y_{t_2}) = a\mu_{x_{t_1}} + b\mu_{y_{t_2}} \tag{2.3.16}$$

Equation (2.3.16) is known as the *distribution law*. It is meaningful if $t_1 = t_2$.

Conclusion: The expectation of the linear combination of x and y is the linear combination of the expectation of x and the expectation of y or of their mean values.

For the special case $a = 1$, $b = 1$,

$$E(x_{t_1} + y_{t_1}) = E(x_{t_1}) + E(y_{t_1}) = \mu_{x_{t_1}} + \mu_{y_{t_1}} \tag{2.3.17}$$

The distribution law applies equally to higher-order moments or expectations. A case of particular importance is:

$$E[(x_{t_1} + y_{t_1})^2] = E(x_{t_1}^2 + 2x_{t_1}y_{t_1} + y_{t_1}^2)$$

$$= E(x_{t_1}^2) + 2E(x_{t_1}y_{t_1}) + E(y_{t_1}^2) \tag{2.3.18}$$

If $\{x_{t_1}\}$ and $\{y_{t_1}\}$ are orthogonal,

$$E(x_{t_1}^2 + y_{t_1}^2) = E(x_{t_1}^2) + E(y_{t_1}^2) \tag{2.3.19}$$

Equation (2.3.17) can be generalized for a number of processes $\{x_i\}$, $i = 1, 2,\ldots,n$. Then, if orthogonality $E(x_j x_k) = 0$, $\forall j,k$ can be assumed, we have

$$E\left[\left(\sum_{i=1}^{n} x_i\right)^2\right] = \sum_{i=1}^{n} E\left(x_i^2\right) \tag{2.3.20}$$

or for generalized moments $E[f(x)]$,

$$E\left[f\left(\sum_{i=1}^{n} x_i\right)\right] = \sum_{i=1}^{n} Ef(x_i)] \tag{2.3.21}$$

The expression (2.3.19) is of considerable practical importance, particularly in analyzing and evaluating error budgets in systems that consist of numerous subsystems, each contributing its own error.

EXAMPLE 2.3—2: Combined error due to environmental interferences

Consider the error due to environmental interferences as defined in (1.6.5).

$$y = \sum_{k=1}^{n} G_{I_k} i_k$$

G_{I_k} are the sensitivity coefficients to the variations in the environmental components i_k. We assume they are statistically independent and we seek the expected RMS error $\{E(y^2)\}^{1/2}$. From (2.3.19) and (2.3.15),

$$E(y^2) = E\left\{\left[\sum_{k=1}^{n} G_{I_k} i_k\right]^2\right\} = \sum_{k=1}^{n} G_{I_k}^2 E(i_k^2)$$

If there is statistical dependence between variables i_l, i_m, the appropriate cross terms $E(i_l i_m)$ must be incorporated in the sum. These terms may be either positive or negative. Therefore, without close examination, it is not possible to predict whether statistical dependence between interferences yields larger or smaller errors than in the case of statistical independence.

We define the *central joint expectation* of x_{t_1} and y_{t_1} as the covariance $V_{x_{t_1}y_{t_1}}$

$$V_{x_{t_1}y_{t_1}} = \sigma^2_{x_{t_1}y_{t_1}} \triangleq E[(y_{t_1} - \mu_{x_{t_1}})(y_{t_1} - \mu_{y_{t_1}})]$$

$$= E(x_{t_1}y_{t_1}) - \mu_{x_{t_1}}\mu_{y_{t_1}} \tag{2.3.22}$$

If $x = y$, (2.3.22) reduces to (2.2.19).

Autocorrelation

We now reconsider (2.3.8), which defined the cross correlation function and we reduce it to the case $x \equiv y$. In other words, we seek the correlation of $\{x_k(t)\}$ with itself. The ensemble $\{x_k(t)\}$ is shown in Figure 2.9 cut at the instances t_1, t_2. In accordance with (2.3.17),

$$C_{xx}(t_1,t_2) \triangleq E(x_{t_1}x_{t_2}) \tag{2.3.23}$$

Definition 2.3—4: The joint expectation of x_{t_1} and x_{t_2}, being a measure of the correlation of the process with itself at the instants t_1 and t_2, $t_2 - t_1 = \tau$, is a function of these variables, and it defines the *autocorrelation function* of the process.

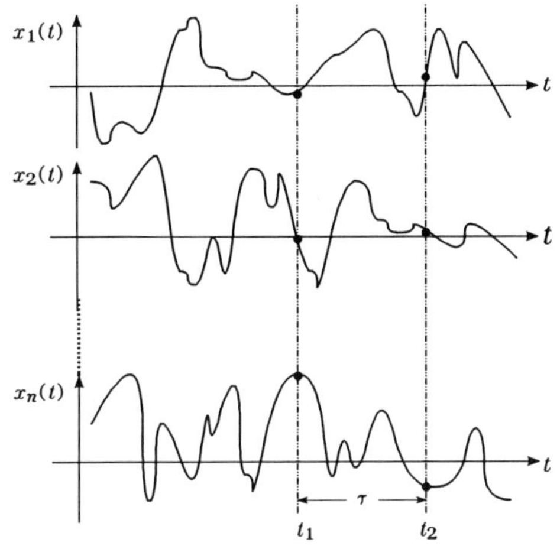

Figure 2.9. Ensemble $\{x_k\}$ sampled at two time instances.

Definition 2.3−5: For a process to be stationary in autocorrelation it is sufficient that with a given τ, $p_{x_{t_1}x_{t_2}}(\alpha_1,\alpha_2) = p_{x_{t_i}x_{t_j}}(\alpha_1,\alpha_2)$, $\forall\ i,j$.

Following the definition of $p_x(\alpha)$ and (2.2.4), the meaning of the joint PDF, $p_{x_{t_1}x_{t_2}}(\alpha_1,\alpha_2)d\alpha_1 d\alpha_2$ is the probability that $x(t)$ at $t=t_1$ is in the slice $\{\alpha_1,\alpha_1+d\alpha_1\}$ and that $x(t)$ at $t=t_2$ is in the slice $\{\alpha_2,\alpha_2+d\alpha_2\}$.

Definition 2.3−6: A process that is stationary in autocorrelation is defined as a wide-sense stationary process.

Since $t_2 = t_1 + \tau$ and $p_{x_{t_1}x_{t_1+\tau}}(\alpha_1,\alpha_2)$, is time-invariant, it is not a function of t_1 and it can be rewritten as $p_{xx\tau}(\alpha_1,\alpha_2)$, which is a function of τ only. Therefore, for a wide-sense stationary process, the autocorrelation function takes the form:

$$C_{xx}(\tau) \triangleq E[x(t)x(t+\tau)] \tag{2.3.24}$$

Equation (2.3.24) means the following: Cut through the ensemble at an instant of time t, cut again at $t+\tau$, multiply all the pairs $x(t)$ and $x(t+\tau)$, and determine the expectation by averaging these products. Repeat for a number of values of τ. This average as a function of τ constitutes the autocorrelation function $C_{xx}(\tau)$.

The properties of the autocorrelation and cross-correlation functions will be discussed in further detail in Section 2.4.

2.4 Correlation coefficient and functions

The correlation coefficient

The correlation coefficient is an important statistical descriptor. In this section, we develop explicit expressions that are measures of this statistical relation, namely, the correlation between the variables x and y. In particular we shall study the principal properties of correlation functions, which are important in understanding random processes and their role in the design of aerospace sensor systems. Correlation has been formally introduced in Section 2.3 as the joint expectation and it has been indicated that as a measure, it can assume negative, zero, or positive values. However, before proceeding developing analytical expressions, some insight into the actual meaning of correlation will be given by a simple example.

Let the variable x represent the age of humans from infancy to old age. Consider first the period from early childhood to young adulthood from age zero to sixteen in increments of one year. Let the variable y be weight. Clearly there is a relation between age and weight. Older children are usually heavier, although quite a few older children may be lighter than younger ones. But we know that as a rule, namely, in a statistical sense, or on the average, a plot of y against x would demonstrate a monotonic increase of weight y against age x. Let this plot be approximated by a straight up-sloping line. We would also speak of a positive correlation between height and age. We sense that the slope of the line represents the correlation and is proportional to it. The same mental exercise, performed for the age group of 20 to 40, would reveal that on the average, weight does not increase with age. We would say that in this age group the correlation is zero. In the age group 40 to 60 there is normally another positive correlation and later, beyond the age of 70, we would observe a decrease in weight with age, i.e., a negative correlation between age and weight. A representation of this age related variation $y = f(x)$ is shown in Figure 2.10.

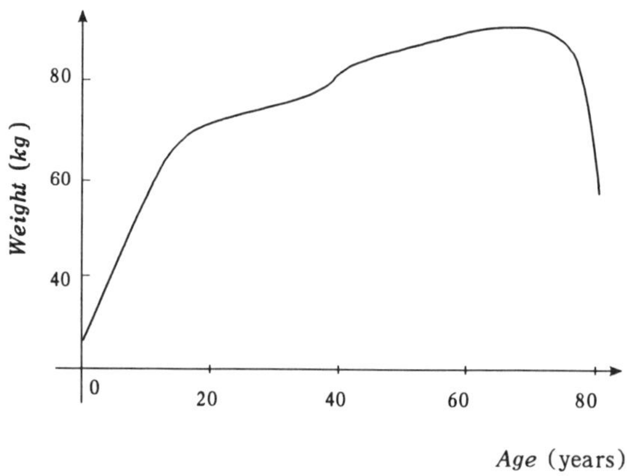

Figure 2.10. Relation beween weight and age in humans.

Recalling the notion of linear regression described in Section 1.8, the statistical approach can be applied in approximating a sample plot of $y = f(x)$ by the linear regression line $y = \hat{m}x + \hat{b}$. Such a sample plot for $x = 0, 1, 2, \ldots, 16$ involving 16 youngsters is illustrated in Figure 2.11, along with its regression line.

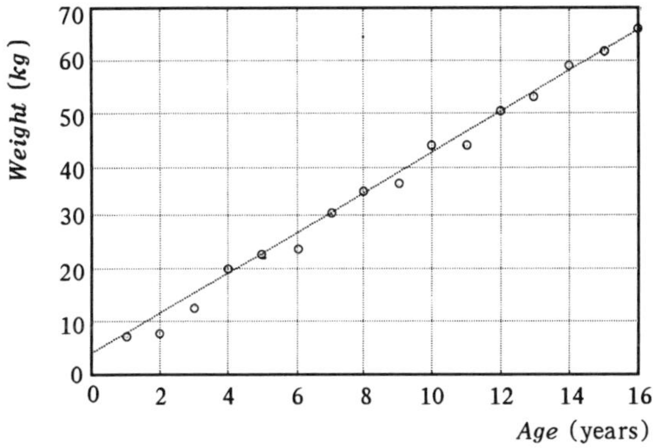

Figure 2.11. Regression line relating weight and age in a sample of youngsters.

This line has a random slope $\hat{m}$ and a random zero-point offset because the sample consists of only 16 elements. Assume now that countless samples $y = f(x)$ of 16 youngsters have been processed in the same manner and averaged. Clearly, the finite averages $A(\cdot)$ in (1.8.19) and (1.8.20) can be replaced by the expectations $E(\cdot)$. Thus, and in view of (2.2.19), and (2.3.22), we have

$$\hat{m} = \frac{E(xy) - E(x)E(y)}{E(x^2) - [E(x)]^2} = \frac{V_{xy}}{\sigma_x^2} \tag{2.4.1}$$

In accordance with (2.3.21), the slope $\hat{m}$ is the covariance of x and y or, the correlation between them.

$$\hat{b} = \frac{E(x^2)E(y) - E(xy)E(x)}{E(x^2) - [E(x)]^2}$$

$$= \frac{(\sigma_x^2 + \mu_x^2)\mu_y - (V_{xy} + \mu_x\mu_y)\mu_x}{\sigma_x^2} = \mu_y - \mu_x\hat{m} \tag{2.4.2}$$

Substituting these solutions of $\hat{m}$ and $\hat{b}$ into $y = \hat{m}x + \hat{b}$ we have,

$$y = \frac{V_{xy}}{\sigma_x^2}x + \mu_y - \mu_x\frac{V_{xy}}{\sigma_x^2} = \mu_y + (x - \mu_x)\frac{V_{xy}}{\sigma_x^2} \tag{2.4.3}$$

Equation (2.4.3) is the normalized regression line expressed in terms of the basic statistical descriptors μ_x, μ_y, σ_x^2, and V_{xy}. This normalized regression line is sketched in Figure 2.12. We now center this line around μ_x and μ_y by defining the following dimensionless variables:

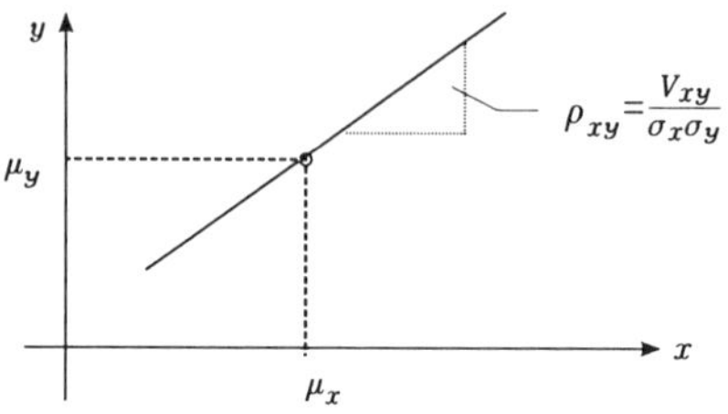

Figure 2.12. Normalized and centered regression line showing correlation coefficient ρ_{xy}.

$$\xi \triangleq \frac{x - \mu_x}{\sigma_x} \tag{2.4.4}$$

$$\eta \triangleq \frac{y - \mu_y}{\sigma_y} \tag{2.4.5}$$

$$\rho_{xy} \triangleq \frac{V_{xy}}{\sigma_x \sigma_y} \tag{2.4.6}$$

With these definitions and from (2.4.3), the centralized normalized regression line is

$$\eta = \rho_{xy}\, \xi \tag{2.4.7}$$

Where ρ_{xy} is the dimensionless correlation coefficient relating the weight y to age x and represented by the slope of the correlation line in the ξ,η plane. We can now express the normalized joint expectation

$$E(\xi\eta) = E(\xi \rho_{xy} \xi) = \rho_{xy}\, E(\xi^2) \tag{2.4.8}$$

and,

$$E(\xi^2) = E\{\frac{(x - \mu_x)^2}{\sigma_x^2}\} = \frac{\sigma_x^2}{\sigma_x^2} = 1 \tag{2.4.9}$$

Equation (2.4.8) yields

$$E(\xi\eta) = \rho_{xy} \tag{2.4.10}$$

Thus, the normalized joint expectation of x and y is equal to the correlation coefficient ρ_{xy}.

The higher the slope ρ_{xy}, the higher the correlation. Observing (2.4.1), and if x and y are statistically independent, we have from (2.3.12),

$$E(xy) = E(x)E(y) = \mu_x\mu_y \tag{2.4.11}$$

Implying that $\hat{m} = 0$ and therefore $\rho_{xy} = 0$.

Conclusion: Statistically independent processes are uncorrelated.

It is of interest to determine the domain of existence of ρ_{xy}. For this purpose we determine the following:

$$E[(\xi \pm \eta)^2] = E(\xi^2) \pm 2E(\xi\eta) + E(\eta^2) \geq 0 \tag{2.4.12}$$

Since $E(\xi^2) = E(\eta^2) = 1$, and $E(\xi\eta) = \rho_{xy}$, it follows that

$$1 \pm 2\rho_{xy} + 1 \geq 0 \tag{2.4.13}$$

or

$$2(1 \pm \rho_{xy}) \geq 0 \tag{2.4.14}$$

and

$$-1 \leq \rho_{xy} \leq 1 \tag{2.4.15}$$

This result indicates that perfect correlation, positive or negative, exists when $\rho_{xy} = |1|$. In other words, correlation cannot be better than perfect.

Conclusion: The range of existence of the correlation coefficient is $\rho_{xy} \leq |1|$. This result is sketched in Figure 2.13.

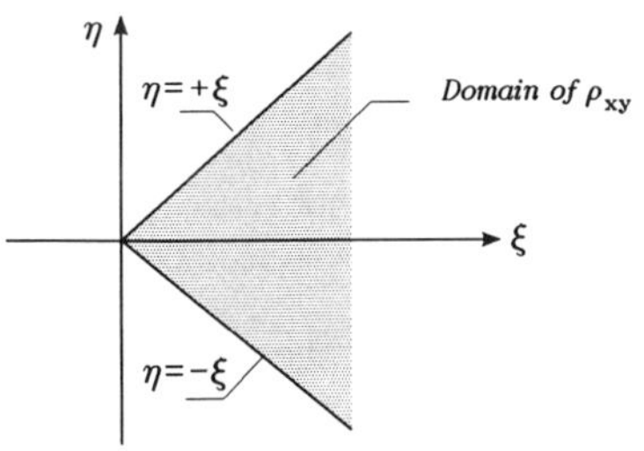

Figure 2.13. Domain of existence of the normalized correlation coefficient.

The autocorrelation function

Correlation functions were introduced in (2.3.7) as joint expectations for different processes x and y and termed crosscorrelation functions If $x \equiv y$, we refer to it as the *autocorrelation function*.

We repeat here the definition of the autocorrelation function $C_{xx}(t_1,t_2)$:

$$C_{xx}(t_1,t_2) \triangleq E(x_{t_1}x_{t_2}) = \int_{-\infty}^{\infty} \alpha_1\alpha_2 p_{x_{t_1}x_{t_2}}(\alpha_1,\alpha_2)\mathrm{d}\alpha_1\mathrm{d}\alpha_2 \qquad (2.4.16)$$

This function is the expectation of x_{t_1}, at t_1, being in the slice α_1, $\alpha_1+\mathrm{d}\alpha_1$ *and* x_{t_2} at t_2 being in the slice α_2, $\alpha_2+\mathrm{d}\alpha_2$; this statistical descriptor is a measure of the statistical similarity of $\{x\}$ with itself at the instants t_1 and t_2, given by $\tau \triangleq t_2 - t_1$. Clearly, if the process $\{x(t)\}$ varies slowly, this similarity, or correlation, will be high. Conversely, if it varies rapidly, the correlation will be low. For example, if $\{x(t)\}$ represents the temperature at a given location, it is expected that the temperature will be similar after about one hour. In dynamical processes this correlation is a result of one form or other of inertia. In the present example, it is thermal inertia. This rapidity, or slowness of variation is closely related to the frequency content of the process. Therefore, the autocorrelation function is a descriptor of the frequency content of the process, as stated in Section 2.1. Clearly, if for a given process τ becomes infinitely large, all autocorrelation vanishes.

We now recall (2.4.6) and substitute $x = x_{t_1}$, $y = x_{t_2}$,

$$\rho_{x_{t_1}x_{t_2}} \triangleq \rho_{xx}(t_1, t_2) = \frac{V_{x_{t_1}x_{t_2}}}{\sigma_{x_{t_1}}\sigma_{x_{t_2}}}$$

$$= \frac{E[(x_{t_1} - \mu_{x_{t_1}})(x_{t_2} - \mu_{x_{t_2}})]}{\sigma_{x_{t_1}}\sigma_{x_{t_2}}}$$

$$= \frac{E(x_{t_1}x_{t_2}) - \mu_{x_{t_1}}E(x_{t_2}) - \mu_{x_{t_2}}E(x_{t_1}) + E(\mu_{x_{t_1}}\mu_{x_{t_2}})}{\sigma_{x_{t_1}}\sigma_{x_{t_2}}}$$

$$= \frac{C_{xx}(t_1,t_2) - \mu_{x_{t_1}}\mu_{x_{t_2}}}{\sigma_{x_{t_1}}\sigma_{x_{t_2}}} \qquad (2.4.17)$$

Equation (2.4.17) is a special form of the correlation coefficient ρ_{xy}, now a function of t_1, t_2; it is in fact the centralized and normalized autocorrelation function $C_{xx}(t_1,t_2)$. As formulated in (2.4.17), some or all four statistical descriptors may be time-varying. Such correlation functions are difficult to analyze and do not lend themselves easily to the establishment of useful rules and explicit mathematical expressions. We therefore focus on processes that are stationary in correlation in accordance with (2.3.24). Thus,

$$\rho_{xx}(t_1,t_1+\tau) = \rho_{xx}(t,t+\tau) = \rho_{xx}(\tau) = \frac{C_{xx}(\tau) - \mu_x^2}{\sigma_x^2} \qquad (2.4.18)$$

Properties of autocorrelation functions can now be studied with the aid of (2.4.18). Following (2.3.24), the order of the product $x(t)x(t+\tau)$ can be reversed. Since the process is stationary, it can be shifted back and forth in time without affecting the expectation. Thus,

$$E[x(t)x(t+\tau)] = E[x(t+\tau)x(t)] = E[x(t)x(t-\tau)] \tag{2.4.19}$$

$$C_{xx}(\tau) = C_{xx}(-\tau) \tag{2.4.20}$$

$$\rho_{xx}(\tau) = \rho_{xx}(-\tau) \tag{2.4.21}$$

Property I: The autocorrelation function of a process that is stationary in autocorrelation, is symmetrical in its argument τ.

The value of $\rho_{xx}(\tau)|_{\tau=0}$ is obtained from (2.4.18),

$$\rho_{xx}(0) = \frac{C_{xx}(0) - \mu_x^2}{\sigma_x^2} = \frac{P_x - \mu_x^2}{\sigma_x^2} = \frac{\sigma_x^2}{\sigma_x^2} = 1 \tag{2.4.22}$$

Property II: The value of the normalized autocorrelation function at the origin $\tau = 0$ is identical to the correlation coefficient and is equal to unity.

From (2.4.22) we have,

$$C_{xx}(0) = \sigma_x^2 + \mu_x^2 = E(x^2) \triangleq P_x \tag{2.4.23}$$

Property III: The value of the autocorrelation function at the origin $\tau = 0$ is equal to the power P_x of the process $\{x(t)\}$.

The value of the autocorrelation function as $\tau \to \infty$ is evaluated by the following argument: Heuristically, we have reasoned that as τ becomes infinitely large, the autocorrelation vanishes implying that $x(t)$ and $x(t+\tau)$ become statistically independent. Therefore

$$E[x(t)x(t+\tau)] = E[x(t)]E[x(t+\tau)] \tag{2.4.24}$$

But since the process is stationary in correlation, it is also stationary in the mean, therefore

$$E[x(t)] = E[x(t+\tau)] = \mu_x \tag{2.4.25}$$

Therefore, (2.4.24) reduces to

$$C_{xx}(\tau)|_{\tau\to\infty} = \mu_x^2 \tag{2.4.26}$$

Property IV: The autocorrelation function converges to the square of the mean value as τ becomes infinitely large.

In view of (2.4.18) and (2.4.26), we have

$$\rho_{xx}(\tau)|_{\tau\to\infty}=0 \tag{2.4.27}$$

Property V: The normalized and centered autocorrelation function converges to zero as $\tau\to\infty$.

Since $\rho_{xx}(\tau)$ is bounded by ± 1, namely, $-1\le\rho_{xx}(\tau)\le 1$, there may be some value or values τ_m for which

$$\rho_{xx}(\tau_m) = -1 \tag{2.4.28}$$

Applying (2.4.18), we have

$$\rho_{xx}(\tau_m) = \frac{C_{xx}(\tau_m)-\mu_x^2}{\sigma_x^2} = -1 \tag{2.4.29}$$

and, in view of (2.4.22), we have

$$C_{xx}(\tau_m)=\mu_x^2-\sigma_x^2 \tag{2.4.30}$$

$$\mu_x^2-\sigma_x^2\le C_{xx}(\tau)\le\mu_x^2+\sigma_x^2 \tag{2.4.31}$$

Property VI: The autocorrelation function $C_{xx}(\tau)$ of a stationary process is bounded by $\mu_x^2-\sigma_x^2$ and $\mu_x^2+\sigma_x^2$.

The foregoing properties are summarized in Figures 2.14 and 2.15.

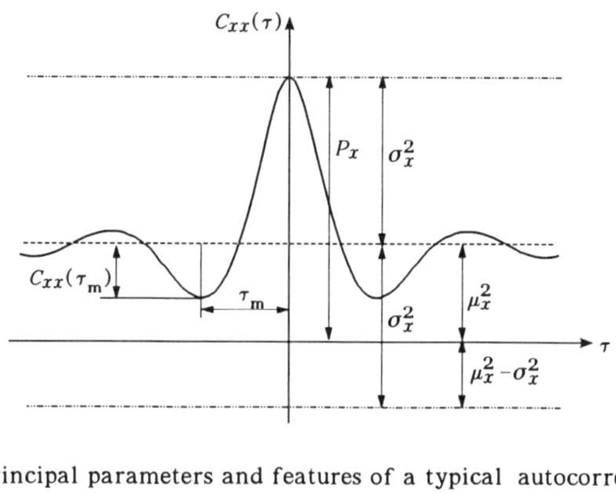

Figure 2.14. Principal parameters and features of a typical autocorrelation function.

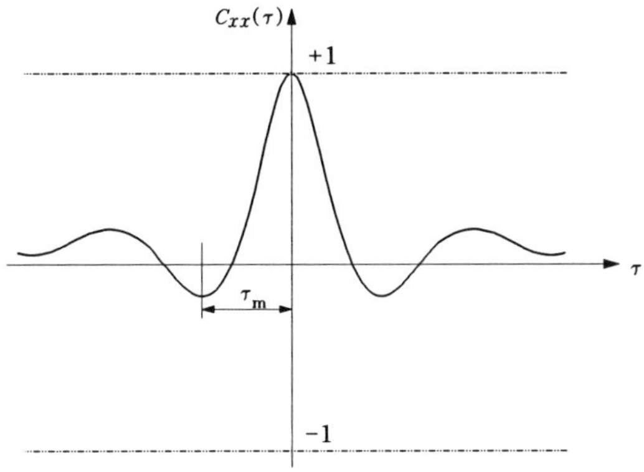

Figure 2.15. Principal parameters and features of the normalized autocorrelation function.

From the discussion following (2.4.16), the following property is evident:

Property VII: A slowly varying process $\{x(t)\}$ has a wide autocorrelation function. The converse is true.

An example of an extremely slow varying process is random bias, i.e., an ensemble $\{x\}$ in which each sample is a constant. The corresponding autocorrelation function is also a constant, as will be illustrated in Example 2.4–1. An extreme example for a rapidly varying process is white noise. Its autocorrelation function is a Dirac function centered at the origin $\tau=0$.

There is one class of random processes for which the minimum of $C_{xx}(\tau)$ reaches $\mu_x^2 - \sigma_x^2$. This is the case of stationary periodic functions, such as $\sin x$, $\cos x$, or sign($\sin x$) as illustrated in Example 2.4–2.

EXAMPLE 2.4–1: Autocorrelation function of a slowly varying process

An extreme example of a slowly varying process is one that does not vary at all. For example, $\{x(t)\}$ is an ensemble of six samples of random biases illustrated in Figure 2.16. The ensemble of biases is given as: $\{x\} = 1.23, -0.51, 0.75\ \ -0.99, 0.22, 0.83$.

It is required to provide an estimate $\hat{C}_{xx}(\tau)$ on the basis of this small number of samples. We replace $E(\cdot)$ by $A(\cdot) = (1/6)\Sigma(\cdot)_i$ and cut the ensemble at the arbitrary times t and $t+\tau$. Clearly, both t and τ are immaterial in this case. Since $x(t) = x(t+\tau)$,

$$\hat{C}_{xx}(\tau) = \frac{1}{6}\sum_{i=1}^{6} x_i(t)x_i(t+\tau) = \frac{1}{6}\sum_{i=1}^{6} [x_i(t)]^2 = \hat{P}_x$$

From the given elements of $\{x\}$, we have

$\{x_i^2\} = 1.51,\ 0.26,\ 0.56,\ 0.98,\ 0.048\ 0.69;$

$\hat{P}_x = \hat{C}_x = 0.674;$

$\hat{\mu}_x = 0.255;$

$\hat{V}_x = \hat{P}_x - \hat{\mu}_x^2 = 0.674 - 0.255^2 = 0.609;$

$\hat{\sigma}_x = \sqrt{0.609} = 0.78.$

The estimate $\hat{C}_{xx}(\tau)$ has been determined with an uncertainty of 10%.

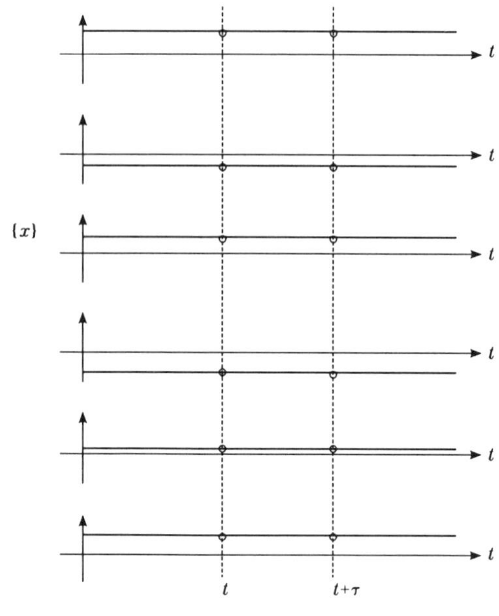

Figure 2.16. Ensemble of random constants underlying constant autocorrelation function.

EXAMPLE 2.4–2: Analytical determination of autocorrelation function

The following is an example in which $C_{xx}(\tau)$ is determined analytically. The process is an ensemble of sinusoidal functions of time with constant amplitude A and constant frequency ω. The random variable is the phase angle θ, which implies that the phases of the various sample functions in the ensemble $\{x\}$ are randomly distributed, for example, between $\theta = -\pi$ and $\theta = +\pi$:

$$x(t) = A\sin(\omega t + \theta)$$

The ensemble $\{x\}$ is illustrated in Figure 2.17.

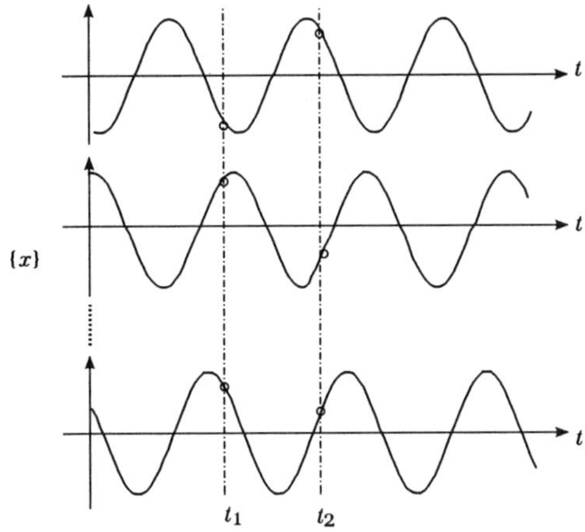

Figure 2.17. Random process consisting of ensemble of sine-waves with random phase shifts.

In accordance with (2.3.21),

$$C_{xx}(t_1,t_2) = E\{A^2\sin(\omega t_1+\theta)\sin(\omega t_2+\theta)$$

$$= A^2\frac{1}{2}E\{\cos\omega(t_1-t_2)-\cos[\omega(t_1+t_2)+2\theta]\}$$

Defining $t_1-t_2=\tau$, noting that $\omega\tau=$ const., and denoting $\omega(t_1+t_2)$ by $\alpha=$ const., we have

$$C_{xx}(\tau) = \frac{A^2}{2}\cos\omega\tau - \frac{A^2}{2}E[\cos(\alpha+2\theta)]$$

The second term is expanded, i.e.,

$$E[\cos(\alpha+2\theta)] = E[\cos\alpha\cos2\theta - \sin\alpha\sin2\theta]$$

$$= \cos\alpha\, E(\cos2\theta) - \sin\alpha\, E(\sin2\theta)$$

The problem is to evaluate $E(\cos2\theta)$ and $E(\sin2\theta)$. In order to do so, we assume a uniform PDF for θ, namely,

$$p_\theta(\beta) = \frac{1}{2\pi};\quad -\pi\le\beta\le\pi$$

$$= 0;\quad \text{elsewhere}$$

We now recall (2.4.13) for generalized moments, namely,

$$E[f(\theta)] = E[\cos2\theta] \text{ or } E[\sin2\theta]$$

which can be explicitly expressed by

$$E[f(\theta)] = \int_{-\infty}^{\infty} f(\beta)p_\theta(\beta)\mathrm{d}\beta = \frac{1}{2\pi}\int_{-\pi}^{\pi}\cos 2\beta \mathrm{d}\beta = \frac{1}{4\pi}\Big|_{-\pi}^{\pi}\sin 2\beta = 0$$

or

$$\frac{1}{2\pi}\int_{-\pi}^{\pi}\sin 2\beta \mathrm{d}\beta = \frac{1}{4\pi}\Big|_{-\pi}^{\pi} -\cos 2\beta = 0$$

Thus,

$$C_{xx}(\tau) = \frac{A^2}{2}\cos\omega\tau$$

$A^2/2$ is the well-known expression for the power of a sinusoidal process. This autocorrelation function displays all the properties except property IV, or V because in this example $\{x(t)\}$ is not strictly random. It is completely predictable for each individual sample function but not for the ensemble. The periodicity of the process implies that statistical dependence prevails even for infinitely large values of τ. Functions of this nature are known as *deterministic random processes*. Other examples in the same class are stationary random square wave, saw-tooth or triangular functions or any superposition of them.

EXAMPLE 2.4–3: The exponential autocorrelation function

We now formally introduce the exponential autocorrelation function. It is defined by the nonanalytic expression

$$C_{xx}(\tau) = \sigma_x^2 e^{-a|\tau|} + \mu_x^2$$

or in normalized form

$$\rho_{xx}(\tau) = e^{-a|\tau|}$$

This function consists of two different branches:

$$\rho_{xx}^{+}(\tau) = e^{-a\tau}; \qquad \tau>0$$

$$\rho_{xx}^{-}(\tau) = e^{a\tau} = e^{-a(-\tau)}; \qquad \tau<0$$

It is graphically illustrated in Figure 2.18. Thus it obeys property I, i.e., symmetry.

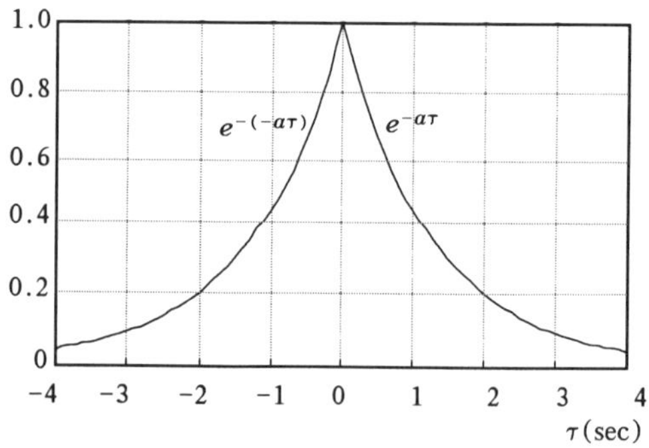

Figure 2.18. Exponential autocorrelation function of a first-order Markov process.

This type of autocorrelation function is particularly important because many actual processes can be approximated by it. Thanks to the simplicity using this model, considerable computational effort can be saved while providing clear insight into apparently complex processes.

Further inspection shows that if a is small, $\rho_{xx}(\tau)$ is wide. In other words, it decays slowly in τ, implying a slowly varying process. In the limit, if $a \to 0$, $\rho_{xx}(\tau)$ represents a constant as in Example 2.4-2. In contrast, if a is large, it decays rapidly, indicating a rapidly varying process. Rapid decay means that even for relatively small values of τ, the correlation of $x(t)$ with itself vanishes. Disregarding μ_x, without loss of generality, we have

$$C_{xx}(\tau) = \sigma_x^2 e^{-a|\tau|} = \frac{\sigma_x^2 a}{\frac{a}{2}\,2} e^{-a|\tau|}$$

Since the exponent $a\tau$ is dimensionless and τ is of dimension (sec), a must be of dimension (sec^{-1}). In other words, a represents frequency. It follows that the ratio $2\sigma_x^2/a$ represents power density, i.e., the amount of power per unit frequency, defined for example as volt^2/(rad/sec) or volt^{2}sec.

The term $(a/2)e^{-a|\tau|}$ has the properties of a Dirac function $\delta(\tau)$, centered at the origin $\tau=0$. The area under the delta function is 1, half in the region $\tau>0$ and half in $\tau<0$, and it is infinity at $\tau \to 0$. Evaluating the appropriate integral of $(a/2)e^{-a\tau}$ yields

$$\int_0^\infty \frac{a}{2}e^{-a\tau}d\tau = \frac{a}{2}\left(-\frac{1}{a}e^{-a\tau}\right)\Bigg|_0^\infty = \frac{1}{2}$$

and

$$\lim_{\substack{\tau \to 0 \\ a \to \infty}} ae^{-a\tau} = \lim_{\substack{\tau \to 0 \\ a \to \infty}} \frac{1}{\tau}a\tau(1-a\tau+\ldots) = \lim_{\substack{\tau \to 0 \\ a \to \infty}} \frac{1}{\tau}a\tau = \lim a = \infty$$

In the double limit, a is an arbitrary large constant while $\tau \to 0$. This justifies the Taylor series expansion and permits the assumption $a\tau \ll 0$ so that τ can cancel out.

Thus, an infinitely large a indicates frequency components approaching infinity. In the language of optics this is known as *white color*. In analogy, a process with $a \to \infty$ is known as a *white process*, often labeled *white noise*. This representation is of major importance in modeling random processes.

The cross-correlation function

Cross-correlation was first defined in (2.3.7). We now examine its properties in detail. Two processes $\{x\}$ and $\{y\}$ illustrated in Figure 2.8. $E(x_{t_1}y_{t_2})$ in (2.3.7) is a measure of the correlation between the two processes and it is formally equivalent to the correlation coefficient given in (2.4.6) for $t_1 = t_2$. In order to derive some explicit properties, we now assume that both $\{x\}$ and $\{y\}$ are stationary in cross-correlation. As such, we define

$$E(x_t y_{t+\tau}) \triangleq C_{xy}(\tau) \tag{2.4.32}$$

$$E(x_t y_{t-\tau}) \triangleq C_{xy}(-\tau) \tag{2.4.33}$$

Contrary to the autocorrelation function $C_{xx}(\tau)$, symmetry with respect to τ cannot be presumed. In fact, except for the trivial case $y = kx$, where k is a constant, we have

$$C_{xy}(\tau) \neq C_{xy}(-\tau) \tag{2.4.34}$$

Reversing the order of the product, however, yields the following equality:

$$E(x_{t-\tau}\, y_t) = E(y_t x_{t-\tau}) \tag{2.4.35}$$

By assuming stationarity, this joint expectation can be shifted in time, namely,

$$E(x_{t-\tau}\, y_t) = E(x_t\, y_{t+\tau}) \tag{2.4.36}$$

The right-hand term in (2.4.35) is $C_{yx}(-\tau)$, and the right-hand term in (2.4.36) is $C_{xy}(\tau)$. Consequently,

$$C_{xy}(\tau) = C_{yx}(-\tau) \tag{2.4.37}$$

Clearly, for $\tau=0$ and in compliance with (2.3.22),

$$C_{xy}(0) = E(x_t y_t) = V_{xy} + \mu_x \mu_y \tag{2.4.38}$$

If at least one of the mean values is zero, $C_{xy}(0) = V_{xy}$.

We can now determine the bounds of $C_{xy}(\tau)$ by expanding the centered, normalized, non-negative expression

$$E\left\{\left(\frac{x_{t_1}-\mu_{x_{t_1}}}{\sigma_{x_{t_1}}} \pm \frac{y_{t_2}-\mu_{y_{t_2}}}{\sigma_{y_{t_2}}}\right)^2\right\}$$

$$= E\left\{\frac{x_{t_1}^2-2\mu_{x_{t_1}}x_{t_1}+\mu_{x_{t_1}}^2}{\sigma_{x_{t_1}}^2}\right\} \pm 2E\left\{\frac{(x_{t_1}-\mu_{x_{t_1}})(y_{t_2}-\mu_{y_{t_2}})}{\sigma_{x_{t_1}}\sigma_{y_{t_2}}}\right\}$$

$$+ E\left\{\frac{y_{t_2}^2-2\mu_{y_{t_2}}y_{t_2}+\mu_{y_{t_2}}^2}{\sigma_{y_{t_2}}^2}\right\} \geq 0 \tag{2.4.39}$$

The numerators of the first and third terms are clearly $\sigma_{x_{t_1}}^2$ and $\sigma_{y_{t_2}}^2$, respectively, so that these terms are equal to one. The second term is readily shown to be $\rho_{xy}(t_1,t_2)$. Substituting $t_2=t_1+\tau$ and recalling that

$$\rho_{xy}(\tau) \triangleq \frac{C_{xy}(\tau)-\mu_x\mu_y}{\sigma_x\sigma_y} \tag{2.4.40}$$

Equation (2.4.40) leads to

$$1 \pm \frac{C_{xy}(\tau) - \mu_x \mu_y}{\sigma_x \sigma_y} \geq 0 \tag{2.4.41}$$

Explicitly, the bounds on $C_{xy}(\tau)$ are

$$C_{xy}(\tau) \leq \sigma_x \sigma_y + \mu_x \mu_y \tag{2.4.42}$$

$$C_{xy}(\tau) \geq -\sigma_x \sigma_y + \mu_x \mu_y \tag{2.4.43}$$

These bounds are equivalent to the bounds on $C_{xx}(\tau)$ given in (2.4.31).

Clearly, if $\{x\}$ and $\{y\}$ are totally unrelated processes, their cross-correlation must be zero for all τ. Hence, a nonzero cross-correlation implies that there must be some linear or nonlinear relationship between $\{x\}$ and $\{y\}$ defined by an operator $O(x)$. Indeed, it will be shown in Section 2.6 that the cross-correlation function is the statistical input-output relationship of linear and nonlinear systems. These are known as *describing functions*, or *equivalent gain* [10].

2.5 Equivalence of time and ensemble averages, ergodicity

In the preceding sections, random processes were defined as ensembles $\{x\}$ or $\{y\}$, and all the statistical descriptors were developed accordingly. Strictly, this is entirely within the realm of probability theory. *Time* was introduced as a parameter, t_1 or t_2, as shown in Section 2.2. In engineering problems, however, situations that call for analysis and evaluation in terms of statistical ensembles are not typical. Most situations deal with a single sample, which may represent a laboratory experiment, a simulation, or a field flight test. The available data are often from this single sample or at best, from a limited number of them. Yet, it is essential to learn the utmost about the entire ensemble in terms of statistical information from these few samples.

In Section 2.1, the four statistical descriptors were introduced on the basis of such a sample random function $x(t)$. Consider any sample function $x_i(t)$ of the ensemble $\{x(t)\}$.

Conditions for equivalence of time and ensemble averages

The sample average was denoted by μ_{x_i}, although it actually stands for the statistical mean of the ensemble. In this section, we recall the definition of the time average given in (2.1.1), but in order to distinguish it conceptually from the ensemble mean μ_x, it will be denoted by

$$\overline{x_i(t)} \triangleq \lim_{T \to \infty} \left\{ \frac{1}{T} \int_{t'-T}^{t'} x_i(\sigma) d\sigma \right\} \tag{2.5.1}$$

or, in more general terms, for $y = f(x)$,

$$\overline{y_i(t)} = \overline{f[x_i(t)]} = \lim_{T\to\infty}\left\{\frac{1}{T}\int_{t'-T}^{t'} f[x_i(\sigma)]d\sigma\right\} \tag{2.5.2}$$

The individual time averages in the ensemble, $i = 1, 2,...,n$, are not necessarily equal to each other. The left-hand side in (2.5.2) represents a sample in the ensemble of time averages. These samples are indicated by the dashed lines in Figure 2.19a.

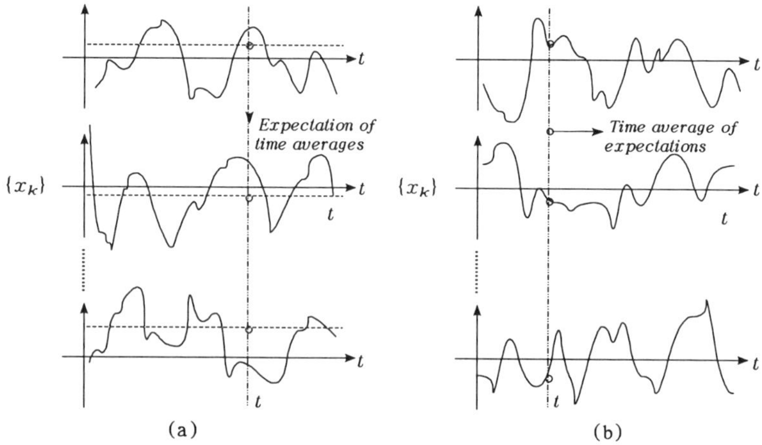

Figure 2.19. Description of (a) expectation of time averages compared to (b) time average of expectations.

Assume a cut through this ensemble of time averages at time t and define the expectation

$$E\{\overline{f[x_i(t)]}\} = \text{expectation of time averages} \tag{2.5.3}$$

where $i = 1, 2,..,n$. The integral in the right-hand side of (2.5.2) can be replaced by the arithmetic average of the sample, namely, by a sum of a large number N of discrete samples $f[x_i]_j$, $j = 1, 2,...,N$ along the time axis divided by N, namely,

$$\overline{f[x_i(t)]} \Leftrightarrow \lim_{N\to\infty}\left\{\frac{1}{N}\sum_{j=1}^{N}[f(x_i)]_j\right\} \tag{2.5.4}$$

The expectation defined in (2.5.3) can therefore be expressed by

$$E\{\overline{f[x_i(t)]}\} = E\left\{\lim_{N\to\infty}\frac{1}{N}\sum_{j=1}^{N}[f(x_i)]_j\right\} \tag{2.5.5}$$

On the basis of the distribution law presented in (2.3.15), we can write the equality

$$E\{\lim_{N\to\infty}\left\{\frac{1}{N}\sum_{j=1}^{N}[f(x_i)]_j\right\} = \lim_{N\to\infty}\frac{1}{N}\left\{\sum_{j=1}^{N}E\{[f(x_i)]_j\right\} \tag{2.5.6}$$

Equation (2.5.6) is equivalent to

$$E\{\overline{f[x_i(t)]}\} = \overline{E\{f[x_i(t)]\}} \tag{2.5.7}$$

where

$$\overline{E\{f[x_i(t)]\}} = \text{Time average of the Expectations at any instant } t \tag{2.5.8}$$

Expression (2.5.8) is illustrated in Figure 2.19(b). The meaning of (2.5.7) is:

The expectation of an ensemble of time averages equals the time average of the expectation at any instant t.

Ergodicity

In the foregoing, nowhere has it been assumed that the process is stationary. Therefore, (2.5.7) is valid whether the process is stationary or not.

If $\{x(t)\}$ is strictly stationary, $E\{f[x_i(t)]\}$ = const. Then for the right-hand side of (2.5.7), we have

$$\overline{E\{f[x_i(t)]\}} = E\{f[x_i(t)]\} \tag{2.5.9}$$

If all the time averages in the left-hand side of (2.5.7) are equal to each other, we have

$$E\{\overline{f[x_i(t)]}\} = \overline{f[x_i(t)]} \tag{2.5.10}$$

Equating the right-hand sides of (2.5.9) and (2.5.10), we have the following important equality:

$$\overline{f[x_i(t)]} = E\{f[x_i(t)]\} \tag{2.5.11}$$

Equation (2.5.11) states the following property:

Property I: A process that is strictly stationary and in which all the sample time averages in the ensemble are equal to each other has the property that the time average of any of the sample functions is equal to the ensemble average.

Definition 2.5–1: *A process that has Property I is an ergodic process.*

The meaning of *ergodicity* is: A process for which every sequence or a sizable sample of it is statistically the same and therefore equally representative of the whole, is *ergodic*. It relates to the probability that any state of the process will reoccur.

Special cases of $f[x_i(t)]$ yield corresponding definitions of ergodicity. Thus, if $f[x_i(t)] = x_i(t)$, the process is *ergodic in the mean*, i.e.,

$$\overline{x_i(t)} = E(x) = \mu_x \tag{2.5.12}$$

Namely, the time average of any sample equals the expectation or the ensemble mean.

If $f[x_i(t)] = x_i^2(t)$, the process is *ergodic in power*. i.e.,

$$\overline{x_i^2(t)} = P_x \tag{2.5.13}$$

The same reasoning applies to other statistical descriptors, such as the correlation function. Recalling the definition of the sample autocorrelation function in (2.1.8), *ergodicity in autocorrelation* means

$$\overline{x_i(t)x_i(t+\tau)} = E(x_t x_{t+\tau}) \tag{2.5.14}$$

or

$$\phi_{xx}(\tau) = C_{xx}(\tau) \tag{2.5.15}$$

Similarly, we can refer to ergodicity in *cross correlation*, in *variance* and in *covariance*. Thus

$$\phi_{xy}(\tau) = C_{xy}(\tau) \tag{2.5.16}$$

In practical terms ergodicity means that if we process the information from a single sample, it is representative of the entire ensemble from which it might have been selected. This raises the immediate question: How do we know that the two conditions, namely, stationarity and equality of averages are being met? The answer is that we do not strictly know. In reality, ergodicity is not a provable property. It should rather be viewed as a hypothesis. For example, given a bag of dice, we want to determine the average number of the throw of the dice in the bag. We can accomplish this by two approaches. We can use the *sample* approach, by which one die is thrown many times, each throw is recorded and the expected average 3.5 is determined.

The other approach is the *ensemble* approach where the whole bag is thrown at once and the number of each die is recorded. If the dice process is ergodic, the ensemble experiment will also yield 3.5, or at least, a number very close to it. A valid question is, why should it not? There seems to be no reason to expect otherwise, provided that all the dice are identical, i.e., totally unloaded or equally loaded, so as to meet the requirement that the averages of all dice be equal, and that they do not deform or wear in the course of the long sequence of throws.

A similar line of reasoning applies in the testing of a guided missile. Its performance evaluation is often determined on the basis of a single experimental test, and, in the case of a destructive test, on the basis of a few tests. The assumption of ergodicity in this case must be justifiable by the knowledge that each missile is manufactured, assembled, and inspected under identical conditions and procedures.

The assumption of ergodicity, if justifiable, permits the representation of dynamical physical random processes by single samples of real-time functions and the application of them to mathematical operations such as integration, differentiation, and filtering. Thus the study of the properties and performance of aerospace sensor systems in environments of random inputs, interferences, or noise is made possible.

For more detailed formulations of stationarity and ergodicity and for broader definitions of random variables as vector quantities, see, for example, [8].

2.6 Mathematical operations on random processes

This section is devoted to the development of mathematical operations on random sample functions $x(t)$ by means of the statistical descriptors presented in Section 2.1.

Autocorrelation of the sum of random functions

Given the autocorrelation function $\phi_{xx}(\tau)$ of $x(t)$ and $\phi_{yy}(\tau)$ of $y(t)$, the autocorrelation function $\phi_{zz}(\tau)$ of $z(t) = x(t) + y(t)$ is derived as follows:

$$\phi_{zz}(\tau) = \overline{[x(t) + y(t)][x(t+\tau) + y(t+\tau)]}$$

$$= \overline{x(t)x(t+\tau)} + \overline{x(t)y(t+\tau)} + \overline{y(t)x(t+\tau)} + \overline{y(t)y(t+\tau)}$$

$$= \phi_{xx}(\tau) + \phi_{xy}(\tau) + \phi_{yx}(\tau) + \phi_{yy}(\tau) \tag{2.6.1}$$

The expansion of (2.6.1) uses the distribution law as applied to time averages. If $x(t)$ is orthogonal to $y(t)$, the two cross-correlation functions are zero, resulting in

$$\phi_{zz}(\tau) = \phi_{xx}(\tau) + \phi_{yy}(\tau) \tag{2.6.2}$$

Result: The autocorrelation function of the sum of two orthogonal functions is the sum of the individual autocorrelation functions.

This result can be directly extended to the sum of any number of functions under the condition of orthogonality.

EXAMPLE 2.6–1: Signal analysis by autocorrelation function

$x(t)$ is a stationary random process characterized by the autocorrelation function $\phi_{xx}(\tau) = 4e^{-|\tau|}$. $y(t)$ is periodic and it has the autocorrelation function $\phi_{yy}(\tau) = \cos \omega\tau$, where $\omega = \pi$ rad/sec. $x(t)$ and $y(t)$ are orthogonal. It is required to determine the autocorrelation of $z(t) = x(t) - y(t)$. It is obvious that the negative sign of $y(t)$ causes the cross correlation terms in (2.6.1) to be negative. But due to the orthogonality, these terms are zero, and the result is as in (2.6.2). Thus,

$$\phi_{zz}(\tau) = 4e^{-|\tau|} + \cos \pi\tau$$

A plot of $z(t)$ generated by simulation is shown in Figure 2.20a. The simulation was performed by PC–Mat lab using the M–file lsim.m. Mere inspection of this plot does not readily disclose the existence of a periodic component in this random noise signal. However, processing $z(t)$ by the operation $\overline{z(t)z(t+\tau)}$ for 1000 data points provides $\hat{\phi}_{zz}(\tau)$, which reveals this periodic component very clearly along with the exponential component, as shown in Figure 2.20b. This processing was performed by PC–Matlab file xcorr.m. This example demonstrates that the autocorrelation function can be a very efficient method of data processing for the purpose of signal diagnostics.

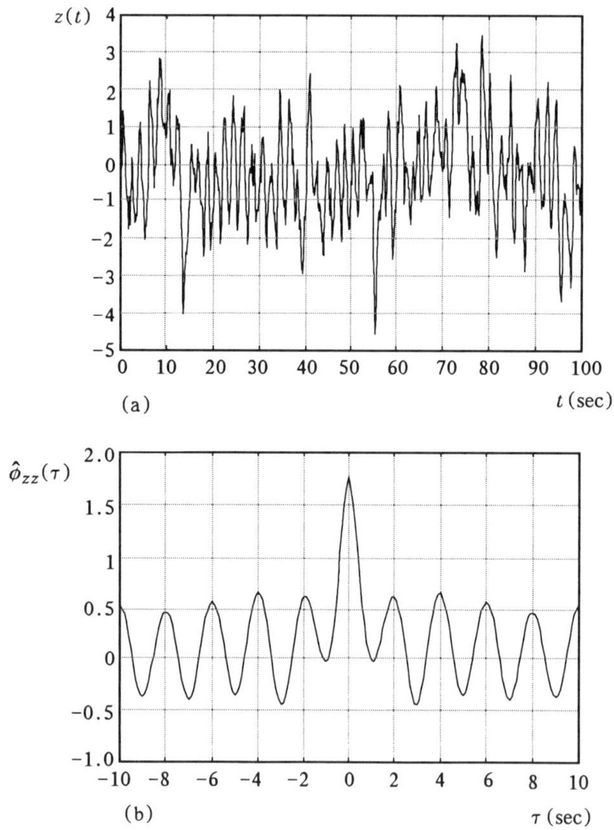

Figure 2.20:–(a) Sample of the sum of first order Markov process and sinusoidal signal;
–(b) Autocorrelation function of their sum.

Cross-correlation between a random function and its time derivatives

The expressions of the correlation functions yield useful relationships for analysis, estimation, and filtering in sensor system design. We are concerned with the correlation between a random sample function $x(t)$ and its *nth* derivative denoted by $x^{(n)}(t)$, namely,

$$\phi_{xx^{(n)}}(\tau) \triangleq \overline{x(t)x^{(n)}(t+\tau)} \tag{2.6.3}$$

The right-hand side of (2.6.3) can be rewritten as follows:

$$\overline{x(t)x^{(n)}(t+\tau)} = \overline{x(t)\frac{d^n x(t+\tau)}{dt^n}} = \overline{x(t)\frac{d^n x(t+\tau)}{d\tau^n}}$$

$$= \frac{d^n}{d\tau^n}\overline{x(t)x(t+\tau)} = \frac{d^n}{d\tau^n}\phi_{xx}(\tau) \tag{2.6.4}$$

Since τ is a parameter in time average over t, in the last step, the derivative $d^n(\cdot)/d\tau^n$ can be placed outside the averaging operator $\overline{(\cdot)}$. We therefore have the following.

Result: The cross-correlation between $x(t)$ and its *n-th* derivative is equal to the *n-th* derivative of the autocorrelation of $x(t)$.

Due to the symmetry of $\phi_{xx}(\tau)$, we have for positive integers $k = 0, 1, 2, 3, \ldots$

$$\frac{d^k}{d\tau^k}\phi_{xx}(-\tau) = (-1)^k \frac{d^k}{d\tau^k}\phi_{xx}(\tau) \tag{2.6.5}$$

(2.6.4) can be generalized for the cross correlation between the *m-th* and the *n-th* derivatives of $x(t)$:

$$\phi_{x^{(m)}x^{(n)}}(\tau) = \overline{x^{(m)}(t)x^{(n)}(t+\tau)} = \overline{x^{(m)}(t)\frac{d^n x(t+\tau)}{dt^n}}$$

$$= \overline{x^{(m)}(t)\frac{d^n x(t+\tau)}{d\tau^n}} = \frac{d^n}{d\tau^n}\overline{x^{(m)}(t)x(t+\tau)}$$

$$= \frac{d^n}{d\tau^n}\overline{x(t+\tau)x^{(m)}(t)} = \frac{d^n}{d\tau^n}\overline{x(t)x^{(m)}(t-\tau)} =$$

$$= \frac{d^n}{d\tau^n}\phi_{xx^{(m)}}(-\tau) = \frac{d^n}{d\tau^n}\left[\frac{d^m}{d\tau^m}\phi_{xx}(-\tau)\right]$$

$$= \frac{d^n}{d\tau^n}\left[(-1)^m \frac{d^m}{d\tau^m}\phi_{xx}(\tau)\right] = (-1)^m \frac{d^{(m+n)}}{d\tau^{(m+n)}}\phi_{xx}(\tau) \tag{2.6.6}$$

Result: The cross correlation between the m-th and the n-th derivative of x(t) is equal to the (m+n)-th derivative of the autocorrelation function $\phi_{xx}(\tau)$. Its sign is given by $(-1)^m$.

We can now consider particular cases of special practical value:

1. k=1; m=0; n=1:

$$\phi_{x\dot{x}}(\tau) = -\frac{d}{d\tau}\phi_{xx}(-\tau) = \frac{d}{d\tau}\phi_{xx}(\tau) \tag{2.6.7}$$

Result: The cross correlation between a process $x(t)$ and its time derivative $\dot{x}(t)$ is equal to the derivative of the autocorrelation function of $x(t)$ with respect to its argument τ.

2. k=2; m=1; n=1:

$$\phi_{\dot{x}\dot{x}}(\tau) = -\frac{d^2}{d\tau^2}\phi_{xx}(\tau) \tag{2.6.8}$$

Result: The autocorrelation function of the time derivative $\dot{x}(t)$ is equal to minus the second derivative of the autocorrelation function of $x(t)$ with respect to its argument τ.

A case of special interest is:

$$\phi_{x\dot{x}}(\tau)\Big|_{\tau=0} = \frac{d}{d\tau}\phi_{xx}(\tau)\Big|_{\tau=0} = 0 \tag{2.6.9}$$

Because of the symmetry of $\phi_{xx}(\tau)$, its derivative at the origin τ=0 must be zero. However, due to the assumed ergodicity of $x(t)$, we recall that $\phi_{x\dot{x}}(\tau) = E(x_t\dot{x}_{t+\tau})$, so that for τ=0, we have,

$$E(x_t\dot{x}_t) = 0 \tag{2.6.10}$$

Result: For an ergodic random process, x(t) is orthogonal to its time derivative $\dot{x}(t)$.

Orthogonality was presented in Section 2.3 under the assumption that two processes are statistically independent. In the present case of $x(t)$ and $\dot{x}(t)$, their statistical independence is not a necessary condition for orthogonality. On the contrary, it is easily demonstrated that $x(t)$ and $\dot{x}(t)$ are in general statistically dependent.

EXAMPLE 2.6–2: Autocorrelation of the derivative of a noisy process

The range measurement $z(t)$ of a radar is accompanied by additive noise $n(t)$

$$z(t) = y(t) + n(t)$$

where $y(t)$ is the true range. The range error $e_z(t) = w(t) - y(t) = n(t)$. Processing the noise $n(t)$ reveals that its autocorrelation function is of the type

$$\phi_{nn}(\tau) = \sigma_n^2(1+a|\tau|)e^{-a|\tau|}$$

which is a second-order Markov process to be formally defined later. It is required to derive the range rate $\dot{z}(t)$ from the range measurements $z(t)$ by differentiation and to determine the range rate error $\dot{e}_z(t) = \dot{n}(t)$ in statistical terms, i.e., $\sigma_{\dot{n}}$. Applying (2.6.8) and taking the positive branch of $\phi_{nn}(\tau)$ only, we have

$$\phi_{\dot{n}\dot{n}}(\tau) = -\frac{d^2\phi_{nn}(\tau)}{d\tau^2} = \sigma_n^2 a^2 e^{-a\tau}\,(1-e^{-a\tau})$$

$$\sigma_{\dot{n}}^2 = \phi_{\dot{n}\dot{n}}(\tau)\big|_{\tau=0} = \sigma_n^2 a^2$$

or

$$\sigma_{\dot{n}} = \sigma_n a$$

We conclude from this result that the standard deviation of the range rate error is equal to the standard deviation of the range error σ_n times the bandwidth a. We also conclude that the larger the bandwidth of the process, the larger the standard deviation of its time derivative. This means that the differentiation of wideband noise in a process may cause very large errors.

In the present radar example, assume that $\sigma_n = 4$ m and that $a = 6$ rad/sec. The range rate error will be

$$\sigma_{\dot{n}} = 4\text{ m} \times 6\text{ rad/sec} = 24\text{ m/sec}$$

Because of this noise amplification caused by differentiation, it is commonly attempted to apply low-pass filtering to attenuate the noise error when differentiation of a process is required.

Filtering of white noise, Markov processes

On the basis of the preceding properties of correlation functions we can now develop the expressions for low-pass filtering of white noise. This is done here for a specific example of a simple first-order filter, but the result is of general importance.

We have the relationship

$$x(s) = H(s)w(s) \tag{2.6.11}$$

where

$$H(s) = \frac{a}{a+s} \tag{2.6.12}$$

and $w(s)$ represents Gaussian white noise. The reason for assuming Gaussian noise is that the process remains Gaussian under any arbitrary linear operation such as the filtering discussed here. This important property and others are addressed in [12]. We formulate the relationship (2.6.12) and (2.6.13) by the corresponding differential equation and its shift in time by τ:

$$\dot{x}(t) + ax(t) = aw(t) \tag{2.6.13}$$

$$\dot{x}(t+\tau) + ax(t+\tau) = aw(t+\tau) \tag{2.6.14}$$

Multiplying the corresponding sides of the equations and time averaging, we have

$$\overline{\dot{x}(t)\dot{x}(t+\tau)} + a\overline{x(t)\dot{x}(t+\tau)} + a\overline{x(t+\tau)\dot{x}(t)} + a^2\overline{x(t)x(t+\tau)}$$

$$= a^2\overline{w(t)w(t+\tau)} \tag{2.6.15}$$

By virtue of the previous definitions of correlation functions, the last equation is rewritten as

$$\phi_{\dot{x}\dot{x}}(\tau) + a\phi_{x\dot{x}}(\tau) + a\phi_{\dot{x}x}(\tau) + a^2\phi_{xx}(\tau) = a^2\phi_{ww}(\tau) \tag{2.6.16}$$

For white noise, the right-hand side in (2.6.16) is given by

$$a^2\phi_{ww}(\tau) = a^2 S_w\,\delta(\tau) \tag{2.6.17}$$

where $S_w = 2\sigma_w^2/b$ is the power density, b is the very large bandwidth of the white noise and $\delta(\tau)$ is the Dirac or delta function. Making use of (2.6.7), (2.6.8), and (2.4.37), we have

$$-\frac{d^2}{d\tau^2}\phi_{xx}(\tau) + a\frac{d}{d\tau}\phi_{xx}(\tau) + a\frac{d}{d\tau}\phi_{xx}(-\tau) + a^2\phi_{xx}(\tau)$$

$$= a^2 S_w\,\delta(\tau) \tag{2.6.18}$$

The second and third terms on the left-hand side obviously cancel each other, and the equation reduces to

$$\frac{d}{d\tau^2}\phi_{xx}(\tau) - a^2\phi_{xx}(\tau) = -a^2 S_w\,\delta(\tau) \tag{2.6.19}$$

Applying Laplace transformation, we have

$$\Phi_{xx}(s)(a^2 - s^2) = a^2 S_w \tag{2.6.20}$$

and

$$\Phi_{xx}(s) = \frac{a^2}{a^2 - s^2} S_w = \frac{a^2 S_w}{2a}\left(\frac{1}{a+s} + \frac{1}{a-s}\right) \tag{2.6.21}$$

The operation $\mathcal{L}\{\phi_{xx}(\tau)\} = \Phi_{xx}(s)$ is the two-sided Laplace transform, which is needed because $\phi_{xx}(\tau)$ consists of two symmetrical branches in the regions $\tau > 0$ and $\tau < 0$. $\Phi_{xx}(s)$ will be addressed in further detail later and will be shown to be the power spectrum of the process $x(t)$. We define

$$\Phi_x^+(s) \triangleq \frac{a^2 S_w}{2a}\frac{1}{a+s} \tag{2.6.22}$$

and

$$\Phi_x^-(s) \triangleq \frac{a^2 S_w}{2a}\frac{1}{a-s} \tag{2.6.23}$$

With these definitions, (2.6.21) is inverse-transformed so that

$$\phi_{xx}(\tau) = \frac{a^2 S_w}{2a}[\phi_x^+(\tau) + \phi_x^-(\tau)]$$

$$= \frac{a^2 S_w}{2a}[e^{-a\tau} + e^{-a(-\tau)}] = \frac{aS_w}{2}e^{-a|\tau|} \tag{2.6.24}$$

Result: First-order low-pass filtering of white noise provides a process $x(t)$, known as a first-order Markov process, which is characterized by an exponential autocorrelation function having an exponent coefficient a equal to the filter bandwidth.

Comment: In interpreting (2.6.24), recall that the two exponential branches exist in different exclusive regions.

This exponential autocorrelation function was formally introduced in Example 2.4–3 without the physical interpretation or justification provided here.

The coefficient $aS_w/2$ is the power P_x. Setting $P_x = 1$ for normalization, we have $aS_w/2 = 1$ and $S_w = 2/a$. Substituting this into (2.6.22), we have

$$\Phi_{xx}(s) = \frac{2}{a}\frac{a^2}{a^2-s^2} = \frac{2a}{a^2-s^2} \tag{2.6.25}$$

This form of a power spectrum is of major importance and is commonly used in modeling and analysis of random signals and noise. It should be noted that $\Phi_{xx}(s)$ is symmetrical in the s plane. The properties of the power spectrum will be discussed in further detail in Section 2.8.

At this stage, it suffices to note that an infinite variety of autocorrelation functions can be generated by an appropriate choice of low-pass filters operating on white noise. This operation is also known as *coloring* white noise, and it is indeed analogous to the effect of color filtering of white light.

EXAMPLE 2.6–3: Estimation of autocorrelation revealing the nature of the process

Gaussian white noise $w(t)$ with unit power density $S_w = 1$ is applied to a lightly damped second-order low-pass filter establishing the relationship

$$x(s) = \frac{5}{s^2 + s + 5}w(s)$$

It is required to determine an estimate $\hat{\phi}_{xx}(\tau)$ of $\phi_{xx}(\tau)$ by computer simulation.

The simulation is performed using PC-Matlab, M-files lsimul.m and xcorr.m. The simulated time plot of $x(t)$ is shown in Figure 2.21, and the corresponding $\hat{\phi}_{xx}(\tau)$ is shown in Figure 2.22. The quasi-periodic $\sqrt{5}$ rad/sec nature of the process is somewhat noticeable in $x(t)$, but it is very clearly pronounced in the plot of $\phi_{xx}(\tau)$. This example demonstrates that the autocorrelation function is a very powerful statistical descriptor, revealing important frequency and damping information of a random process.

Having established that any Gaussian process can be viewed as filtered white noise, it follows that correlation signal processing is an important tool in identifying linear dynamical systems [12]. This means that if an unknown dynamical system is driven by wide-band noise that approximates white noise, the shape of $\phi_{xx}(\tau)$ reveals the nature of the unknown dynamical model of the system.

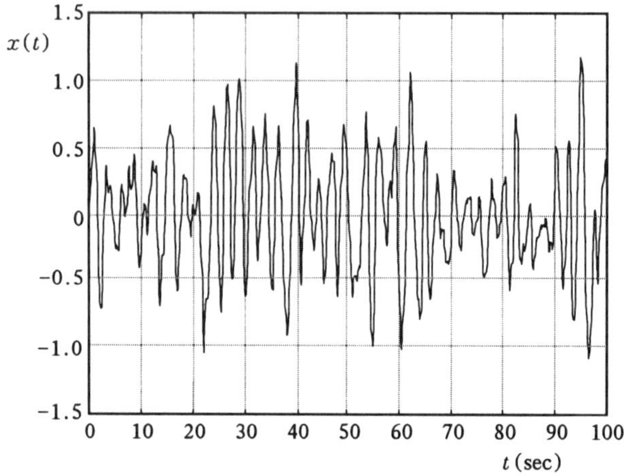

Figure 2.21. Simulation of lightly damped filtering of white noise; oscillatory nature is noticeable.

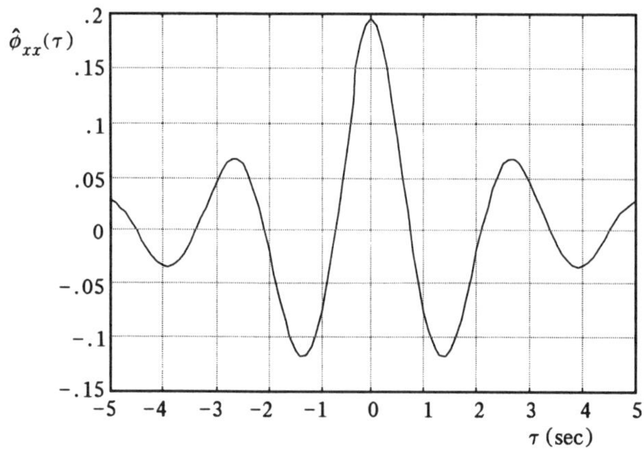

Figure 2.22. Autocorrelation of lightly damped, filtered white noise; frequency and damping clearly revealed.

Further properties of correlation functions

It is not accidental that in Example 2.6-2 the noise model was chosen as a second order Markov process and not the simpler exponential first order Markov process. Let us examine what would occur if a first order process $x(t)$

is differentiated to provide $\dot{x}(t)$. We have seen that this operation requires twice differentiation of $\phi_{xx}(\tau)$ and inversion of its sign. This examination is readily performed by the graphical representation Figure 2.23 where (a), (b), and (c) illustrate the respective derivatives.

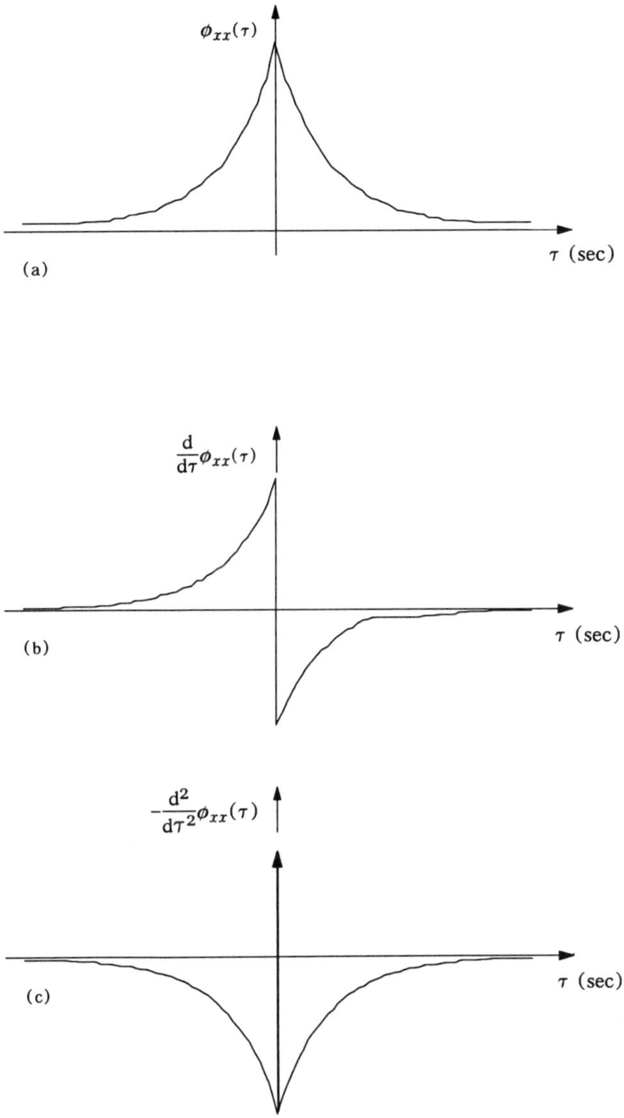

Figure 2.23. Derivatives of exponential autocorrelation function: (a), (b), and (c) (a) zeroth derivative; (b) first derivative; second derivative.

Figure 2.23(c) reveals that

$$-\frac{\mathrm{d}^2\phi_{xx}(\tau)}{\mathrm{d}\tau^2}\bigg|_{\tau=0} = \phi_{\dot{x}\dot{x}}(0) = P_{\dot{x}} = \infty \tag{2.6.26}$$

meaning that the power of the derivative of a first order Markov process is infinite. This clearly results from the fact that the exponential autocorrelation function is not analytical at the origin $\tau=0$ and therefore not differentiable. Thus, in realistic modeling of random processes when involving differentiation, the first order Markov process is inadequate, and consequently higher-order processes must be chosen.

Another way to interpret this property is by the frequency response approach. The spectrum of a first-order Markov process has been shown to be

$$\Phi_{xx}(s) = \frac{a^2}{a^2 - s^2} S_w = \frac{a}{a+s}\frac{a}{a-s} S_w = H(s)H(-s)S_w \tag{2.6.27}$$

$\Phi_{xx}(s)$ can be regarded as the input to the filter $F(s) = s$, i.e., the differentiation operator. Denote the output of this differentiator by $y(t)$. Then, in analogy with (2.6.27),

$$\Phi_{yy}(s) = F(s)F(-s)\Phi_{xx}(s) = (s)(-s)\frac{a}{a+s}\frac{a}{a-s}S_w$$

$$= \frac{-a^2 s^2}{a^2 - s^2} S_w \tag{2.6.28}$$

This expression shows that

$$\lim_{s\to\infty} \Phi_{yy}(s) = a^2 S_w = \text{const.} \tag{2.6.29}$$

indicating that the power density is constant out to infinite frequency. In other words, the power, which is the integral of the power density out to infinitely high frequencies, is infinite. This agrees with the analysis of the autocorrelation function $\phi_{\dot{x}\dot{x}}(\tau)$. Equation (2.6.29) indicates that for the power to be finite, the power density must approach zero as $s \to \infty$. This means that the highest power in the denominator of the spectrum must be larger than the power of the numerator, or, in other words, the relative degree must be greater than zero. By *relative degree* we mean the difference between the highest powers in the denominator and the numerator.

Elementary prediction

In Section 2.1 the point was made that the future of a random process is unpredictable, that is, as indicated in Figure 2.1, one cannot tell the value of $y(t+\lambda)$, of the process $y(t)$, where t is the present time and $\lambda > 0$. Strictly speaking, this statement is correct. However, what it actually implies is that one cannot predict $y(t+\lambda)$ precisely. However, if $y(t)$ is reasonably smooth, it is differentiable. This implies that the length of its autocorrelation function is not vanishingly small, so that $y(t+\lambda)$ can be expanded into a converging Taylor series:

$$y(t+\lambda) = y(t) + \dot{y}(t)\lambda + \ddot{y}(t)\frac{\lambda^2}{2} + \ldots \tag{2.6.30}$$

This expression is a zero, first, or second order extrapolation, in accordance with the number of terms retained in the series expansion, which is an approximation to the unrealizable operation of pure prediction. With the first three terms, for example, we have second order extrapolation and the prediction error is,

$$\epsilon_\lambda(t) = y(t+\lambda) - y(t) - \dot{y}(t)\lambda - \ddot{y}(t)\frac{\lambda^2}{2} - \ldots \tag{2.6.31}$$

Assuming ergodicity in autocorrelation of $y(t)$, the mean square error of $\epsilon_\lambda(t)$ is given by

$$\overline{\epsilon_\lambda^2} = \overline{[y(t+\lambda) - y(t) - \dot{y}(t)\lambda]^2} = \overline{y^2(t+\lambda)} + \overline{y^2(t)}$$

$$+ \overline{\dot{y}^2(t)}\lambda^2 - 2\overline{y(t+\lambda)y(t)} - 2\overline{y(t+\lambda)\dot{y}(t)}\,\lambda + 2\,\overline{y(t)\dot{y}(t)}\,\lambda \tag{2.6.32}$$

The last term on the right-hand side of (2.6.32) is zero because of the orthogonality of y and $\dot{y}$, so it reduces to

$$\overline{\epsilon_\lambda^2} = 2P_y + \lambda^2 P_{\dot{y}} - 2\phi_{yy}(\lambda) - 2\lambda \frac{d}{d\lambda}\phi_{yy}(\lambda) \tag{2.6.33}$$

Expressing $\phi_{yy}(\lambda)$ in normalized form, expressing $P_{\dot{y}}$ explicitly assuming $\mu_y = 0$, and rearranging, we have

$$\overline{\epsilon_\lambda^2} = 2\sigma_y^2 - 2\sigma_y^2\rho_{yy}(\lambda) - \sigma_y^2\lambda^2 \frac{d^2}{d\lambda^2}\rho_{yy}(\lambda)\Big|_{\lambda=0} - \sigma_y^2 2\lambda \frac{d}{d\lambda}\rho_{yy}(\lambda)$$

$$= 2\sigma_y^2\left\{1 - \rho_{yy}(\lambda) + \lambda\frac{d}{d\lambda}\rho_{yy}(\lambda) - \frac{1}{2}\lambda^2\frac{d^2}{d\lambda^2}\rho_{yy}(\lambda)\Big|_{\lambda=0}\right\} \tag{2.6.34}$$

The prediction error clearly vanishes as $\lambda \to 0$. Alternatively, for a given value of λ, the error is smaller, when $\rho_{yy}(\lambda)$ is wider or in other words, the process $y(t)$ is slower.

The foregoing rudimentary analysis disregards any possible a priori knowledge of the process statistics, specifically its PDF. If such knowledge were used, substantially smaller prediction errors would be achieved.

EXAMPLE 2.6–4: Elementary prediction

The process $y(t)$ is described by a second order Markov process with $\sigma_y = 1$ and $\mu_y = 0$ given by

$$\rho_{yy}(\lambda) = (1 + a\lambda)e^{-a\lambda} \; ; \qquad \lambda \geq 0$$

The first and second derivatives are,

$$\frac{d}{d\lambda}\rho_{yy}(\lambda) = -a^2\lambda e^{-a\lambda} \; ; \qquad \lambda \geq 0$$

$$\frac{d^2}{d\lambda^2}\rho_{yy}(\lambda) = -a^2 e^{-a\lambda}(1 - a\lambda); \quad \lambda \geq 0$$

Substituting these expressions into (2.6.34) and rearranging, we have

$$\overline{\epsilon_\lambda^2} = 2[1 - \{1 + a\lambda)e^{-a\lambda} + \frac{a^2\lambda^2}{2}(1-2e^{-a\lambda})]$$

It is easily seen that the expression up to the second term only corresponds to zero-order extrapolation, and inclusion of the third term corresponds to first-order extrapolation as formulated in (2.6.32).

A plot of $\overline{\epsilon_\lambda^2}$ as a function of $a\lambda$ is shown for zero order and first-order extrapolation in Figure 2.24. It shows, as expected, that first-order extrapolation yields much smaller errors for small values of $a\lambda$, while the opposite is true for large values of $a\lambda$. The inclusion of the quadratic term in (2.6.31) would yield still smaller prediction errors for small values of $a\lambda$. However, the higher-order terms would eventually lead to diminishing returns because of the high-order differentiation of inevitable wide-band additive noise encountered in real systems.

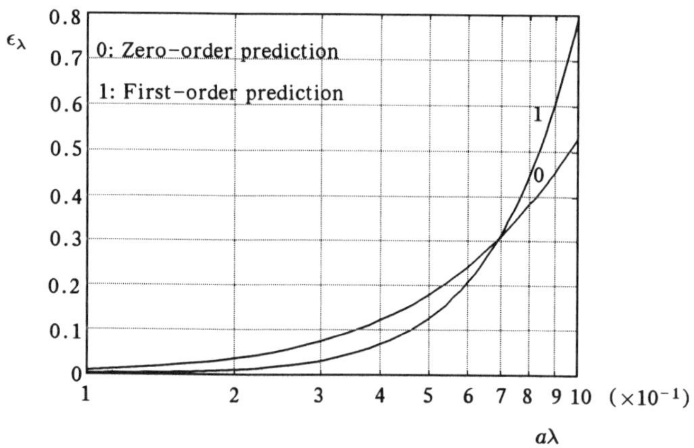

Figure 2.24. Zero-order and first-order rms prediction error as a function of the normalized prediction time $a\lambda$.

One might ask what the results for ϵ_λ would be if $\lambda < 0$. Clearly, without noise this is a trivial proposition because this case refers to the past which, as pointed out in Section 2.0, is precisely known. Therefore ϵ_λ is, disregarding noise, identically zero.

2.7 Input–output relationships

Convolution

This section addresses the general topic of linear operations on random processes. The previous sections have dealt with the special cases of low-pass filtering of white noise and the differentiation and extrapolation of ergodic processes. The basic approach here will be in the time domain by means of the convolution integral, which defines the relationship between the input $x(t)$ and a linear dynamical system modeled by its time invariant impulse response $h(t)$ and its corresponding output $y(t)$. The subject matter focuses primarily on asymptotically stable and causal systems [11].

The notion of the convolution integral will first be reviewed. We relate the formulation to the concept of real time t, as indicated in Section 2.0. This input–output relationship is mathematically defined by the basic form of the convolution integral

$$y(t) = \int_{\tau=-\infty}^{\tau=t} h(t-\tau)x(\tau)d\tau \tag{2.7.1}$$

A graphical interpretation of (2.7.1) is given in Figure 2.25.

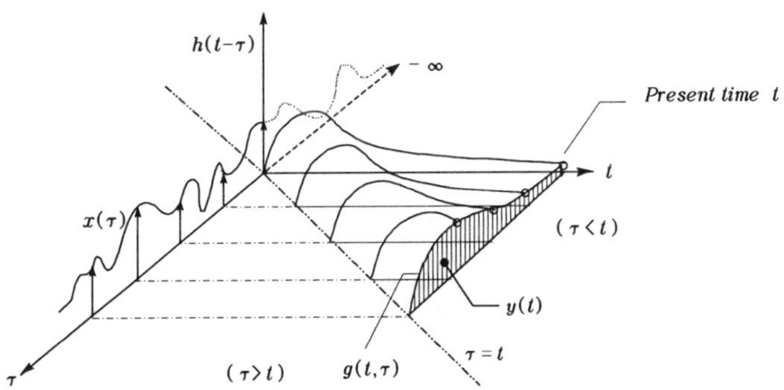

Figure 2.25. Graphical interpretation of the convolution integral.

$h(t-\tau)$ is a function of the two variables t and τ and is described in the t,τ plane. t is the "real-time" axis and τ is the time axis along which the input $x(\tau)$ has been evolving from the distant past $\tau = -\infty$ up to the present $\tau = t$. $x(\tau)$ can be interpreted as a sequence of pulses, as indicated in Figure 2.25. The line $\tau = t$ divides the t,τ plane into the two regions, $\tau > t$ and $\tau < t$. A *response* is an outcome of cause and effect. In a physically realizable system the cause must always precede the effect. Therefore $h(t-\tau)$ cannot exist in the region $\tau > t$. Thus, the sequence of pulses $x(\tau)$ excites the sequence of asymptotically stable impulse responses $h(t-\tau)$ originating on the line $\tau = t$. The line parallel to the τ axis at the instant t is the present. The earlier responses $h(t-\tau)$, i. e., those with the smaller values of τ, retain the smaller residual values at the present time t and the more recent ones, i.e., larger τ, retain the larger residual values. The sum of all these residual values, defined by $g(t,\tau)=h(t-\tau)x(\tau)$ is described by the shaded area under the curve $g(t,\tau)$, namely, the integral in (2.7.1), which is the output $y(t)$.

In the application to input–output relationships for random processes, it is often preferable to express the convolution integral in alternative forms by transformation of variables as follows:

$$t-\tau=\sigma; \Rightarrow \tau=t-\sigma; \Rightarrow \mathrm{d}\tau=-\,\mathrm{d}\sigma \tag{2.7.2}$$

The limits of the convolution integral are accordingly,

$$\tau=-\infty, \Rightarrow \sigma=\infty\ ; \tau=t, \Rightarrow \sigma=0 \tag{2.7.3}$$

Substituting these variables and limits into (2.7.1), we obtain

$$\begin{aligned} y(t) &= \int_{\infty}^{0} h(\sigma)x(t-\sigma)(-\mathrm{d}\sigma) \\ &= \int_{0}^{\infty} h(\sigma)x(t-\sigma)\mathrm{d}\sigma \end{aligned} \tag{2.7.4}$$

In many applications, the origin of $x(\tau)$ is set at $\tau=0$ so that (2.7.1) would assume the form

$$y(t)=\int_{0}^{t} h(t-\tau)x(\tau)\mathrm{d}\tau \tag{2.7.5}$$

On applying the transformation of variables of (2.7.2), the limits of the integral become

$$\tau=0, \Rightarrow \sigma=t; \tau=t, \Rightarrow \sigma=0 \tag{2.7.6}$$

so that,

$$\begin{aligned} y(t) &= \int_{t}^{0} h(\sigma)x(t-\sigma)(-\mathrm{d}\sigma) \\ &= \int_{0}^{t} h(\sigma)x(t-\sigma)\mathrm{d}\sigma \end{aligned} \tag{2.7.7}$$

Since $h(t-\tau)=0$ for $\tau>t$, the upper limit t in (2.7.1) can be changed to ∞ without altering the value of the integral. Thus, we have an additional form for the convolution integral:

$$y(t)=\int_{-\infty}^{\infty} h(t-\tau)x(\tau)\mathrm{d}\tau = \int_{-\infty}^{\infty} h(\sigma)x(t-\sigma)\mathrm{d}\sigma \tag{2.7.8}$$

These forms of the convolution integral are particularly useful in expressing

$y(t)$ as a stationary process, thus permitting the formulation of the output auto-correlation $\phi_{yy}(\tau)$ as follows:

$$\phi_{yy}(\tau) \triangleq \overline{y(t)y(t+\tau)}$$

$$= \overline{\int_{-\infty}^{\infty} h(\sigma)x(t-\sigma)d\sigma \int_{-\infty}^{\infty} h(\sigma')x(t+\tau-\sigma')d\sigma'}$$

$$= \overline{\int_{-\infty}^{\infty}\int_{-\infty}^{\infty} h(\sigma)h(\sigma')x(t-\sigma)x(t+\tau-\sigma')d\sigma d\sigma'} \tag{2.7.9}$$

Since the average of the product of a linear operator and a random process is the product of the linear operator and the average of the random process, (2.7.9) can be rewritten as

$$\phi_{yy}(\tau) = \int_{-\infty}^{\infty}\int_{-\infty}^{\infty} h(\sigma)h(\sigma')\overline{x(t-\sigma)x(t+\tau-\sigma')}d\sigma d\sigma' \tag{2.7.10}$$

The averaged shifted product in the integrand is easily shown to be

$$\overline{x(t-\sigma)x(t+\tau-\sigma')} = \phi_{xx}(\tau+\sigma-\sigma') \tag{2.7.11}$$

Thus, we have the input-output relationship between x and y in terms of the input and output correlation functions,

$$\phi_{yy}(\tau) = \int_{-\infty}^{\infty}\int_{-\infty}^{\infty} h(\sigma)h(\sigma')\phi_{xx}(\tau+\sigma-\sigma')d\sigma d\sigma' \tag{2.7.12}$$

It is of particular practical engineering interest to express this relationship for the output power $P_y = \overline{y^2(t)}$ by setting $\tau = 0$ in (2.7.12),

$$\overline{y^2(t)} = \int_{-\infty}^{\infty}\int_{-\infty}^{\infty} h(\sigma)h(\sigma')\phi_{xx}(\sigma-\sigma')d\sigma d\sigma' \tag{2.7.13}$$

Conclusion: The output power $y(t)$ is determined by the double convolution of the impulse response $h(\sigma)$ and the autocorrelation function of the input $x(t)$.

Nonstationary processes

We now recall (2.7.7), which defines the input-output relationship from the instant $\tau = 0$ to t. The output $y(t)$ now has the initial condition $y(0) = 0$. A sketch of a typical response is shown in Figure 2.26.

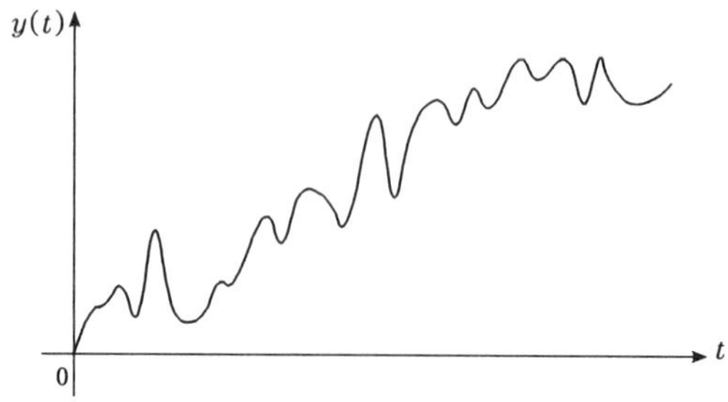

Figure 2.26. A sample of a nonstationary process with zero initial conditions.

Clearly, in this case $y(t)$ is not ergodic and the time averages used above do not apply. Therefore, in order to establish an input–output relationship for the transient phase, application of the formulation of the *ensemble average* is required in accordance with (2.3.23), thus

$$C_{yy}(t,\tau) \triangleq E[y(t)y(t+\tau)]$$

$$= E\left\{\int_0^t h(\sigma)x(t-\sigma)d\sigma \int_0^{t+\tau} h(\sigma')x(t+\tau-\sigma'd\sigma'\right\}$$

$$= E\left\{\int_0^t\int_0^{t+\tau} h(\sigma)h(\sigma')x(t-\sigma)x(t+\tau-\sigma')d\sigma d\sigma'\right\}$$

$$= \int_0^t\int_0^{t+\tau} h(\sigma)h(\sigma')E[x(t-\sigma)x(t+\tau-\sigma')]d\sigma d\sigma'$$

$$= \int_0^t\int_0^{t+\tau} h(\sigma)h(\sigma')C_{xx}(\tau+\sigma-\sigma')d\sigma d\sigma' \qquad (2.7.14)$$

The output power P_y is again determined by setting $\tau = 0$ so that

$$P_y(t) = C_{yy}(t,\tau)\Big|_{\tau=0} = E[y^2(t)]$$

$$= \int_0^t\int_0^t h(\sigma)h(\sigma')C_{xx}(\sigma-\sigma')d\sigma d\sigma' \qquad (2.7.15)$$

Conclusion: The output power of a linear system with zero initial conditions excited by a stationary random input is determined by the double convolution of the impulse response with the correlation function of the input, and is a function of time t.

A case of special interest is that of pure integration, namely, $h(\sigma)=1$, which is the impulse response of an integrator. The output power in this case reduces to the expression

$$E[y^2(t)] = \int_0^t \int_0^t C_{xx}(\sigma-\sigma')d\sigma d\sigma' \tag{2.7.16}$$

The double integral in the last expression is not necessarily bounded even though the autocorrelation $C_{xx}(\sigma)$ is bounded, namely,

$$\lim_{\sigma \to \infty} C_{xx}(\sigma) = 0 \tag{2.7.17}$$

Thus, the power of the time integral of a stationary bounded process $x(t)$, even if zero-mean, will, in general, diverge with time t without bound. It will be referred to as *stochastic divergence.* This property manifests itself in numerous physical processes and in sensor systems. For example, an ascending meteorological balloon, when subjected to zero-mean random crosswinds will invariably develop a lateral deviation from its nominal vertical ascent path. This phenomenon is also known as *random walk*, and it is closely associated with Brownian motion in gas dynamics. Familiarity with the properties of stochastic divergence is important in analyzing and modeling error processes in sensor systems, particularly in inertially stabilized platforms.

EXAMPLE 2.7–1: Determination of time–varying output power of an ergodic process

The autocorrelation function of a zero–mean ergodic process $x(t)$ with $\sigma_x = 1$ is given by

$$\phi_{xx}(\tau) = e^{-a|\tau|}$$

It is required to derive an explicit expression for the power $P_y(t) = \mathrm{E}[y^2(t)]$ in accordance with (2.7.16):

$$P_y(t) = \int_0^t \int_0^t e^{-a|\sigma-\sigma'|}d\sigma' d\sigma = \int_0^t d\sigma' \int_0^t e^{-a|\sigma-\sigma'|}\, d\sigma$$

The integrand in the inner integral is not analytic because it consists of two different, positive and negative exponential branches, which must be integrated separately, in accordance with the sketch shown in Figure 2.27, as follows:

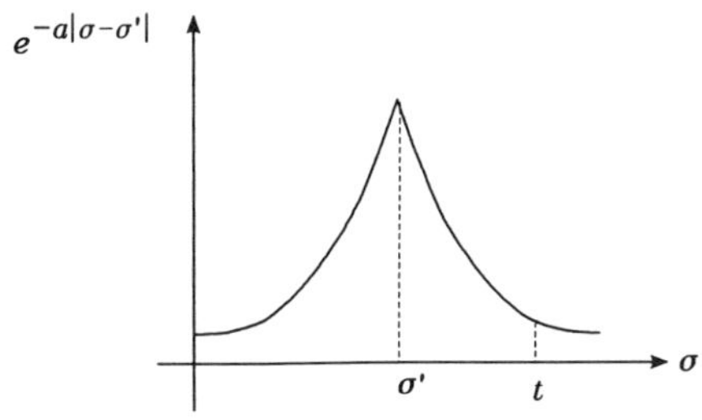

Figure 2.27. Nonanalytic integrand of the double convolution.

$$P_y(t) = \int_0^t d\sigma' \left\{ \int_0^{\sigma'} e^{a(\sigma-\sigma')} d\sigma + \int_{\sigma'}^t e^{-a(\sigma-\sigma')} d\sigma \right\}$$

$$= \int_0^t d\sigma' \left\{ \frac{1}{a} e^{a(\sigma-\sigma')} \Big|_0^{\sigma'} - \frac{1}{a} e^{-a(\sigma-\sigma')} \Big|_{\sigma'}^t \right\}$$

$$= \frac{1}{a} \int_0^t d\sigma' \left\{ 1 - e^{-a\sigma'} - (e^{-(t-\sigma')} - 1) \right\}$$

$$= \frac{1}{a} \int_0^t d\sigma' \{ 2 - e^{-a\sigma'} - e^{-a(t-\sigma')} \}$$

$$= \frac{1}{a} \left\{ 2t - \left(-\frac{1}{a} e^{-a\sigma'}\right)\Big|_0^t - \left(\frac{1}{a} e^{-a(t-\sigma')}\right)\Big|_0^t \right\}$$

$$= \frac{1}{a} \left[2t - \frac{1}{a}(1 - e^{-at}) - \frac{1}{a}(1 - e^{-at}) \right]$$

Finally,

$$P_y(t) = \frac{2}{a^2}(at - 1 + e^{-at})$$

The normalized power with respect to $2/a^2$ is plotted in Figure 2.28.

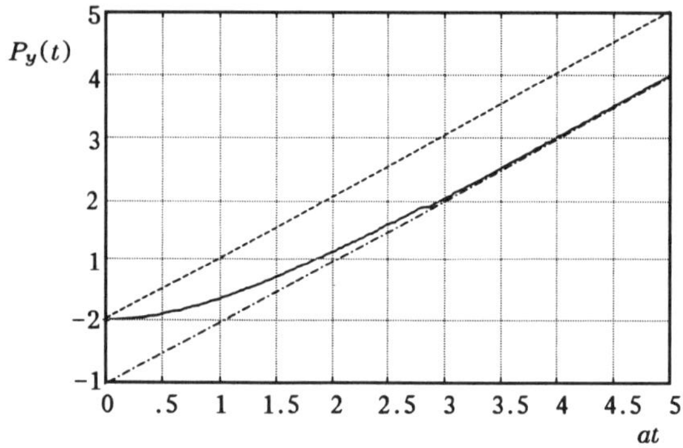

Figure 2.28. Normalized output power $P_y(t)$ as a function of normalized time at.

For a wide-band process or a relatively long time, $at \to \infty$, and $P_y(t)$ is approximately

$$P_y(t) \cong \frac{2}{a} t$$

For a narrow band process or a relatively short time, $at \to 0$, it is readily shown by expanding the exponential term e^{-at} that

$$P_y(t) \cong \frac{2}{a^2} \frac{a^2 t^2}{2} = t^2$$

Conclusion: The standard deviation σ_y of a wide-band process diverges with the square root of time, and σ_y, for a narrow-band process approaching a constant diverges linearly with time, as expected from the integration of a constant.

It is often surprising that the integration of a zero-mean stationary random process diverges from zero. One should bear in mind that this divergence is conceived and visualized for a sample $y(t)$ of the ensemble $\{y(t)\}$. Indeed, any sample can diverge from zero, but either in the positive or the negative direction. Recall that the variable that diverges is the power $P_y(t) = E[y^2(t)]$ in accordance with

$$P_y(t) = \int_{-\infty}^{\infty} \beta^2 p_{yt}(\beta)\mathrm{d}\beta \tag{2.7.18}$$

as illustrated in Figure 2.29 which indicates that the PDF spreads out with time. However, the mean value of the ensemble stays at zero in accordance with

$$\mu_y(t) = \int_{-\infty}^{\infty} \beta p_{yt}(\beta)\mathrm{d}\beta = 0 \tag{2.7.19}$$

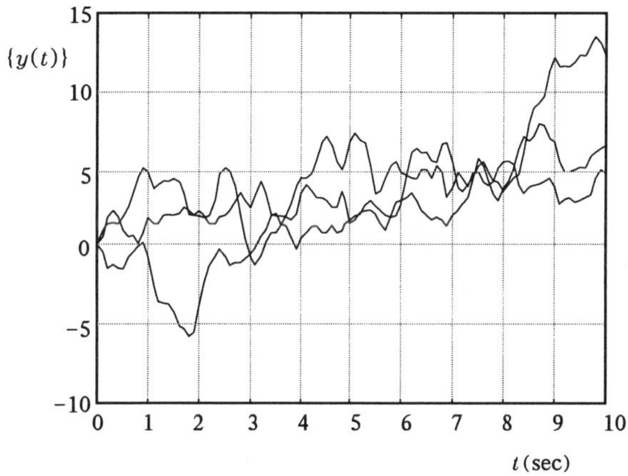

Figure 2.29. Illustration of ensemble power divergence of the integral of a random stationary process.

Output power in the steady state

The power at the system output can be determined by recalling (2.7.12). We now introduce the following change of variables:

$$\sigma - \sigma' = -\lambda; \Rightarrow \sigma' = \sigma + \lambda; \Rightarrow \mathrm{d}\sigma' = \mathrm{d}\lambda$$

so that

$$\phi_{yy}(\tau) = \int_{-\infty}^{\infty}\int_{-\infty}^{\infty} h(\sigma)h(\sigma+\lambda)\phi_{xx}(\tau-\lambda)\mathrm{d}\sigma\mathrm{d}\lambda \tag{2.7.20}$$

We now define the autocorrelation function of the impulse response $h(\sigma)$,

$$\phi_{hh}(\lambda) = \int_{-\infty}^{\infty} h(\sigma)h(\sigma+\lambda)d\sigma \tag{2.7.21}$$

which clearly is symmetrical in λ as indicated by the sketches of $\phi_{hh}(\lambda)$ and $\phi_{hh}(-\lambda)$ represented by the identical shaded areas shown in Figures 2.30 (a) and (b).

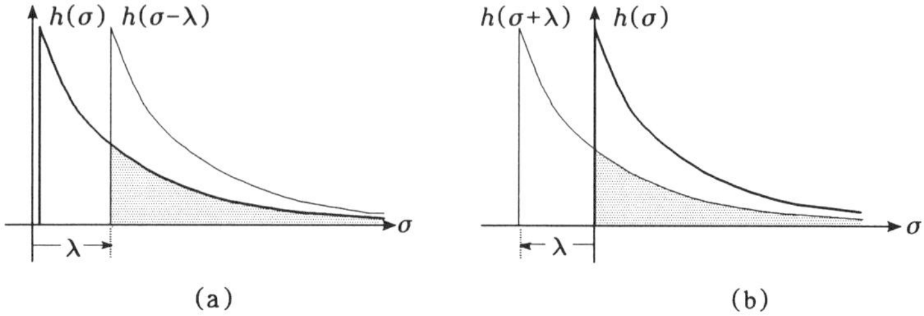

Figure 2.30. Illustration of the two-sided impulse response autocorrelation function: (a)− $\lambda>0$; (b)−$\lambda<0$.

Substituting (2.7.21) into (2.7.20), we have

$$\phi_{yy}(\tau) = \int_{-\infty}^{\infty} \phi_{hh}(\lambda)\phi_{xx}(\tau-\lambda)d\lambda \tag{2.7.22}$$

The output power is determined by setting $\tau = 0$, thus

$$\begin{aligned} P_y = \phi_{yy}(\tau)\Big|_{\tau=0} &= \int_{-\infty}^{\infty} \phi_{hh}(\lambda)\phi_{xx}(-\lambda)d\lambda \\ &= \int_{-\infty}^{\infty} \phi_{hh}(\lambda)\phi_{xx}(\lambda)d\lambda \\ &= 2\int_{0}^{\infty} \phi_{hh}(\lambda)\phi_{xx}(\lambda)d\lambda \end{aligned} \tag{2.7.23}$$

The power P_y is bounded if the integrand in (2.7.23) meets the condition:

$$\lim_{\lambda\to\infty}\{\phi_{hh}(\lambda)\phi_{xx}(\lambda)\} = 0 \tag{2.7.24}$$

This condition is met if $h(t)$ is asymptotically stable [11] and if $\phi_{xx}(\tau)$ is bounded.

EXAMPLE 2.7–2: Smoothing of a stationary random process

In this example $h(t)$, is defined as a smoothing operator as follows:

$$\begin{aligned} h(t) &= \frac{1}{T}; \quad t>0 \\ &= 0\ ; \quad t<0 \end{aligned}$$

The input $x(t)$ is a zero-mean first-order Markov process

$$\phi_{xx}(\lambda) = e^{-a|\lambda|}$$

with unity power $P_x = 1$. $h(t)$ and the corresponding graphical description for the determination of $\phi_{hh}(\lambda)$ are shown in Figures 2.31 a and b respectively.

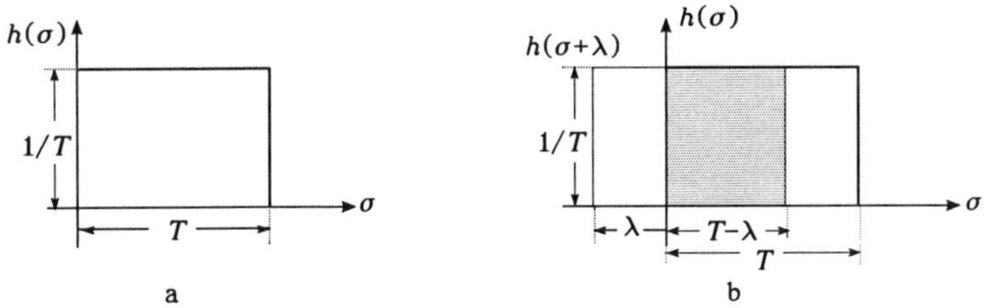

Figure 2.31. Determination of impulse response autocorrelation function for smoothing; (a) $h(t)$; (b) shifted $h(t)$ and representation of $\phi_{hh(\lambda)}$.

$$\phi_{hh}(\lambda) = \frac{1}{T}\frac{1}{T}(T-\lambda) = \frac{1}{T}(1-\frac{\lambda}{T})\;; \qquad \lambda > 0$$

In accordance with (2.7.23),

$$\phi_{yy}(\tau)\Big|_{\tau=0} = P_y = 2\int_0^\infty \frac{1}{T}(1-\frac{\lambda}{T})e^{-a\lambda}\,d\lambda$$

$$= 2\int_0^T \frac{1}{T}e^{-a\lambda}\,d\lambda - \frac{2}{T^2}\int_0^T \lambda e^{-a\lambda}\,d\lambda$$

$$= \frac{2}{aT}\{1 + \frac{1}{aT}(e^{-aT}-1)\}$$

Examination of the last expression shows that for $aT \to 0$, $P_y = 1$ and that for $aT \to \infty$, $P_y = 0 \cong 2/aT$. P_y as a function of aT is described in Figure 2.32. aT is the *smoothing factor* determined by the correlation length and the smoothing interval. Clearly, with small values of aT, $P_y/P_x \cong 1$, implying that there is essentially no smoothing, while for large values of aT, the smoothing is inversely proportional to aT, or the rms value of y, σ_y is inversely proportional to $\sqrt{aT}$. Thus, for example, if it is required to smooth the standard deviation of the process $x(t)$ by a factor of 10, aT must be $\sim$ 100.

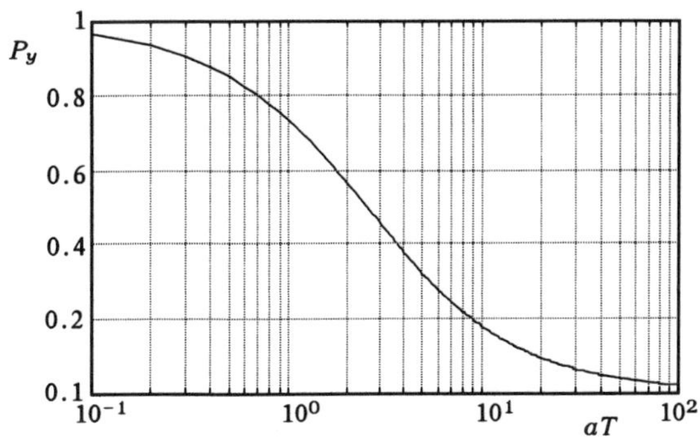

Figure 2.32. Power of first-order Markov process smoothed over the normalized time interval aT.

Response of linear systems to white noise

A special case of interest is that of linear operations on a white noise process. The autocorrelation function of the input $x(t)$ in this case becomes

$$C_{xx}(\tau) = \delta_{xx}(\tau) \tag{2.7.25}$$

Substituting this Dirac function into (2.7.15), we have

$$C_{yy}(t,\tau)\Big|_{\tau=0} = \int_0^t \int_0^t h(\sigma)h(\sigma')\delta_{xx}(\sigma - \sigma')$$

$$= \int_0^t h(\sigma)\left\{\int_0^t h(\sigma')\delta_{xx}(\sigma-\sigma')\mathrm{d}\sigma'\right\}\mathrm{d}\sigma \tag{2.7.26}$$

The inner integral is readily shown to be $h(\sigma)$. The result therefore is

$$C_{yy}(t,\tau)\Big|_{\tau=0} = P_y(t) = \int_0^t h^2(\sigma)\mathrm{d}\sigma \tag{2.7.27}$$

Since $h^2(\sigma)$ is a non-negative function, $P_y(t)$ is a nondecreasing function of t. For example, assume that $h(t) = \sin\omega t$. This yields the result,

$$P_y(t) = \int_0^t \sin^2(\omega\sigma)d\sigma = \int_0^t \frac{1-\cos 2\omega\sigma}{2}d\sigma$$

$$= \frac{1}{2}\left(t - \frac{\sin 2\omega t}{2\omega}\right) = \frac{\omega t}{2\omega}\left(1 - \frac{\sin 2\omega t}{2\omega t}\right) \tag{2.7.28}$$

A plot of this expression is shown in Figure 2.33, which clearly demonstrates the nondecreasing, diverging nature of the process.

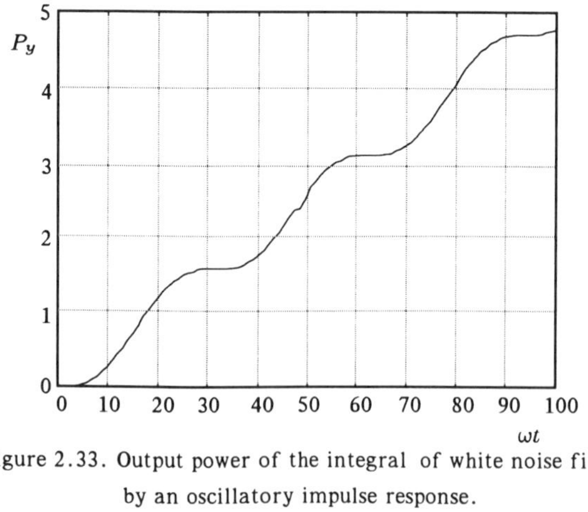

Figure 2.33. Output power of the integral of white noise filtered by an oscillatory impulse response.

Response of linear systems to slowly varying input

In the case of a slowly varying input the autocorrelation function of $x(t)$ can be approximated by a constant C, so that

$$C_{xx}(\tau) = C^2 \tag{2.7.29}$$

Substituting into (2.7.15), we have

$$P_y(t) = C^2\int_0^t\int_0^t h(\sigma)h(\sigma')\mathrm{d}\sigma\mathrm{d}\sigma' = C^2\left\{\int_0^t h(\sigma)\mathrm{d}\sigma\right\}^2 \tag{2.7.30}$$

This is a non-negative function of time, but not necessarily nondecreasing. For the previous example $h(t) = \sin\omega t$, resulting in

$$P_y(t) = \frac{C^2}{\omega^2}(1 - \cos\omega t)^2 \tag{2.7.31}$$

In realistic situations, $x(t)$ is neither strictly white nor strictly "black", but rather intermediate, demonstrating a divergence process which is intermediate between (2.7.27) and (2.7.31).

Single, double, and triple integration

An application of major importance is the integration of a random process $x(t)$ with initial conditions $y(0) = 0$, as illustrated in Example 2.7–1 for $\phi_{xx}(\tau) = e^{-a|\tau|}$. As shown in Example 2.4–3, and for relatively large values of a, $\phi_{xx}(\tau)$ can be approximated by $(\sigma_x^2/a/2)\delta(\tau)$. We denote the power density by $S_{xx} = \sigma_x^2/a/2$. In conjunction with (2.7.27) and assuming $\mu_x = 0$, we have

$$\sigma_y^2(t) = S_{xx}\int_0^t h^2(\sigma)\mathrm{d}\sigma \tag{2.7.32}$$

In dynamical systems, $h(t)$ often represents single, double or triple integration, as in the single integration of acceleration to determine velocity or the triple integration of specific force if it varies with time to obtain position. This situation arises, for example, if an accelerometer is mounted on an airborne stabilized platform in horizontal flight. This platform, in general, undergoes an angular drift so that the sensed component of the D'Alembert force opposing gravity varies with time, say, in a linear fashion. In order to derive explicit expressions, recall that

$$I_1 = \int(\cdot)\mathrm{d}t \rightarrow \frac{1}{s} \rightarrow h(t) = 1 \tag{2.7.33}$$

$$I_2 = \int\int(\cdot)\mathrm{d}t\mathrm{d}t' \rightarrow \frac{1}{s^2} \rightarrow h(t) = t \tag{2.7.34}$$

$$I_3 = \int\int\int(\cdot)dtdt'dt'' \rightarrow \frac{1}{s^3} \rightarrow h(t) = \frac{t^2}{2} \tag{2.7.35}$$

Substituting the corresponding expressions of $h(t)$, we have,

$$\sigma_y^2(t) = S_{xx}\int_0^t 1^2 d\sigma = S_{xx}t \tag{2.7.36}$$

$$\sigma_y^2(t) = S_{xx}\int_0^t \sigma^2 d\sigma = S_{xx}\frac{t^3}{3} \tag{2.7.37}$$

$$\sigma_y^2(t) = S_{xx}\int_0^t \frac{\sigma^4}{4} d\sigma = S_{xx}\frac{t^5}{20} \tag{2.7.38}$$

The corresponding rms values $\sigma_y(t)$ are given by

$$\sigma_y(t) = S_{xx}^{1/2}\, t^{1/2} = \frac{\sigma_x}{a^{1/2}} t^{1/2} \tag{2.7.39}$$

$$\sigma_y(t) = S_{xx}^{1/2}\, t^{3/2} = \frac{\sigma_x}{(3a)^{1/2}} t^{3/2} \tag{2.7.40}$$

$$\sigma_y(t) = S_{xx}^{1/2}\, t^{5/2} = \frac{\sigma_x}{(20a)^{1/2}} t^{5/2} \tag{2.7.41}$$

Simulated examples of samples of the ensemble processes for the last three expressions are shown in Figures 2.34 (a), (b), and (c). The input is Gaussian white noise. I_1, I_2, and I_3 are its single, double, and triple integrations, in accordance with (2.7.33), (2.7.34), and (2.7.35) respectively.

So far, the input-output relationships in linear systems have been addressed by means of the output autocorrelation function as related to the input auto correlation function. In particular, attention has been given to the output power, i. e., $\phi_{yy}(\tau)|_{\tau=0}$.

Another important relationship is the cross correlation $\phi_{xy}(\tau)$ between the input $x(t)$ and the output $y(t)$. Reconsider (2.7.8). In accordance with (2.5.11),

$$\begin{aligned}\phi_{xy}(\tau) &= \overline{x(t)y(t+\tau)} = \int_{-\infty}^{\infty} \overline{x(t)h(\sigma)x(t+\tau-\sigma)}d\sigma \\ &= \int_{-\infty}^{\infty} h(\sigma)\overline{x(t)x(t+\tau-\sigma)}d\sigma \\ &= \int_{-\infty}^{\infty} h(\sigma)\phi_{xx}(\tau-\sigma)d\sigma \end{aligned} \tag{2.7.42}$$

Result: The input-output cross-correlation function is given by the convolution of the input autocorrelation function and the impulse response of the system.

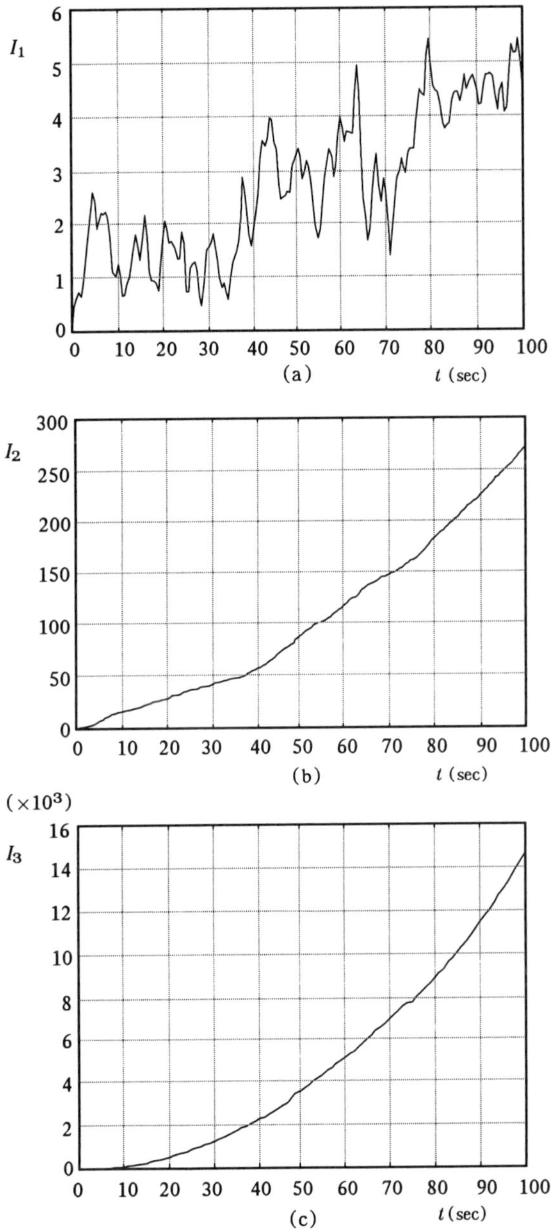

Figure 2.34. Simuation of a sample of:

(a)– Single integration of zero–mean white noise;

(b)– Double integration of zero–mean white noise;

(c) – Triple integration of zero–mean white noise.

2.8 Spectral analysis

The self-spectrum

In the foregoing section it was shown how the output power of a linear system driven by a stationary random input $x(t)$ can be determined both for ergodic and nonergodic outputs. The time-domain approach demonstrated so far, has the advantage that it treats the ergodic and non-ergodic cases by the same analytical tools. However, in many cases, the time domain approach is not the most convenient method because it often involves unwieldy mathematical expressions.

The alternative approach for determining output power is by spectral integration. In principle, the expression for spectral integration is derived by Laplace transforming (2.7.22). Spectral integration is particularly advantageous if $h(t)$ and $\phi_{xx}(\tau)$, when transformed into the frequency domain, are rational functions. If they are, they can be factored and described by a finite number of poles and zeros. Thus, the integration in the frequency domain reduces to the summation of residues [11], [14], which is a simple procedure. If, on the other hand, $h(t)$ or $\phi_{xx}(\tau)$ does not transform into a rational expression e.g., when they involve pure time delays, spectral integration is next to impossible, and it is necessary to perform the integration in the time domain, as shown in Section 2.7. Spectral analysis, for the special case of white noise input and a first-order lag, was initially introduced in (2.6.25). However, the power spectrum of a stationary random process is a fundamental statistical descriptor determined by the Wiener-Kintchin equations [8], which relate the power spectrum to the autocorrelation function in accordance with the two-sided Laplace transform, [11].

$$\Phi_{xx}(s) = \mathcal{L}\{\phi_{xx}(\tau)\} \triangleq \int_{-\infty}^{\infty} \phi_{xx}(\tau)e^{-s\tau}d\tau \tag{2.8.1}$$

$$\phi_{xx}(\tau) = \mathcal{L}^{-1}\{\Phi_{xx}(s)\} \frac{1}{2\pi j}\int_{-j\infty}^{j\infty} \Phi_{xx}(s)e^{s\tau}ds \tag{2.8.2}$$

The justification for applying the Laplace transform to random processes is given, among others, in [12], [13]. An important example is $\phi_{xx}(\tau) = \sigma_x^2 e^{-a|\tau|}$. In accordance with (2.8.1),

$$\begin{aligned}\Phi_{xx}(s) &= \sigma_x^2\left(\int_{-\infty}^{0} e^{a\tau}e^{-s\tau}d\tau + \int_{0}^{\infty} e^{-a\tau}e^{-s\tau}d\tau\right) \\ &= \sigma_x^2\left(\int_{-\infty}^{0} e^{(a-s)\tau}d\tau + \int_{0}^{\infty} e^{(a+s)\tau}d\tau\right) \\ &=\sigma_x^2\left(\frac{1}{a-s}+\frac{1}{a+s}\right) = \frac{2\sigma_x^2 a}{a^2 - s^2} = \frac{2\sigma_x^2 a}{(a-s)(a+s)}\end{aligned} \tag{2.8.3}$$

Equation (2.8.3) is identical to the expression for the spectrum given by (2.6.26), except for the variance σ_x^2 which was disregarded there.

The case of special interest is when $a \to \infty$, namely,

$$\Phi_{xx}(s) \to \frac{2\sigma_x^2}{a} = \frac{\sigma_x^2}{a/2} \tag{2.8.4}$$

which is the expression of a uniform power spectral density (PSD), or white noise. The following cases are sometimes of special interest:

1) The PSD at the origin of s: Setting $s=0$ in (2.8.1):

$$\Phi_{xx}(s)\Big|_{s=0} = \int_{-\infty}^{\infty} \phi_{xx}(\tau)d\tau \tag{2.8.5}$$

Result: The power spectral density at the origin $s = 0$ is given by the area under the respective autocorrelation function.

2) From (2.8.1):

$$\phi_{xx}(\tau)\Big|_{\tau=0} = \frac{1}{2\pi j}\int_{-j\infty}^{j\infty} \Phi_{xx}(s)ds = P_x \tag{2.8.6}$$

Result: The value of the autocorrelation function at $\tau=0$, which is the power of the process, is equal to the integral of the PSD over the entire frequency domain.

Since $\phi_{xx}(\tau)$ is symmetrical in τ, it is evident from (2.8.1) that $\Phi_{xx}(s)$ is symmetrical in s and that it factorizes into

$$\Phi_{xx}(s) = \Phi_x(-s)\Phi_x(s) \tag{2.8.7}$$

as demonstrated by (2.8.3). By the same token, the PSD $\Phi_{yy}(s)$ of the output y of a linear system must be symmetrical in s as well. Recalling (2.7.22), which is formally the convolution of $\phi_{hh}(\lambda)$ and $\phi_{xx}(\lambda)$, its Laplace transform is

$$\Phi_{yy}(s) = \Phi_{hh}(s)\Phi_{xx}(s) \tag{2.8.8}$$

With the aid of (2.7.21), we have

$$\Phi_{yy}(s) = H(-s)H(s)\Phi_{xx}(s) \tag{2.8.9}$$

Therefore, the output power, in analogy with (2.8.6), is determined by

$$P_y = \frac{1}{2\pi j}\int_{-j\infty}^{j\infty} H(-s)H(s)\Phi_{xx}(s)ds \tag{2.8.10}$$

Parseval's theorem

Equating the last expression with (2.7.23), we have

$$\int_{-\infty}^{\infty} \phi_{hh}(\lambda)\phi_{xx}(\lambda)d\lambda = \frac{1}{2\pi j}\int_{-j\infty}^{j\infty} \Phi_{hh}(s)\Phi_{xx}(s)ds$$

$$= \frac{1}{2\pi j}\int_{-j\infty}^{j\infty} H(s)H(-s)\Phi_{xx}(s)ds \tag{2.8.11}$$

which is a form of Parseval's theorem [8] and [12] specialized for statistical input-output relationships. More formal derivations of the foregoing relationships can be found, e.g., in [8].

Spectral decomposition

We now proceed to specialize the preceding PSD expressions for rational functions [13]. As stated earlier, this class readily lends itself to analytical power spectrum integration yielding closed-form solutions. We recall (2.8.1) and rewrite it explicitly as a two-sided Laplace Transform:

$$\Phi_{xx}(s) = \mathcal{L}\{\phi_{xx}(\tau)\} = \int_{-\infty}^{\infty} \phi_{xx}(\tau)e^{-s\tau}d\tau = \int_{-\infty}^{0} \bar{\phi}_x(\tau)e^{-s\tau}d\tau$$

$$+\int_{0}^{\infty} \bar{\phi}_x(\tau)e^{-s\tau}d\tau \triangleq \bar{\Psi}_x(s) + \overset{+}{\Psi}_x(s) \tag{2.8.12}$$

where $\bar{\phi}_x(\tau)$ and $\overset{+}{\phi}_x(\tau)$ represent the negative and positive branches of $\phi_{xx}(\tau)$ corresponding to $\tau<0$ and $\tau>0$.

Substituting $\tau = -u$, the first term of the right-hand side can be rewritten as follows:

$$\int_{\tau=-\infty}^{0} \bar{\phi}_x(\tau)e^{-s\tau}d\tau = \int_{-u=-\infty}^{0} \bar{\phi}_x(-u)e^{-s(-u)}d(-u)$$

$$= \int_{u=\infty}^{0} \bar{\phi}_x(-u)e^{-(-s)u}d(-u) = -\int_{0}^{\infty} \bar{\phi}_x(-u)e^{-(-s)u}d(-u)$$

$$= \int_{0}^{\infty} \bar{\phi}_x(-u)e^{-(-s)u}d(u) \tag{2.8.13}$$

In the last integral, $\bar{\phi}_x(-u)$ can be replaced by $\overset{+}{\phi}_x(u)$ since they are equal. We therefore have

$$\int_{0}^{\infty} \overset{+}{\phi}_x(u)e^{-(-s)u}du \triangleq \overset{+}{\Psi}_x(-s) \tag{2.8.14}$$

In conjunction with (2.8.12), it follows that

$$\bar{\Psi}_x(s) = \overset{+}{\Psi}_x(-s) \tag{2.8.15}$$

indicating that the mirror image $\overset{-}{\phi}_x(\tau) = \overset{+}{\phi}_x(-\tau)$ of $\overset{+}{\phi}_x(\tau)$ in the τ domain corresponds to the mirror image $\Psi_x^-(s)=\Psi_x^+(-s)$ of $\Psi_x^+(s)$ in the s domain. We now express $\Psi_x^-(s)$ and $\Psi_x^+(s)$ in (2.8.12) as rational functions in s,

$$\Psi_x^-(s) = \frac{b(-s)}{d(-s)}\ ;\ \Psi_x^+(s) = \frac{b(s)}{d(s)} \tag{2.8.16}$$

where $b(s)$ and $d(s)$ are polynomials in s. From (2.8.12) and (2.8.16),

$$\Phi_{xx}(s) = \frac{b(-s)}{d(-s)} + \frac{b(s)}{d(s)} = \frac{b(-s)d(s) + b(s)d(-s)}{d(-s)d(s)} \tag{2.8.17}$$

With

$$b(s) = \sum_{i=0}^{n} b_i\, s^i\ ; \qquad d_j(s) = \sum_{j=1}^{m} d_j\, s^j \tag{2.8.18}$$

it is readily shown that the numerator in (2.8.17) is an even function in s and that it assumes the form $2b_0d_0 + 2(b_0d_2 - b_1d_1)s^2 + \ldots$, which can be expressed as the product

$$\left(\sum_{i=0}^{n} c_i s^i\right)\left(\sum_{i=0}^{n} c_i(-s)^i\right)$$

Therefore the power spectrum factorizes into the symmetrical branches

$$\Phi_{xx}(s) = \Phi_x(s)\Phi_x(-s) = \left\{\frac{\displaystyle\sum_{i=0}^{n} c_i s^i}{\displaystyle\sum_{j=0}^{m} d_j s^j}\right\}\left\{\frac{\displaystyle\sum_{i=0}^{n} c_i(-s)^i}{\displaystyle\sum_{j=0}^{m} d_j(-s)^j}\right\} = \frac{\left|\displaystyle\sum_{i=0}^{n} c_i s^i\right|^2}{\left|\displaystyle\sum_{j=0}^{m} d_j s^j\right|^2} \tag{2.8.19}$$

This is a standard form for expressing rational power density spectra. Expressions for m up to 10 can be found, among others, in [13]. This method of integration, however, becomes extremely unwieldy for $m>2$. $\Phi_{xx}(s)$ can also be expressed in factored form in terms of poles and zeros, as follows:

$$\Phi_{xx}(s) = \left(\frac{\displaystyle\prod_{i=0}^{n}(z_i+s)}{\displaystyle\prod_{j=0}^{m}(p_j+s)}\right)\left(\frac{\displaystyle\prod_{i=0}^{n}(z_i-s)}{\displaystyle\prod_{j=0}^{m}(p_j-s)}\right) \tag{2.8.20}$$

The factored form in (2.8.20) is particularly suitable for the integration of the power spectrum by summation of its residues. These are determined by the partial fraction expansions or by the Laurent series expansions. Details of this procedure can be found in [11] and [13]. Explicit expressions for the PSD integral I_m for the rational function form in (2.8.19) up to $m = 10$, can be found in [13] and are given in Appendix A2. These computational algorithms, however, become extremely unwieldy for powers $m > 2$. The determination of residues and, moreover, the complete integration algorithm are readily determined by modern commercial software such as PC–Matlab [15], also presented in Appendix A2 .

Residues related to poles on the imaginary axis are infinite. Therefore, the power P_y resulting from the integration of such power density spectra is infinite. These cases correspond to the nonergodic examples of a periodically diverging process $y(t)$ described earlier in this section.

From Section 2.6, it follows that for the integral of the PSD, namely, for the output power P_y to be bounded, the relative degree in (2.8.18) must be greater than one. Therefore, from the foregoing, the following two conditions must be met in order to ensure the boundedness of P_y:

1. $m \geq n+1$

2. $\mathrm{Re}\{p_j\} \leq 0$

If condition 1 is not met, the spectrum spreads out to infinity so that the system is not strictly physically realizable. If condition 2 is not met, it only means that P_y diverges with time to infinity in accordance with Section (2.7) and that the integration of (2.8.10) to determine the output power P_y must be performed in the time domain.

EXAMPLE 2.8–1: Filtering of white band noise in a FBS sensor

Filtering of wideband noise: Recall Example 1.11–3 dealing with the noise output of a force–balanced sensor described in Chapter 1, p. 51. The output to the noise transfer function $H_v(s)$ was given by

$$z(s) = H_v(s)v(s) = \frac{(a+s)A}{a(1+Ar)+s}v(s)$$

The relative degree of $H_v(s)$ in this example is zero. The spectral input–output relationship is

$$\Phi_{zz}(s) = H_v(s)H_v(-s)\Phi_{vv}(s)$$

If $v(s)$ is represented by the ideal white noise model, then

$$\Phi_{vv}(s) = \frac{2\sigma_v^2}{b} = N_v = \text{const.}$$

where N_v is a flat PSD and b derives from the exponential model of the noise autocorrelation function,

$$\phi_{vv}(\tau) = \sigma_v^2 \, e^{-b|\tau|}$$

Therefore,

$$P_z = \frac{1}{2\pi j}\int_{-j\infty}^{j\infty} \Phi_{zz}(s)ds \to \infty$$

Thus, the output power P_z of this idealized model of the force-balanced sensor with a white noise input would be infinite.

Reconsidering this example with the more realistic noise model having an exponential autocorrelation function, its corresponding PSD is given by

$$\Phi_{vv}(s) = \frac{2\sigma_v^2 b}{b^2 - s^2}$$

With this noise model we have

$$\Phi_{zz}(s) = \frac{2\sigma_v^2 A^2 b(a+s)(a-s)}{[a(1+Ar)+s](b+s)[a(1+Ar)-s](b-s)}$$

$$= 2\sigma_v^2 A^2 b \left| \frac{(a+s)}{s^2+[a(1+Ar)+b]s+ab(1+Ar)} \right|^2$$

With the notations of (2.8.19), we have

$$\Phi_{zz}(s) = 2\sigma_v^2 A^2 b \frac{|c_1 s+c_0|^2}{|d_2 s^2+d_1 s+d_0|^2}$$

where $c_1 = 1$; $c_0 = a$; $d_2 = 1$; $d_1 = a(1+Ar)+b$; and $d_0 = ab(1+Ar)$.

Since in this example, $m = 2$, I_2 is given by

$$I_2 = \frac{c_1^2 d_0 + c_0^2 d_2}{2d_0 d_1 d_2}$$

and P_z is

$$P_z = \sigma_v^2 A^2 \frac{b(1+Ar)+a}{(1+Ar)[a(1+Ar)+b]}$$

Assume the following numerical values: $\sigma_v = 3$ mv; $A = 100$; $a = 1\ \text{sec}^{-1}$; $b = 600\ \text{sec}^{-1}$; and $r = 1$. With these values, P_z can be approximated by

$$P_z = \sigma_v^2 A^2 \frac{b}{a(1+Ar)+b} = 3^2 \times 100^2 \times \frac{600}{101+600} = 77033\ (\text{mv})^2$$

and

$$\sigma_z = \sqrt{P_z} = 277\ \text{mv}$$

This numerical example quantitatively confirms the observations in Section 1.11 which predict high noise amplification.

Cross-spectrum

In accordance with (2.8.1), the Laplace transformation of (2.7.42) yields the following relationship, which defines the input-output cross-spectrum,

$$\Phi_{xy}(s) = H(s)\Phi_{xx}(s) \qquad (2.8.21)$$

This relationship prescribes, among others, the algorithm by which a dynamical model of an unfamiliar system $H(s)$, often referred to as the *system describing function*, can be identified from its input-output relationships, namely,

$$H(s) = \frac{\Phi_{xy}(s)}{\Phi_{xx}(s)} \tag{2.8.22}$$

Equation (2.8.22) is also known as the *Wiener solution* [13]. If $\Phi_{xy}(s)$ derives from the actual response $y(t)$ of a physical causal system, it is clear that the right hand pole in $\Phi_{xx}(s)$ cancels out as readily seen from (2.8.21) and (2.8.22).

The self-spectrum and the cross-spectrum in the last expression are usually estimated on the basis of a finite length of data which can provide the estimates $\hat{\phi}_{xx}(\tau)$ and $\hat{\phi}_{xy}(\tau)$. On applying the fast Fourier transform (FFT) [16] to these estimates, we obtain the corresponding estimates of the spectra. Thus, the estimated describing function is

$$\hat{H}(s) = \frac{\hat{\Phi}_{xy}(s)}{\hat{\Phi}_{xx}(s)} \tag{2.8.23}$$

The last expression indicates that it is advantageous to excite the system $H(s)$ by white-noise. In this case, $\Phi_{xx}(s) = \text{const.} = N$, and $H(s)$ is uniquely determined by $\Phi_{xy}(s)$.

Problems

Problem 2.1—Tactical missile guidance: Effect of accelerometer errors

A short-range tactical missile flying a shallow trajectory is inertially guided by body-mounted accelerometers A_y and A_z having identical error characteristics. y and z are the horizontal and vertical body axes. The error of e.g., A_y consists of white (band-limited) noise having σ_{a_y} mg, a bandwidth $\alpha = 30$ Hz and a random bias $\mu_{a_y} = 0.1\sigma_{a_y}$.

The average velocity of the missile is $v = 200$ m/sec and the maximum range is 4 km. The largest permitted $2\sigma_{\Delta_y}$ miss distance at this range is 5 m.

a) What are the largest permissible σ_{a_y} and μ_{a_y}?

b) What would these values be for half the maximum range?

Hint: In the computations, take advantage of the trajectory shallowness.

Problem 2.2—Autocorrelation function: Surface roughness

The vertical cross-section of a runway surface and a wheel of an undercarriage is shown in the following sketch:

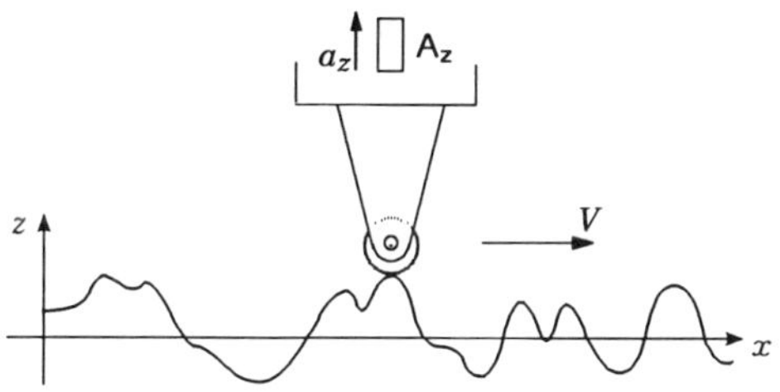

The spatial autocorrelation function of the surface is modeled as a third-order Markov process:

$$\phi_{zz}(\lambda) = \sigma_z^2\left[1 + \left|\frac{\lambda}{l}\right| + \frac{1}{3}\left|\frac{\lambda}{l}\right|^2\right]e^{-\left|\frac{\lambda}{l}\right|}$$

where l is the autocorrelation length of the surface roughness measured by z.

a) Find the expression for the temporal autocorrelation function of z resulting from the travel at speed V.

b) Find the temporal autocorrelation function $\phi_{a_z a_z}(\tau)$.

c) Why is the surface roughness modeled as a third-order Markov process? Is a lower order not proper?

c) What is the expression that relates the rms value σ_{a_z} with travel speed V?

d) Give a graphical description of $\sigma_{a_z}(V)$ and discuss the result.

f) Compute σ_{a_z} for $\sigma_z = 5$ cm; $l = 10$ m; and $V = 54$ Km/hr.

Note: Observe consistency of dimensions.

Problem 2.3— Autocorrelation function: Radar range prediction

A fire control system incorporates a radar that measures the range to target $R(t)$. The range measurement R_m is accompanied by additive noise

$$R_m(t) = R(t) + n(t)$$

where $n(t)$ is described by a third-order Markov process having the autocorrelation function

$$\phi_{nn}(\tau) = \sigma_n^2\left[1 + a|\tau| + \frac{a^2}{3}|\tau|^2\right]e^{-a|\tau|}$$

It is required to predict the range $R(t+\theta)$ at a future instant of time $t+\theta$ on the basis of the present time range measurement $R_m(t)$ and the measured range rate $\dot{R}_m(t)$ in accordance with

$$\hat{R}(t+\theta) = R_m(t) + \dot{R}_m(t)\theta$$

The prediction error is defined by

$$\epsilon(t) \triangleq \hat{R}(t+\theta) - [R(t) + \dot{R}(t)\theta]$$

a) Assuming that $n(t)$ is ergodic, does this imply that $\epsilon(t)$ is also ergodic? Explain.

b) Compute the temporal autocorrelation function $\phi_{\epsilon\epsilon}(\tau)$.

c) It is given that $\sigma_n = 10$ m; and $a = 0.3\ \text{sec}^{-1}$. What is the standard deviation σ_ϵ for $\theta = 0.5$ sec and for $\theta = 1$ sec?

d) Is it imperative to model the noise as a third-order Markov process? Is a lower order admissible in this problem? Explain.

Problem 2.4— Autocorrelation function: Its information content

The autocorrelation function of a stationary random process $x(t)$ is given by

$$C_{xx}(\tau) = 4\cos(0.3|\tau|)e^{-|\tau|}$$

a) Is this $C_{xx}(\tau)$ a legitimate autocorrelation function? If so, explain and justify.

b) Can you determine the following from the preceding expression:
1. The power P_x ?
2. The mean value μ_x ?
3. The standard deviation σ_x ?
2. The power spectral density and effective bandwidth?
3. The probability density function $p_x(\alpha)$?

Give the numerical values wherever appropriate.

Problem 2.5— White noise: rms of smoothed process

$x(t)$ is a band limited zero-mean white noise process with power spectral density $S_x = \sigma_x^2/a$, where a is the band limit. $x(t)$ is smoothed throughout the period $0-T$. The smoothed signal is denoted by $y(t)$.

a) Determine the expression for σ_y in the steady state.

b) How does T relate to a in terms of smoothing effectiveness?

Problem 2.6—Low–pass filtering: Of colored noise

The output of a sensor is corrupted by colored noise defined by the autocorrelation function

$$\phi_{nn}(\tau) = \sigma_n^2 e^{-a|\tau|}$$

To attenuate the noise, the output is low–pass filtered by $H(s) = 1/(1+sT)$. Determine the following:

a) The autocorrelation function and standard deviation of the filter output y.

b) Plot σ_y as a function of aT.

c) Determine an aT that assures $\sigma_y/\sigma_n \leq 0.01$

Problem 2.7— Noisy radar measurements: Position and velocity errors

A fire control radar tracks a target flies flies a straight and level path at an altitude h=3000 m above ground and at a constant speed V=150 m/sec. Noisy range $R(t)+\tilde{R}(t)$ and elevation $\lambda(t)=\tilde{\lambda}(t)$ measurements are fed into the fire control computer.

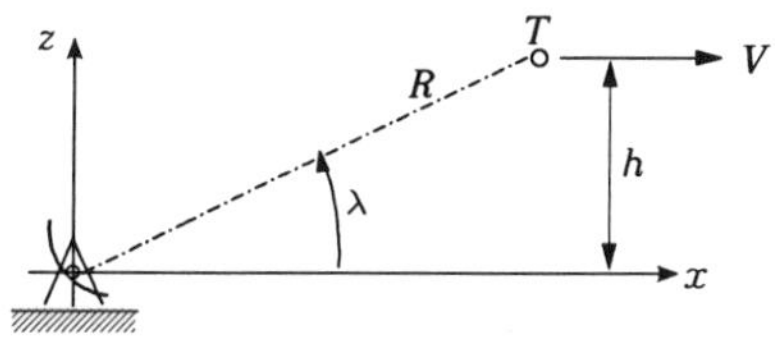

The noisy measurement errors $\tilde{R}(t)$ and $\tilde{\lambda}(t)$ are assumed to be statistically independent and are characterized by the statistics

$$\phi_{\tilde{\lambda}\tilde{\lambda}}(\tau)=\sigma_{\tilde{R}}^2(1+\alpha|\tau|)e^{-\alpha|\tau|}$$

$$\phi_{\tilde{R}\tilde{R}}(\tau)=\sigma_{\tilde{\lambda}}^2(1+\beta|\tau|)e^{-\beta|\tau|}$$

$$\sigma_{\tilde{\lambda}} = 5 \text{ mrad}; \ \sigma_{\tilde{R}} = 10 \text{ m}; \ \alpha = 1 \text{ sec}^{-1}; \ \beta = 0.2 \text{ sec}^{-1}$$

a) Find the rms values of the position errors $\tilde{x}$ and $\tilde{z}$ in the Cartesian coordinates x,z.

b) Compute the rms value $\sigma_{\tilde{v}}$ of the velocity error $\tilde{v}$ for λ=30 deg and λ=90 deg.

Problem 2.8— Determination of rate of change: Altitude rate from altimeter

The measurement error $\tilde{h}$ of altitude is described by an exponential autocorrelation function having an exponent coefficient $a = 0.05$ sec^{-1} and a standard deviation $\sigma_{\tilde{h}} = 10$ m.

It is required to derive $\dot{h}$ by the approximate differentiation $H(s) = s/(s+1)$.

a) Derive the expression for $\sigma_{\dot{h}}$ and its numerical value.

b) Since the differentiation operation is not ideal, can you compare it with the correct value provided by ideal differentiation $H(s) = s$? Explain with theoretical arguments.

c) How would you modify the statistical model of $\tilde{h}$ to enable such a comparison? Give a numerical example.

Problem 2.9—Correlation Coefficient: Of two random stationary processes

Given are two random processes: $x(t)$ and $y(t) = cx(t) + n(t)$; $x(t)$ and $n(t)$ are uncorrelated and c is a constant.

a) Determine the central normalized correlation coefficient ρ for $c > 0$ and $c < 0$.

b) Investigate the effect of $n(t)$ on ρ for two situations: $n \to 0$; and $n \to \infty$.

Given are two stationary random processes:

$$z_1(t) = x(t)+y(t)$$

$$z_2(t) = x(t)-y(t)$$

c) Express the variances V_{z_1} V_{z_2} for

1) $x(t)$ and $y(t)$ are statistically independent.

2) Their correlation coefficient ρ is known.

Problem 2.10—Processing of statistical data: Evaluation of expectation and power

A series of random numbers represents the result of an experiment. It is required to determine $E(x)$, and $E(x^2)$.

a) In view of the relationships

$$E(x) = \int_{-\infty}^{\infty} \alpha p_x(\alpha) d\alpha$$

$$E(x^2) = \int_{-\infty}^{\infty} \alpha^2 p_x(\alpha) d\alpha$$

what would be your approach?

1. Start by evaluating an estimate of $p_x(\alpha)$ and then perform the two integrals?
2. Process the random numbers directly?

b) Express the two approaches explicitly with the discrete equivalents and compare in regard to computational efficiency.

c) If, by theoretical insight, $p_x(\alpha)$ can be determined analytically, how, if at all, would you make use of the series of random numbers?

Problem 2.11— Evaluation of the PDF: Using theoretical insight

Continue with Problem 2.10, considering the following: A missile is guided along a flat horizontal trajectory toward a target having a height h_T. We treat here the vertical dimension only so that the width is of no consequence. The actual trajectory is a random function $z(t)$. By tracking the target in a flight test, it has been observed that the peak-to-peak vertical excursion z_{pp} of the missile from its nominal straight line trajectory is $z_{pp} = 2h_T$. From the nature of the guidance system, it is known that its performance is very close to a second-order dynamical system with a very small damping factor $\varsigma = 0.1$.

a) What does this a priori information convey with regard to the nature of the flight trajectory?

b) What can you deduce with regard to an approximate PDF of $z(t)$, and $p_z(\alpha)$?

c) What is the hit probability of the missile?

Appendix A2 Integration of power density spectra

1. Integration of spectra from rational transfer functions.

$$I_m = \frac{1}{2\pi j}\int_{-j\infty}^{j\infty}\left|\frac{c_{m-1}s^{m-1}+\cdots c_0}{d_m s^m+\cdots d_0}\right|^2 ds \qquad \text{(A2.1)}$$

$$I_1 = \frac{c_0^2}{2d_1d_2} \qquad \text{(A2.2)}$$

$$I_2 = \frac{c_1^2d_0 + c_0^2d_2}{2d_0d_1d_2} \qquad \text{(A2.3)}$$

$$I_3 = \frac{c_2^2d_0d_1 + (c_1^2 - 2c_0c_2)d_0d_3 + c_0^2d_2d_3}{2d_0d_3(-d_0d_3 + d_1d_2)} \qquad \text{(A2.4)}$$

$$I_4 = \frac{c_3^2(d_0^2d_3 + d_0d_1d_2) + (c_2^2 - 2c_1c_3)d_0d_1d_4 + (c_1^2 - 2c_0c_2)d_0d_3d_4 + c_0^2(-d_1d_4^2 + d_2d_3d_4)}{2d_0d_4(-d_0d_3^2 - d_1^2d_4 + d_1d_2d_3)} \qquad \text{(A2.5)}$$

Integrals up to I_{10} are given in [13].

2. Integration by state variable techniques with Matlab Control Toolbox algorithms.

num $\triangleq$ numerator coefficients in descending powers of s.
den $\triangleq$ denominator coefficients in descending powers of s.

$[A,B,C,D]$ = tf2ss(num,den)

where A,B,C,D are state space and measurement matrices in accordance with,

$$\dot{x} = Ax + Bu \qquad \text{(A2.6)}$$

$$y = Cx + Du \qquad \text{(A2.7)}$$

The special form of the general Lyapunov matrix equation $AX + XB + C = 0$ if $B = A'$ is

$$AX + XA' + C = 0 \qquad \text{(A2.8)}$$

If R is the amplitude of white noise input to a state space system,

$$E[w(t)w(\tau')] = R(t)\delta(t-\tau), \quad \dot{x} = Ax + Bw$$

where δ is the Dirac delta function. Then the state covariance

$$Q = E[xx'] \tag{A2.9}$$

is the solution to a Lyapunov equation of the form

$$AQ + QA' + BRB' = 0 \tag{A2.10}$$

The solution Q is found with the algorithm

$$\mathrm{Q = lyap(A,B{*}R{*}B')} \tag{A2.11}$$

The output covariace $P = E[yy']$ is found with

$$\mathrm{P = C{*}Q{*}C' + D{*}R{*}D'} \tag{A2.12}$$

which is the system output power.

References

1. Cannon, Robert H., *Dynamics of Physical Systems*, McGraw Hill, New York, 1967.

2. Liung, Lennart, *System Identification: Theory for the User*, Prentice-Hall, Englewood Cliffs, NJ, 1987.

3. Franklin G.F., Powell, J.D. and Workman, M.L., *Digital Control of Dynamic Systems*, Addison-Wesley, Reading, 1990.

4. Lawrence, P.D., and Mauch, K., *Real-Time Microcomputer System Design: An Introduction*, Book, McGraw-Hill, New York, 1987.

5. Kepler, Johannes, *Magnum Opus, a New Astronomy Based on Causation or, a Physics of the Sky derived from Investigation of the Motion of the Star Mars founded on the Observations of the Noble Tycho Brahe*, 1609.

6. Newton Isaac, *Principia*, 1687.

7. Stewart, Ian, *Does God Play Dice? The Mathematics of Chaos*, Basil Blackwell Inc., Cambridge, MA, 1989.

8. Melsa, J.L., and Sage, A.P. *An Introduction to Probability and Stochastic Processes*, Prentice-Hall, Englewood Cliffs, NJ, 1973.

9. Davenport, W.B., and Root, William L., *An Introduction to Random Signals and Noise*, Book, McGraw-Hill, New York, 1958.

10. Merhav, S.J., "Equivalent Gain of Single Valued Nonlinearities with Multiple Inputs," *IRE Transactions on Automatic Control*, Vol. AC-7, No. 5, October 1962.

11. Truxal, J.G., *Automatic Feedback Control System Synthesis*, McGraw-Hill, New York, 1955.

12. Peebles, P.Z., *Probability, Random Variables and Random Signal Principles*, McGraw-Hill, New York, 1987.

13. Seifert, W.W., and Steeg, C.W., Ed. *Control Systems Engineering*, McGraw-Hill, New York, 1960.

14. McRuer, D.T., Ashkenas, I., and Graham, D., *Aircraft Dynamics and Automatic Control*, Princeton University Press, Princeton, 1973.

15. The Mathworks Inc., "PC-Matlab, Control Systems Toolbox," tf2ss and lyap tools.

16. Brigham, E.O., *The Fast Fourier Transform*, Prentice-Hall, Englewood Cliffs, NJ, 1974.

3

Inertial Force Sensors – Accelerometers

3.0 Introduction

This chapter introduces specific force sensors, or, as traditionally known in the technical literature, *accelerometers*. The subject is treated at both the sensor and the system level, that is, it considers both *how*, and *what* it measures on moving vehicles. It ranges from traditional spring-mass devices to modern vibrating beam quartz or silicon technologies. The effects of sensor imperfections in various applications are also addressed.

A fundamental requirement in aerospace system design and operation is the measurement of vehicular acceleration with respect to inertial space. These measurements are provided by sensors commonly known as *accelerometers*. However, as the title of this Chapter suggests, we shall often use the term *force sensor* or *specific force sensor* (SFS). Specific force, denoted by the vector **a**, implies the total net Newtonian force **F** acting on the vehicle divided by its mass M, i.e.,

$$\mathbf{a} = \frac{\mathbf{F}}{M} \text{ m sec}^{-2}$$

a is commonly resolved along the body-axes of the vehicle x, y, and z, so that $\mathbf{a} \triangleq [a_x, a_y, a_z]^T$. Since SFS's are single input-axis devices, each axis requires the dedication of at least one SFS. The x, y, and z subscripts will therefore be dropped in the analysis or design of the sensor, unless otherwise specified.

The term *SFS* is used because this sensor does not actually measure acceleration. This is best verified by considering a vehicle in free fall equipped with an SFS. While the vehicle clearly accelerates at $g = 9.81$ m sec^{-2}, the SFS reading is zero. The reason, of course, is that the gravitational pull acts equally on each mass particle of the SFS so that no relative displacement takes place between the proof mass m and the SFS casing. Specific force and acceleration are only identical in a gravitation-free environment which, strictly speaking, does not exist in our planetary system, or for that matter in the universe. Thus, in every situation, in order to establish the reading of an SFS, one has to identify *inertial* (*Newtonian*) *forces only* disregarding gravitational forces or their projections caused by the tilting of the vehicle with respect to the local horizon. In aerospace vehicles, such inertial forces include thrust, aerodynamic lift or drag, centripetal forces, present in curved flight paths, or D'Alembert reaction forces acting vertically upwards in reaction to the gravitational pull. These forces manifest themselves as $-g$ and *are* sensed by the SFS. They should not be confused with the gravitational pull g, which is not sensed by the

SFS. A simple example is an SFS A_y mounted on-board a boat with its input axis aligned with the y axis of the boat as shown in Figure 3.1(a). The boat is rolled by the angle ϕ. The vertical flotation force is $F_Z = -W$. The specific force is $a_z = F_Z/M = -W/(W/g) = -g$. Its projection along the input axis is $a_y = -g \sin\phi$. The negative sign indicates that the proof mass m has moved in the $+y$ direction, implying that the projected component of F_Z is in the $-y$ direction. This demonstrates how the SFS responds to the tilt angle ϕ.

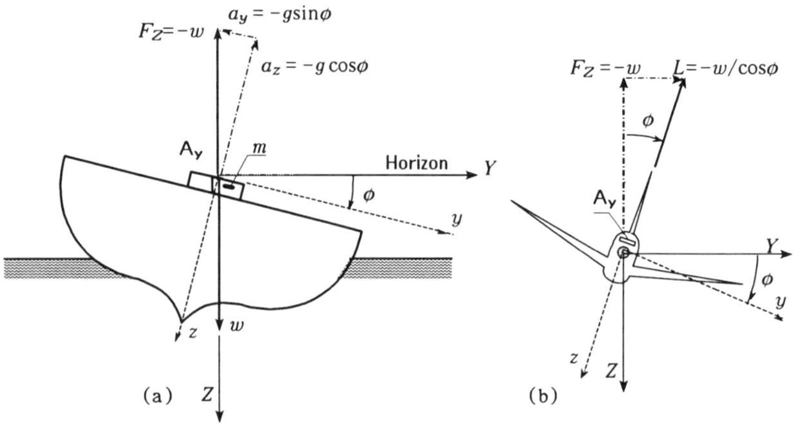

Figure 3.1. (a): g component measured on board a surface vehicle;
(b): g component measured on board a rolled aircraft.

On the other hand, consider A_y mounted in a rolled aircraft during a side-slip-free turn in the horizontal plane, as indicated in Figure 3.1(b). The vertical force that keeps the aircraft in this plane is again $F_Z = -w$. However, the lift force $L = -w/\cos\phi$ is always perpendicular to the y axis and thus to the input axis of A_y. Therefore, the reading of the SFS is zero.

Conclusion: It is impossible to obtain tilt information from accelerometers mounted on board conventional bank-to-turn or pitch-to-loop aircraft.

The complete set of three degrees-of-freedom equations that determine the readings a_x, a_y, a_z of the platform-mounted accelerometer triad under the influence of gravity reaction force and vehicular attitude angles ϕ, θ, ψ, is given in Section 4.1.

The usual operational requirements of SFS's are:

- *Frequency response*: Flat out to ≈ 50 Hz in relatively crude applications and out to ≈ 1000 Hz in inertial navigation applications or vibration measurements.

- Dynamic range: $D_r = 10^2 - 10^3$ in flight control, $10^3 - 10^4$ in tactical guidance applications and $10^4 - 10^8$ in inertial navigation systems.

- Linearity: 1%—0.1% of full scale in conventional flight control, and 10^{-4} to 10^{-5} in guidance and inertial navigation applications.

- *Wide-band noise*: In the order of 0.1% rms of full scale in medium class, and 0.01% rms of full scale in high grade SFS's spread over the previously mentioned bandwidths. The PSD's in high-grade SFS's are substantially lower.

3.1 Specific force readings on moving platforms

Consider the vehicle as a rigid body with a mass M which moves with the inertial velocity **V** and angular velocity **Ω**. From Newton's second law,

$$\mathbf{F} = \frac{d}{dt}(M\mathbf{V}) = \frac{dM}{dt}\mathbf{V} + M\frac{d\mathbf{V}}{dt} \tag{3.1.1}$$

We assume that M = const. so that the first term is zero. Thus, **F** is determined by the second term only, namely,

$$\mathbf{F} = M\frac{d\mathbf{V}}{dt} = \mathbf{1}_V M\frac{d\mathbf{V}}{dt} + M(\mathbf{\Omega}\times\mathbf{V}) \tag{3.1.2}$$

Dividing through by M, the acceleration vector is

$$\mathbf{a} = \mathbf{1}_V\frac{d\mathbf{V}}{dt} + \mathbf{\Omega}\times\mathbf{V} \tag{3.1.3}$$

The first term represents the tangential acceleration. The second term, often termed the *Coriolis acceleration* is actually the centripetal acceleration resulting from the angular rate **Ω** of the velocity vector **V** in inertial space. $\mathbf{1}_V \triangleq (\mathbf{i}, \mathbf{j}, \mathbf{k})^T$ is the unit vector with its components **i**, **j**, **k** along the x, y, z body-axes. Thus,

$$\mathbf{V} \triangleq \mathbf{i}U + \mathbf{j}V + \mathbf{k}W \tag{3.1.4}$$

$$\mathbf{\Omega} \triangleq \mathbf{i}P + \mathbf{j}Q + \mathbf{k}R \tag{3.1.5}$$

With these notations, we have

$$\mathbf{1}_V\frac{d\mathbf{V}}{dt} = \mathbf{i}\dot{U} + \mathbf{j}\dot{V} + \mathbf{k}\dot{W} \tag{3.1.6}$$

and

$$\mathbf{\Omega}\times\mathbf{V} = \begin{vmatrix} \mathbf{i} & \mathbf{j} & \mathbf{k} \\ P & Q & R \\ U & V & W \end{vmatrix} = \mathbf{i}(QW-RV) + \mathbf{j}(RU-PW) + \mathbf{k}(PV-QU) \tag{3.1.7}$$

We denote

$$\mathbf{F} = \mathbf{i}F_x + \mathbf{j}F_y + \mathbf{k}F_z \tag{3.1.8}$$

Combining (3.1.2) with (3.1.5) and (3.1.4) and dividing through by M, the Newtonian sensed readings $\mathbf{a}^n$ of the SFS's are

$$\begin{bmatrix} a_x^n \\ a_y^n \\ a_z^n \end{bmatrix} = \begin{bmatrix} \dot{U} + QW - RV \\ \dot{V} + RU - PW \\ \dot{W} + PV - QU \end{bmatrix} \tag{3.1.9}$$

They consist of the linear acceleration components $\dot{U}$, $\dot{V}$, $\dot{W}$ and the components of centripetal acceleration due to the curved path of the vehicle. We initially assume that the earth is flat and nonrotating. This assumption is valid for short-term flight trajectories. Later, it will be removed.

A case of special interest is a vehicle with the resultant net local vertical force $F_Z + W = 0$, namely, $F_z = -W$, and we determine the projections of the specific force $a_z = F_z/M = -W/W/g = -g$ on the x, y, z body axes by means of the Euler transformation $[E]$ (Appendix A4),

$$\begin{bmatrix} a_x^g \\ a_y^g \\ a_z^g \end{bmatrix} = [E] \begin{bmatrix} 0 \\ 0 \\ -g \end{bmatrix}$$

$$= \begin{bmatrix} c\psi c\Theta & s\psi c\Theta & -s\Theta \\ (c\Psi s\Theta s\Phi - s\Psi c\Phi) & (s\Psi s\Theta s\Phi + c\Psi c\Phi) & c\Theta s\Phi \\ (c\Psi s\Theta c\Phi + s\Psi s\Phi) & (s\Psi s\Theta c\Phi - c\Psi s\Phi) & c\Theta c\Phi \end{bmatrix} \begin{bmatrix} 0 \\ 0 \\ -g \end{bmatrix}$$

$$= \begin{bmatrix} g s\Theta \\ -g c\Theta s\Phi \\ -g c\Theta c\Phi \end{bmatrix} \tag{3.1.10}$$

where $s(\cdot) \triangleq \sin(\cdot)$ and $c(\cdot) \triangleq \cos(\cdot)$.

Combining (3.1.9) and (3.1.10), we have for the total specific force vector,

$$\begin{bmatrix} a_x \\ a_y \\ a_z \end{bmatrix} = \begin{bmatrix} a_x^n \\ a_y^n \\ a_z^n \end{bmatrix} + \begin{bmatrix} a_x^g \\ a_y^g \\ a_z^g \end{bmatrix} = \begin{bmatrix} \dot{U} + QW - RV \\ \dot{V} + RU - PW \\ \dot{W} + PV - QU \end{bmatrix} + \begin{bmatrix} g s\Theta \\ -g c\Theta s\Phi \\ -g c\Theta c\Phi \end{bmatrix} \tag{3.1.11}$$

Again, recall that the second vector term on the right-hand side in (3.1.10) is not due to gravity but rather to the D'Alembert force $F_z = -W$.

It is important to realize that the readings of on-board SFS's are, in general, not identical to the vehicular acceleration as seen by an outside observer. This is illustrated by the following hovercraft example.

Elementary strapdown mechanization for vehicle guidance

Equations (3.1.10), in conjunction with the Euler transformation $[E]$ and the Euler differential equations given in (7.0.1)–(7.0.4), constitute the basis for the mechanization of *attitude-heading reference systems* (AHRS) and strapdown inertial navigation systems, as schematically described in Figure 3.2.

The diagram concisely summarizes all the foregoing inertial, dynamical and kinematic equations which provide vehicle attitude angles ϕ, θ, and ψ, inertial velocity V_X, V_Y, V_Z and vehicle position R_X, R_Y, R_Z in Earth coordinates. The diagram, however, by no means represents a mechanization of a complete inertial strapdown navigation system. Additional information and functions, such as Earth rate and sphericity, alignment by leveling, gyro–compassing and Schuler tuning, are required to accomplish this.

The system functions in Figure 3.2 consist of three categories:

1. measurements by the rate gyros G_x, G_y, G_z and by the specific force sensors A_x, A_y, A_z.
2. modeling based on prior knowledge. In the present case, the knowledge of $g \cong 9.81\,\mathrm{m\,sec^{-1}}$.
3. computations by coordinate transformations and integrations.

These three basic categories can be identified in all aerospace sensor systems. The actual implementation of the computational process is iterative to comply with real time requirements and is based on quaternions rather than directly on Euler angles, to avoid singularities. These issues are addressed in further detail in Chapter 4.

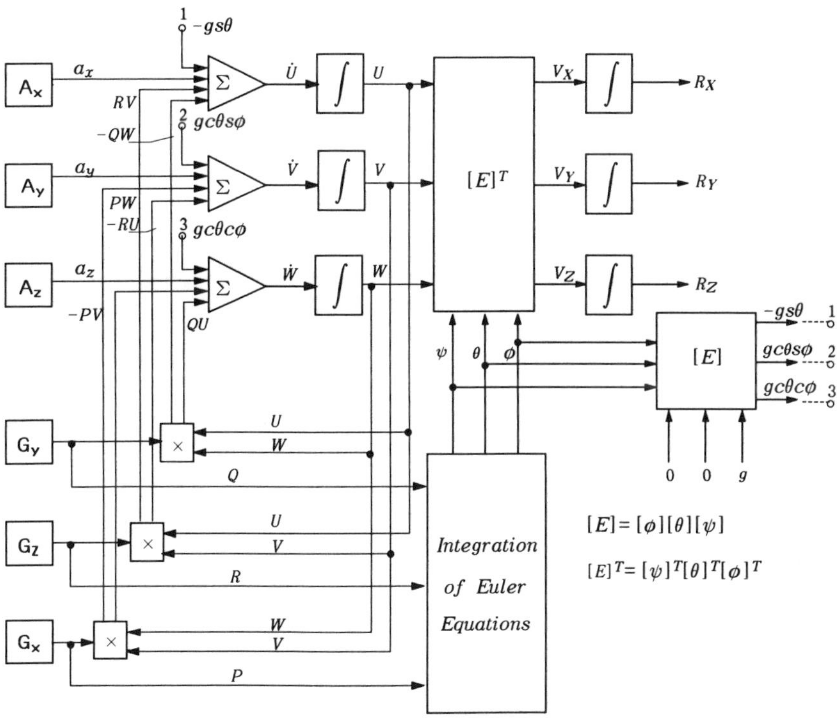

Figure 3.2. Basic strapdown mechanization for computation of ground speed and vehicle position based on measured a and Ω.

EXAMPLE 3.1–1: Specific force measurements on a hovercraft

A hovercraft H weighing w kg is schematically described in Figure 3.3.

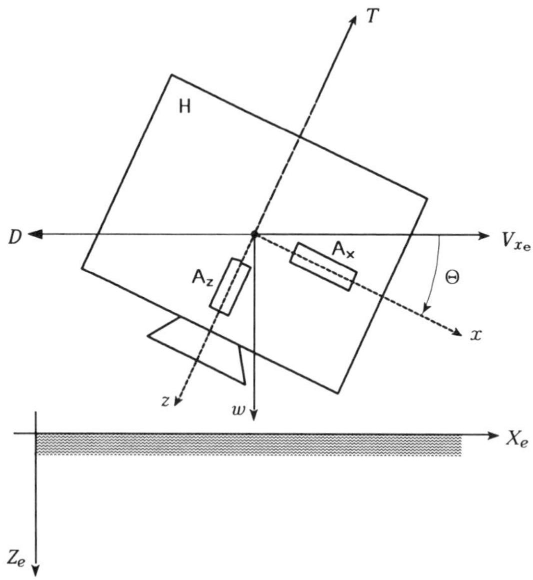

Figure 3.3. Specific force measurements on board a tilted and accelerating hovercraft.

A control system, not shown here, provides control of the pitch angle Θ and the magnitude of the thrust vector T, which is always perpendicular to the x axis and adjusted to counteract w. Two SFS's A_x and A_z, are mounted on H and are aligned with the x and z axes, respectively. Motion is with respect to the earth axes X_E and Z_E. The angular rate Ω of H, its roll angle Φ, and its yaw angle Ψ are always assumed to be zero.

1. What are the readings of A_x and A_z while H hovers over a fixed point?

Since the angular rate and all the attitude angles, including Θ during hover, are zero, we have

$$a_x = \dot{U} + QW - RV + gs\Theta = \dot{U}$$

$$a_z = \dot{W} + PV - QU - gc\Theta c\Phi = \dot{W} - g$$

Since T is perpendicular to x, $a_x = 0$. Therefore, $\dot{U} = 0$. Since there is no vertical motion during hover, $\dot{W} = 0$ and $a_z = -g$.

2. What are the readings of A_x and A_z when Θ is tilted as shown in the sketch? Initially, when the horizontal velocity of H is small, the drag force D is still negligible. What are the expressions for $\dot{U}$ and $\dot{W}$?

Since T is perpendicular to x, $a_x = 0$. Consequently, $\dot{U} = -gs\Theta$. To hold the altitude, the thrust must be increased to $T/c\Theta = -w/c\Theta$. Therefore, $a_z = -g/c\Theta$, and $-g/c\Theta = \dot{W} - gc\Theta$; $\dot{W} = gc\Theta - g/c\Theta = -g\tan\Theta s\Theta$. Note that $g\tan\Theta$ is the specific force along the X_E axis and $-g\tan\Theta s\Theta$ is its component along the z axis.

3. Eventually, D increases and V_{XE} reaches a constant value, $\dot{V}_{XE} = 0$, which implies that the net horizontal force $-w\tan(-\Theta) + D = 0$, i.e., $D = -w\tan\Theta$. What are the readings a_x and a_z of A_x and A_z?

The remaining vertical Newtonian force component is $T = -w$. Its projections along the x axis are $-ws(-\Theta)$ so that $a_x = gs\Theta$, and its projection along the z axis is $-gc\Theta$, as can be verified from Figure 3.4.

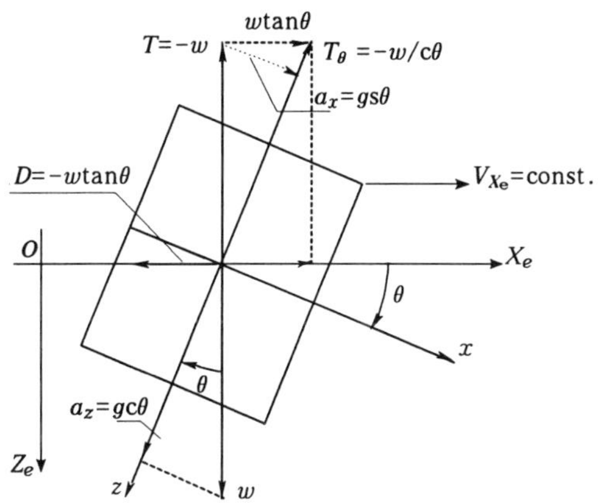

Figure 3.4. Force components on a tilted and accelerating hovercraft.

The results of this example are initially counterintuitive in the sense that when H accelerates the reading of the longitudinal SFS is zero, whereas when traveling at a constant velocity, its reading is not zero. The example reemphasizes that one must consider only the Newtonian and D'Alembert force components and never the gravitational forces. These, of course, play their role in the formulation of the equations of vehicular motion alongside other forces such as thrust, lift, or drag.

3.2 Leveling the supporting platform

The foregoing examples and expressions, in particular (3.1.11), pertain to geospheric motion of an aircraft in nominally horizontal flight, ships, or land vehicles. The second term on the right-hand side indicates the fundamental dilemma that it is impossible to separate Newtonian forces from the supporting D'Alembert force component, which may be aerodynamic, hydrostatic, or simply mechanical reaction forces according to Newton's first law. The crucial factor is the precision with which the platform that supports the SFS's can be aligned with the local level, i.e., $\Theta \to 0$ and $\Phi \to 0$. The simplest method would apparently be to suspend the platform pendulously. However, it is obvious that under motion, the pendulous platform will be perturbed by horizontal acceleration components which excite Θ and Φ and introduce parasitic errors in the measurement of a_x, a_y and a_z. Consequently, a platform which is insensitive to horizontal accelerations is sought.

Mathematical Schuler pendulum

We will now develop the concept of the mathematical Schuler pendulum and study its properties. The following analysis is based on [1]. In Figure 3.5 a mathematical pendulum P of length ρ and point mass m is suspended above the surface of the earth E so that the distance of its pivot p from the center of the earth O is $R = R_e + h$, where $h \ll R_e$. R points to O, i. e., to the local vertical LV. θ_{pv} is the angular deviation from the LV. The pivot p, of P, is accelerated by a_p. X_i, Z_i are the axes of an arbitrary inertial reference frame (i). θ_v is the angle of the LV with respect to (i), and θ_{p_i} is the direction of ρ with respect to X_i.

$$a_p = \ddot{\theta}_v R \tag{3.2.1}$$

$$\theta_v + \theta_{pv} = \theta_{pi} \tag{3.2.2}$$

$$\ddot{\theta}_v + \ddot{\theta}_{pv} = \ddot{\theta}_{pi} \tag{3.2.3}$$

The equation of moments for small θ_{p_v} is

$$I_p \ddot{\theta}_{p_i} + m\rho g \theta_{p_v} = m\rho R \ddot{\theta}_v \quad (LV \,||\, g) \tag{3.2.4}$$

Substituting (3.2.2), we have

$$I_p(\ddot{\theta}_v + \ddot{\theta}_{p_v}) + m\rho g \theta_{p_v} = m\rho R \ddot{\theta}_v \tag{3.2.5}$$

$$I_p \ddot{\theta}_{p_v} + m\rho g \theta_{p_v} = (\rho Rm - I_p)\ddot{\theta}_v \tag{3.2.6}$$

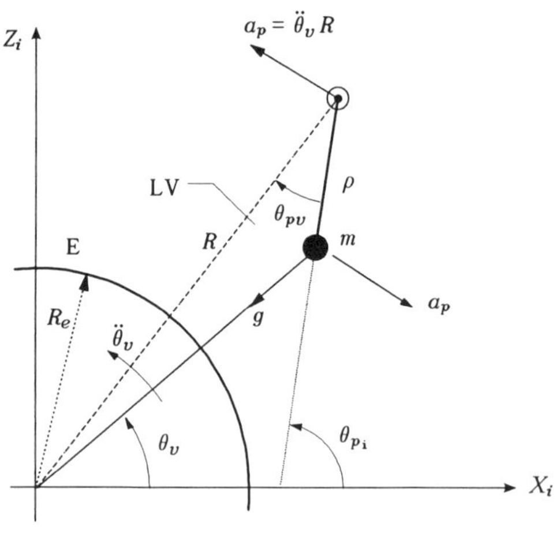

Figure 3.5. Mathematical pendulum defined in the radial gravitational field of the planet.

The forcing moment on the right hand side must be zero if P is to be acceleration-insensitive, i.e.,

$$\rho Rm - I_p = 0 \tag{3.2.7}$$

since $I_p = m\rho^2$, $\rho Rm = m\rho^2$, resulting in

$$\rho = R \tag{3.2.8}$$

Thus, for the pendulum to be acceleration insensitive, its length must equal the radius of Earth. On substituting this result into (3.2.4), the pendulum equation takes the form

$$\ddot{\theta}_{Pv} + \frac{g}{R}\theta_{Pv} = 0 \tag{3.2.9}$$

where the angular frequency is given by

$$\omega_s = \sqrt{g/R} \tag{3.2.10}$$

ω_s is known as the *Schuler frequency*. Using the approximation $h \ll R_e$, we have $R \simeq R_e$. Given the constants on earth for g and R, $g = g_e = 9.81\ \mathrm{m\,sec^{-2}}$, $R = R_e = 6.381 \times 10^6$ m results in ω_s = 4.44 rad/hr.

Conclusion: An Earth-based platform that has been made insensitive to horizontal accelerations must have a natural frequency of ω_s = 4.44 rad/hr or a period $T_s = 2\pi/\omega_s$ = 84.4 min. The length ρ of this pendulum, however, must be equal to $R_e = 6.381 \times 10^6$ m. Clearly, its realization is totally unfeasible.

Physical Schuler pendulum

Another attempt to implement the Schuler pendulum might be a physical pendulum with acceptable small dimensions, as described in Figure 3.6.

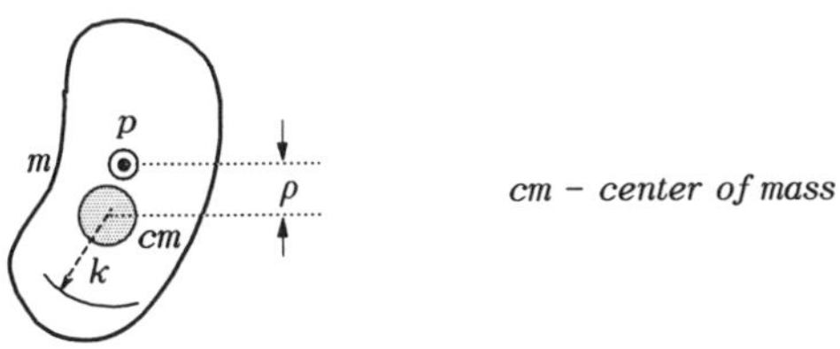

Figure 3.6. Physical pendulum with minute effective length ρ.

The moment of inertia is $I_p = mk^2 + m\rho^2$, where k is the radius of gyration of the pendulum. Since $\rho \ll k$, the term $m\rho^2$ can be neglected. Again, setting $\rho R_e m = mk^2$, we have

$$\rho = \frac{k^2}{R_e} \tag{3.2.11}$$

Assume a reasonable dimension for k, e. g., $k = 0.1\text{m}$. With $R_e = 6.38 \times 10^6$ m we have

$$\rho = \frac{10^{-2}}{6.38} \times 10^{-4} \cong 1.5\,\text{nanometer} \tag{3.2.12}$$

which is about the wavelength of X-rays, again a totally infeasible solution. It appears that the realization of a Schuler pendulum by elementary mechanical methods is next to impossible. It has been a long-standing question since the turn of the century, when this principle was discovered by the German scientist Maximilian Schuler. Later in this book, we shall see, that with the advent of modern high-precision gyroscopes, accelerometers, and servomechanism techniques, it is indeed practical to realize such a pendulum, which is also known in the literature as the *Schuler tuned platform.* When tilted from the horizontal by the angle θ_0, it will indeed swing indefinitely in accordance with

$$\theta(t) = \theta_0 \cos \omega_s t \tag{3.2.13}$$

The Schuler tuned platform has so far been described for Θ only. Clearly, a true platform must be realized by a three-axis mechanization, namely, for Φ and Ψ as well, to provide complete isolation from the angular motions of the flight vehicle. At this point, it is of interest to examine the effect of Schuler tuning on the positional error propagation $\Delta x(t)$ resulting from leveling error θ, which, in accordance with example 3.1−1, is $\Delta a_x = g\sin\theta$. From (3.2.13), and assuming $\theta_0 \ll 1$,

$$\Delta a_x = g\sin(\theta_0 \cos\omega_s t) \cong g\theta\cos\omega_s t \tag{3.2.14}$$

The velocity error is accordingly,

$$\Delta v_x(t) = g\theta_0 \int_0^t \cos\omega_s \sigma\, d\sigma = \frac{g\theta_0}{\omega_s}\sin\omega_s t \tag{3.2.15}$$

The position error is

$$\Delta x(t) = \int_0^t \Delta v_x(\sigma)\, d\sigma = \frac{g\theta_0}{\omega_s^2}(1-\cos\omega_s t) = \theta_0 R_e (1-\cos\omega_s t) \tag{3.2.16}$$

where R_e is the radius of Earth. This result demonstrates the important effect of Schuler tuning in that it bounds the position error Δx to within zero and $2R_e\theta_0$. Thus, for example, if θ_0 = 100 μrad,

$$\Delta x \le 2 \times 6380 \times 10^3 \times 10^{-4} = 1276\ \text{m}$$

This result should be compared to $\Delta x = g\theta_0 t^2/2$ without Schuler tuning, which diverges with the square of time.

3.3 Schuler frequency on other planets

We now examine the essential geophysical parameters that determine the Schuler frequency on other planets. From (3.2.9), we focus on the ratio g/R. Let m_p be the mass of the planet, m, the mass of the pendulum, and F_p the pull of the planet and R_p its radius. Then according to Newton's law of gravitation,

$$F_p = K\frac{m_p m}{R^2} \tag{3.3.1}$$

where K is the universal constant of gravitation. Let d_p be the mass density of the planet. Then, its mass is given by

$$m_p = \frac{4\pi}{3}R_p^3 d_p \tag{3.3.2}$$

Substituting (3.3.1) into (3.3.2), we have

$$F_p = K\frac{4\pi}{3}\frac{R_p^3}{R^2}d_p m \tag{3.3.3}$$

The gravitational specific force near the surface of the planet, i. e., $R = R_p$ is given by

$$g_p = \frac{F_p}{m} = K\frac{4\pi}{3}R_p d_p \tag{3.3.4}$$

Finally, the planetary Schuler frequency ω_{s_p} is determined by,

$$\omega_{sp}^2 = \frac{g_p}{R_p} = K\frac{4\pi}{3}d_p \tag{3.3.5}$$

Conclusion: The Schuler frequency ω_{s_p} on any planet is determined solely by the mass density of the corresponding planet.

If all the planets were made of the same material, their respective Schuler frequencies would be the same.

3.4 Force balance accelerometers

Pickoff considerations

In Section 1.11 the relationships between the principal design parameters, resolution y_r, dynamic range D_r, and bandwidth ω_n, were briefly addressed. It was shown there that ω_n and y_r are strongly conflicting factors in elementary open-loop deflection-type accelerometers. For a given required maximum acceleration a_{max}, an increase, by a factor of say, 10 in bandwidth ω_n, would require, for a given D_r, an increase of 100 in resolution y_r. Alternatively, for a given pickoff technology, y_r is given, so that the dynamic range would have to be decreased by a factor of 100 (Equations (1.11.19–20)).

Clearly, contact pick-offs such as potentiometers must be ruled out for high resolution accelerometers. Non-contact pick-offs such as differential transformers or synchros, are superior, but they may not be sufficiently linear for the relatively large excursions of the proof mass typical in open loop sensors. Furthermore, the spring may not be sufficiently linear to satisfy the requirements of a high precision accelerometer. In Section 1.10, the basic properties and advantages of the *forced balance sensor* (FBS) were introduced. We shall now apply its principle of operation to the realization of a forced balance accelerometer.

Spring-mass accelerometer model

The elementary spring-mass mechanization of Section 1.11 is depicted in Figure 3.7. The deflection of the proof mass m, denoted here by x, in response to the force to which it is subjected is,

$$\frac{x(s)}{m\ddot{X}(s)} = \frac{\frac{1}{m}}{s^2 + \frac{D}{m}s + \frac{k}{m}} \tag{3.4.1}$$

$\ddot{X}(s)$ is the Newtonian acceleration of the SFS casing or the vehicle due to specific forces. A block diagram describing the dynamical model representing (3.4.1) is shown in Figure 3.7. A pick-off is placed at the output x to provide an electrical signal x_m to represent $\ddot{X}$. At this point we assume that the pick-off is totally friction free and that it has infinite resolution. Also, in order to assure its linearity, we now assume that the deflection x is vanishingly small. Noise is presently disregarded. Under these assumptions, the pick-off, denoted by G_x, is in effect a detector in accordance with the definitions given in Chapter 1. Since friction-free operation and small size is required, a capacitive pick-off is a preferred choice.

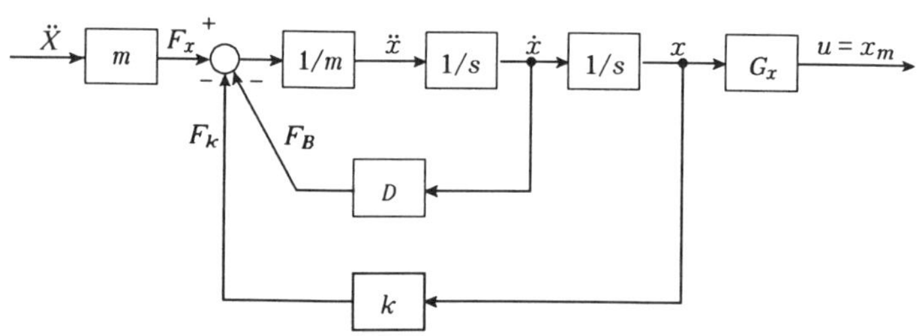

Figure 3.7. Block diagram of the basic damped spring-mass accelerometer.

The capacitive detector

A technical arrangement of a variable capacitance bridge is shown in Figure 3.8.

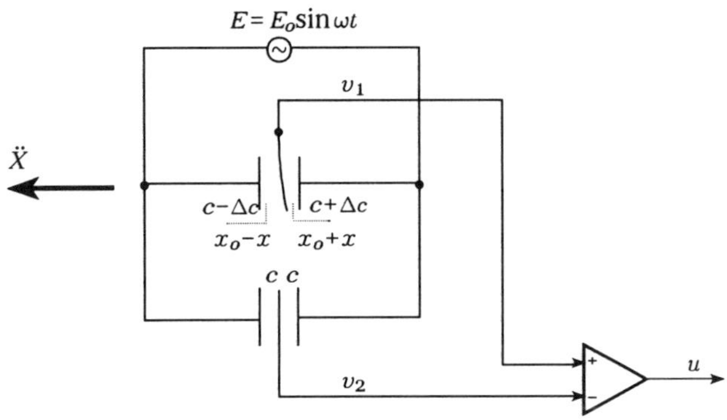

Figure 3.8. Capacitance bridge for measurement of proof mass displacement.

The capacitance of a single plate pair in vacuum with surface A in^2 and a gap of x in, is given by

$$C = \frac{0.225A}{x} \text{ pF} \tag{3.4.2}$$

1 pF = 10^{-12} Farad. The variable capacitor C is arranged in a capacitance Wheatstone bridge as shown in Figure 3.8. It is driven by a high-frequency oscillator generating the voltage $E = E_0 \sin\omega t$. The variation of each capacitance is denoted by $C+\Delta C$ and $C-\Delta C$. Clearly, $V_2 = E/2 = (E_0/2)\sin\omega t$. The current in the upper branch is

$$i_1 = \frac{E}{\frac{1}{j\omega C_1} + \frac{1}{j\omega C_2}} = \frac{E}{\frac{1}{j\omega(C+\Delta C)} + \frac{1}{j\omega(C-\Delta C)}}$$

$$= -\frac{\omega(C^2 - \Delta C^2)}{2jC} E \tag{3.4.3}$$

where V_1 is given by

$$V_1 = \frac{1}{j\omega C_2} i_1 = -\frac{\omega(C^2 - \Delta C^2)}{2jCj\omega(C-\Delta C)} E = \frac{C^2 - \Delta C^2}{2C(C-\Delta C)} E \tag{3.4.4}$$

The output u of the bridge is

$$u = V_2 - V_1 = \left\{ \frac{1}{2} - \frac{1}{2} \frac{C^2(1 - \frac{\Delta C^2}{C^2})}{C^2(1 - \frac{\Delta C}{C})} \right\} E \tag{3.4.5}$$

Reduction of (3.4.5) leads to the simple expression

$$u = -\frac{E}{2}\frac{\Delta C}{C} \tag{3.4.6}$$

In accordance with (3.4.2), the nominal value of C, C_0, is given by

$$C_0 = \frac{0.225A}{x_0} pF \tag{3.4.7}$$

Let

$$C_x \triangleq \left.\frac{\partial C}{\partial x}\right|_{x=x_0} = -\frac{0.225}{x_0^2}A = -\frac{C_0}{x_0} \tag{3.4.8}$$

$$\Delta C = C_x x = -\frac{C_0}{x_0}x \tag{3.4.9}$$

and

$$\frac{\Delta C}{C_0} = -\frac{x}{x_0} \tag{3.4.10}$$

From (3.4.4) and (3.4.10), we have

$$u = \frac{Ex}{2x_0} \triangleq G_x x \tag{3.4.11}$$

where G_x is defined by,

$$G_x \triangleq \frac{\mathrm{d}}{\mathrm{d}x}u = \frac{E}{2x_0} \tag{3.4.12}$$

Comments:

1. The larger E_0, and the smaller x_0, the higher the sensitivity G_x.
2. The frequency ω has dropped out in the foregoing analysis. However, this does not imply that the capacitive detector can be realized with DC excitation. u is typically very small, and subsequent DC amplification would entail large noise-to-signal ratios.

Recalling that E is an alternating signal, we actually have

$$u = \frac{E_0}{2x_0}\sin\omega t\, x = G_x \sin\omega t\, x \tag{3.4.13}$$

Clearly, whenever x changes sign, the phase of u changes by 180 deg. Also, to make the detector a practical device, the periodically alternating output u must be rectified in such a way that the correct sign of u is retained. This operation is accomplished by means of synchronous detection. u, as given in (3.4.13), is multiplied by sgn(ωt) derived from the high-frequency oscillator. The sign-sensitive rectification shown in Figure 3.9 is self-explanatory. The full-wave rectified signal $\bar{u}$ is subsequently low-pass filtered to attenuate the residual ripple to an acceptable level, denoted by $\bar{u}_f$:

$$\bar{u} = \frac{E_0}{2x_0} \sin\omega t \, \mathrm{sgn}\,\omega t\,; \qquad x > 0 \tag{3.4.14}$$

$$\bar{u} = -\frac{E_0}{2x_0} \sin\omega t \, \mathrm{sgn}\,\omega t\,; \qquad x < 0 \tag{3.4.15}$$

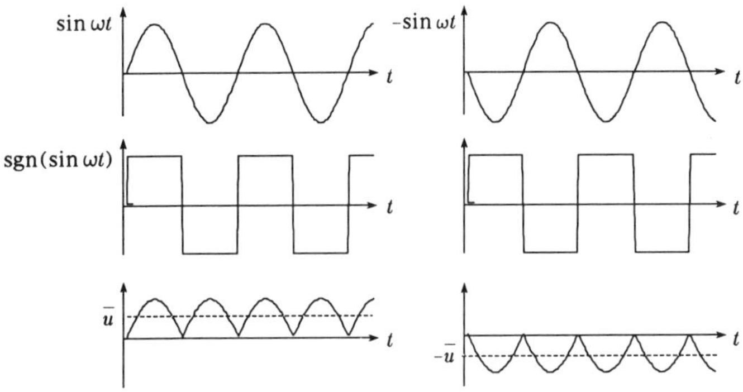

Figure 3.9. Signal processing of capacitive bridge output.

I_r, flowing through the variable output resistor R_0 provides the output voltage $V_{out} = I_r R_0$.

Implementation of loop closure

The incorporation of $F_r = G_x Arx$ modifies the force equation to

$$m\ddot{X} = m\ddot{x} + B\dot{x} + kx + G_x Arx \tag{3.4.16}$$

The last two terms combine to form the equivalent gain K_{eq},

$$K_{eq} = k + G_x Ar \tag{3.4.17}$$

The realization of the accelerometer as a *force balanced sensor* is shown in Figure 3.10.

The closed loop-relationships are

$$F_r = G_x Arx \tag{3.4.18}$$

$$x = \frac{1}{s^2 + \frac{B}{m}s + \frac{K_{eq}}{m}} \ddot{X} \tag{3.4.19}$$

$$V_0 = G_x A R_0 x = \frac{G_x A R_0}{s^2 + \frac{B}{m}s + \frac{k + G_0 A r}{m}} \ddot{X} \tag{3.4.20}$$

The complete closed loop of the forced balance sensor is shown in Figure 3.10.

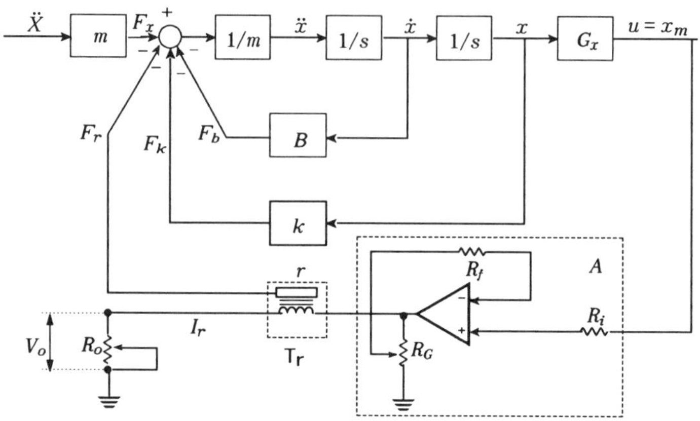

Figure 3.10. Technical diagram of closed loop force balance accelerometer with scale and gain adjustment.

In the last expression, in the steady state, as $s \to 0$, we have

$$V_0 \cong \frac{mR_0}{r} \ddot{X} \tag{3.4.21}$$

The precision, repeatability and long-term stability of this FBS are thus determined entirely by the precision and stability of the output resistor R_0 and the scale factor r of the force transducer T_r. The mass m can, of course, be considered absolutely stable. The large value of K_{eq} dictates that special provisions must be made to ensure an acceptable damping coefficient $\varsigma = 0.5 - 0.8$. This can be achieved by gas trapped in a very narrow gap enclosing the moving mass m. The dynamics of this FBS are tuned primarily by varying A. The scale factor is adjusted by varying R_0. The sensitivity is defined in terms of mA/g. A drawing of a FBS indicating the mechanical structure and the principal components, as outlined in this section, is shown in Figure 3.11. This FBS, **Sundstrand Model 305B**, is a linear high-precision sensor that responds only to the specific force along its longitudinal input axis. This is assured by the triad of parallelogram-type flexor arms, which are almost ideally rigid along the cross axes. This structure also prevents sensitivity to angular acceleration and vibro-pendulous rectification effects encountered in pendulous mechanizations. Overall precision is on the order of 0.01% of full scale. All the electronics are encapsulated in the sensor casing.

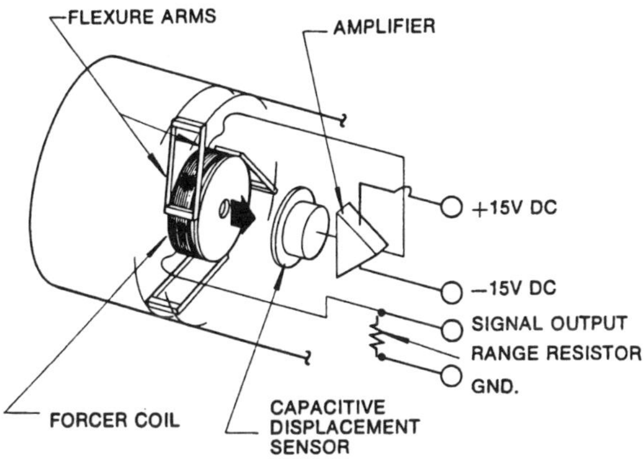

Figure 3.11. Exploded view of a force balance accelerometer using metal parallelogram springs (Source: Sundstrand Data Control publication).

A recording of low-level vibrations from DC to 100 Hz of a building on a quiet day as measured by the 305B is shown in Figure 3.12. The recording is evidence of its very low noise level and high resolution.

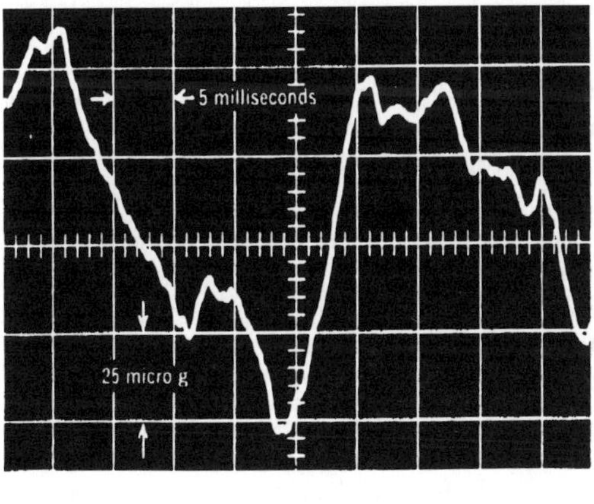

Figure 3.12. Recording of Sundstrand 305B force balance accelerometer output demonstrating low noise level and high sensitivity.

The Q-flex accelerometer

A more recent development of the FBS SFS's is the *Sundstrand Q-Flex*. Its special feature is the quartz pendulous flexible membrane which provides outstanding long-term stability and reliability and eliminates hysteresis and fatigue, which exist even in the best metal alloy suspensions. An exploded view of its construction is shown in Figure 3.13. For the past twenty years the Q-flex has, been, and still is, an important and much used component in a variety of aerospace and other applications. Therefore, its structure and its technology is described below in some detail.

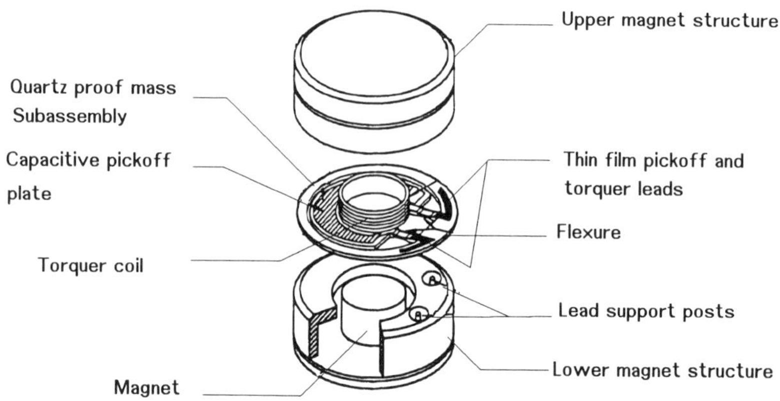

Figure 3.13. Exploded view of Q-flex accelerometer subassembly (based on Sundstrand Doc. 979-0147-801).

To begin forming the flexor-mass assembly, an annular slot is cut in the quartz blank so as to form a central disk connected to a continuous outer rim by two thin narrow bridges that serve as hinges. The central disc serves as part of the proof mass. The thin bridges serve as flexors, and the outer rim is the supporting structure for the flexors and the attached pendulum. The two critical flexor Sections are formed by chemical milling in order to provide and control the desired physical characteristics. A portion of the central disc is made conductive by vapor depositing metallic films to provide electrical surfaces required for the balanced capacitive pickoff. Conducting leads for the pick-off signals and torquer drive current are carried across both sides of the two flexors by similar vacuum deposited metallic films. Finally, torque coils arebonded symmetrically to the sides of the central disc and connected electrically to complete the assembly of the proof mass, flexors and flexor support. This subassembly is then clamped between the symmetrical magnetic structures. The balanced capacitive bridge pickoff is formed by the small gaps between the metalized portions of the fused silica proof mass and the fixed reference plates of the magnetic structure. The small, precise gap between the magnet housing and the proof mass assembly also provides damping through the motion of the gas trapped in the gap. The high level of damping is electronically augmented by conventional servo techniques. The payoff of the refined technology and the solid principle of operation described previously is demonstrated by the data in Table 3.1. Especially note the typical noise frequency distribution which is in accordance with the noise characteristics predicted in Chapter 2, the large bandwith, the excellent linearity, and the very low sensitivity to environmental interferences.

Table 3.1. Partial Data List (2123/578)
of Sundstrand Q-Flex Accelerometers

Functional Description	Numerical Value
Performance	
Range, maximum full scale, QA 1000–1300	± 25–40g
Scale factor	1.33–3.00 mA/g±10%
Frequency response	300 Hz ±5%
Natural frequency	>800 Hz
Damping ratio	0.3–0.7
Noise	
0–10 Hz	<10 Nanoamps rms
0–50 Hz	<100 Nanoamps rms
500–10 kHz	<1 Microamp rms
Supply voltage	±13 VDC–28 VDC
Scale factor sensitivity to supply voltage	0.002 %/V
Bias shift sensitivity to supply voltage	10 micro-g/V
Resolution (DC)	0.000001 g
Threshold	0.000001 g
Linearity (DC)	20 $\mu g/g^2$
Hysteresis	0.001 % of Full scale
Repeatability	0.003 % of Full scale
Scale factor temperature coefficient	180 ppm/deg C (max)
Axis alignment/traverse sensitivity	0.002 g/g
Vibration rectification coefficient	75 $\mu g/g^2$

The common feature of the class of accelerometers described so far is that they all are analog devices, i.e., their output signal is either a current or a voltage proportional to the applied specific force. When used within a computerized flight control system, the analog output must be converted into digital format involving high-precision voltage-to-frequency conversion f/V [7], which generates a train of pulses whose rate is proportional to the output voltage V. These accumulate by an up–down so as to provide vehicular velocity.

3.5 Measurement of angular acceleration

Possible approaches

In the design of aerospace attitude control systems it is sometimes required to measure the components of the angular acceleration vector of the controlled vehicle

$$\dot{\boldsymbol{\Omega}} = [\dot{p}, \dot{q}, \dot{r}]^T \tag{3.5.1}$$

These measurements can be obtained using the following three approaches:

1. angular accelerometers.

2. pairs of linear accelerometers.

3. determining the time derivative of the measured angular rates p_m, q_m, r_m provided by a triad of rate gyros.

These methods are briefly reviewed here.

Angular accelerometers

The basic angular accelerometer is a straightforward analog of the linear spring loaded accelerometer described earlier. The principle of operation is explained by the sketch in Figure 3.14.

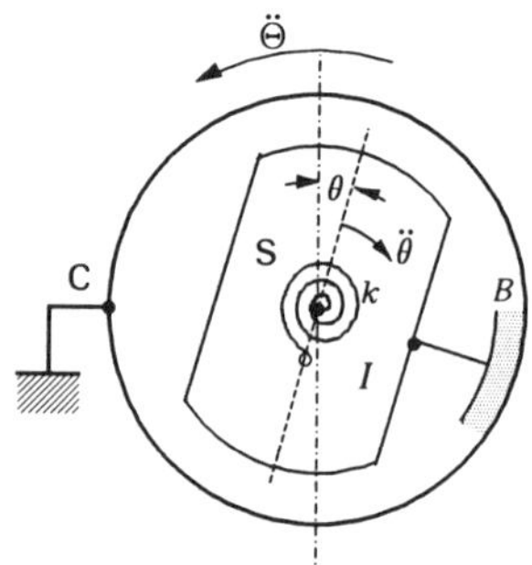

Figure 3.14. Schematic drawing of a rotational accelerometer.

A mass m having a moment of inertia I is centrally supported by a pivot and bearings in a vehicle-mounted casing C. It is balanced by a torsional spring with the constant k and is viscously damped with the constant D. At the onset of an inertial vehicular angular acceleration $\ddot{\Theta}$, m tends to retain its orientation in inertial space so that the steady-state angular deflection θ between m and C that develops is a measure of $\ddot{\Theta}$. The torque equation is

$$I(\ddot{\Theta} - \ddot{\theta}) = D\dot{\theta} + k\theta \tag{3.5.2}$$

Rearranging and Laplace transforming yields

$$\theta(s)=\frac{1}{s^2+\frac{B}{I}s+\frac{k}{I}}\ddot{\Theta}(s) \tag{3.5.3}$$

In the steady state, as $s\to 0$, the deflection θ is,

$$\lim_{s\to 0}\theta(s)=\frac{I}{k}\ddot{\Theta}(s)=\frac{1}{\omega_n^2}\ddot{\Theta}(s) \tag{3.5.4}$$

$$\omega_n \triangleq \sqrt{\frac{k}{I}}\,\text{rad/sec} \tag{3.5.5}$$

The angular accelerometer can, in principle, be mechanized as a FBS sensor to improve precision and linearity. However, there has not been a technical justification to do this because for short-term attitude determination applications, the open loop precision of $\sim 1\%$ is sufficient. On the other hand, if long-term attitude determination is required, even for a force balance mechanization, the required precision becomes prohibitively costly. This can be ascertained by the following numerical example. In order to determine the attitude angle Θ from θ, double integration is required. For example, if a precision of 1 deg in Θ is required after 1 hr of flight, the permissible bias, or resolution error $\epsilon_{\ddot{\Theta}}$, would be determined by

$$\epsilon_{\ddot{\Theta}}\frac{t^2}{2}=\epsilon_{\ddot{\Theta}}\frac{3400^2}{2}=1\,\text{deg} \tag{3.5.6}$$

Thus, the permissible error is

$$\epsilon_{\ddot{\Theta}}=\frac{2\times 1}{1.294\times 10^7}=1.543\times 10^{-7}\ \text{deg sec}^{-2} \tag{3.5.7}$$

Assume now that the required maximum vehicle angular acceleration is $\ddot{\Theta}_{max}=150$ deg sec^{-2}. Thus, the required dynamic range D_r of this sensor would be on the order of

$$D_r=\frac{\ddot{\Theta}_{\max}}{\epsilon_{\ddot{\Theta}}}=\frac{150}{1.543\times 10^{-7}}=9.74\times 10^8\cong 10^9 \tag{3.5.8}$$

This requirement is in the class of costly high-grade inertial navigation technology. It must therefore be concluded that angular accelerometers, as such, are not compatible with the requirements for attitude determination, even for limited time durations. This task must be left to gyro technology for the obvious reason that only one integration of a rate gyro output, e.g., $\dot{\Theta}$, is required to determine Θ.

Another field of application of angular accelerometers is the measurement of rotational vibrations. Modern sensors for this purpose employ piezoresistive transduction. They are designed to measure angular accelerations up to several hundred Hz. The measurement range is up to several thousand rad sec^{-2}.

Typical null point uncertainty is on the order of 2–3% of full scale. Typical examples are the high range ENDEVCO MODELS 7302B/7302M1 having a range of ± 50.000 rad sec^{-2} and medium-range MODEL 7302BM2 with a range of ± 5000 rad sec^{-2}.

Accelerometer pairs

A pair of parallel mounted accelerometers A_1 and A_2, as shown in Figure 3.15, can provide the measurement of vehicular angular acceleration along with a linear acceleration component.

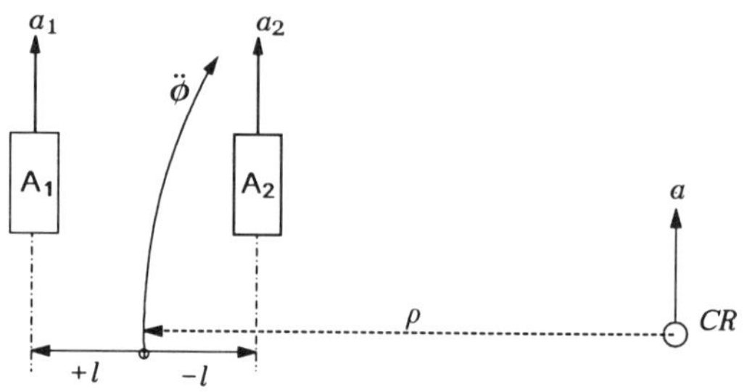

Figure 3.15. A linear accelerometer pair arranged to operate as an angular accelerometer along with linear acceleration.

CR is the instantaneous center of rotation of the vehicle, which is subjected to the linear acceleration a and the angular acceleration $\ddot{\phi}$.

The individual readings of the accelerometers are a_1 and a_2:

$$a_1 = a + (\rho + l)\ddot{\phi} \tag{3.5.9}$$

$$a_2 = a + (\rho - l)\ddot{\phi} \tag{3.5.10}$$

From the difference

$$\Delta a = a_1 - a_2 = 2l\ddot{\phi} \tag{3.5.11}$$

we have

$$\ddot{\phi} = \frac{\Delta a}{2l} \tag{3.5.12}$$

and from the sum

$$\Sigma a = a_1 + a_2 = 2(a + \rho\ddot{\phi}) = 2(a + \rho\frac{a_1 - a_2}{2l}) \tag{3.5.13}$$

from which

$$a = \frac{\Sigma a}{2} - \frac{\rho}{2l}\Delta a = \frac{a_1 + a_2}{2} - \frac{\rho}{l}\frac{a_1 - a_2}{2} \tag{3.5.14}$$

Conclusion: A pair of parallel placed accelerometers provides the information on both the vehicular angular acceleration component $\ddot{\phi}$ and the corresponding linear acceleration component a. Since l is a perfectly known constant, $\ddot{\phi}$ is determined with the precision of A_1 and A_2. The precision by which a is determined, is, to a substantial extent, dictated by the precision of the knowledge of ρ.

The technique of using accelerometer pairs for measuring angular accelerations, as described earlier, has the advantage that it can use the broad range of well developed and proven linear accelerometers technologies while providing components of the vehicle's translatory accelerations.

A block diagram describing the mechanization of (3.5.12) and (3.5.14) is shown in Figure 3.16.

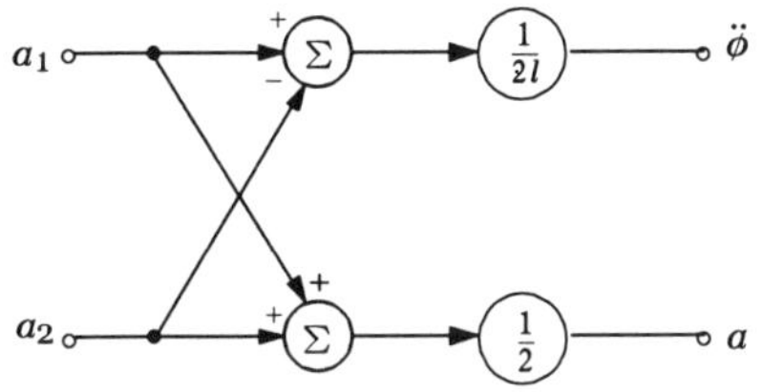

Figure 3.16. Schematic for the measurement of angular and linear acceleration.

EXAMPLE 3.5–1: Determination of roll acceleration of an aircraft

It is required to determine the angular roll acceleration $\ddot{\phi}$ from the readings of an accelerometer pair a_{1m} and a_{2m} attached to wing tips of an aircraft. The input axes of the accelerometers are perpendicular to the wing of the plane. The wing span is b m. It is assumed that the displacement of the center of rotation from the aircraft axis of symmetry ρ, is negligibly small compared to b; it is therefore set to zero. The null point offsets of the accelerometers are denoted by δa_1 and δa_2, respectively. Under an angular acceleration $\ddot{\phi}$, the accelerometer readings are assumed to be a_{1m}= +0.1g and a_{2m}= −0.1g, respectively.

It is required to determine the actual values of $\ddot{\phi}$ and a from the accelerometer readings. From (3.5.12), we have

$$\ddot{\phi}=\frac{a_1-a_2}{b}=\frac{a_{1m}-\delta a_1-(a_{2m}-\delta a_2)}{b}=\frac{\Delta a_m+(\delta a_2-\delta a_1)}{b}$$

The last expression shows that if the two accelerometer offsets are equal, their effect cancels out by virtue of the common mode rejection. δa_1 and δa_2 can also be viewed as components of the D'Alembert lift forces $L_\phi = -g(1-\cos\phi)$, which, as shown here, fortunately cancel out.

Recalling (3.5.14) and setting $\rho=0$, we have

$$a=\frac{a_{1m}-\delta a_1+a_{2m}-\delta a_2}{2}=\frac{a_{1m}+a_{2m}}{2}-\frac{\delta a_1+\delta a_2}{2}$$

Thus, for the measurement of the acceleration a, the effective offset is the mean value of the individual accelerometer offsets.

Differentiating angular rates

On the assumption that on-board angular rate sensors are available, angular acceleration can be derived by differentiation. In spite of the enhanced noise that accompanies this operation, it is fairly common practice. This noise depends on the quality of the angular rate pickoff. Potentiometers are relatively noisy and have finite resolution. Microsyn output is typically accompanied by high-frequency ripple, and digital encoders, in conjunction with pulse counters, may have significant quantization noise. However, in spite of these potential shortcomings, differentiation is a fairly common practice. This is true for attitude control design, in particular where the actually experienced noise levels are effectively filtered by the relatively slow control loop dynamics.

3.6 Integrating accelerometers

Measurement of acceleration, or specific force per se, is not necessarily an end but rather a means, in particular in guidance and navigation applications. The variables representing the essential information are actually inertial velocity or ground speed and position. A cursory description of the principle of operation of an inertial velocimeter that inherently integrates specific force is given in Section 1.9. However, for technical reasons, that mechanization cannot be considered a practical solution. In the 1930s and '40s, analog electronic integration in inertial navigation technology could not provide the required precision because of the substantial drift rates of electronic amplifiers and the difficulty in assuring the required scale factor precision and firmness. Those would have caused unacceptable offset errors even for relatively short integration times. Sufficiently precise analog-to-pulse rate conversion was not yet available, so integration by pulse counting was also a premature concept. For this reason, a mechanization based on an unbalanced angular rate gyroscope was considered a plausible solution in the 1940s and '50s. A major application was the determination of the rocket motor cut-off instant for ballistic missiles as the velocity reached a predetermined value for the required missile ground range. The principle involved is the utilization of the gyro dynamic property by which a torque, caused by an intended mass imbalance, applied at right angles to the spin axis, causes a proportional angular precession rate at a right angle to the applied torque. This torque is the product of the applied acceleration and the pendulosity of the gyro. Therefore, the accumulated precession angle is proportional to the integral of the acceleration, namely the velocity. This device has been known as the *gyro pendulum* or *pendulous integrating gyro accelerometer* (PIGA) [5]. Since the principle of operation is essentially gyro dynamics, it will be briefly addressed in Chapter 4 in relation to rate gyro applications.

With the advent of modern solid-state electronics, analog-to-digital and analog-to-frequency conversion and very high-precision interpolating vernier pulse counting, these World War II gyro dynamical bulky and costly techniques have become obsolete.

3.7 Vibrating beam accelerometers

Background

Vibrating beam accelerometer (VBA) technology is emerging as a major force in guidance systems. The frequency output of the vibrating accelerometer can be processed exclusively by digital electronics without causing additional delta velocity errors typical in analog-to-digital conversion. This compatibility with high-speed digital navigation and flight control systems, its excellent precision, and other cost, power, and weight advantages have spurred continued research and development of the VBA, which has matured as a revolutionary class of products of first-rate importance. A good overview of the-state-of-the-art is given in [2].

Principle of operation

The VBA is a class of instruments that utilizes an elastic beam as the resonating element in a piezoelectrically or electromagnetically excited electromechanical oscillator. The resonant frequency of the beam varies as a function of the applied acceleration by attaching one end of the rigid beam to the seismic mass and the other end to the body-mounted casing. The seismic mass is constrained by flexures that permit a small motion of the mass along the desired sensitive axis. This type of accelerometer is open-loop. However, the high axial stiffness of the beam results in very high fundamental natural frequency of the seismic mass system, thereby providing broad response bandwidth. Single resonating beam force transducers have an inherent nonlinear force-to-frequency conversion function. In any precision accelerometer, two force transducers, namely, the vibrating beams, are arranged in a push-pull loading mode that effectively cancels this nonlinearity.

Crystalline quartz is the most common material for the resonating beam element. However, silicon is rapidly gaining in importance, especially with the advent of silicon microfabrication technology. The vibrating beam accelerometer has the fundamental advantage over the earlier vibrating string accelerometer [3] in that it requires no pre load tension in order to measure positive and negative acceleration. Furthermore, it needs no delicate cross-supports for the proof mass attached to the string. The elastic restoring force of the relatively high thickness-to-length ratio beam, sketched in Figure 3.17, results in a relatively high resonant frequency f_0, which increases with tension and decreases under compression. The variation in this zero-load frequency as a function of applied load is dictated primarily by the Euler buckling load of the resonant element under compression. A sketch of the single VBA is shown in Figure 3.18. a is the applied acceleration pointing to the left and $F = Ma$ is the inertial force applied to the mass M, which either stretches or compresses the beam. The joint J between the beam and the body causes substantial energy dissipation, which must be replenished in order to sustain the beam vibration.

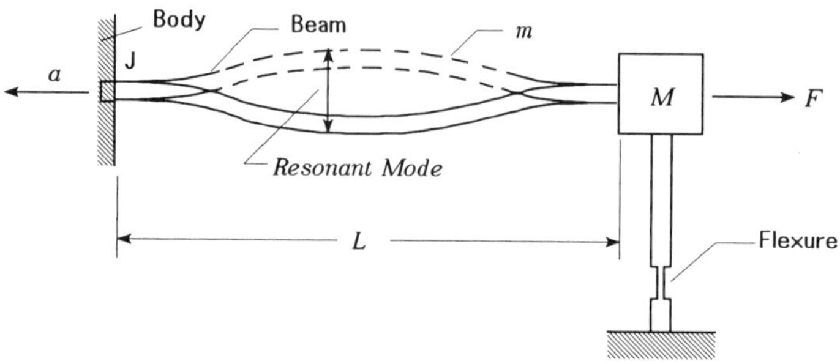

Figure 3.17. Principle of operation of the vibrating beam accelerometer.

Acceleration-to-frequency conversion

The following parameters and variables determine the relationships between the input acceleration and the resonant frequency.

K — beam stiffness (largely determined by L);
m — equivalent mass of the beam;
M — suspended mass;
F — inertial force acting on M;
f_0 — zero-load frequency;
f_n — Resonant frequency under load;
a — input acceleration;
C — Euler buckling load.

The following relationships apply:

$$f_0 = \frac{1}{2\pi}\sqrt{\frac{K}{m}} \tag{3.7.1}$$

$$f_n = \frac{1}{2\pi}\sqrt{\frac{K+\alpha Ma}{m}} = \frac{1}{2\pi}\sqrt{\frac{K}{m}}\sqrt{\frac{1+\alpha Ma}{K}} = f_0\sqrt{1+\frac{Ma}{C}} \tag{3.7.2}$$

$$C \triangleq \frac{K}{\alpha} \tag{3.7.3}$$

For physical reasons, $|Ma/C| \ll 1$. A positive a will be associated with the stretching of the beam, while a negative a implies its compression. The relationship between f_n and a follows from (3.7.2), namely,

$$a_c = \frac{C}{M}\left(\frac{f_n^2}{f_0^2} - 1\right) \tag{3.7.4}$$

where the subscript $(\cdot)_c$ stands for the computed value. It follows that the nonlinear relationship between the acceleration and the resonant frequency is quadratic. This nonlinear relationship is obviously highly undesirable for a high-precision instrument. This problem has been elegantly resolved by placing the proof mass between two coaxial beams such that upon application of the axial acceleration, one beam is compressed, while the other is stretched. A sketch of this arrangement is shown in Figure 3.18 in which the vibrating beams are double-ended tuning forks (DETFs). These are presented in greater detail later. This type of VBA, specifically those with quartz DETFs are known as the *Accelerex* series, developed and manufactured by Sundstrand Data Control (SDC). The reason for using tuning forks is that their vibration is virtually not transmitted to the vehicle body so that very little energy is lost. This implies that only minute energy is required to sustain the oscillation and a very high Q factor can be achieved. This feature is essential because, as will be discussed, a high Q assures high resolution of frequency variations in f_n. The reason for using DETFs is technically obvious because otherwise stretching and compressing are impossible.

The rigidity of the structure implies a relatively high natural frequency f_B of the beam mass assembly (not to be confused with f_0), which typically is in excess of 1 kHz. Consequently, the vibro-pendulous effect is minute because the energy of external vibration is typically confined to lower frequency regions.

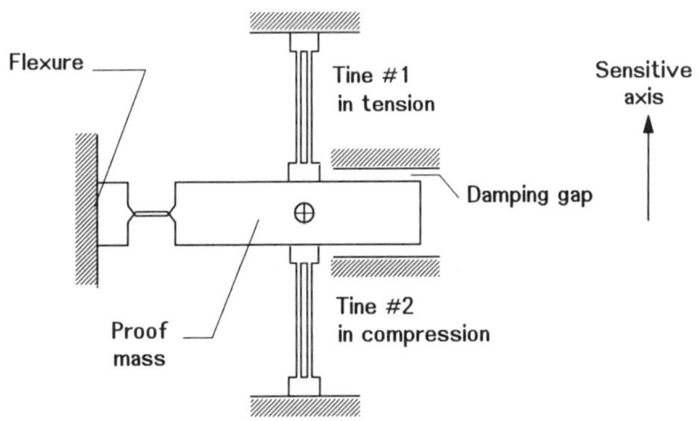

Figure 3.18. Push-pull arrangement of a pair of double-ended tuning forks (based on Ref. 2).

Thus, a push-pull operation providing an almost ideal linear relationship is implemented. Subtracting the corresponding outputs $a_1 - a_2$, and assuming that C in the two beams is identical, the following relationships between the applied acceleration difference a and the resonance frequencies f_{n_1} and f_{n_2} are obtained as follows:

$$a = a_1 - a_2 = \frac{C}{M}\left(\frac{f_{n1}^2}{f_0^2} - 1\right) - \frac{C}{M}\left(\frac{f_{n2}^2}{f_0^2} - 1\right) = \frac{C}{M}\left(\frac{f_{n1}^2 - f_{n2}^2}{f_0^2}\right)$$

$$= \frac{C}{M}\left(\frac{f_{n1} + f_{n2}}{f_0}\right)\left(\frac{f_{n1} - f_{n2}}{f_0}\right) \tag{3.7.5}$$

Let $f_{n1} = f_0 + \Delta f_1$; $f_{n2} = f_0 - \Delta f_2$, which implies that the frequency increase in beam 1 is not assumed to be precisely equal to the frequency decrease in beam 2. With these notations, we have from (3.7.5),

$$a = \frac{C}{M}\left(\frac{f_0 + \Delta f_1 + f_0 - \Delta f_2}{f_0}\right)\left(\frac{f_0 + \Delta f_1 - f_0 + \Delta f_2}{f_0}\right)$$

$$= \frac{C}{M}\left(\frac{2f_0 + \Delta f_1 - \Delta f_2}{f_0}\right)\frac{\Delta f_1 + \Delta f_1}{f_0} = \frac{4C}{M}\left(1 + \frac{\Delta^2 f}{f_0}\right)\frac{\overline{\Delta f}}{f_0}$$

$$\cong \frac{4C}{M}\frac{\overline{\Delta f}}{f_0} \tag{3.7.6}$$

where $\Delta^2 f \triangleq \Delta f_1 - \Delta f_2$ is negligibly small compared to f_0 and can be ignored; $\overline{\Delta f} \triangleq (\Delta f_1 + \Delta f_2)/2$ may deviate slightly from linearity especially for very high accelerations, but this error can be calibrated in view of the excellent repeatability of crystalline materials, such as quartz or silicon, used in the fabrication of the vibrating beams. It therefore follows that (3.7.4) establishes an almost perfect linear relationship between the input acceleration a and the total relative frequency deviation $\overline{\Delta f}/f_0$, which can be on the order of 10%. The essence of the VBA operation therefore, is the precise measurement of the frequency deviation. Since two of the principal requirements are high resolution and a large dynamic range, the ability to resolve minute frequency deviations is a major issue which is discussed in further detail later.

EXAMPLE 3.7–1: Required Q-factor in a vibrating beam accelerometer

Let $f_0 = 35$ kHz and $\overline{\Delta f}/f_0 = 0.1$. The desired dynamic range of the accelerometer is $a_{max}/a_{th} = 10^4$.

This requirement implies $\overline{\Delta f}/f_{th} = 0.1 \times 35000/f_{th} = 10^4$, from which it follows that $f_{th} = 3500/10^4 = 0.35$ Hz. This result implies that the resonating beam must have a very high-quality factor Q and high-performance digital processing algorithm to provide this accurate readout.

Double Ended Tuning Fork

The dual beam resonator design, often referred to as the *double ended tuning fork* (DETF) resonator utilizes two coupled beams vibrating 180 deg out of phase. This dynamic motion causes the shear and moment reactions at the end of the beams to cancel. The stresses approach zero within the distance of a few beam widths from the root of the beam. With the use of this feature, bulky end isolation elements are not required. Consequently, the resonators are small, simple and relatively easy to manufacture.

Early quartz dual beam resonators [4] such as those shown in Figure 3.19 were cut by mechanical methods such as a Sheffield Cavitron, which yielded good surface finish and approximately 100-microinch dimensional accuracy. Lapping and polishing were required to achieve a high Q resonance. This fine machining and tuning considerably added to the cost and sensitivity to the individual workmanship. The end conditions were still slightly imperfect and allowed energy loss to the surrounding structure. The performance characteristics, nonetheless, were very good and encouraging.

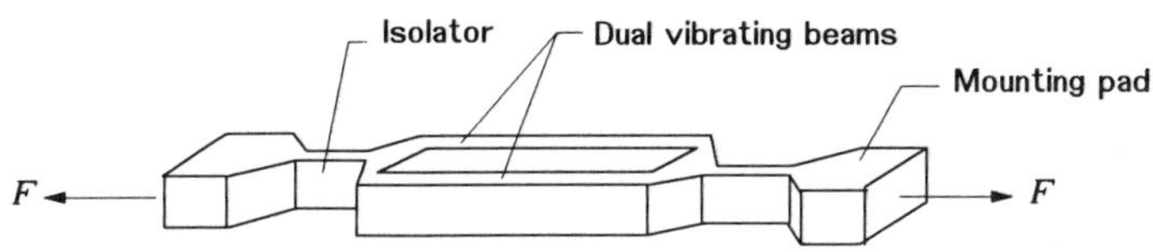

Figure 3.19. Dual-beam resonator with isolating mounting pads (based on [2]).

Photholitographic masking technology, which, was developed in the semiconductor industry in the early 1980s, was adapted for crystal fabrication using wet chemical etching. The photolithographic process is capable of submicron total position accuracy over a 4-inch silicon wafer. The feature-to-feature matching, critical for beam balancing, is at least an order of magnitude better than the absolute accuracy, i.e., better than 10 microinches. The consequence of this development is the capability of manufacturing dozens of perfectly matched resonators from a single wafer at a very low cost. This made the dual beam resonator not only practical but also highly cost-effective. It also improved the matching of crystal pairs for greatly enhanced common mode error rejection. As a result, the production volume of VBAs and pressure sensors using DETF resonators has become massive. A sketch of a DETF resonator is shown in Figure 3.20. The pair of slender vibrating beams, referred to as *tines*, are 180 deg out of phase. In principle, they are similar to a guitar string, except that they can be compressed as well as stretched. The shear and bending stresses, which cancel out within a few beam widths, permit the mechanical connections to be made directly with mounting pads without resorting to bulky isolators. The unique geometry of the crystal, as shown in Figure 3.20, allows the achievement of very high-performance quality from an extremely small accelerometer. The overall length of the crystal depicted is less than 0.8 mm.

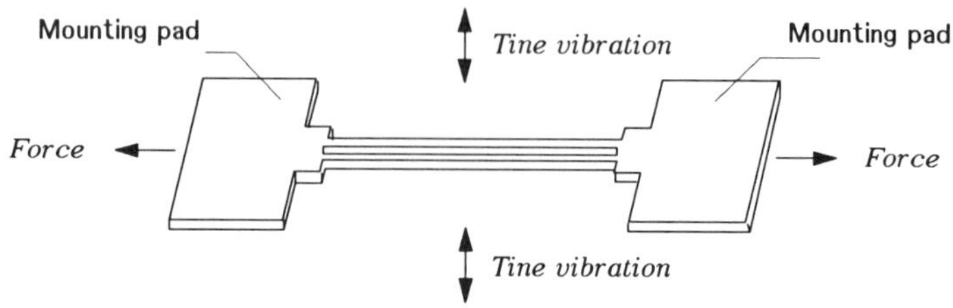

Figure 3.20. Miniature crystalline vibrating VBA with monolithic mounting pads and tines (based [2]).

Crystalline quartz is an ideal material for the VBA. The zero-load resonant frequency depends almost entirely on the dimensions of the crystal and the mechanical and physical properties of the quartz, which are highly stable. Quartz also has extremely low internal losses under oscillation, which makes it ideal for high Q resonators. Its piezoelectric properties provide the coupling between the mechanical resonance and the oscillator electronics by the use of a surface electrode pattern. The frequency of a quartz resonator is relatively insensitive to temperature variations.

The quartz VBA should not be confused with the traditional quartz piezoid charge-coupled type accelerometer, which responds to AC inputs only. On the other hand, the VBA, while having a wide bandwidth, fully responds down to DC. The carrier frequency f_0 continues to provide information even if the acceleration is zero or unchanging.

Crystal controlled oscillation

The equivalent circuit of a vibrating crystal, shown in Figure 3.21, is useful for explaining the basic concepts governing its function. The crystal is a two-port network with mechanical drive and motion-sensing functions performed by the same electrodes. The L_1, R_1, C_1 branch of the equivalent circuit is known as the *motional arm.* The motional inductance L_1 represents the active mass in the tines. The motional capacitance C_1 represents the elastic constraint of this mass. It consists of a constant portion, created by the highly stable elastic modulus of the quartz, and a variable portion, as indicated in (3.7.1) and (3.7.2), created by the axial forces on the tines and resulting in additional restoring forces on the mass.

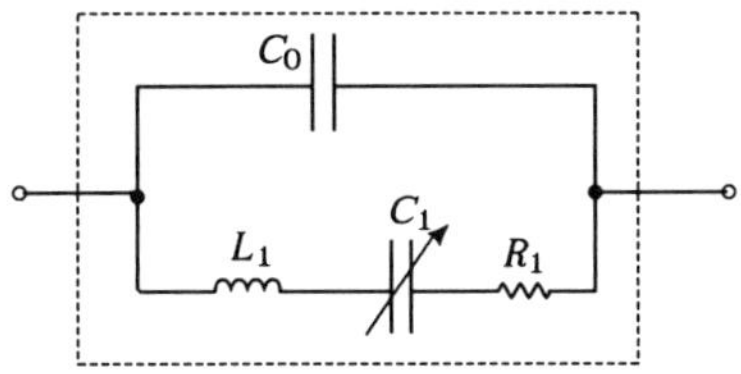

C_0 - *Static capacitance (electrodes)*

L_1 - *Motional inductance (mass)*

C_1 - *Motional capacitance (elasticity)*

R_1 - *Equivalent resistance (losses)*

Figure 3.21. Equivalent circuit of oscillating crystal

The variation in capacitance modulates the resonant frequency as a function of the force applied. R_1 represents the bulk crystal losses created by all the sources including mechanical losses, acoustic energy losses, and electrode resistances. The static capacitance C_0 is the sum of the capacitance between the electrodes and that added by the wiring.

The quartz crystal feedback oscillator is arranged to provide the feedback at the desired output frequency. Oscillation occurs when the circuit has a phase angle of $n \times 360$ deg, $n = 1, 2, 3, \ldots$ and a gain of 1. Crystals can be forced to oscillate in either series (antiresonance) or parallel resonance frequencies. The series mode is preferable because it is less sensitive to external component parameter variations. Frequency stability is therefore essentially dictated by the highly stable quartz parameters. The oscillation frequency is given by

$$f_s = \frac{1}{2\pi\sqrt{L_1 C_1}} \tag{3.7.7}$$

This frequency is nominally designed to 35 kHz and can vary by $\pm 10\%$ due to changes in C_1 caused by the sensed acceleration.

The crystal quality factor Q is related to the series resonant frequency f_s, the equivalent series resistance R_1, and the motional inductance L_1 as given by

$$Q = \frac{2\pi f_s L_1}{R_1} = \omega_s \frac{L_1}{R_1} = \varsigma^{-1} \tag{3.7.8}$$

where ς is the damping factor. Since R_1 represents internal energy losses, it is clear from (3.7.8) why the inherent favorable properties of quartz are associated with high levels of Q. It should be kept in mind, however, that a very high Q is associated with a small negative real part a of the complex pole pair given in the s plane. Since $a=\omega_s\varsigma$ and $\varsigma=1/2Q$, it follows from (3.7.8) that

$$a=\frac{\omega_s}{2Q}=\frac{\pi f_s}{Q} \tag{3.7.9}$$

The associated time constant of the oscillation envelope build-up after energizing the oscillator is

$$\tau=a^{-1}=\frac{Q}{2\pi f_s} \tag{3.7.10}$$

For example, for $\pi f_s=\pi\times 35000 \cong 110.000\ \text{sec}^{-1}$, and $Q=10^5$, $\tau \cong 1$ sec. In many applications, such as navigation, this build-up time constant is acceptable, but it may be excessive in some missile guidance problems. On the other hand the high Q assures very low power consumption, drawing only about 5 mA from a 10 volt power supply. The output stage provides a robust, buffered zero to five volt square wave output.

Signal processing system

The signal processing of the accelerometer output is performed entirely with digital electronics without analog to digital conversion. A block diagram of a typical system implementation is shown in Figure 3.22.

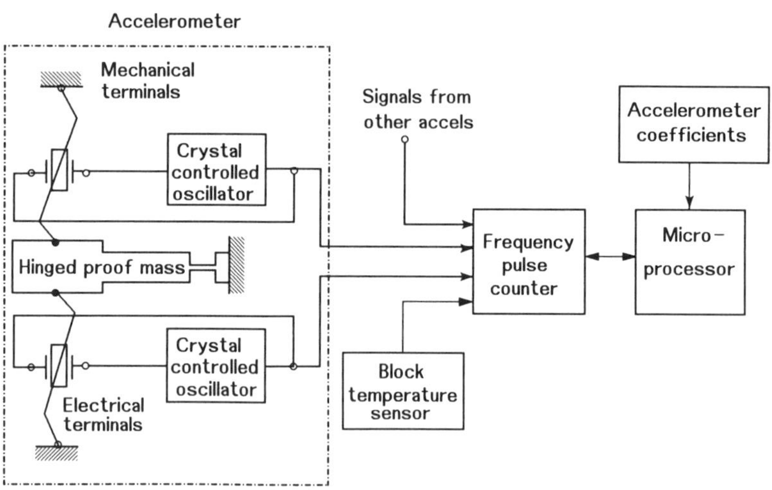

Figure 3.22. Push-pull vibrating beam accelerometer including compensation and signal processing.

The purpose of this system is to provide a numerical output that is proportional to the integral of the sensed acceleration, to eliminate residual deviations from linearity, and to compensate for the effects of temperature variations. The philosophy of the entire signal processing and compensation open loop design approach is based on the excellent stability and repeatability of the crystals, which permit the abandonment of the closed-loop principles such as in forced-balance sensors. The accelerometer output is a pair of buffered 0—5 volt square waves. These represent the resonant frequency of the VBA crystals. The difference between these two frequencies is closely proportional to the acceleration. With zero g acceleration, both crystals output 35 kHz. Each crystal frequency changes by ±3500 Hz at the ±100 g full-scale range. The sensitivity of the accelerometer is therefore 35 Hz/g. This 10% change in frequency provides a large scale factor which contributes to accuracy and resolution. The algorithm coefficients characterize the output parameters of the completed sensor without resorting to the costly process of matching the individual characteristics of individual crystals. Temperature sensitivities are also characterized and removed by thermal modeling instead of using bulky and costly temperature regulation methods. The accelerometer does not have an individual built-in temperature sensor. Instead, a single block-mounted temperature sensor is used to provide temperature information for compensating the entire triad of accelerometers. The temperature sensor itself is a quartz crystal resonator, which provides the temperature information in the desired format of frequency changes. The push-pull arrangement, which provides inherent common mode rejection, removes most of the effects of temperature variation so that the additional temperature compensation serves as a fine adjustment.

Frequency-to-acceleration conversion algorithm

Digital counters and sampling are used to measure the crystal frequencies. The high frequency system master clock operating at 10 MHz more is used as a vernier to enhance resolution by interpolating between the crystal cycles. The frequency to acceleration conversion algorithm is defined as

$$q = a_0 + a_1 f_1^2 + a_2 f_2^2 \tag{3.7.11}$$

where f_1 = frequency of crystal #1 and f_2 = frequency of crystal #2. The term a_0 accounts for the bias, i.e., the difference between f_1 and f_2 for zero acceleration input. The coefficients a_1 and a_2 are weighted contributions from f_1 and f_2, respectively, to cancel the residual effect of second-order nonlinearity. The three constants are modeled as third-order polynomials in temperature T, namely,

$$a_i = a_{io} + a_{i_1}\mathrm{T} + a_{i_2}\mathrm{T}^2 + a_{i_3}\mathrm{T}^3;\ i = 0, 1, 2, 3 \tag{3.7.12}$$

a_{ij} change slowly with temperature. Therefore they can be updated at a much slower rate than that required for the acceleration conversion algorithm.

The principal signal processing algorithm converts frequency to delta velocity, which typically is the desired format for inertial navigation. The digital measurement of frequency is given by the number of zero crossings accumulated over a predetermined sampling time interval T. This is equivalent to integrating the frequency, which implies a change in phase over that period. But this phase change represents the change of velocity ΔV over the time interval T, which is formulated as follows:

$$\Delta V = A_1(\Delta\phi + A_2\Sigma\phi + A_0 T) \tag{3.7.13}$$

where A_0, A_1, and A_2 are constants that are modeled functions of temperature.

$\Delta\phi = \phi_1 - \phi_2$;

$\Sigma\phi = \phi_1 + \phi_2$;

$\phi_1 =$ crystal #1 phase change over one sample period;

$\phi_2 =$ crystal #2 phase change over one sample period;

$$\phi_i = \int_{t_0}^{t_0+T} f_i \mathrm{d}t; \quad i = 1, 2 \tag{3.7.14}$$

ϕ_1 is assumed to be positive while ϕ_2 is assumed to be negative. Clearly, $\Delta\phi$ is the principal term representing the velocity increment over T. In (3.7.13), A_1 is the scale factor. A_0 is a constant bias due to a zero input acceleration offset at the frequency that accumulates a bias in ΔV over the sampling period T. $\Sigma\phi$ should ideally be zero, but it may contribute additional bias plus some residual even-order nonlinearity effects. The three coefficients A_0, A_1, and A_2 are modeled as third-order polynomials in temperature similarly to (3.7.13),

$$A_i = A_{i0} + A_{i1}\mathrm{T} + A_{i2}\mathrm{T}^2 + A_{i3}\mathrm{T}^3\ ; \quad i = 1, 2, 3 \tag{3.7.15}$$

Again, these constants change slowly with temperature and consequently can be updated at a slow rate.

Average acceleration output can be obtained by modifying the scale factor A_1 to

$$A_1' = \frac{A_1}{g_0 T} \tag{3.7.16}$$

where $g_0 = 9.81$ m sec^{-2} = 32.2 ft sec^{-2}, and T again is the sampling period. he two linearization models (3.7.11) and (3.7.13) are essentially equivalent, except that the latter is computationally more efficient.

Resolution and dynamic range

Resolution of the output incremental velocity ΔV depends on the measurements of the signal phases as derived in the foregoing. Using a very high-frequency clock as a vernier to interpolate between signal cycles in consecutive samples can greatly improve the precision of the phase measurements. The signal phase resolution will be inversely proportional to the number of clock cycles accumulated within one signal cycle of frequency f_0, which is the nominal zero input VBA oscillation frequency. The resolution of the delta velocity is determined by

$$r_{\Delta V}=A_1\frac{f_0}{f_{\text{clock}}} \tag{3.7.17}$$

With f_{clock} = 10 MHz and a typical value of $A_1=0.15$ m sec^{-1},

$$r_{\Delta V}=0.15\times\frac{35{,}000}{10{,}000000}=5.25\times10^{-4}\ \text{m sec}^{-1} \tag{3.7.18}$$

The resolution r_a of acceleration is determined by division by T. Normalizing by g defines the resolution in units of g,

$$r_a=\frac{r_{\Delta V}}{gT} \tag{3.7.19}$$

With a sampling rate of 500 Hz, namely a sampling interval of 2 m sec, the preceding example yields

$$r_a=\frac{5.25\times10^{-4}}{2\times10^{-3}\times9.81}=24.7\ \text{mg per sample} \tag{3.7.20}$$

We now assume that the counters are perfect, i.e., no counts are missing between samples. Consider the resolution r_a^N for N consecutive samples. This is equivalent to the increase of the sampling interval to NT. Namely, in view of (3.7.19),

$$r_a^N=\frac{r_{\Delta V}}{gNT}=\frac{r_a}{N} \tag{3.7.21}$$

The foregoing is true because the +1 or −1 count errors at the juncture of two consecutive samples cancel so that for the entire N samples, the count error is still +1 or −1. Thus, for $N=500$, the resolution is

$$r_a^{500}=\frac{24.7}{500}=53\ \mu\text{g}\cong50\ \mu\text{g} \tag{3.7.22}$$

Since we have assumed a sampling rate of 500 Hz, this resolution is obtained for the smoothing period $NT=1$ sec. This resolution competes very well with high-grade force balance accelerometers. The delay of ~ 1 sec caused by the smoothing operation is of minor significance in inertial navigation applications where high resolution is required. For the full-scale acceleration input of ± 100 g assumed earlier, this class of VBA has a dynamic range of

$$D_r = \frac{100}{50} \times 10^4 = 2 \times 10^4 \tag{3.7.23}$$

This is a suitable figure for high-grade tactical applications. The principal factor which decides the resolution, as seen in (3.7.17), is f_{clock}. For example, a 50 MHz clock would improve the resolution to the level of 10 μg and the dynamic range to $D_r = 10^7$. This improved performance level is of significance in special applications of the VBA, such as in the Coriolis rate sensor described in Chapter 6.

Sensitivity to clock frequency variations

Errors in the clock frequency f_{clock} have the same effect as errors in crystal frequency f_0. They produce apparent crystal frequency deviations, which appear to be bias errors. With the push-pull pair arrangement, shifts in f_0 are primarily common mode so that they mutually cancel out. Yet, in view of (3.7.5) and (3.7.4), the net shift depends on how closely the crystal buckling loads C are matched. We denote this mismatch by ΔC. For crystals from the same wafer, the Cs are expected to be within $\pm 1\%$ of each other. For example, $C = 500$ g, it follows that $\Delta C = 5.0$ g. Denoting the shift in f_{clock} by Δf_{clock}, the shift in bias Δb would be

$$\Delta b = \Delta C \frac{\Delta f_{\text{clock}}}{f_{\text{clock}}} \tag{3.7.24}$$

For a relative clock shift of one per million,

$$\Delta b = 5.0 \text{ g} \times 10^{-4} = 5\ \mu\text{g} \tag{3.7.25}$$

This expected bias shift is normally negligible in comparison with other larger bias sources, which are an order of magnitude larger.

Frequency response and sensitivity to extraneous vibration

The moving subassembly of the VBA is constrained by the high spring rate of the crystal tines, and it is damped by the squeeze film gas in the damping gap, as indicated in Figure 3.23. The dynamics is modeled as a conventional second-order system with two complex poles and a damping factor denoted by the transfer function:

$$G_s(s) = \frac{\omega_s^2}{s^2 + 2\omega_s\varsigma s + \omega_s^2} = \frac{\omega_s^2}{s^2 + \frac{\omega_s}{Q}s + \omega_s^2} \tag{3.7.26}$$

For a typical SDC Tactical Grade Accelerex (TGA) accelerometer, the peak resonance response is at $f_s = \omega_s/2\pi = 1450$ Hz and the damping factor ς, which equals 0.15, provides a Q factor of 3.4, in accordance with the relationship $Q = 1/2\varsigma$. The corresponding frequency response plot is shown in Figure 3.23.

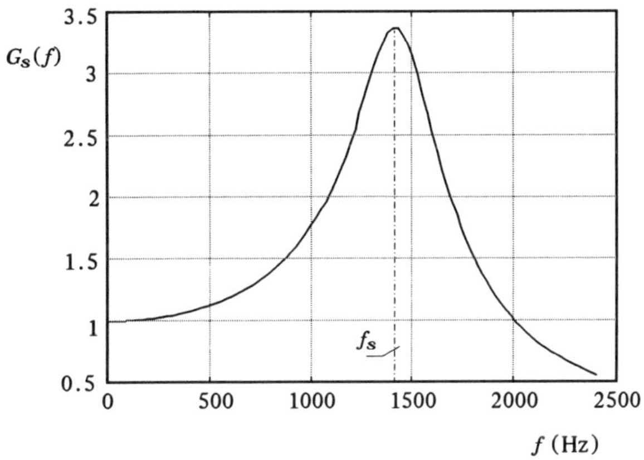

Figure 3.23. Crystal frequency response in an accelerex-type VBA (source: Sundstrand Data Control internal report).

This broad bandwidth assures small phase shifts in the useful operating frequency range out to ~700 Hz. This performance characteristic, which is important in flight control and navigation, compares well with that of the FBS Q-flex accelerometer.

Vibration rectification is a major aspect of an accelerometer, the consideration of which strongly influences choices of the VBA mechanical design. Individual force crystals have a substantial even-order nonlinearity. The method of using matched crystal pairs made from the same wafer in a push-pull mode causes all the even order nonlinearities to cancel out exactly. Moreover, the use of a single pendulum with dual crystals causes the dynamic responses of the pair to be nearly identical because they are exposed to the same input. As a result, the vibration rectification coefficient (VRC) is minute out to about 1000 Hz. Typical values, for the TGA in the region 100–500 Hz are substantially less than 10 μg/g^2. However, in the region of the resonance, around 1500 Hz, the vibration is about 3.5 times more intense, thus causing an accordingly increased VRC.

3.8 Piezo and capacitive transducers ⊕

Piezo sensors/transducers

Besides the specialized high-performance sensors described in the foregoing sections, there exists a large family of piezo sensors or transducers. The term *piezo* derives from Greek; its meaning is *pressure*. The common property of piezo devices is that they all react to pressure in the sense that they respond through different physically measurable quantities to pressure or tension. As such, they are known as *piezovoltaic*, *piezoresistive*, *piezocapacitive*, or *piezo-magnetic* devices. Strictly, the piezo response acts by virtue of a unique natural property of the material, such as quartz, or other types of crystalline structures.

A unique property of truly physical piezoelectric devices is that they are bilateral. This means that inasmuch as their response to pressure is a corresponding electrical output, they also respond to an electrical input by generating a force or torque. This property underlies, for example, the operation of quartz crystal oscillators. This bilateral property is obviously not shared by piezoresistive materials which are unilateral. Recent research indicates that changes in the polarization of light passing through optical fibers caused by their stretching can yield a large family of miniature transducers by means of tension sleeves that react to changes of physical quantities, such as temperature, humidity or acceleration. More generally, man-made piezo properties can be provided by the design of special mechanisms that deform under pressure and cause a change in a physical parameter such as the gap between the plates of a capacitor, thus changing capacitance, or in a different arrangement, inductance or resistance. In the modern development of this family of sensors, quartz and silicon play a prominent role because of their uniquely favorable properties of stability, repeatability, and suitability to microminiaturization.

By and large, piezotransducers are low-cost moderate precision commercial devices. Their most common purpose is to measure, monitor, and/or control structural vibrations in aerospace vehicles. As such, in contrast to the precision accelerometers described so far, they are generally required to provide measurements of shock and high-intensity vibrations up to thousands of g's and out to high frequencies in the 10—20 kHz range, but not necessarily in the very low-frequency and dc range. Piezoelectric devices naturally fulfill this requirement.

A major emerging application of piezotransducers is the artificial rigidization of structures by closed-loop control [7], in which both the sensor, which measures a deformation, and the forcer, which is intended to attenuate it, are piezoelectric devices that naturally interface with electrical servo technology.

Piezoelectric (PE) accelerometers use a spring-mass unit to generate a force proportional to the amplitude of a vibration. This force is applied to a PE element that produces a charge on its terminals that is proportional to the mechanical motion. An advantage of PE materials is that they are self-generating and therefore do not require an external power source. Crystals such as quartz are naturally piezoelectric if properly cut, but they have relatively low sensitivity. Ferroelectric materials are more commonly used. They are man–made ceramics that are given piezoelectric properties by a process of artificial polarization. By means of suitable ceramics composition, much higher charge sensitivity and operating temperature can be attained. The shape of the material can be easily configured to obtain the desirable transducer characteristics, with special emphasis on output stability achieved by *aging*. The overview in this Section draws upon technical information of Endevco data on accelerometers and other principal leading manufacturers in the area of miniature piezo—force transducers.

Piezoelectric accelerometers

A variety of mechanical designs are available. The most common is the single–ended compression (SEC) whose high sensitivity and high resonance frequency render it a useful device for general–purpose applications. The device has a center post on which the mass, which is fitted with a center hole, is torqued down onto the sensitive element to prestress it. The Isobase is a version of the SEC that minimizes the effects of base–coupled inputs. It isolates against nonvibration environments by using a specially contoured base. It is more suitable for low–level measurements or those involving strain at the mounting surface possibly caused by thermal transients. Designs that use shear forces have the advantage that they permit low proof mass, thus resulting in miniature sizes. As the sensing element is well isolated from the base, shear accelerometers provide the best protection against pickup from base bending and acoustic noise.

The Isoshear allows the use of multiple crystals and passive compensators so that a much broader range of sensitivity and temperature is possible. The Isoshear design allows a broad range of piezoelectric materials thus enabling optimization for temperature extremes. It offers excellent temperature stability and extremely high signal–to–noise ratios in a great variety of operating conditions. Sketches of SEC, Isobase, and Isoshear accelerometers are shown in Figure 3.24(a), (b), and (c).

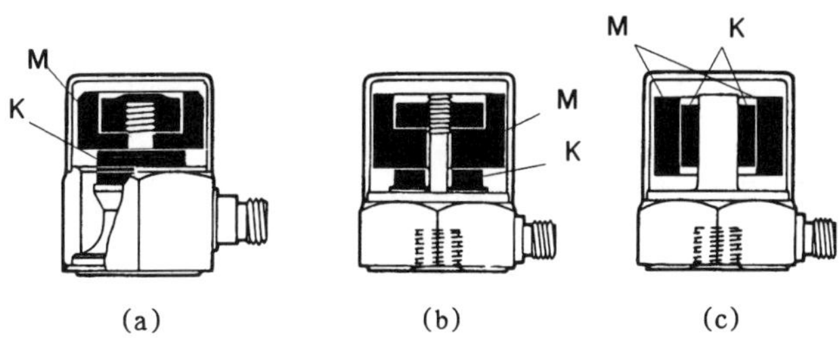

Figure 3.24. (a)–Single ended compression accelerometer;
(b)– Isoshear accelerometer;
(c)–Isobase–type accelerometer.

ISOTRON accelerometers are piezoelectric with internal electronics. They enable increased design flexibility. Electronic gain can be used to reduce the transducer size and increase its sensitivity. Frequency response can be tailored to specific applications by electronic filtering. The mechanical resonance of the accelerometer can be suppressed. Because of the low output impedance provided by the integrated output amplifier, output cabling is not a critical issue. ISOTRON transducers can drive long cables without loss of effective sensitivity and without causing an increase in noise. Moreover, low impedance devices are less sensitive to contamination and moisture.

Piezoresistive sensors

Piezoresistive (PR) elements are solid-state silicon resistors that change their electrical resistance in proportion to applied mechanical stress. They are mechanically connected to cantilever beams and electrically connected to a Wheatstone bridge to produce an electrical output that is proportional to the vibratory acceleration is determined by its amplitude and the square of its angular frequency. Compared to wire gauges, PR gauges are virtually free of hysteresis and their sensitivity is several orders of magnitude greater. They offer the advantage of dc response. This renders them suitable for static measurements or pulses of long duration. Since they utilize an external power source, they can have very low output impedance. Because of this and their inherent high sensitivity, preamplification is not necessary. They can be made in miniature sizes, a feature that increases the diversity of their usage. A sketch of a piezoresistive gage is shown in Figure 3.25. It is made from a single piece of silicon. Large mounting pads with their electrical contacts are connected via a compliant neck, which is the active part of the gauge.

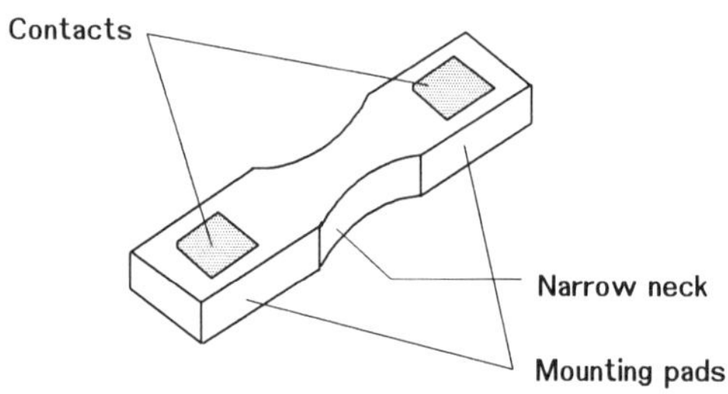

Figure 3.25. Sketch of a piezoresistive gauge fabricated from a single piece of silicon.

Variable capacitance accelerometers

The piezo devices just described are primarily intended for very high g inputs. A variable capacitance device known as the MICROTRON is particularly suitable for low-level g measurements. The individual sensor element is micro-machined from a single silicon crystal and then electrostatically bonded to form a parallel plate capacitive device. In conjunction with the electronic circuitry, the result is a low-noise sensor which responds to accelerations down to the dc, range up to 400 Hz and provides stable damping characteristics and a large bandwidth. The integral electronics provides a high level output signal over a wide temperature range at a very low output impedance. The MICROTRON is a low-g high-sensitivity accelerometer intended for applications to monitoring of trajectories, structural flutter phenomena, motion studies in man-machine systems, and biodynamic interference. Its design features a pair of silicon elements arranged in a differential bridge configuration that uses gas damping to maximize frequency response. The gas damping permits a stable damping coefficient over a wide temperature range. Full-scale ranges are typically ± 10 g or ± 30 g. A partial data list for model 7290 is given in Table 3.2.

Table 3.2. Performance Data of a Variable Capacitance Wideband Accelerometer (Endevco *Microtron*)

Parameter	Performance	
Range	g	±10
Sensitivity (at 10 VDC excitation)	mV/g	200±5
Nonlinearity & Hysteresis (% of reading, max to full range)	%	±1
Frequency response (±5% max, ref 100 Hz)	Hz	1 to 400
Mounted Resonant frequency	Hz	2700, typical
damping ratio	0.7 ±0.2	
Transverse Sensitivity	%	1
Thermal sensitivity shift (-25 deg to +75 deg C)	%	±2
Zero measurand output (Typical/maximum)	mV	±20/±40
Thermal zero shift (-25 deg C to +75 deg C)	% F.S.	±2
Zero shift due to 10000 g shock (max)	% F.S.	0.1
Noise (0 to 600 Hz)	mV rms	3
Excitation voltage	VDC	13.0 to 18.0
Input current (max)	mA	15
Warm-up time	μsec	1

The physical dimensions of the MICROTRON type 7290 are given in Figure 3.26.

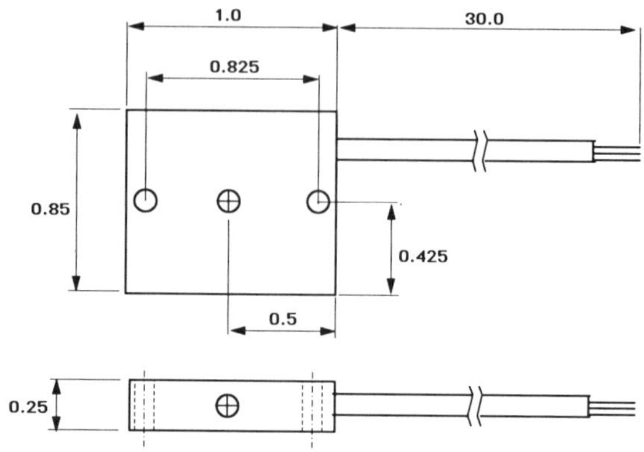

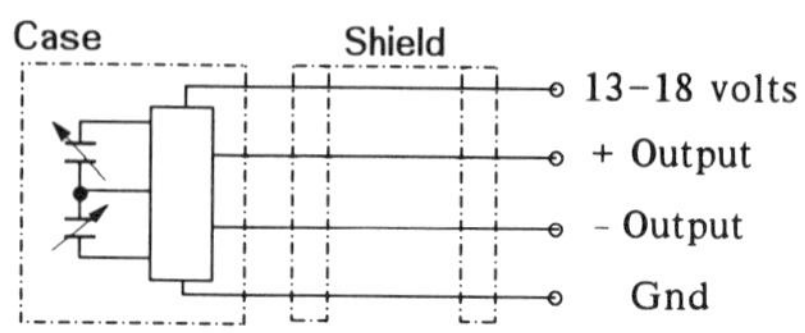

All dimensions in inches

Figure 3.26. A variable capacitance bridge accelerometer (Endevco MICROTRON type).

Problems

Problem 3.1— Accelerometer bandwidth: Effect on estimated range by integration

An accelerometer aligned with the longitudinal axis of a missile flying a horizontal flat trajectory is intended for determining its range by double integration of the measured acceleration $\ddot{X}_m$. The dynamics of the accelerometer is modeled as a first-order lag such that $\ddot{X}_m = 1/(1+s\tau)\ddot{X}$. $\ddot{X}$ is described by a step acceleration lasting Tsec.

a) Formulate the dynamic error $\epsilon_d = \ddot{X} - \ddot{X}_m$ and the expression for the range error $\Delta X(T)$ just after the Tsec boost period. Does it increase during $t > T$?

b) The initial acceleration $\ddot{X} = 40$ g and the boost duration $T = 10$ sec. What are the estimated range and range error after t=40 sec ?

Problem 3.2— Accelerometer error: Aircraft roll angle estimation

A pair of nominally identical accelerometers are mounted on the wing tips of an aircraft with their input axes normal to the wing plane.

a) Describe the appropriate schematic for extracting the normal acceleration a_z and the angular acceleration $\ddot{\phi}$ for two mounting versions:. 1) The input axes are parallel. 2) The input axes are in opposition.

b) Is there any advantage of version 1 over 2, or vice versa, with respect to feasibility or errors?

c) Is it necessary to compensate for the effect of gravity?

d) The respective scale factor errors of the accelerometers are ϵ_1 and ϵ_2. What is the nature of their effect on the estimate of ϕ?

d) The roll angle is determined by double integration of $\ddot{\phi}$. To what value of Δa must the pair be matched so that the error $\Delta\phi$ does not exceed 1 deg after 10 min of flight? In light of the results, discuss the practicality of the concept considering the fact that the null point uncertainty of a good inertial grade accelerometer is ~50 μg.

Problem 3.3— Accelerometer reading: On pendulum

An accelerometer A_x is attached to a thin rigid rod of length l, pivoted to act as a pendulum as shown in the following figure.

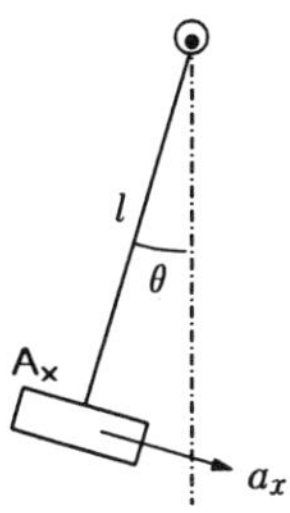

The pendulum swings in accordance with its length l with initial condition $\theta(0)=\theta_0$ and $\dot{\theta}(0)=0$.

a) Develop the expression for the accelerometer reading $a_x = a_x(t)$.
b) Give an intuitive interpretation of the result.

Problem 3.4– Accelerometer controlled platform: Properties

An accelerometer A_x is mounted on a gimbaled single axis platform. It responds to both an inclination angle θ and an acceleration component a_x. Its output is amplified by a gain factor A and fed into the corresponding torque motor M_y having a torque constant b_y. (The index y indicates that M_y is mounted on the y–axis).

a) Draw a block diagram of this servoed system indicating all the parameters and variables involved.

b) Formulate the differential equation that defines the dynamics of the system. Present its solution with initial conditions $\theta(0)=\theta_0$ and $\dot{\theta}(0)=0$. Assume small angles for θ.

c) Present the solution and interpret the result. What is the response $\theta(t)$ of a horizontal acceleration a_X in Earth axes.

d) How does the platform respond to a constant accelerometer bias Δa?

e) Can you set a suitable value for A so that the platform becomes insensitive to a_x? Discuss and interpret the result in physical terms.

Problem 3.5– Accelerometer noise: Prefiltering

The noise characteristics of an accelerometer are like those of the Microtron model 7290 as given in Table 3.2. The data given there determine the power spectral density (PSD) of the noise.

a) Compute its value using the data given in Table 3.2.

It is necessary to integrate the output of the accelerometer in order to estimate the velocity of the vehicle.

b) What is the velocity error due to the noise after one hour of flight?

The error is found to be twice as large as the tolerable error. An apparently obvious remedy is to prefilter the accelerometer output e.g., by a first–order low–pass filter $a/(a+s)$.

c) Investigate this possibility with a as a parameter and decide if, and to what extent, this noise prefiltering idea is effective.

d) Does the prefiltering affect the precision of the estimated velocity? Develop the expression for a given acceleration impulse $a\Delta t$ = 100 m/sec.

e) Give a qualitative interpretation for your findings.

Problem 3.6– Accelerometer design: Open–and closed–loop versions

The principal data of an accelerometer are:

Proof mass m = 1 grf

Natural frequency ω_n = 200 $\sec^{-1}$

a) What is the deflection of m if the accelerometer is placed on the table with its input axis vertical?

b) What is the value of the spring constant K_S?

It is required to increase the natural frequency of the accelerometer ten fold by placing it in a force balance loop.

c) What is the required loop gain that includes all the elements, namely, the detector sensitivity, the amplifier gain, and the transducer?

d) Now what is the mass deflection under the same situation as in a)?

e) This accelerometer is mounted in an aircraft that accelerates at 2 g along a straight climb trajectory at 30 deg. The input axis is along the climb trajectory.

f) What is the mass deflection under these conditions?

Problem 3.7– Accelerometer reading: Under steady climb

An accelerometer is mounted in an aircraft along its longitudinal axis. The aircraft climbs at a steady speed V m/sec flight path angle of γ rad.

a) What is the reading of the accelerometer?

b) What is the effect of variations in V on the accelerometer reading?

c) What is the effect of variations in vehicle weight?

Problem 3.8– Accelerometer reading: under free fall in the atmosphere

An aerodynamically shaped load is released from a stationary balloon toward a vertical drop. The mass m of the load and its drag coefficient D are given. The load is equipped with an accelerometer aligned with the longitudinal axis of the load.

a) Calculate the accelerometer reading versus time. Explain intuitively the nature of the result. Compare with the actual acceleration history of the load.

b) What is the accelerometer reading history if the drag is essentially zero? Compare with the actual acceleration history.

Problem 3.9– Accelerometer noise: effect on velocity and position estimation

Processing the output noise n of an accelerometer reveals a PSD of the general nature shown in the linear frequency scale sketch:

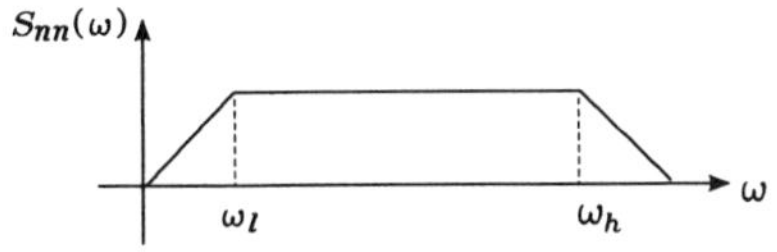

a) Formulate a continuous Markov-type PSD that roughly represents this sketch. It is implied that $\omega_h \gg \omega_l$.

b) Determine the expression for the power P_n. Inspect the effects of ω_l and ω_h on P_n.

The accelerometer output is integrated once to estimate the velocity v.

c) What is the nature of the velocity error $\tilde{v}$ due to n? Is it a stationary process? What is the value of $P_{\tilde{v}}$?

d) What would be the nature of the corresponding position error $\tilde{x}$? Is it a stationary process? Express $P_{\tilde{x}}(t)$. If not, what is the law of its divergence process?

Problem 3.10 – Null point errors: Dead space versus bias

An accelerometer may be deficient as a result of 1) a bias or 2) as a result of a dead space. An aircraft flies through a turbulent atmosphere which excites random lateral accelerations modeled to a good approximation as a first order Markov process. The accelerometer output is integrated twice to provide a position estimate. Clearly, both error sources cause a position error.

a) Develop models for the error that results from cases 1 and 2.

b) Discuss the nature of the error as a function of time and argue which one is more detrimental than the other and under what assumptions on the error time history.

Problem 3.11 – Null point offset: High-pass filtering

A null-point offset is often considered to be a principal error source. In principle, it is possible to block the bias by placing a high pass filter at the output of the accelerometer.

a) Is this idea sound? If so, under what assumptions and circumstances?

b) What would be the effect of the high-pass filter in INS applications?

Discuss these issues and draw a conclusion.

References

1. Åström, K.J., and Hector, N.F., "Vertical Indication with a Physical Pendulum Based on Electromechanical Synthesis of High Moment of Inertia", Royal Institute of Technology, Stockholm 1959.

2. Norling, B.L., "An Overview of the Evolution of Vibrating Beam Accelerometer Technology," Symposium on Gyro Technology, Stuttgart, Germany, 1991.

3. Hayes, H.C., "Method and Apparatus for Determining the Force of Gravity," U. S. Patent 1,995,305, March 1933.

4. Serra, N.R. "Technical Report on the Quartz Resonator Gigital Accelerometer," *Inertial Navigation Systems and Components Proceedings*, AGARD Conference Proceeding, AGARD-CP-43, Advisory Group for Aerospace R&D (AGARD of NATO, May 1968), pp. 487–516.

5. Macomber, G.R., and Fernandez, M., *Inertial Guidance Engineering*, Book, Prentice-Hall, Englewoods Cliffs, 1962.

6. Rao, S.S., and Sunar, M., "Piezoelectricity and Its Use in Disturbance Sensing and Control of Flexible Structures: A Survey," *Applied Mechanics Review*, Vol. 47, No. 4, April, 1994, pp. 113–123.

7. Doebelin, E.O., *Measurement Systems, Applications and Design*, fourth edition, McGraw-Hill, New York, 1990.

4

Inertial Rotation Sensors

4.0 Introduction

This chapter discusses some of the most important gyroscopic rotation sensors that have been in use throughout this century. The central issues are their mathematical model and characteristic error sources, which allow the prediction of their performance in various applications and environments. The knowledge of the angular attitude of a flight vehicle with respect to an external reference axis system is of paramount importance in the realization of flight control, attitude estimation and control, and automatic navigation systems. The typical quantities to be measured are the vehicle attitude Euler angles Φ, Θ, and Ψ, defined with respect to the local level and local north and the inertial vehicular angular rates P, Q, R. These quantities are depicted in Figure 4.1. The transformation of any vector from Earth axes X_e, Y_e, Z_e to body axes x_b, y_b, z_b is performed by the Euler transformation $[E] = [\Phi][\Theta][\Psi]$, formulated in detail in Appendix A4.

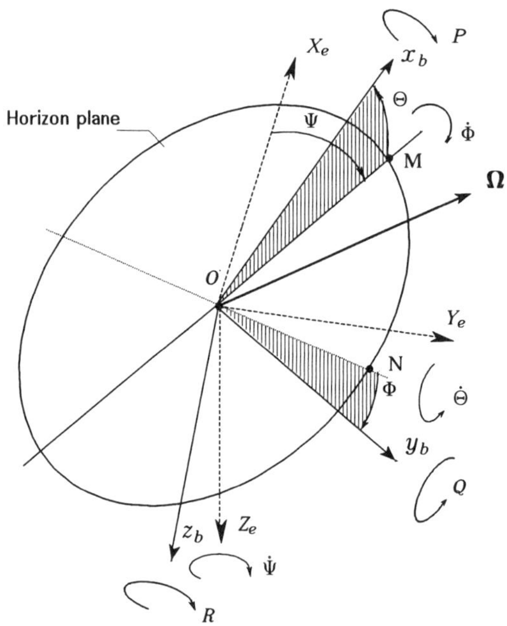

Figure 4.1. Relation of body axes and Earth axes.

It is important to realize at the outset that the inertial angular rates P, Q, R are not identical with the Euler angular rates $\dot{\Phi}$, $\dot{\Theta}$, $\dot{\Psi}$. The former triplet represents the projections of the total inertial angular rate vector $\boldsymbol{\Omega}$ on the vehicle body axes x_b, y_b, z_b while the latter triplet represents the Euler angular rates $\dot{\Phi}$, $\dot{\Theta}$, $\dot{\Psi}$ around the axes OM, ON and Z_e respectively. The kinematic relationships between P, Q, R and $\dot{\Phi}$, $\dot{\Theta}$, $\dot{\Psi}$ are known as the Euler differential equations presented in, among others, [1] or [2]. They are repeated here for the sake of completeness.

$$P = \dot{\Phi} - \dot{\Psi}\sin\Theta \tag{4.0.1}$$

$$Q = \dot{\Theta}\cos\Phi + \dot{\Psi}\cos\Theta\sin\Phi \tag{4.0.2}$$

$$R = -\dot{\Theta}\sin\Phi + \dot{\Psi}\cos\Theta\cos\Phi \tag{4.0.3}$$

Alternatively, solving for $\dot{\Phi}$, $\dot{\Theta}$, $\dot{\Psi}$ in terms of P, Q, R yields

$$\dot{\Phi} = P + Q\sin\Phi\tan\Theta + R\cos\Phi\tan\Theta \tag{4.0.4}$$

$$\dot{\Theta} = Q\cos\Phi - R\sin\Phi \tag{4.0.5}$$

$$\dot{\Psi} = Q\frac{\sin\Phi}{\cos\Theta} + R\frac{\cos\Phi}{\cos\Theta} \tag{4.0.6}$$

The last three nonlinear equations are of particular technical importance because they permit the computation of the Euler angles Φ, Θ, Ψ given their initial conditions Φ_0, Θ_0, Ψ_0 and using the measurements P, Q, R provided by the triad of rate gyros. Thus,

$$\Theta = \Theta_0 + \int_0^t \dot{\Theta}dt \tag{4.0.7}$$

$$\Phi = \Phi_0 + \int_0^t \dot{\Phi}dt \tag{4.0.8}$$

$$\Psi = \Psi_0 + \int_0^t \dot{\Psi}dt \tag{4.0.9}$$

The Euler angles can be derived simultaneously from the measured inertial angular rate components P, Q, R and the computational algorithm based on (4.0.4) to (4.0.9). However, the measured angular rates P, Q, R invariably incorporate errors known as drift rates. Therefore, the integrations of (4.0.5) to (4.0.9) result in time-diverging angular errors. For this reason, the method based on the integration of the Euler equations is applicable only in either high-precision and costly inertial guidance technology or in very short-term tactical missile guidance applications.

Traditionally, the roll and pitch angles Φ, Θ are measured by a vertical free gyroscope, while the heading angle Ψ is measured by a directional free gyroscope, a compass, or a magnetometer. The inertial angular rates *P*, *Q*, *R* are measured by an orthogonal triad of restrained angular rate gyroscopes. Accordingly, we distinguish between these two classes of gyroscopes, namely, *attitude gyros* and *rate gyros*. The first class, ideally freely suspended, operates on the principle of conservation of angular momentum which, is rooted in Newton's first law, namely that the angular momentum vector of a spinning rotor will be maintained in magnitude and direction in inertial space, unless perturbed by external torques. Ideally, this would be accomplished with a perfectly friction-free suspension of the spinning rotor such as electrostatic or electromagnetic suspension in a perfect vacuum (Autonetics or Honeywell electrostatically suspended gyro (ESG)). In reality, however, residual torques, such as those due to bearing friction or mass imbalance, are inevitable. These bring about a change in the angular momentum vector,which manifests itself by the precession of the rotor spin axis. Consequently, the direction in inertial space of a free gyro drifts away from its original angular orientation.

In the second class, namely the mechanical rate gyros, the rotor assembly is restrained to the vehicle frame and is not freely suspended. It therefore is subject to the vehicle angular rates *P*, *Q*, or *R*, which induce changes in the angular momentum of the rotor. In analogy to Newton's second law, these rates of change in angular momentum manifest themselves as torques proportional to their magnitude. As will be shown later, these torques are balanced by springs that deflect proportionally so that the corresponding angular rate components *P*, *Q*, *R* are measured and displayed. Parasitic torques resulting from friction in the suspension bearings clearly add to the spring-balancing torques and introduce drift rates much the same as in the free gyros. In order to reduce these drift rates, especially in high-precision inertial navigation systems, formidable efforts have been invested in the past four decades in the development of friction-free suspensions. These yielded a large variety of ingenious technologies, some of which will be addressed in the course of this chapter. The problem of determining the angular attitude of a flight vehicle has been a challenge for decades. The traditional common, but disliked, use of the vertical gyro operates satisfactorily under ordinary flight conditions, i.e., shallow flight paths with relatively small values of Θ. This restriction is evident from (4.04)–(4.06), which indicates singularity or gimbal lock if $\Theta \rightarrow \pm\pi/2$. Its advantage is low cost and bounded error. However, it is relatively bulky, consists of numerous moving parts, and is not sufficiently reliable. Typical values of mean time between failure (MTBF) are 300–500 hours. On the other hand, an attitude measurement unit based on rate gyros and angular rate integration has better inherent reliability, but it requires very costly rate gyros, which are prohibitive in many aerospace applications. A very good and detailed account of traditional electromechanical gyro technology, covering the period up to the early 1970s is given in [12]. Ongoing research [4], [5] attempts to provide improved performance, reliability and more cost-effective solutions. Some of these will be addressed in Chapters 6 and 9. Spacecraft control poses a different set of challenges both at the mission and technology levels. A modern comprehensive account of spacecraft sensors and control is given in [6].

4.1 The free gyroscope

Basic mechanization and dynamics

The mechanization of the vertical gyro is schematically shown in Figure 4.2.

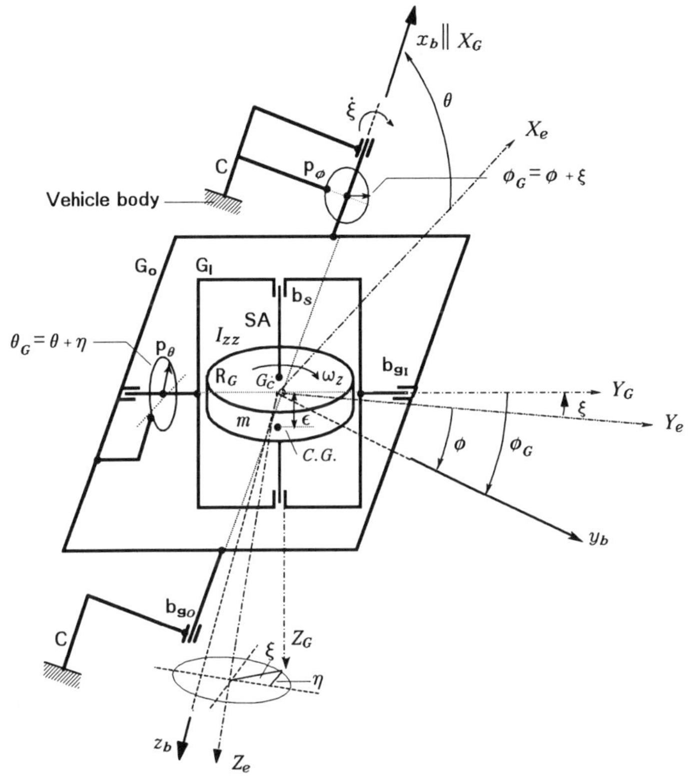

Figure 4.2. Schematic of free vertically oriented gyroscope.

The rotor R_G, vertically suspended by the spinner bearing b_s in the inner gimbal G_I, spins at the angular rate of ω_z rad/sec. G_I, in turn, is suspended in the outer gimbal G_0 by the inner gimbal bearing b_{gI}. G_0 is suspended by the outer gimbal bearing b_{g0}, which is affixed to the vehicle body-mounted casing C. The gimbal mount center G_c is the intersection of the two gimbal axes as indicated in Figure 4.2. Ideally, the center of gravity C.G. of the entire mount of mass m should coincide with G_c. In reality, it is displaced by ϵ, shown as vertical only, and causes a mass imbalance. This two-axis mechanization should ideally isolate R_g from possible angular vehicle body motion P, Q, R. As stated in Section 4.0, if b_{gI} and b_{go} are absolutely friction-free, and if the suspended rotor-gimbal assembly is perfectly balanced and no other external torques are present, the spin axis SA of R_g will maintain the initial vertical or any other initial orientation indefinitely. The pitch and roll angles Θ and Φ

are measured by the angular position pickoffs p_θ and p_ϕ, respectively. These too, must be friction-free. Therefore they are invariably realized by synchros. In those applications where Θ and Φ are used for Euler transformations, the linear pickoffs are replaced by resolvers. This case will be demonstrated by Example 4.1–1.

Torque and precession relationships

Denote the moment of inertia tensor of R_G by

$$J=\begin{bmatrix} I_{xx} & J_{xy} & J_{xz} \\ J_{yx} & I_{yy} & J_{yz} \\ J_{zx} & J_{zy} & I_{zz} \end{bmatrix} \tag{4.1.1}$$

where I_{xx} etc. and J_{xy} are the principal and cross moments of inertia, respectively. Denote the total angular rate of R_G by

$$\Omega_G \triangleq [P_G, Q_G, R_G]^T \tag{4.1.2}$$

Its total angular momentum is

$$H=J\Omega_G \tag{4.1.3}$$

and its vectorial time rate of change which determines the corresponding torque vector, is

$$T=\dot{H}+\Omega_G\times H \tag{4.1.4}$$

Due to their symmetry with respect to the x, y, z axes, all the cross inertias in J are zero, which reduces to the diagonal matrix

$$J=\begin{bmatrix} I_{xx} & 0 & 0 \\ 0 & I_{yy} & 0 \\ 0 & 0 & I_{zz} \end{bmatrix} \tag{4.1.5}$$

Due to the mechanical isolation of R_G, only the R_G component of Ω along its z axis is of significance. Denoting $R_G=\omega_z$, (4.1.3) specializes to

$$H=\begin{bmatrix} I_{xx} & 0 & 0 \\ 0 & I_{yy} & 0 \\ 0 & 0 & I_{zz} \end{bmatrix}\begin{bmatrix} 0 \\ 0 \\ \omega_z \end{bmatrix}=\begin{bmatrix} 0 \\ 0 \\ I_{zz}\omega_z \end{bmatrix} \tag{4.1.6}$$

Since I_{xx} = const., and so is ω_z by its constant speed motor drive, we have in (4.1.4),

$$\dot{H}=0 \tag{4.1.7}$$

Denoting $I_{zz}\omega_z \triangleq H_z$, the vector product $\Omega\times H$ is given by

$$T=\Omega_G\times H=\begin{vmatrix} \mathrm{i} & \mathrm{j} & \mathrm{k} \\ P_G & Q_G & R_G \\ 0 & 0 & H_z \end{vmatrix}$$

$$=\mathrm{i}(Q_G H_z-0.R_G)-\mathrm{j}(P_G H_z-0.R_G)+\mathrm{k}(P_G 0-0.Q_G)$$

$$=H_z(\mathrm{i}Q_G-\mathrm{j}P_G) \tag{4.1.8}$$

where, i, j, k are unit vectors in the corresponding x, y, z directions. Thus, the torque vector is

$$T\triangle\begin{bmatrix} T_x \\ T_y \end{bmatrix}=\begin{bmatrix} H_z Q_G \\ -H_z P_G \end{bmatrix} \tag{4.1.9}$$

from which

$$Q_G=\frac{T_x}{H_z} \tag{4.1.10}$$

$$P_G=-\frac{T_y}{H_z} \tag{4.1.11}$$

The last two equations disclose that a torque around the x axis, T_x invokes a precession rate component Q_G around the y axis, while a torque T_y around the y axis invokes a precession rate component P around the x axis. Conversely, an external torque exerted around the x axis causes a precession rate Q around the y axis, causing the reaction torque $H_z Q_G$ which balances the external torque T_x. Clearly, Q_G, T_x, H_z and P_G, T_y, H_z constitute orthogonal vector components, as indicated in Figure 4.3 (a) and (b) respectively.

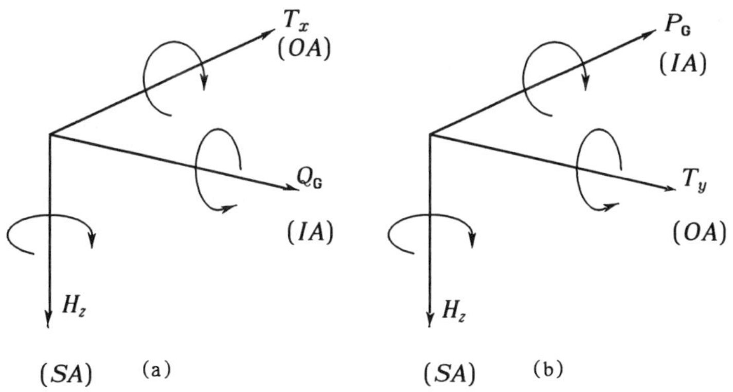

Figure 4.3. Orientation of spin, input, and output axes and corresponding precession torques.

Effects of mechanical imperfections

Equations (4.1.10) and (4.1.11) provide the basis for the gyro performance in terms of its long term drift rate and expresses the significance of the principal parameters that determine it. These parameters are:

- the angular momentum H_z;
- gimbal bearing friction;
- mass unbalance of the gimbal-rotor suspension.

Clearly, the tendency should be to make the rotor moment of inertia I_z and the spin rate ω_z as large as possible while keeping the bearing friction and the mass imbalance as small as possible. Some of these requirements are mutually contradictory. Thus, increasing the rotor mass with a given suspension technology causes an increase in the bearing friction torque, and with a given level of manufacturing precision increases the mass unbalance torque as well. Thus, just increasing the rotor mass does not ensure improved performance. Clearly, the dominant parameters contributing to long term stability are the spin rate ω_z and the radius of gyration of the rotor r_g which increases I_z by its square. This observation indicates that the rotor design should permit the largest concentration of its mass close to its outer rim. This is indeed the case in gyroscopic rotor design.

EXAMPLE 4.1–1: Attitude determination by vertical gyro and magnetometer

As stated earlier, the principal task of the VG is to provide pitch and roll information θ and ϕ. For the most part, this information is used directly as the essential measurement for aircraft pitch and roll control by an autopilot. In order to provide complete attitude information, the third angle, namely, heading ψ, must be independently measured. A widely used method is by means of a compass, a flux valve, or a body fixed (strapdown) magnetometer. A compass, or flux valve, however, is inferior if the aircraft is intended for intense maneuvers, in which case θ or ϕ can be quite large so that conventional gimbaling schemes become ineffective. We therefore consider the strapdown magnetometer as the preferred sensor. The method is based on the measurement of the projections of the local Earth magnetic field vector H by an orthogonal triad of magnetic sensors. The method is complicated by the fact that H is inclined downwards by an angle λ, which varies across the surface of the globe. In intermediate latitude zones, a typical value for λ is -40°. The vector representation of H in Earth coordinates in the X_e, Z_e plane coincides with the magnetic north and the local vertical. In accordance with the following sketch,

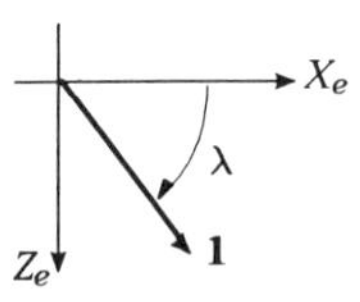

1 is defined by

$$(c\lambda \;\; 0 \;\; s\lambda)^T = \mathbf{1}$$

The measured components of H in the rolled and pitched magnetometer axes are,

$$\begin{bmatrix} H_x \\ H_y \\ H_z \end{bmatrix} = [E] \begin{bmatrix} c\lambda \\ 0 \\ s\lambda \end{bmatrix} = \begin{bmatrix} c\psi\, c\theta & s\psi\, c\theta & -s\theta \\ c\psi\, s\theta\, s\phi - s\psi c\phi & s\psi s\theta s\phi + c\psi c\phi & c\theta s\phi \\ c\psi s\theta c\phi + s\psi s\phi & s\psi s\theta c\phi - c\psi s\phi & c\theta c\phi \end{bmatrix} \begin{bmatrix} c\lambda \\ 0 \\ s\lambda \end{bmatrix}$$

$$= \begin{bmatrix} c\psi c\theta c\lambda - s\theta s\lambda \\ c\psi s\theta s\phi c\lambda - s\psi c\phi c\lambda + c\theta s\phi s\lambda \\ c\psi s\theta c\phi c\lambda + s\psi s\phi c\lambda + c\theta c\phi s\lambda \end{bmatrix} \tag{4.1-1}$$

The naively computed value ψ_c, which considers only the measured values of H_x and H_y, as if the aircraft were precisely horizontal, is given by

$$\tan\psi_c = -\frac{H_y}{H_x} = \frac{s\psi c\phi c\lambda - c\theta s\phi s\lambda - c\psi s\theta s\phi c\lambda}{c\psi c\theta c\lambda - s\theta s\lambda} \tag{4.1-2}$$

The minus sign follows from the definitions of axes and signs of angles as shown:

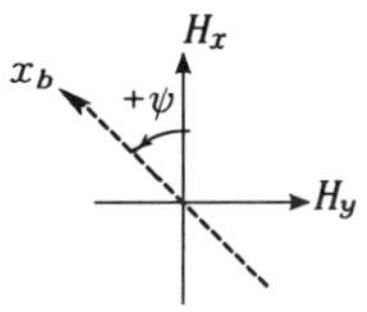

Assume the following numerical values: $\theta = 10^0$; $\phi = 20^0$; $\lambda = 40^0$; Thus, $s\psi = c\psi = 0.707$; $\psi = 45^0$; $s\phi = 0.342$; $c\phi = 0.942$; $s\theta = 0.174$; $c\theta = 0.984$; $s\lambda = 0.64$; $c\lambda = 0.766$. Substituting these values into (4.1–2), we have

$$\tan\psi_c = 0.615;\ \psi_c = 31.5^0 \tag{4.1-3}$$

Therefore, the error is $\Delta\psi = 45^0 - 31.5^0 = 14.3^0$. This is a substantial error even though θ and ϕ are quite moderate. Indeed, if $\theta = \phi = 0$, from (4.1–2), we would have

$$\tan\psi_c = -\frac{H_x}{H_y} = \frac{s\psi c\lambda}{c\psi c\lambda} = \tan\psi \tag{4.1-4}$$

which is exactly correct. In order to derive the correct answer from the magnetometer readings, it is necessary to derotate the aircraft by θ and ϕ. This is done by inverse transforming $(H_x, H_y, H_z)^T$ as follows:

$$[\theta]^{-1}[\phi]^{-1} \begin{bmatrix} H_x \\ H_y \\ H_z \end{bmatrix} = [\theta]^T[\phi]^T \begin{bmatrix} H_x \\ H_y \\ H_z \end{bmatrix} = \begin{bmatrix} H_x' \\ H_y' \\ H_z' \end{bmatrix} \tag{4.1-5}$$

θ and ϕ are assumed to be available from the VG pickoff readings. $(H_x', H_y', H_z')^T$ are the derotated magnetometer readings visualized by the following piogram:

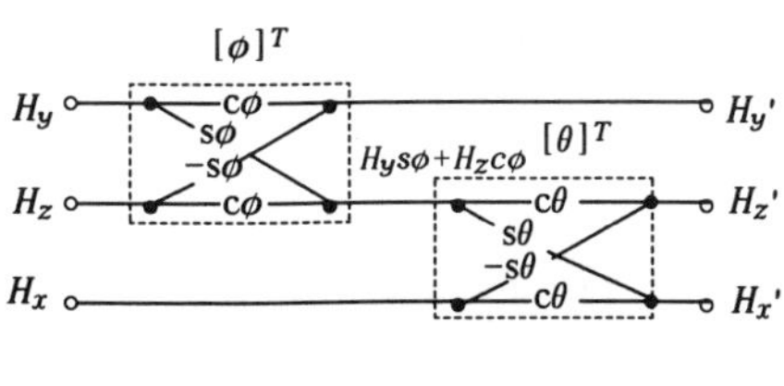

It is evident that in addition to H_x, H_y, the vertical component H_z is required in order to implement the transformation algorithm.

The algorithm that provides the complete set of Euler angles ϕ, θ, ψ was described in Figure 3.6. The folllowing diagram describes the Euler transformation operation by means of electromechanical resolvers that replace the linear pickoffs p_θ, p_ϕ. This subsystem demonstrates that the combination of a vertical gyro and a three-axis magnetometer can provide complete attitude information even for intensely maneuvering aircraft where large values of θ and ϕ are often involved.

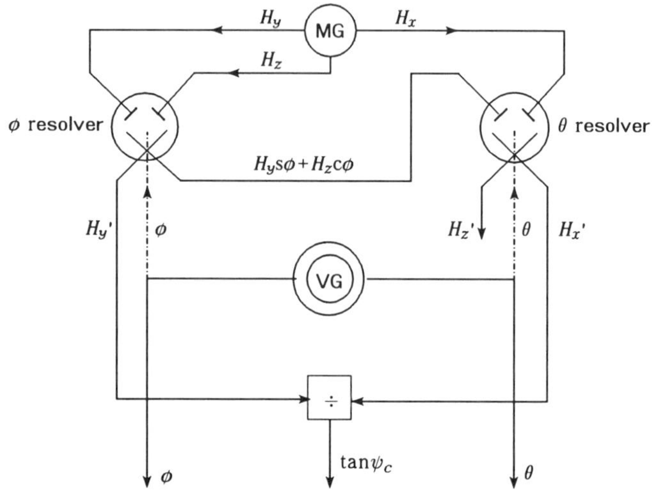

Figure 4.4. Implementation of axes rotation ϕ, θ by means of a vertical gyro and resolvers.

EXAMPLE 4.1–2: Free gyro in a satellite, effect of linear acceleration on spin axis

A free gyro is mounted in a satellite: The weight of the rotor and shaft is $w_r = 400$ gr-f. Its radius of gyration r_g is 2 cm. Its spin rate is $n = 24000$ rpm. The weight of the gimbals is $w_g = 200$ gr-f; The center of gravity of the rotor-gimbal assembly C_{G_0} is below $C.G.$ by $\epsilon = 0.1$ mm. The satellite uniformly changes its inertial speed along X by 100 m sec^{-1} over a time interval of 1 min. What is the magnitude and direction of the angular drift at the end of the acceleration period?

The mass of the rotor-gimbal assembly is

$$m = \frac{w_r + w_g}{981} = \frac{400+200}{981} = 0.611 \text{ gr cm}^{-1}\text{ sec}^2;$$

$$I_z = \frac{w_r}{g} r_g^2 = \frac{400}{981} 2^2 = 1.631 \text{ gr cm sec}^2;$$

$$\omega_z = \frac{2\pi n}{60} = \frac{6.28 \times 24000}{60} = 2512 \text{ rad sec}^{-1};$$

$$H_z = I_z \omega_z = 1.631 \times 2512 = 4095 \text{ gr cm sec};$$

The satellite acceleration is $a_x = 120/60 = 2$ m sec^{-2} $= 200$ cm sec^{-2}; the torque due to mass imbalance T_y is

$$T_y = -ma_x\epsilon = -0.611 \times 200 \times 0.01 = -1.222 \text{ gr cm};$$

In accordance with (4.1.11), the resulting roll rate $\dot{\xi}$ is

$$\dot{\xi} = -\frac{T_y}{H_z} = -\frac{-1.222}{4095} = 2.98 \times 10^{-4} \text{ rad sec}^{-1} = 1.51 \times 10^{-2} \text{ deg/sec};$$

By the end of the 60 sec interval, the corresponding drift angle ξ is

$$\xi = 1.51 \times 10^{-2} \times 60 = 1.025^0;$$

EXAMPLE 4.1–3: Free gyro in an aircraft – coning motion due to linear acceleration

The same gyro is now mounted in an aircraft that flies a geospherical trajectory at a constant altitude. The Newtonian torque T_y is the same as in Example 4.1–1. However, due to the angular deviation ξ and the action of the weight of the gimbal mount $w = w_r + w_g$, a restoring torque around the X_G axis, $T_x = w\epsilon\sin\xi$ exists. We therefore have

$$T_x = gm\epsilon\sin\xi$$

$$T_y = -a_x m\epsilon + gm\sin\eta\epsilon$$

Substituting into (4.1.10) and (4.1.11), assuming small angles, $\sin\xi \cong \xi$ and $\sin\eta \cong \eta$, and observing the conventions of a right-hand axis system,

$$\dot{\eta} = \frac{gm\epsilon\xi}{H_z} = \frac{gm\epsilon\int\dot{\xi}dt}{H_z}$$

$$\dot{\xi} = \frac{am\epsilon\eta}{H_z} - \frac{gm\epsilon\int\dot{\eta}dt}{H_z}$$

Laplace transforming these equations yields

$$\dot{\eta}(s) = \frac{gm\epsilon}{H_z}\frac{\dot{\xi}}{s} = \alpha\epsilon\frac{\dot{\xi}}{s}$$

$$\dot{\xi}(s) = \frac{a_x m\epsilon}{H_z} - \frac{gm\epsilon}{H_z}\frac{\dot{\eta}}{s} = \frac{a_x}{g}\frac{gm\epsilon}{H_z} = \delta\alpha\epsilon - \alpha\epsilon\frac{\dot{\eta}}{s}$$

where $\frac{gm}{H_z}\Delta\alpha$ and $\frac{a_x}{g}\Delta\delta$. Multiplying through by s and rearranging, we have,

$$\dot{\xi}(s) = \delta(s)(\alpha\epsilon)\frac{s^2}{s^2 + \alpha^2\epsilon^2}$$

$$\dot{\eta}(s) = \delta(s)(\alpha\epsilon)^2\frac{s}{s^2 + \alpha^2\epsilon^2}$$

Retaining the assumptions that ξ and η are small angles, for the corresponding transfer functions, we have

$$\xi(s) = \delta(s)(\alpha\epsilon)\frac{s}{s^2 + \alpha^2\epsilon^2}$$

$$\eta(s) = \delta(s)(\alpha\epsilon)^2\frac{1}{s^2 + \alpha^2\epsilon^2}$$

The corresponding impulse responses are

$$\xi(t) = \delta(\alpha\epsilon)\cos(\alpha\epsilon)t$$

$$\eta(t) = \delta(\alpha\epsilon)\sin(\alpha\epsilon)t$$

Describing the onset of a_x as a step function, the corresponding step responses are

$$\xi(t) = \delta \sin(\alpha\epsilon)t$$

$$\eta(t) = \delta(1 - \cos(\alpha\epsilon)t)$$

The interpretation of this result is a conical circular motion of amplitude δ and of frequency $\alpha\epsilon$ rad/sec around the mean angular deviation δ. Clearly, $\xi(0)=0$ and $\eta(0)=0$. The frequency $\alpha\epsilon$, which should be made as small as possible, is directly proportional to ϵ and inversely proportional to H_z. With the given numerical values in this example, we have

$$\delta = \frac{a_x}{g} = \frac{200}{981} = 0.204 \text{ rad} = 11 \text{ deg}$$

$$\alpha\epsilon = \frac{gm\epsilon}{H_z} = \frac{(w_r + w_g)\epsilon}{H_z} = \frac{600 \times 0.01}{2050} = 0.00292 \text{ rad/sec.}$$

The corresponding period is $T = 2\pi/0.00292 = 2150$ sec.

It is easily verified that if the eccentricity ϵ indicated in Figure 4.2 changes its sign, the resulting solution is a hyperbolic cosine indicating a periodic divergence of ξ and η.

EXAMPLE 4.1–4: Effect of gimbal bearing friction

The effect of gimbal bearing friction can be assessed by the following exemplary assumptions: The radius of the inner gimbal bearing b_{g_1} is $r_{b_1} = 1$ cm. The friction coefficient is $\mu_b = 0.001$; the corresponding friction torque T_f, with the assembly weight of 600 gr is

$$T_f = (w_r + w_g) r_{b_1} \mu_b = 600 \times 1 \times 0.001 = 0.6 \text{ gr cm.}$$

As a result of, for example, random angular pitch motion, T_f behaves like a binary zero-mean random process with an exponential autocorrelation time of $T = 4$ sec. In accordance with Chapter 2, the resulting angular roll drift is given by

$$\sigma_\xi(t) = \frac{T_f T}{H_z}\sqrt{t};$$

By the end of 60 sec, the accumulated expected roll error σ_ξ would be

$$\sigma_\xi(120) = \frac{0.6 \times 4}{2050}\sqrt{60} = 9 \times 10^{-3} \text{ rad} = 0.52 \text{ deg.}$$

The preceding example demonstrates that the realization of a free gyroscope, either vertical or directional, by means of conventional electromechanical technology, would result in intolerable drift errors in ξ and η easily amounting to many degrees per hour. Furthermore, for regular geospherical flight, the relevant angular reference is the local vertical, which varies substantially with respect to inertial space both by virtue of Earth rotation and the motion of the aircraft. Similarly, the spin axis of a directional gyroscope, designed to measure azimuth with respect to the geographical north, must be aligned with the local horizontal plane. Again, due to Earth rotation, motion of the aircraft, or both, this alignment cannot be maintained for any length of time. Therefore, additional control elements must be incorporated to render the free gyroscope as a useful device. This issue is discussed in the following section.

4.2 The vertical gyroscope

Description and imposing the vertical

The obvious approach to enforce the direction of the local vertical (LV) on the spin axis is to bring to bear the gravitational pull of Earth. The simplest and least expensive method would be a bob weight attached to the inner gimbal G_I, that is, to intentionally introduce a mass imbalance. Obviously, this approach renders the gyroscope sensitive to horizontal components of acceleration and may therefore cause undamped precession of the spin axis and lead to substantial deviations from the local vertical, as indicated by the angles ξ and η in Figure 4.2. The conventional method for imposing the vertical is described in Figure 4.5. The inner gimbal G_I is equipped with two perpendicular level sensors l_ξ and l_η that sense the deviations ξ and η. The corresponding signals $s_{l_1}\xi$ and $s_{l_2}\eta$ are fed into the amplifiers A_y and A_x, which drive the torque motors M_y and M_x. The torque coefficients of the motors are b_y and b_x. s_{l_1} and s_{l_2} are the scale factors of the level sensors l_ξ and l_η in volt/deg.

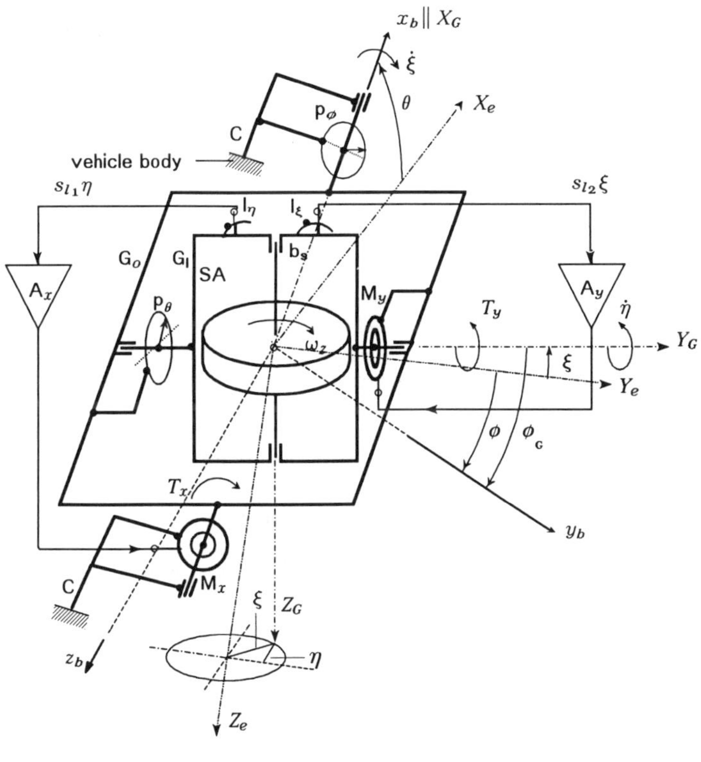

Figure 4.5. Imposing the direction of the vertical by level sensors and servo-control of the gimbal system.

The corresponding torques T_y and T_x induce the precession rates $\dot{\xi}$ and $\dot{\eta}$ in the direction opposite to the deviations ξ and η so as to annul them. This mechanization imposes a gravity controlled two–axis *erection mechanism.* The dominant reaction torques to T_y and T_x are $H_z\dot{\xi}$ and $H_z\dot{\eta}$ respectively. The relatively small Newtonian reaction torques $I_x\ddot{\xi}$ and $I_y\ddot{\eta}$ are disregarded in the following analysis. Under this assumption, the dynamics around the X_G and Y_G are not coupled, and each can be treated separately. For this purpose, the rotor R_g is depicted in Figure 4.6 as viewed along the X_G axis.

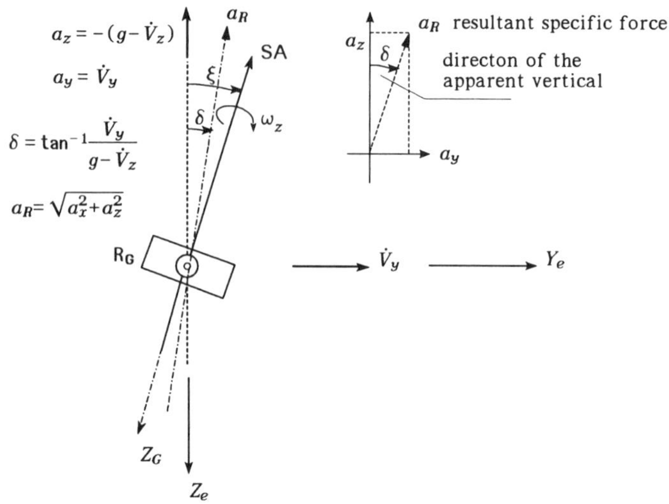

Figure 4.6. Principal variables and parameters defining spin axis dynamics.

Dynamical model and erection process

In the following, we refer to the direction of the spin axis SA. For modeling the gyro dynamics, the following notations are used:

$\dot{\xi}_I$ – angular rate with respect to inertial space.

$\dot{\xi}_v$ – angular rate of the local vertical with respect to inertial space.

$\dot{\xi}$ – angular rate with respect to the local vertical.

$\dot{\xi}_d$ – drift rate of the spin axis due to extraneous torques.

ε – null point error of the level sensor.

β – reading of the leveling sensor.

δ – angle of the apparent vertical.

The leveling sensor is a bubble level filled with a conducting fluid, as depicted in Figure 4.7. The position of the bubble, measured by electrodes to be described later, is affected both by the tilt ξ and a_y, namely, δ. The reading, e.g., of l_x under the influence of ξ and δ is indicated in Figure 4.7. The position of the bubble affects the resistance of the paths **a–c** and **c–b**, where the electrode **c** plays the role of the sliding contact of a potentiometer.

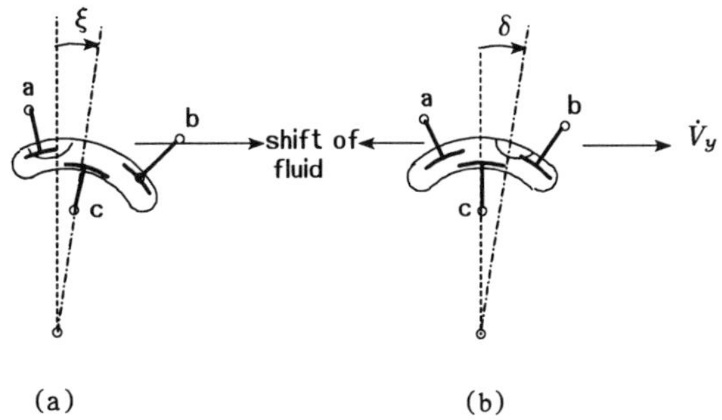

Figure 4.7. Bubble position under influence of: (a) tilt ξ; (b) force angle δ.

The total angular displacement of the bubble in l_ξ is therefore $\xi-\delta$, and its reading is

$$\beta = s_l(\xi - \delta - \varepsilon) \tag{4.2.1}$$

In principle, l_ξ is similar to a pendulous device in the sense that it responds both to Newtonian acceleration, and tilt with respect to the local horizontal. However, unlike an accelerometer, which has its own internal restoring force, it becomes totally ineffective in the outer space environment.

The torque exerted by motor M_y is

$$T_y = -\beta b_y A_y \tag{4.2.2}$$

$\dot{\xi}_l$ is given by

$$\dot{\xi}_l = \dot{\xi} + \dot{\xi}_v \tag{4.2.3}$$

The total torque is given by

$$H_z\dot{\xi}_l = T = T_y + T_d \; (+I_y\ddot{\eta}_l) \tag{4.2.4}$$

where T_d is the disturbance torque due to bearing friction and mass imbalance and $I_y\ddot{\eta}_l$ is disregarded. Dividing through by H_z, we have

$$\dot{\xi}_I = \frac{T_y}{H_z} + \frac{T_d}{H_z} = -\frac{\beta b_y A_y}{H_z} + \dot{\xi}_d \tag{4.2.5}$$

where $\dot{\xi}_d \triangleq T_d/H_z$. Define

$$K \triangleq \frac{s_l b_y A_y}{H_z} \tag{4.2.6}$$

and substitute (4.2.1) and (4.2.3) into (4.2.5):

$$\dot{\xi} + \dot{\xi}_v = -K(\xi - \delta - \varepsilon) + \dot{\xi}_d \tag{4.2.7}$$

Rearranging and solving for ξ by Laplace transforming while considering $\dot{\xi}_v$ and $\dot{\xi}_d$ as constants, we have

$$\xi(s)(s+K) = K[\delta(s) + \varepsilon(s)] - \dot{\xi}_v + \dot{\xi}_d \tag{4.2.8}$$

$$\xi(s) = \frac{K}{s+K}[\delta(s) + \varepsilon(s)] - \frac{\dot{\xi}_v}{s+K} + \frac{\dot{\xi}_d}{s+K} \tag{4.2.9}$$

In the steady state, i.e., for $s \to 0$,

$$\xi = \delta + \varepsilon - \frac{\dot{\xi}_v}{K} + \frac{\dot{\xi}_d}{K} \tag{4.2.10}$$

Disregarding the errors due to the last three terms, we conclude that *SA* aligns itself with the apparent vertical given by δ and not with the true vertical.

Similar expressions can be readily obtained for the Y_G axis, namely, η. The deviation ξ consists of four principal components due to:

1. horizontal acceleration.
2. null-point error.
3. rotation of the LV in inertial space due to rotation of Earth and motion of the aircraft with respect to Earth.
4. drift of the gyro *SA* due to mechanical imperfections.

We observe that the implementation of the gravity-control system causes the closed-loop system to behave as a first-order dynamical system with the erection time constant

$$\tau_e = K^{-1} \triangleq \frac{H_z}{s_l b_y A_y} \tag{4.2.11}$$

Its effect, as seen in (4.2.10), is to bound the errors in ξ due to $\dot{\xi}_v$ and $\dot{\xi}_d$ to finite values as determined by K in the last two terms. Design practice shows that the various parameters that determine τ_e can be chosen so that the total steady-state deviation ξ does not exceed $\sim 1^\circ$. In the choice of K, we have two conflicting requirements:

1. In the expression for the response to δ, disregarding ε, we have

$$\xi(s) = \frac{K}{s+K}\delta(s) \tag{4.2.12}$$

The corresponding time response is

$$\xi(t) = \delta(1-e^{-Kt}) = \delta(1-e^{-t/\tau_e}) \tag{4.2.13}$$

It is desirable to select the smallest possible value for K in order to achieve effective low-pass filtering of $\delta(s)$ due to sudden changes in horizontal acceleration or due to atmospheric turbulence.

2. On the other hand, to keep the steady-state errors due to (4.2.10) as small as possible, K should be made as large as possible. Thus, the choice of K must be a compromise between these conflicting requirements.

The process of erection to the local vertical can be directly derived from (4.2.9). Assume that δ, $\dot{\xi}_v$, $\dot{\xi}_d$ are all zero. Let the initial value of ξ be $\xi(0)=\xi_o$ and $\dot{\xi}_o=0$. Thus,

$$\xi(s) = \frac{\xi_o}{s+K} \tag{4.2.14}$$

and

$$\xi(t) = \xi_o e^{-Kt} = \xi_o e^{-t/\tau_e} \tag{4.2.15}$$

The erection time constant τ_e is commonly chosen in the range of 1—5min. Yet, for the initial erection process, τ_e can be reduced to a substantially smaller value and then adjusted to the larger nominal value for normal operation. In reality, as seen by (4.2.9), $\xi(t)$ converges to the constant residual errors

$$\begin{aligned}\lim_{t\to\infty}\{\xi(t)\} &= \lim_{t\to\infty}\{\xi_o e^{-t/\tau_e} + \delta(1-e^{-t/\tau_e}) + (\dot{\xi}_v+\dot{\xi}_d)\tau_e\} \\ &= \delta + (\dot{\xi}_v+\dot{\xi}_d)\tau_e \end{aligned} \tag{4.2.16}$$

4.3 Error sources in the vertical

Errors due to gyro drift rate

Gyro drift rate components $\dot{\xi}_d$ and $\dot{\eta}_d$ cause deviations from the vertical as shown in the following example.

EXAMPLE 4.3-1: Determination of drift error in a vertical gyro

The technical data of a vertical gyroscope are:
- weight of the rotor w_r=400 gr-f;
- weight of the gimbal mount w_g = 200 gr-f;
- radius of gyration of the rotor: r_g=2 cm;
- spin rate n = 24000 rpm = 400 rps = 2512 rad/sec;
- moment of inertia of the rotor $I_z = (w_r/g)r_g^2 = (400/981)\times 2^2 = 1.63$ gr-f cm sec^2;
- angular momentum $H_z = I_z\omega_z = 1.63\times 2512 = 4094$ gr-f cm sec;

- horizontal mass eccentricity $\varepsilon = 0.1$ mm;
- torque coefficient of $\mathbf{M}_x$ and $\mathbf{M}_y$, $b_y = 350$ gr-f cm/amp;
- sensitivity of the levels l_ξ and l_η $s_l = 20$ mv/deg $= 1.146$ volt/rad
- The required erection time constant $\tau_e = 2\,\text{min} = 120$ sec;

It is required to determine the amplifier gain A_y.

$$\tau_e = \frac{H_z}{s_l b_y A_y}$$

$$A_y = \frac{H_z}{s_l b_y \tau_e} = \frac{4094}{1.146 \times 350 \times 120} = 85 \text{ mA/volt;}$$

The mass unbalance torque due to ε,

$$T_d = (w_r + w_g)\varepsilon = 600 \times 0.01 = 6 \text{ grf cm}$$

The resulting gyro drift rate is

$$\dot{\xi}_d = \frac{T_d}{H_z} = \frac{6}{4094} = 1.46 \times 10^{-3} \text{ rad/sec} = 5.04 \text{ deg/min}$$

The angular deviation due this drift rate is,

$$\xi_d = \dot{\xi}_d \tau_e = 5.04 \times 2 = 10.08 \text{ deg;}$$

Essentially, identical expressions can be derived for the error η_d and the corresponding design parameters around the Y_G axis.

This result demonstrates substantial deviation from the true vertical and illustrates the high sensitivity of the gyro precision to a seemingly small mechanical imperfection ε. In accordance with (4.2.10), the angular rate of the local vertical contributes to the angular deviation as well. Its magnitude depends on the flight path direction and the vehicle ground speed, which adds vectorially to the tangential velocity of Earth at any given point on its surface. It may manifest itself either in η_v or ξ_v or both. This example indicates that the precision of the gyro is not solely determined by its electromechanical precision and design parameters but also by its flight conditions such as position, heading, and flight velocity. Recalling (4.2.9), through δ, its precision is even more affected by its acceleration and atmospheric turbulence.

Elimination of drift rate errors by integral control

It is possible to eliminate the effects of $\dot{\xi}_d$ and $\dot{\xi}_v$ in the steady state by introducing integral control in the erection loop. A pure integrator is placed in parallel with the gain as shown in the following block diagram. The subscript y in A_y is dropped here because the same considerations apply to η.

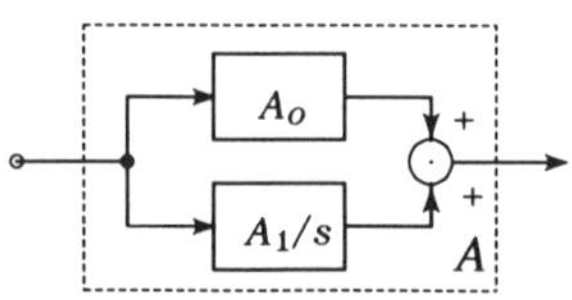

Thus, the substitution of $A = A_o + A_1/s$ into (4.2.6) modifies K to

$$K = K_o + \frac{K_1}{s} = K_o \frac{s + K_1/K_o}{s} = K_o \frac{s+z}{s} \tag{4.3.1}$$

where $z \triangleq K_1/K_o$. Substituting K from (4.3.1) into (4.2.9), we have

$$\xi(s) = \frac{\dfrac{K_o(s+z)}{s}}{s + \dfrac{K_o(s+z)}{s}}(\delta + \epsilon) - \frac{1}{s + \dfrac{K_o(s+z)}{s}}\dot{\xi}_v + \frac{1}{s + \dfrac{K_o(s+z)}{s}}\dot{\xi}_d$$

$$= \frac{K_o(s+z)}{s^2 + K_o s + K_o z}(\delta + \epsilon) - \frac{s}{s^2 + K_o s + K_o z}\dot{\xi}_v + \frac{s}{s^2 + K_o s + K_o z}\dot{\xi}_d \tag{4.3.2}$$

Conclusion: The dynamics are modified to second order, and the steady- state error due to constant values of $\dot{\xi}_v$ and $\dot{\xi}_d$ vanishes.

In any case, $\dot{\xi}_v$ and $\dot{\xi}_d$ are, at most, very slow variables, and their effect is almost eliminated by the action of the integral control. The well damped low-pass filtering operation on $\delta + \epsilon$ can essentially be retained, provided the zero z is adequately placed.

Errors induced by off-great-circle motion

In addition to the effect of δ due to horizontal centripetal or linear accelerations, additional accelerations exist that are due to the sphericity and rotation of Earth. One such effect, which is purely kinematic, occurs when the flight path of an aircraft is not on a great circle. Consider the spherical geometry described in Figure 4.8.

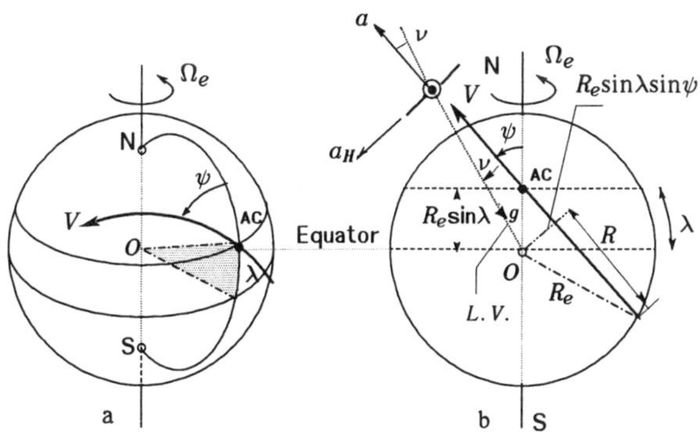

Figure 4.8. Horizontal lateral acceleration induced by non-great-circle flight.

The aircraft **AC** flies a circular path that happens to cross latitude λ at the heading ψ with respect to the local longitude line. Its ground speed is V m sec^{-1} and its altitude is $h \ll R_e$. The rotation of Earth Ω_e is initially disregarded.

The following geometrical relationships apply:

$$R_e^2 - R_e^2 \sin^2\lambda \sin^2\psi = R_e^2(1 - \sin^2\lambda \sin^2\psi) = R^2 \tag{4.3.3}$$

From Figure 4.8, we have

$$\sin\nu = \frac{R_e \sin\lambda \sin\psi}{R_e} = \sin\lambda \sin\psi \tag{4.3.4}$$

The circular path of the aircraft is of radius R. Therefore, the centripetal acceleration is

$$a = \frac{V^2}{R} = \frac{V^2}{R_e(1-\sin^2\lambda \sin^2\psi)^{1/2}} \tag{4.3.5}$$

which is coplanar with the small circle of radius R of the flight trajectory. The component perpendicular to the local vertical is

$$a_H = a \sin\nu \tag{4.3.6}$$

Substituting (4.3.5),

$$a_H = \frac{V^2}{R} \sin\nu = \frac{V^2}{R_e} \frac{\sin\lambda \sin\psi}{(1 - \sin^2\lambda \sin^2\psi)^{1/2}} \tag{4.3.7}$$

If either λ or ψ is zero, the flight path is a great circle, lateral or longitudinal respectively, and $a_H = 0$.

The foregoing provides only the result for a_H due to the displacement $R_e \sin\lambda \sin\psi$ of the circular flight plane from the center of Earth.

Coriolis acceleration

For atmospheric flight, the rotation rate of Earth Ω_e cannot be disregarded. Together with V, it causes a dynamic force in the local horizontal plane known as the Coriolis acceleration, which occurs when a body moves relative to a rotating reference axis system. Since vehicular motion is defined with respect to the local horizontal plane, the components of the Coriolis accelerations referred to it need to be determined. The corresponding geometry is shown in Figure 4.9.

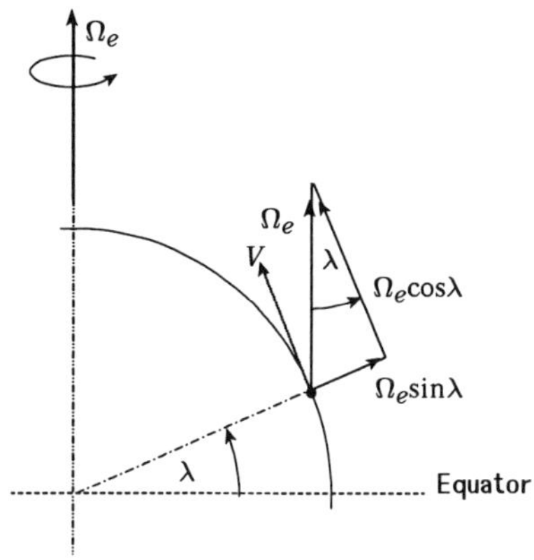

Figure 4.9. Geometry and motion underlying Coriolis acceleration.

In the simple case, if the aircraft flies due north, i. e., $\psi=0$, the Coriolis acceleration is given by the cross product [7] given as

$$a_c = 2\Omega_e \sin\lambda \times V \tag{4.3.8}$$

which is perpendicular to both $\Omega_e\sin\lambda$ and V. According to the conventions of the local axis system, $V>0$, $\lambda>0$ and $\Omega_e>0$, $a_c>0$, pointing to the east. In the more general situation $\psi\neq 0$, the corresponding vector product is determined in the local coordinate frame x, y, z by: $\Omega_e=[\omega_{e_x}, 0, \omega_{e_z}]^T$; $V=[v_x, v_y, v_z]^T$; where $\omega_{e_x}=\Omega_e\cos\lambda$; $\omega_{e_y}=0$; $\omega_{e_z}=\Omega_e\sin\lambda$; $v_x=V\cos\psi$; $v_y=V\sin(-\psi)$; $v_z=0$.

The vector product is:

$$a_c = 2\Omega_e \times V = 2\begin{vmatrix} \mathbf{i} & \mathbf{j} & \mathbf{k} \\ \omega_{e_x} & 0 & \omega_{e_z} \\ v_x & v_y & 0 \end{vmatrix} = 2(-\mathbf{i}\omega_{e_z}v_y + \mathbf{j}\omega_{e_z}v_x + \mathbf{k}\omega_{e_x}v_y) \tag{4.3.9}$$

The three Coriolis components are:

$$a_{c_x} = -2\omega_{e_z}v_y$$

$$a_{c_y} = 2\omega_{e_z}v_x$$

$$a_{c_z} = 2\omega_{e_x}v_y$$

Clearly, a_{c_z} is balanced by the local vertical lift force. It is readily verified that a_{c_x} and a_{c_y} are perpendicular components to v_y and v_x respectively.

EXAMPLE 4.3–2: Error in the vertical due to pseudo acceleration

The heading of an aerospace vehicle is $\psi = -45$ deg northwest and it just crosses the latitude $\lambda = 45$ deg north. Its ground speed is 900 m/sec, approximately Mach 3. What is the angle δ of the apparent vertical with respect to the true vertical?

We have $\sin\lambda = 1/\sqrt{2}$; $\sin\psi = -1/\sqrt{2}$; $R_e = 6380$ Km;

From (4.3.5), we have

$$a_H = \frac{V^2}{R_e} \frac{-1/2}{(1-1/4)^{1/2}} = 0.127 \times (-0.577) = -7.33 \times 10^{-2} \text{ m sec}^{-2}$$

which points in the negative direction of Y.

$$\tan\delta \cong \delta = 7.33 \times 10^{-2}/9.81 = 7.47 \times 10^{-3} \text{ rad} = 0.428 \text{ deg}$$

In principle, navigational data on V, λ, and ψ can be utilized to correct this error if system requirements warrant such a correction.

The Coriolis components are determined as follows.

The angular rate of Earth is,

$$\Omega_e = 360 \text{ deg}/(24 \times 3600 \times 57.3) = 7.27 \times 10^{-5} \text{ rad/sec};$$

$$v_x = 900\cos 45^0 = 636.3 \text{ m/sec}; \; v_y = 900\sin(-45^0) = -636.3 \text{ m/sec};$$

$$\omega_{e_x} = 7.27 \times 10^{-5}\cos 45^0 = 5.14 \times 10^{-5} \text{ rad/sec};$$

$$\omega_{e_z} = 7.27 \times 10^{-5}\sin 45^0 = 5.14 \times 0^{-5} \text{ rad/sec};$$

$$a_{c_x} = 2 \times (-5.14 \times 10^{-5}) \times (-636.3) = 6.54 \times 10^{-2} \text{ m/sec}^2 = 0.00666 \text{ g};$$

$$a_{c_y} = 2 \times 5.14 \times 10^{-5} \times 636.3 = 6.54 \times 10^{-2} \text{ m/sec}^2 = 0.00666 \text{ g}.$$

We therefore have

$$\delta_x = \frac{a_{c_x}}{g} = 0.00666 \text{ rad} = 0.38 \text{ deg}; \quad \delta_y = \frac{a_{c_x}}{g} = 0.00666 \text{ rad} = 0.38 \text{ deg};$$

$$|\delta| = (\delta_x^2 + \delta_y^2)^{1/2} = 0.53 \text{ deg}$$

The vectorial presentation in the local horizontal plane is shown in the following figure:

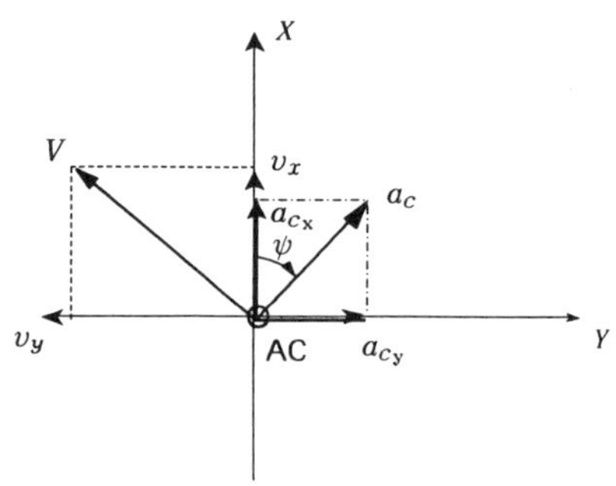

Conclusion: The combined effect of the kinematic and Coriolis accelerations can cause deviations of δ, and thus in ξ, up to ~ 1 deg for high speed vehicles.

The cut-out mechanism

Equations (4.212) and (4.2.13) imply that if a horizontal acceleration persists for a substantial length of time the spin axis *SA* gradually aligns itself with δ so that very substantial errors in the indicated vertical will occur. This can be avoided if the erection mechanism is disabled as long as the horizontal acceleration persists. The scheme that provides this function is known as the *cutout mechanism.* Its principle of operation is best explained by recalling (4.2.12), namely, $\xi(s) = K/(s+K)\delta(s)$, which can be interpreted as a first-order servo loop shown in Figure 4.10.

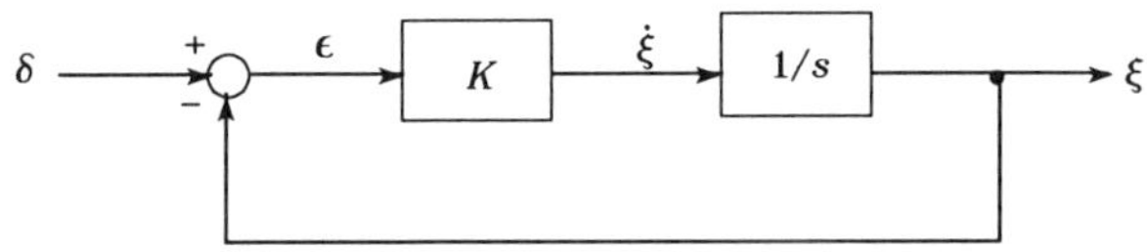

Figure 4.10 . Erection control loop interpreted as a first-order closed-loop system.

From the diagram, we have

$$K\epsilon = K(\delta - \xi) = K\left(1 - \frac{K}{s+K}\right)\delta = \frac{Ks}{s+K}\delta \tag{4.3.10}$$

where $\delta-\xi$ is the position of the bubble in accordance with Figure 4.10. ϵ, the reading of the level sensor l_ξ is determined by

$$\epsilon = \frac{s}{s+K} \tag{4.3.11}$$

For a step function in δ, that is, a sudden onset of horizontal acceleration,

$$\epsilon(t) = \delta e^{-Kt} \tag{4.3.12}$$

Thus, initially, ϵ assumes the full value of δ, but as ξ approaches δ, the bubble gradually returns to its null point position, indicating the apparent vertical. The cutout mechanism, designed to avoid this, is schematically described in Figure 4.11(a). A relay switch SW disconnects the amplifier A_x or A_y from the corresponding torque motor M_x or M_y once the absolute value of the low-pass filtered ϵ, ϵ^* exceeds the disconnect threshold ϵ^*_d of the relay. When the horizontal acceleration recedes, SW recloses once $|\epsilon^*| < \epsilon_c^*$, where ϵ^*_c is the connect threshold of the relay. Thus, for the duration of the acceleration, the gyro operates as a free gyro and is subject to the error due to drift, ξ_d. The history of these functional phases is described in Figures 4.11-(b), (c) and (d).

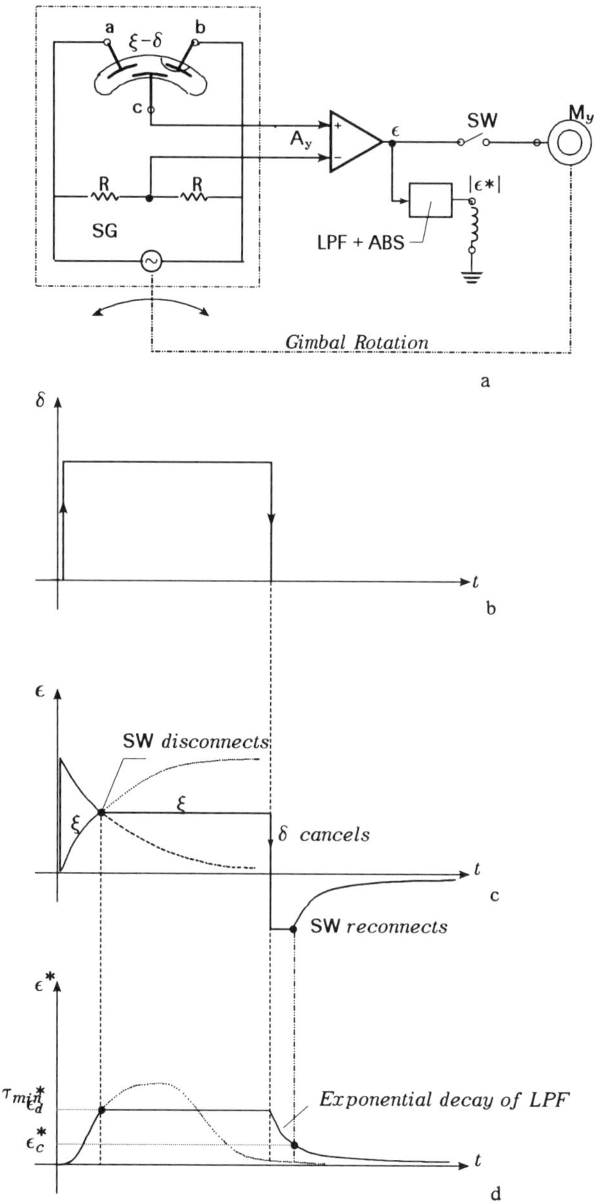

Figure 4.11. (a)– Block diagram of the cutout mechanism;
(b)– Time history of the horizontal acceleration δ;
(c)– Time history of the error ϵ;
(d)– Time history of the relay switching signal ϵ^*.

The LPF is required to prevent sporadic cutouts resulting from short acceleration impulses. The filter is so designed that it will permit ϵ^* to exceed ϵ_d only if δ persists for some minimum time e.g., $\tau_{\min} \cong 10$ sec.

Errors due to atmospheric turbulence

The horizontal components of wind gusts v_{gx} and v_{gy} due to atmospheric turbulence excite accelerations in the horizontal plane a_x and a_y. The conventional mathematical model for the turbulence is a stationary zero mean first order Gauss-Markov process [8] described in terms of its autocorrelation function and self power-spectral density Section 2.8, in 2.8.1—2.8.3. This formulation underlies the Dryden model [9] for atmospheric turbulence which is essentially described as a spatial gust distribution [1], [9]. This model is characterized by its rms intensity σ_g and its spatial exponential autocorrelation length L. At moderate altitudes $L \cong 300$ meters. Along the flight path x, it is modeled by

$$\phi_{v_g v_g}(\xi) = \sigma_{v_g}^2 e^{-|\frac{\xi}{L}|} \tag{4.3.13}$$

An aircraft traveling at ground speed V traverses this gust profile. The distance ξ traversed over a time interval τ is $\xi = V\tau$. Substituting into the right-hand side of (4.3.1), we have

$$\phi_{v_g v_g}(\tau) = \sigma_{v_g}^2 e^{-|\frac{V\tau}{L}|} = \sigma_{v_g}^2 e^{-a|\tau|} \tag{4.3.14}$$

where $a \triangleq V/L$ rad/sec. The corresponding power density spectrum is

$$\Phi_{v_g v_g}(s) = \sigma_{v_g}^2 \frac{2a}{a^2 - s^2} \tag{4.3.15}$$

Thus, the aircraft motion at speed V transforms the spatial correlation model (4.3.1) into a temporal random process with a correlation time constant $T = L/V$ sec. The side gust v_g causes a side specific force, resulting in a lateral velocity v of the aircraft and a corresponding side acceleration $a_g = sv$. v derives from v_g by the lateral stability derivative $Y_v \triangleq (1/m)\partial Y/\partial v$ (sec^{-1}), of the aircraft [1]. $Y_v < 0$ since the force increment ∂Y is opposed to ∂v.

The approximate relationship between v and v_g is

$$v = \frac{-Y_v}{s - Y_v} v_g \tag{4.3.16}$$

Thus, the spectrum of the side acceleration a_g is

$$\Phi_{a_g a_g}(s) = -s^2 \Phi_{vv}(s) = -s^2 \frac{-Y_v}{(-Y_v + s)} \frac{-Y_v}{(-Y_v - s)} \sigma_{v_g}^2 \frac{2a}{(a+s)(a-s)}$$

$$= 2a\sigma_{v_g}^2(-Y_v)^2\left|\frac{s}{(-Y_v+s)(a+s)}\right|^2 \tag{4.3.17}$$

With $\delta = a_g/g$, it is rewritten as

$$\Phi_{\delta\delta} = \frac{2a\sigma_{v_g}^2(-Y_v)^2}{g^2}\left|\frac{s}{(-Y_v+s)(a+s)}\right|^2 \tag{4.3.18}$$

Its integration over s yields (Appendix A2)

$$\sigma_\delta^2 = \frac{1}{2\pi j}\int_{-j\infty}^{j\infty}\Phi_{\delta\delta}(s)ds = \frac{\sigma_{v_g}^2}{g^2}\frac{(-Y_v)^2 a}{-Y_v+a} \tag{4.3.19}$$

Based on (4.2.12), we have for the power spectral density of ξ

$$\Phi_{\xi\xi}(s) = \frac{(-K)^2}{(K+s)(K-s)}\Phi_{\delta\delta}(s)ds$$

$$= \frac{2a\sigma_{v_g}^2(-Y_v)^2(-K)^2}{g^2}\left|\frac{s}{(-Y_v+s)(a+s)(K+s)}\right|^2 \tag{4.3.20}$$

The integration of this power spectrum over s yields the mean square value of ξ, σ_ξ^2

$$\sigma_\xi^2 = \frac{1}{2\pi j}\int_{-j\infty}^{j\infty}\Phi_{\xi\xi}(s)ds = \frac{\sigma_{v_g}^2 a(-Y_v)^2K^2}{g^2(a+K)[-Y_v(-Y_v+a+K)+aK]} \tag{4.3.21}$$

The attenuation provided by the vertical gyro therefore is

$$\frac{\sigma_\xi}{\sigma_{v_g}} = \frac{\sqrt{a}(-Y_v)K}{g\sqrt{(a+K)[-Y_v(-Y_v+a+K)+aK]}} \tag{4.3.22}$$

This result is illustrated by the following numerical example, which permits some simplifying approximations and enhances insight.

EXAMPLE 4.3–3: Error in the vertical due to atmospheric turbulence

An aircraft travels at $V = 300$ m/sec. The correlation length of the atmospheric turbulence is $L = 300$ m and the gust rms value is $\sigma_g = 3$ m/sec. The erection time constant of the gyro, is $\tau_e = 50$ sec; The lateral force stability derivative of the aircraft is $Y_v = -0.2\ \text{sec}^{-1}$.

Determine σ_ξ/σ_{v_g} and the attenuation factor σ_ξ/σ_δ.

$a = V/L = 300/300 = 1$ rad/sec;

$K = 1/50 = 0.02\ \text{sec}^{-1}$;

Since $K \ll a$, (4.3.22) can be simplified and rewritten as follows:

$$\sigma_\xi \cong \frac{\sqrt{a}(-Y_v)K}{g\sqrt{a[-Y_v(-Y_v+a)]}}\sigma_{v_g} = \sqrt{\frac{-Y_v}{-Y_v+a}}\frac{K}{g}\sigma_{v_g} = \sqrt{\frac{-Y_v}{-Y_v+V/L}}\frac{K}{g}\sigma_{v_g}$$

The rms error σ_ξ tends to be smaller if:

1. Y_v, is smaller, i.e. the mass m of the aircraft is greater.

2. V, the ground speed of the aircraft, is higher.

3. K is smaller, i.e., the erection time constant τ_e is larger.

Substituting the numerical values

$$\frac{\sigma_\xi}{\sigma_{v_g}} = \sqrt{\frac{0.2}{0.2+1}} \times \frac{0.02}{9.81} \text{ rad/(m/sec)} = 0.048 \text{ deg/(m/sec)}$$

For $\sigma_{v_g} = 3$ m/sec, we have $\sigma_\xi = 3 \times 0.048 = 0.144$ deg.

From (4.3.19), the rms value of the bubble deviation is,

$$\sigma_\delta = \frac{\sigma_{v_g}(-Y_v)\sqrt{a}}{g\sqrt{-Y_v+a}} = \frac{3}{9.81}\frac{0.2\times 1}{\sqrt{0.2+1}} = 0.056 \text{ rad} = 3.18 \text{ deg}$$

For a Gaussian process, this implies peak-to-peak deviations of ~13 deg. From the foregoing, the angular attenuation factor is:

$$\frac{\sigma_\xi}{\sigma_\delta} = \frac{K}{\sqrt{-Y_v a}} = \frac{0.02}{\sqrt{0.2\times 1}} = 0.045$$

The results of this example illustrate the very effective filtering of the interferences due to turbulence provided by the gyro dynamics of the erection mechanism.

The erection mechanism of vertical gyroscopes does not necessarily involve liquid level sensors. In simpler designs, this sensor is often a mercury switch which makes or breaks the contact of the torque motors M_x and M_y to a suitable power supply. This causes the gimbal system to virtually rotate at all times at the full angular rate to and fro in a bang-bang fashion. Due to hysteresis and stiction effects, this technology, though less costly, is less precise. Another equivalent technology is the use of gravity switches, which have similar qualities.

In summary, since the vertical gyro is controlled by pendulous level sensors, one may wonder if and in what aspect it differs from a two-axis pendulum. Clearly, a pendulum would require heavy damping to prevent sustained coning oscillations, which would continue even after the disturbance recedes. But such damping in the suspension would degrade the isolation between the pendulum axes and the flight vehicle. In the gyroscopic mechanization, however, horizontal accelerations translate into angular rates, which stop once the disturbance recedes.

4.4 The directional gyroscope

Description

As shown in the previous section, the vertical gyroscope provides measurements only of the Euler roll and pitch angles Φ and Θ. Measurement of the azimuth angle Ψ with respect to the geographical north requires an additional nonmagnetic directional sensor commonly realized by a directional gyroscope. In principle, it is analogous to the vertical gyro. However, the spin axis SA of the rotor is designed to lie in the local horizontal plane. It also requires a two-gimbal suspension to assure its isolation from the vehicular angular motion. A schematic of its construction, along with notations, is shown in Figure 4.12.

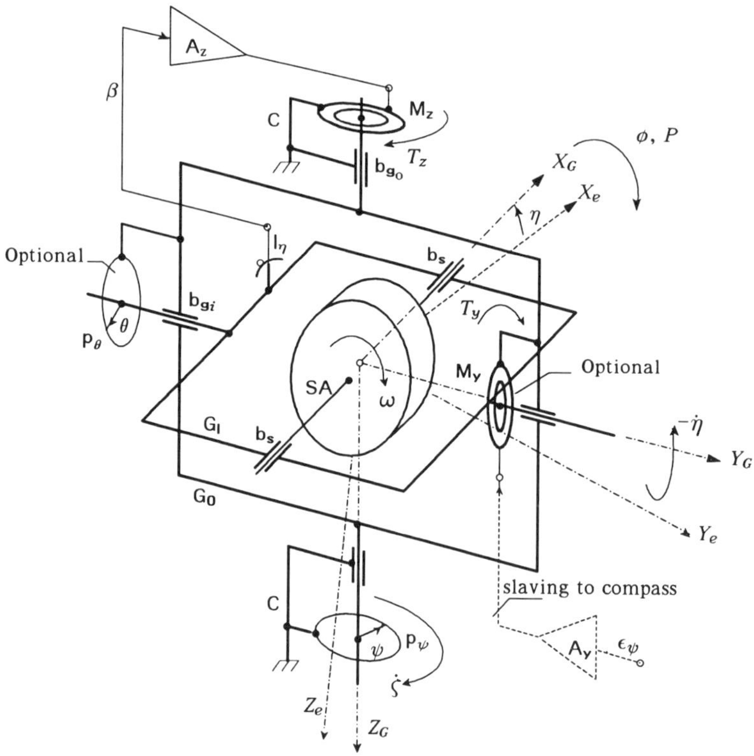

Figure 4.12. Schematic description of the directional gyro, including leveling mechanism and motor M_y for slaving to ψ_m.

Imposing the horizontal

The outer gimbal G_O is suspended from the gimbal bearings b_{go} placed in the vehicle body-mounted casing C. The inner gimbal G_I, supporting the rotor R_G by the spinner bearings b_s, is suspended by b_{gi}. This mechanization clearly isolates the gyro from the aircraft pitch and yaw motions Θ and Ψ but not from roll Φ. Through extraneous torques T_d, a drift rate $\dot{\eta}_d$ occurs, so that the spin axis SA tends to drift from the local horizontal plane by the angle η_d. The level sensor l_η, mounted in G_l, senses the total deviation $\beta = s_l(\eta - \delta - \epsilon)$, which feeds the current amplifier A_x driving the torque motor M_Z. Its torque T_z around the Z_G axis causes a precession rate $-\dot{\eta}$, which tends to annul η_d. The gyro is also quipped with an optional pitch pickoff p_θ and a torque motor M_y between G_o and G_l. Their purpose will be explained later. The basic closed-loop gyro dynamics is similar to that of the vertical gyro described in Section 4.3. With the torque coefficient b_z of M_Z, we have

$$T_z = -\beta b_z A_z \tag{4.4.1}$$

and in analogy to (4.2.3), the total inertial angular rate is

$$\dot{\eta}_I = \dot{\eta} + \dot{\eta}_v \tag{4.4.2}$$

The total torque is determined by

$$H_x \dot{\eta}_I = T = T_z + T_d \ldots \; (+I_z \ddot{\varsigma}) \tag{4.4.3}$$

The term in parentheses is disregarded because in practice, $I_z \ddot{\varsigma} \ll H_x \dot{\eta}$. Dividing through by H_x, we have

$$\dot{\eta} = \frac{T_z}{H_x} + \frac{T_d}{H_x} = -\frac{\beta b_z A_z}{H_x} + \dot{\eta}_d \tag{4.4.4}$$

Defining

$$K \triangleq \frac{s_l b_z A_z}{H_x} \tag{4.4.5}$$

we have, in analogy to (4.2.5),

$$\dot{\eta} + \dot{\eta}_v = -K(\eta - \delta - \epsilon) \tag{4.4.6}$$

which, after Laplace transforming and rearranging, yields the solution

$$\eta(s) = \frac{K}{s+K}[\delta(s) + \epsilon(s)] - \frac{\dot{\eta}_v}{s+K} + \frac{\dot{\eta}_d}{s+K} \tag{4.4.7}$$

Effects of motion and drift rate

The directional gyro, like the vertical gyro, is subject to horizontal accelerations and to the interferences due to the drift rate $\dot{\eta}_d$ and the rotation $\dot{\eta}_v$ of the local vertical in inertial space. To overcome the effect of δ, a cutout mechanism similar to the one shown in Figure 4.11 can be used. $\eta(s)$, which is around the Y_G axis, does not directly cause a deviation ς, so that the precision of Ψ is not directly affected. However, friction torques in b_{g_i} around Y_G and torques due to mass imbalance combine to cause the drift rate $\dot{\varsigma}_d$, which may accumulate to long-term deviations in ς and, thus, directly affect the precision of Ψ.

Effects of aircraft angular rates

Since the gyro mechanization does not isolate the gimbal system from roll variations, additional error sources result from the aircraft roll ϕ in combination with the ensuing rate of turn $\dot{\Psi}$. Moreover, aircraft roll rate P, in combination with η, directly contributes to $\dot{\varsigma}$. The effect due to ϕ is described by means of Figure 4.13-(a) and (b).

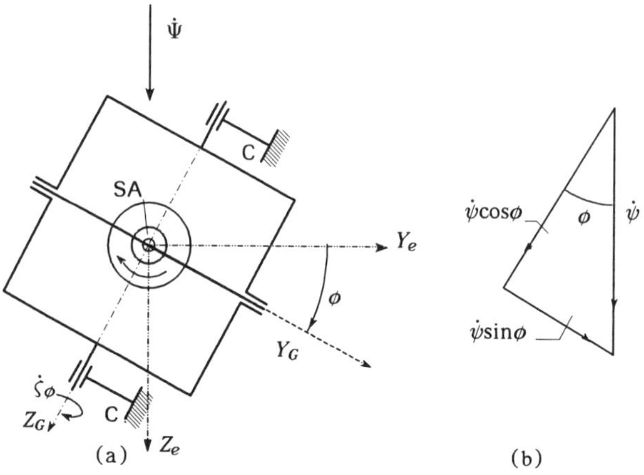

Figure 4.13. Effect of a turn maneuver on the directional gyro
(a) Rear viewof rolled gyro; (b) components of $\dot{\psi}$.

The angular rate component $\dot{\Psi}\sin\phi$ is perpendicular to SA and thus to H_x. Their cross-product causes a precession torque T_{Z_ϕ} around the Z_G axis,

$$T_{Z_\phi} = \dot{\Psi}\sin\phi \times H_x \tag{4.4.8}$$

The resulting precession rate is

$$\dot{\varsigma}_\phi = \frac{T_{Z_\phi}}{H_x} = \dot{\Psi}\sin\phi \tag{4.4.9}$$

Recalling that in a coordinated turn of an aircraft, $\dot{\Psi} = g\tan\phi / V$ [1], this error can be canceled by a compensating current fed into $\mathbf{M}_z$, which produces the torque

$$T_c = \dot{\Psi}\sin\phi \times H_x = \frac{g\tan\phi\sin\phi \times H_x}{V} \tag{4.4.10}$$

where ϕ is obtained from the vertical gyro, H_x is known, and V is provided by the air data system.

An additional error results from the roll rate P of the aircraft in conjunction with the deviation from the horizontal η as described with the aid of Figure 4.14.

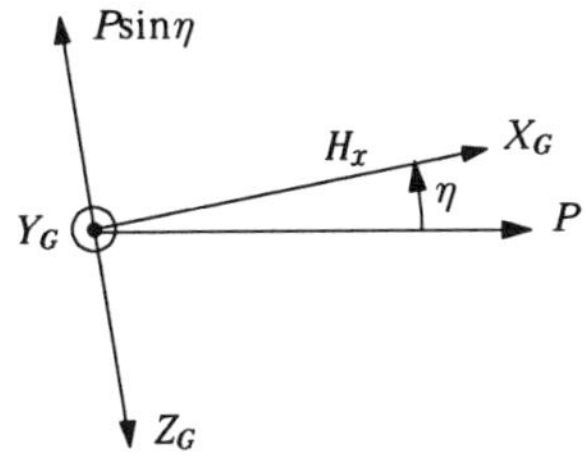

Figure 4.14. Effect of roll rate P on the directional gyro.

The roll component $P\sin\eta$ is perpendicular to H_x and thus produces a torque T_{y_p} around the Y_G axis,

$$T_{y_p} = P\sin\eta \times H_x \tag{4.4.11}$$

The corresponding precession rate $\dot{\varsigma}_p$ around Z_G is T_{y_p}/H_x, thus

$$\dot{\varsigma}_p = P\sin\eta \tag{4.4.12}$$

This error cannot be compensated because η is not measurable; however, roll rates persist only for relatively short time intervals. Thus, if η is on the order of 1 deg or less, the accumulated error can be tolerable in aircraft that exercise only moderate roll rates. For example, for P=0.1 rad/sec, $\eta = 1$ deg and since $\sin\eta \cong \eta$, it follows that $\dot{\varsigma}_p \cong 0.1$ deg/sec. If P persists for, say, 10 sec, thus accumulating a roll angle of $\phi = 1$ rad=55.3 deg, the directional error will be $\varsigma = 0.1$ deg/sec$\times$10 sec = 1 deg. This error is tolerable because eventually the accumulated roll ϕ is annulled by a subsequent negative P to restore level flight. However, this numerical example also indicates that the directional gyro is not applicable in fighter-type aircraft where intense and sustained roll rates P often occur.

Slaving to a compass

From the foregoing, it follows that the error ς in the directional gyroscope is quite susceptible to dynamic and kinematic interferences and that it cannot provide acceptable precision in the autonomous mode described so far. In fact, it is not equipped with an external directional reference measurement as in the case of the vertical gyro which is aided by the gravitation vector sensed by means of its level sensors. To provide the equivalent of such aiding for the directional gyroscope, a magnetic sensor such as a compass or magnetometer is employed, and the directional gyro is slaved to it. Such slaving prevents long-term divergence of ς but limits the precision of Ψ to that of the magnetic device. The method by which this is achieved, is described with the aid of Figure 4.15(a) and (b).

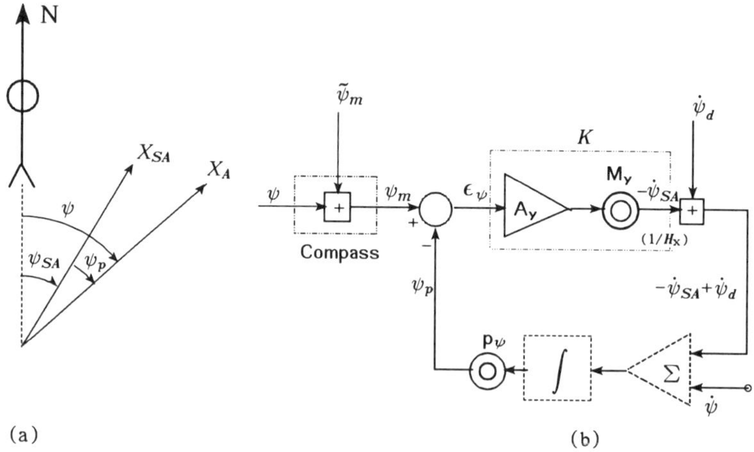

Figure 4.15. Slaving of a directional gyro to a compass
(a) Geometry in the horizontal plane;
(b) Implementation of slaving to compass.

In Figure 4.15(a), ψ is the actual heading angle of the longitudinal aircraft axis X_A, ψ_{SA} is the direction of the gyro spin axis X_{SA}, and ψ_p is the angle between the longitudinal axis X_A of the aircraft and the direction of the spin axis X_{SA} measured by the pickoff p_ψ. The following relationships are evident:

$$\psi_p = \psi - \psi_{SA} \qquad (4.4.13)$$

$$\dot{\psi}_p = \dot{\psi} - \dot{\psi}_{SA} \qquad (4.4.14)$$

$$\psi_m = \psi + \tilde{\psi}_m \qquad (4.4.15)$$

As indicated in Figure 4.15(b), $\tilde{\psi}_m$, is the compass error. It is a function of geographical position, not time. Figure 4.15(b) shows the slaving to the compass by M_y. From (4.4.13), and assuming that $\dot{\varsigma} \cong \dot{\psi}$, we have

$$(-\dot{\psi}_{SA} + \dot{\psi}_d + \dot{\psi})\frac{1}{s} = [(\psi_m - \psi_p)K + \dot{\psi}_d + \dot{\psi}]\frac{1}{s} = \psi_p \qquad (4.4.16)$$

With $\epsilon_\psi A_y b = T_y = -H_x\dot{\psi}_{SA}$; $\epsilon_\psi A_y b/H_x = -\dot{\psi}_{SA}$; $K \triangleq A_y b/H_x$, and expanding Equation (4.4.16),

$$K\psi + K\tilde{\psi}_m - K\psi_p + \dot{\psi}_d + \dot{\psi} = \dot{\psi}_p \tag{4.4.17}$$

Laplace transforming and rearranging gives

$$(s+K)\psi(s) + K\tilde{\psi}_m(s) + \dot{\psi}_d(s) = (K+s)\psi_p(s) \tag{4.4.18}$$

The solution for ψ_p therefore is

$$\psi_p(s) = \psi(s) + \frac{K}{s+K}\tilde{\psi}_m(s) + \frac{\dot{\psi}_d(s)}{s+K} \tag{4.4.19}$$

Conclusions:

1. The measured heading angle ψ_p represents the actual heading ψ without any delay.

2. $\tilde{\psi}_m(s)$ is the magnetic heading error, which varies with the vehicle position X so that $\tilde{\psi} = \tilde{\psi}(X)$. Aircraft motion at ground-speed V translates $\tilde{\psi}$ into a time-dependent variable $\tilde{\psi}(t)$. The higher V, the wider its bandwidth. It is low-pass filtered by $K/(s+K)$. This may be important for the heading display presented to the pilot or for the performance of a heading control autopilot.

3. The drift rate of the gyro $\dot{\psi}_d$ causes a bounded error in ψ. Its steady-state value is $\psi_d = \dot{\psi}_d/K$.

4. The lower K, the lower the effect of $\tilde{\psi}_m$, but the higher the effect of $\dot{\psi}_d$ and vice versa.

5. The result (4.4.19) is a special case of complementary filtering or measurement fusion, discussed in Chapter 8. The precision obtained with the complementary measurements of the compass and the gyro is substantially better than with each individual measurement.

EXAMPLE 4.4–1: Error of magnetically aided directional gyroscope

The gain K for a given directional gyro is $K = 0.01$ sec. The free drift rate is 10deg/hr.

The spatial rms deviation of the magnetic field in a given region is $\sigma_{\tilde{\psi}m} = 2$ deg. The spatial distribution of $\tilde{\psi}_m$ is characterized by the exponential autocorrelation function

$$\phi_{\tilde{\psi}_m\tilde{\psi}_m}(x) = \sigma^2_{\tilde{\psi}m}\, e^{-|\frac{x}{\lambda}|}$$

where λ is the correlation length of $\tilde{\psi}_m(x)$. At the given flight path altitude, $\lambda = 150$ m. The ground speed is $V = 200$ m/sec. In analogy to (4.4.2), the corresponding temporal autocorrelation function is

$$\phi_{\tilde{\psi}_m\tilde{\psi}_m}(\tau) = \sigma^2_{\tilde{\psi}m}\, e^{-\frac{V}{\lambda}|\tau|} = \sigma^2_{\tilde{\psi}m}\, e^{-a|\tau|}$$

where $a \triangleq V/\lambda$. The corresponding power spectrum is

$$\Phi_{\tilde{\psi}_m\tilde{\psi}_m}(s) = \sigma^2_{\tilde{\psi}m}\frac{2a}{a^2 - s^2}$$

In accordance with (4.4.19), this signal is filtered by $K/(s+K)$. The spectrum of the filtered signal therefore is

$$\Phi_{\tilde{\psi}\tilde{\psi}}(s)=\sigma^2_{\tilde{\psi}_m}2a\left|\frac{K}{(a+s)(K+s)}\right|=\sigma^2_{\tilde{\psi}_m}2a\left|\frac{K}{s^2+(a+K)s+aK}\right|^2$$

The variance of the filtered azimuth deviation is therefore

$$\sigma^2_{\tilde{\psi}}=\frac{1}{2\pi j}\int_{-j\infty}^{j\infty}\Phi_{\tilde{\psi}\tilde{\psi}}(s)ds=\sigma^2_{\tilde{\psi}_m}\frac{K}{a+K}$$

$a=V/\lambda=200/150=1.33$ rad/sec. Thus, the attenuation factor is

$$\sigma_{\tilde{\psi}}/\sigma_{\tilde{\psi}_m}=\sqrt{\frac{K}{a+K}}=\sqrt{\frac{0.01}{1.33+0.01}}=0.086$$

This result indicates that slaving the gyro to the compass can attenuate $\tilde{\psi}_m$ by a factor of ~ 10. The residual rms therefore, is $\sigma_{\tilde{\psi}}=2\times 0.086=0.172$ deg. Moreover, the bandwidth of $\tilde{\psi}_m(t)$, which is 1.33 rad/sec, is reduced to ~0.01 rad/sec. This is particularly significant in display applications since it reduces often annoying fluctuations of the heading indicator. The steady-state error due to the free drift $\dot{\psi}_d$ is $\psi_d=\dot{\psi}_d/K=(10\text{ deg}/3600)/(1/100)=0.27$ deg. Clearly, if the correlation length λ is larger, the attenuation factor $\sigma_{\tilde{\psi}}/\sigma_{\tilde{\psi}_m}$ is also larger. On the other hand, if the velocity of the aircraft V is larger, the attenuation factor is smaller.

4.5 Gyrocompassing

Geometry and motion

The most significant property of the level-controlled directional gyroscope is that it automatically aligns its spin axis along the local longitude. That is, the directional gyro seeks the direction of the geographical north. This property is the direct outcome of two factors: 1. the rotation of the Earth; 2. the enforcement of the local level. The gyro is located at latitude λ, and its ground speed V is initially assumed to be zero. The direction of the SA deviates from the north by the angle ψ, as indicated in Figure 4.16(b).

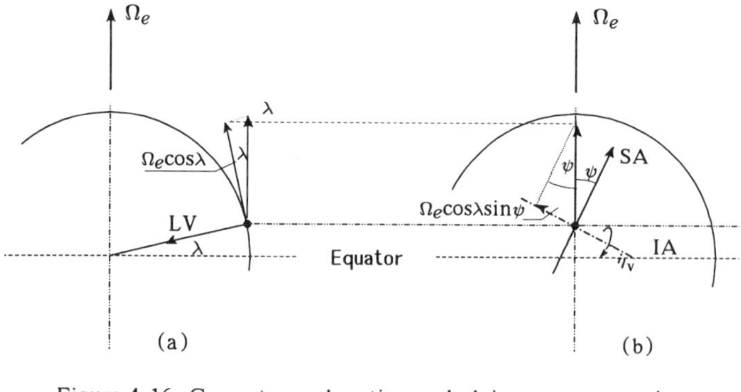

Figure 4.16. Geometry and motion underlying gyrocompassing
(a) Earth rate components due to latitude λ;
(b) Earth rate components due to λ and heading ψ.

From Figure 4.16(a), the rate of turn of the local vertical LV is $\Omega_e \cos\lambda$. It assumes its largest value on the equator and is zero at the geographical poles. From Figure 4.16(b), the gyro input axis IA senses only the component $\Omega_e \cos\lambda \sin\psi$, which is the inertial angular rate $\dot{\eta}_v$

$$\dot{\eta}_v = -\Omega_e \cos\lambda \sin\psi \tag{4.5.1}$$

It is initially assumed that $\epsilon = 0$, $V = 0$, and $\dot{V} = 0$. Thus, in (4.4.5), $\delta = 0$ so that $\beta = \eta$. The torque T_z exerted by $\mathbf{M}_z$ therefore is,

$$T_z = A_z b_z \eta \tag{4.5.2}$$

This torque is in equilibrium with the sum of the reactive precession torque $T_{z_p} = H_\mathcal{I} \dot{\eta}$ and the dynamic reaction torque $T_{z_D} = I_{zG}\ddot{\xi} \cong I_{zG}\ddot{\psi}$. Since $T_{z_D} \ll T_{z_p}$, the dynamic torque will be disregarded. We thus have

$$T_z + T_{z_p} = A_z b_z \eta + H_\mathcal{I} \dot{\eta} = 0 \tag{4.5.3}$$

From (4.4.7), in the steady state $s \to 0$, we have, on substituting $\dot{\eta}_v$ from (4.5.1),

$$\eta = -\frac{\dot{\eta}_v}{K} + \frac{\dot{\eta}_d}{K} = \frac{\Omega_e \cos\lambda \sin\psi}{K} + \frac{\dot{\eta}_d}{K} \tag{4.5.4}$$

Substituting η from (4.5.4) into (4.5.3), recalling that $K \triangleq A_z b_z / H_\mathcal{I}$, assuming that $\dot{\eta}_d = \text{const.}$, and observing from (4.5.4) that $\dot{\eta} = -\ddot{\eta}_v / K$, we have

$$A_z b_z \frac{\Omega_e \cos\lambda \sin\psi}{A_z b_z} H_\mathcal{I} + A_z b_z \frac{H_\mathcal{I}}{A_z b_z} \dot{\eta}_d + \frac{H_\mathcal{I} \Omega_e \cos\lambda \cos\psi\, \dot{\psi}}{A_z b_z} H_\mathcal{I} = 0 \tag{4.5.5}$$

Assuming small ψ so that $\sin\psi \approx \psi$ and $\cos\psi \approx 1$, rearranging, and dividing through by $H_\mathcal{I}$, we have

$$\Omega_e \cos\lambda\, \psi + \dot{\eta}_d + \Omega_e \cos\lambda \frac{H_\mathcal{I}}{A_z b_z} \dot{\psi} = 0 \tag{4.5.6}$$

Thus,

$$\dot{\psi} + \frac{A_z b_z}{H_\mathcal{I}} \psi = -\frac{A_z b_z}{\Omega_e \cos\lambda H_\mathcal{I}} \dot{\eta}_d \tag{4.5.7}$$

or

$$\dot{\psi} + K\psi = \frac{K}{\Omega_e \cos\lambda} \dot{\eta}_d \tag{4.5.8}$$

Directional error in the steady state

In the steady state, as $t \to \infty$ and $\dot{\psi} \to 0$, the solution for ψ is

$$\psi = -\frac{\dot{\eta}_d}{\Omega_e \cos\lambda} \triangleq \Delta\psi_d \tag{4.5.9}$$

Recall (4.4.5), which includes the effects of δ and ϵ. These effects have so far been disregarded in the development leading to (4.5.8). Their inclusion, which is clearly required as part of the performance analysis of the gyrocompassing function, adds the term $A_z b_z(\delta+\epsilon)$ to the right-hand side of (4.5.2). Developing (4.5.4) for the steady-state $s \to 0$ with these terms include results in the extension of (4.5.8), namely,

$$\dot{\psi} + K\psi = -\frac{K}{\Omega_e \cos\lambda}\dot{\eta}_d - \frac{K^2}{\Omega_e \cos\lambda}(\epsilon + \delta) \tag{4.5.10}$$

It therefore follows that the contribution of $\epsilon + \delta$ to the steady-state alignment error, i. e., when $\dot{\psi} \to 0$, is,

$$\psi = -\frac{K}{\Omega_e \cos\lambda}(\epsilon + \delta) \tag{4.5.11}$$

Thus, for example, $\Omega_e \cos\lambda = 10$ deg/hr, $\epsilon = 0.05$deg, $\delta = 0$, and $K = 0.005\,\text{sec}^{-1} =$ 103 deg/hr, we have

$$\psi = \frac{103}{10} 0.05\,\text{deg} = 0.515\,\text{deg}$$

This example signifies that small values of K tend to reduce the influence of the steady-state alignment error due to level sensor bias ϵ. For further aspects, the reader is referred to [7].

Conclusions:

1. If $\dot{\eta}_d = 0$, the directional gyroscope asymptotically aligns itself with the geographical north by driving its spin axis to $\psi = 0$, starting from an arbitrary initial condition $\psi(0) < 90$ deg. This process is known as gyrocompassing.

2. In general, for $\dot{\eta}_d \neq 0$, the gyro aligns within the residual deviation $\Delta\psi_d$, as given in (4.5.8). This result demonstrates that in the northern or southern polar regions, the residual alignment error increases as λ increases. At the poles, $\Delta\psi_d \to \infty$.

3. The function of Gyrocompassing is valid if $\dot{\eta}_v$ is due solely to the rotation of Earth. Vehicular motion introduces an additional term in (4.4.5) at the same level as the drift term due to $\dot{\eta}_d$. This implies that the ground speed must be zero or that V must be known with great precision to permit the compensation of an additional term in $\dot{\eta}_v$ due to vehicular motion.

4. For this reason classical gyrocompassing, as described in this section has been restricted to land-based applications, such as geodesy, and artillery gun-laying, or to those marine operations where the effect of V is relatively small and can therefore be compensated with acceptable precision. Thus, gyrocompassing generally cannot be dissociated from the function of terrestrial navigation, which must be executed with great precision.

5. Once the gyrocompassing process has been completed, i.e., $\psi \to 0$, it follows from (4.5.4) that since $\dot{\eta}_v = 0$, $\eta = 0$ as well. Consequently, except for the small effect of the drift term due to $\dot{\eta}_d$, the north pointing directional gyros aligned with the local level. Therefore, the additional potential errors due to η, given in (4.4.11) and (4.4.12), are canceled by the the gyrocompassing action.

6. For aerospace applications in which V is large, and because of possible horizontal accelerations, gyrocompassing is implemented by means of inertial gimbaled or strapdown platforms. In these platforms the required navigation and gyrocompassing operations are executed by the same instrument and software package.

7. The time constant for gyrocompassing is $\tau_{GC} = K^{-1}$ in accordance with (4.5.8). Thus, the time constant of gyrocompassing is determined by the same gain coefficient K as in the leveling control system. The time to effectively complete the north alignment is on the order of $5\tau_{GC}$.

EXAMPLE 4.5–1: Gyrocompassing error due to gyro drift rate

A directional gyro has a drift rate $\dot{\eta}_d = 0.1$ deg/hr is placed at the latitude $\lambda = 35$ deg N.

1. What is the residual deviation $\Delta\psi_d$?

$$\Omega_e \cos\lambda = (360\,\text{deg}/24\,\text{hr}) \times \cos 35^0 = 15\,\text{deg/hr} \times 0.819 = 12.28\,\text{deg/hr}.$$

Thus,

$$\Delta\psi_d = \frac{\dot{\eta}_d}{\Omega_e \cos\lambda} = \frac{0.1}{12.28} = 8.14 \times 10^{-3}\ \text{rad} = 0.466\ \text{deg}.$$

This result demonstrates that even for the moderate precision of ~ 0.5 deg, a relatively costly inertial grade gyro with a drift rate of $\dot{\eta}_d < 0.1$ deg/hr is required.

The gain of the leveling control loop K is chosen to be $K = 0.005\ \text{sec}^{-1}$.

2. If the initial azimuth deviation is $\psi(t)\big|_{t=0} = \psi(0) = \psi_0 = 30$ deg, disregarding the drift rate $\dot{\eta}_d$, what is the time t_{al} required to complete the north alignment to within $\Delta\psi = \psi(t_{al}) = 0.1$ deg?

According to (4.5.8) we have

$$\dot{\psi} + K\psi = 0; \quad \psi_0 = 30\ \text{deg}$$

The solution is

$$\psi(t) = \psi_0 e^{-Kt}$$

$$\frac{\psi(t_{al})}{\psi_0} = \frac{0.1\,\text{deg}}{30\,\text{deg}} = e^{-0.005 t_{al}}$$

$$t_{al} = 200 \log_e 300 = 1140\ \text{sec} = 19\ \text{min}.$$

Thus, depending on the design parameters of the leveling control loop, the duration of the alignment process to within a fraction of a degree may require a substantial part of an hour.

4.6 The single-axis deflection rate gyro

Description

Recalling Section 4.0, we turn to the second class of gyroscopes, namely, the rate gyro. Initially, we address the single-axis gyro which can measure P, Q, or R, provided its input-axis is aligned with the required angular rate measurement. This sensor is used in a large variety of aerospace applications, such as shaping the dynamic response of an aircraft or spacecraft, inertially stabilizing a payload and controlling its orientation, damping oscillations of space structures, or implementing active aircraft flutter suppression. A schematic description of a traditional spring-loaded rate gyro is shown in Figure 4.17.

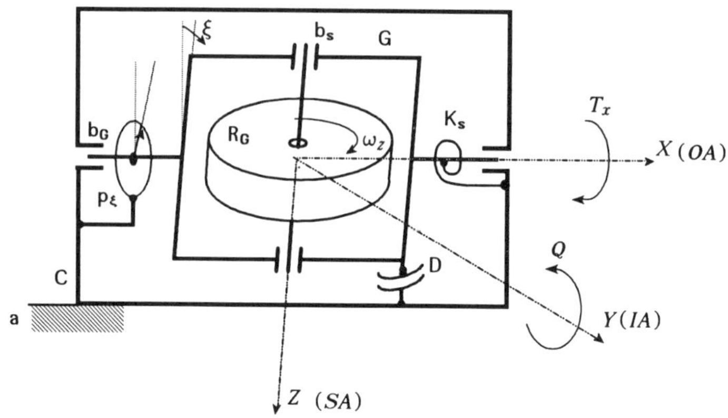

Figure 4.17. Schematic description of the elementary spring-loaded gyro.

The rotor R_G is suspended by the spinner bearings b_s mounted in the gimbal G suspended in the outer gimbal, which is the gyro casing C, fixed to the vehicle body **a**. The gyro is depicted with Y as its input axis and X as its output axis. The gyro, in accordance with (4.1.9), therefore is sensitive to the angular rate component Q. The torque T_x along the X axis is balanced by the spring which is measured by p_ξ. Thus, K_s is the transducer, which provides the measurement of T_x and thus of Q.

Dynamical Model

The following notations are used:

$\mathbf{\Omega}_a \triangleq [P, Q, R]^T$ – The inertial angular rate vector of the axis system a of **a**.

$\mathbf{H} \triangleq [H_x, H_y, H_z]^T$—The total angular momentum vector of the vehicle.

$\boldsymbol{\omega} \triangleq [\omega_x, \omega_y, \omega_z]^T$—The angular rate of the gyro with respect to a.

$\mathbf{H}_a \triangleq J\boldsymbol{\omega}$ — the vector of the gyro angular momentum with respect to a.
$\mathbf{T} \triangleq [T_x, T_y, T_z]^T$— the torque vector exerted on a due to $\mathrm{d}\mathbf{H}_a/\mathrm{d}t$.
In accordance with (4.1.5) and (4.1.6)

$$\mathbf{H}_a \triangleq J\boldsymbol{\omega} = \begin{bmatrix} I_{xx} & 0 & 0 \\ 0 & I_{yy} & 0 \\ 0 & 0 & I_{zz} \end{bmatrix} \begin{bmatrix} \omega_x \\ \omega_y \\ \omega_z \end{bmatrix} = \left[I_{xx}\omega_x,\ I_{yy}\omega_y,\ I_{zz}\omega_z \right]^T \tag{4.6.1}$$

The torque $\mathbf{T}$ between the aircraft a and the gyro is

$$\mathbf{T} = \dot{\mathbf{H}}_a + \boldsymbol{\Omega}_a \times \mathbf{H}_a \tag{4.6.2}$$

Because of the minute inertia of the gyro compared to that of **a**, the effect of **T** on **a** is negligible. However, this is not always the case. For example, the spinning rotor of a jet engine, can exert a very substantial torque on the aircraft while in a turn. Dropping the second subscript in $I_{xx,}$ etc, explicitly, this torque is,

$$\mathbf{T} = \mathbf{i}\frac{\mathrm{d}}{\mathrm{d}t}(I_x\omega_x) + \mathbf{j}\frac{\mathrm{d}}{\mathrm{d}t}(I_y\omega_y) + \mathbf{k}\frac{\mathrm{d}}{\mathrm{d}t}(I_z\omega_z)$$

$$+ \begin{vmatrix} \mathbf{i} & \mathbf{j} & \mathbf{k} \\ P & Q & R \\ I_x\omega_x & I_y\omega_y & I_z\omega_z \end{vmatrix} \tag{4.6.3}$$

Since the gyro is strapped to the vehicle body, $\omega_y = 0$; Furthermore, $\dot{\omega}_z = 0$; Therefore, the $\mathbf{i}$ and $\mathbf{j}$ derivative terms are zero. Thus, on developing the determinant,

$$\mathbf{T} = \mathbf{i}\, I_x\dot{\omega}_x + \mathbf{i}(I_z\omega_z Q - 0) - \mathbf{j}(I_z\omega_z P - I_x\omega_x R) + \mathbf{k}(0 - I_x\omega_x Q)$$

$$= \mathbf{i}\,T_x + \mathbf{j}T_y + \mathbf{k}T_z \tag{4.6.4}$$

Because of the strapdown mounting, the torque components multiplying **j** and **k** are disregarded. The remaining torque component along **i** is

$$T_x = I_x\dot{\omega}_x + I_z\omega_z Q \tag{4.6.5}$$

Because of the angular deviation of the rotor ξ, the torque is actually

$$T_x = I_x\dot{\omega}_x + I_z\omega_z Q\cos\xi \tag{4.6.6}$$

To avoid nonlinearity due to $\cos\xi$, in actual designs, ξ is confined to be small, so that $\cos\xi \approx 1$ and (4.6.5) holds. Denoting $\dot{\omega}_x \triangleq \ddot{\xi}$ and, $I_z\omega_z \triangleq H$, we have

$$T_x = I_x\ddot{\xi} + HQ \tag{4.6.7}$$

where H is assumed to be constant. The restoring torque T_R is

$$T_R = T_S + T_V + T_D \tag{4.6.8}$$

where

T_S — spring torque $K_s\xi$;
T_V — viscous friction torque $D\dot{\xi}$;
T_D — disturbance torque due to friction T_f and mass imbalance $m\epsilon$.

K_s is initially assumed to be constant thus, representing a perfectly linear spring. T_V, which determines the damping factor, is not a highly critical parameter. T_D is governed by a variety of factors and is quite difficult to model mathematically. The sum of all torques around the X axis is zero, thus

$$\sum_i T_i = T_x + T_R = 0 \tag{4.6.9}$$

namely,

$$I_x\ddot{\xi} + HQ + K_s\xi + D\dot{\xi} + T_D = 0 \tag{4.6.10}$$

Laplace transforming (4.6.10) and solving for $\xi(s)$ we have

$$\xi(s) = -\frac{H}{s^2 I_x + Ds + K_s}Q(s) - \frac{T_D(s)}{s^2 I_x + Ds + K_s}$$

$$= -\frac{H/K_s}{\dfrac{s^2}{\omega_n^2} + \dfrac{2\varsigma}{\omega_n}s + 1}Q(s) - \frac{1/K_s}{\dfrac{s^2}{\omega_n^2} + \dfrac{2\varsigma}{\omega_n}s + 1}T_D(s) \tag{4.6.11}$$

where the natural frequency ω_n and the damping coefficient ς are

$$\omega_n \triangleq \sqrt{\frac{K_s}{I_x}} \tag{4.6.12}$$

$$\varsigma \triangleq \frac{D}{2(I_x K_s)^{\frac{1}{2}}} \tag{4.6.13}$$

Equation (4.6.11) expresses the dynamical model of the rate gyro with $Q(s)$ as the input and $T_D(s)$ as the disturbance or noise. D is designed to render ς in the region $\varsigma = 0.6 - 0.8$. In the steady state, as $s \to 0$, we have,

$$\xi(s) = \frac{H}{K_s}Q(s) - \frac{1}{K_s}T_D(s) = \frac{H}{K_s}[Q(s) - Q_D(s)] \tag{4.6.14}$$

where

$$Q_D(s) \triangleq \frac{T_D(s)}{H} \tag{4.6.15}$$

is the drift rate of the gyro. Its presence implies that the gyro outputs the erroneous angular rate Q_D while the actual angular rate Q is zero. Equation (4.6.15) shows that the drift rate can be reduced by making H as large and T_D as small as possible. These goals reflect on size, weight, spin rate, low bearing friction, and mechanical manufacturing and assembly precision, all of which contribute to increased cost. Q_D, which by (4.6.14) has the nature of an additive error, is random and, in general, consists of a time-varying component $Q_D(t)$ and a bias term Q_B. In addition to the above errors due to the gyro mechanical assembly imperfections, additional electrical errors due to the pickoff contribute to the error. p_ς represents ς via a scale factor S (volt/deg). The electrical output of the gyro therefore is,

$$V_0 = S(\xi + \xi_0) \tag{4.6.16}$$

where ς_0 is an electro mechanical offset of the pick-offs. Substituting (4.6.16) into (4.6.14), we have

$$V_0 = \frac{SH}{K_s}[Q(s) - Q_D(s)] + S\xi_0 \tag{4.6.17}$$

SH/K_s is the actual overall scale factor. To translate V_0, which is measured in volts into the required dimension of rad/sec or deg/min, V_0 needs to be scaled by $\sigma \triangleq SH/K_s$. But, none of the parameters constituting σ is exactly known. The best one can do is to rely on prior calibration measurements of S, H and K_s, denoted by S_c, H_c, and K_{sc}. We therefore have at our disposal the calibration coefficient $\sigma_c \triangleq S_c H_c / K_{sc}$. Thus, the calibrated angular rate is,

$$Q_c(s) = \frac{V_0}{\sigma_c} = \frac{\sigma}{\sigma_c}\{[Q(s) - Q_D(s)] + \frac{K_s}{H}\xi_0\} \tag{4.6.18}$$

If we define $\sigma = \sigma_c + \Delta\sigma$, where $\Delta\sigma$ is the scale factor error, then (4.6.18) can be rewritten as

$$Q_c(s) = (1 + \frac{\Delta\sigma}{\sigma_c})\{[Q(s) - Q_D(s)] + \frac{K_s}{H}\xi_0\} \tag{4.6.19}$$

The error of $Q(s)$ is

$$\begin{aligned} \epsilon_Q(s) &= Q_c(s) - Q(s) = \frac{\Delta\sigma}{\sigma_c}Q(s) - (1 + \frac{\Delta\sigma}{\sigma_c})[Q_D(s) - \frac{K_s}{H}\xi_0\} \\ &\cong \frac{\Delta\sigma}{\sigma_c}Q(s) - Q_D(s) + \frac{K_s}{H}\xi_0 \end{aligned} \tag{4.8.20}$$

It is implied that $\Delta\sigma/\sigma_c \ll 1$. The first term is the scale factor error, the second is the gyro mechanical drift term and the third is the electromechanical pickoff offset error. All three terms may consist of constant and random components. The constant term is essentially immaterial since it can be calibrated once for all instances. Its magnitude for each individual gyro is given in the manufacturer's data sheet. A design example of the spring loaded rate gyro is given in the following.

EXAMPLE 4.6–1: Drift rate of a rate gyro due to internal imperfections

The following data are specified:

$H = 10$ gr f cm sec
$n = 24000$ rpm

The requirement for the natural frequency is $\omega_n = 200\,\text{sec}^{-1}$.
The required maximum angular rate is $Q_{\text{max}} = 100$ deg/sec = 1.54 rad/sec;

Thus,

$$\omega_z = \frac{2\pi}{60} n = 2513 \text{ rad/sec}$$

$$I_z = \frac{H}{\omega_z} = \frac{10}{2513} \cong 0.004 \text{ gr f cm sec}^2$$

The gyro mechanical assembly has a moment of inertia $I_x = 0.02$ gr f cm–sec^2. From (4.6.12),

$$\omega_n^2 = \frac{K_s}{I_x}; \quad K_s = I_x \omega_n^2 = 0.02 \times (200)^2 = 800 \text{ (gr–f cm)/rad} = 14 \text{ gr f cm/deg;}$$

The maximum torque T_{max} is,

$$T_{\text{max}} = HQ_{\text{max}} = 10 \times 1.74 = 17.4 \text{ gr f cm;}$$

The maximal angular deflection of the gimbal assembly ξ_{max} is

$$\xi_{\text{max}} = \frac{T_{\text{max}}}{K_s} = \frac{17.4}{14} = 1.24 \text{ deg}$$

This is a reasonable result which assures good linearity. However, such a small angular excursion of the pickoff rotor imposes requirements for high sensitivity and high resolution. These requirements are typically met by microsyns.

The permitted drift rate Q_D due to the horizontal eccentricity ϵ_h of the gimbal mount is 1% of full scale, i.e., 1 deg/sec. The weight of the gimbal mount is $w_g = 40$ gr.

What is the permitted ϵ_h?

$$T_D = w_g \epsilon_h = 40\epsilon_h.$$

$$Q_D = \frac{T_D}{H} = \frac{40\epsilon_h}{10} \text{ rad/sec} = 4 \times 57.3\,\epsilon_h \text{ deg/sec}$$

$$\epsilon_h = \frac{1}{57.3 \times 4} \text{ cm} = 4.36 \times 10^{-3} \text{ cm} = 0.0436 \text{ mm.}$$

This result illustrates the considerable precision required even for a relatively large drift rate of 1deg/sec = 60deg/min.

Imperfections and limitations

The relatively simple low cost technologies applied in the deflection-type rate gyro described earlier a number of significant imperfections and limitations that affect their performance. The most significant of these are the following.

Bandwidth

In commercial rate gyros, the bandwidth denoted by ω_n hardly exceeds 300 rad/sec, or, 50 Hz. Equation (4.6.12) shows that an increase of ω_n by a factor x would involve an increase by x^2 of K_s or a decrease by x^2 of I_x. Increasing K_s implies a corresponding decrease of ξ_{max} thus, increasing the requirement for the resolution of p_ξ or, sacrificing dynamic range by the corresponding factor.
Alternatively, decreasing I_z implies an increase by x^2 of the spin rate ω_z if the magnitude of H has to be maintained. Otherwise, by (4.6.15), the drift rate would increase by the factor x^2.

Drift rate

The principal contributors to drift, which is normally in the 10–50 deg/min range, are gimbal ball bearing friction T_f and mass imbalance of the gimbal suspension. Within state-of-the-art technology, it is impossible to significantly reduce the bearing friction or achieve substantially better machining or assembly methods. Null-point offsets, even if compensated, are a function of temperature and will typically vary even over periods of minutes.

Scale factor errors

The overall scale factor of the rate gyro is determined by $\sigma \triangleq SH/K_s$ as given in (4.6.17). The relative scale factor uncertainty is readily shown to be the sum of the individual relative scale factor uncertainties

$$\frac{\Delta\sigma}{\sigma} = \frac{\Delta S}{S} + \frac{\Delta H}{H} - \frac{\Delta K_s}{K_2}$$

Assuming the individual uncertainties are uncorrelated, the rms value of $\Delta\sigma/\sigma$ is the square root of the sum of squares of the three individual terms as shown in Chapter 1.

Nonlinearity and hysteresis

Linearity, as in all deflection-type sensors, is primarily determined by the spring which normally is a flat torsion bar made of phosphor bronze having relatively low hysteresis. If, for example, the rate gyro is subjected to Q_{max}, the spring "remembers" about 0.01 of the corresponding ξ_{max}, i.e., $0.01Q_{max}$. Thus, for example, $Q_{max} = 100$ deg/sec, the gyro will retain an offset of up to 1deg/sec. It is therefore quite difficult to model the null-point error of a deflection-type rate gyro. In addition, gimbal bearing friction causes a dead space. Before any input rate Q_{min} shows as a gimbal deflection ξ, the dry friction torque T_f must first be exceeded by the precession torque HQ_{min}.

Saturation

The gimbal mount must be protected from excessive input rates which might damage the torsion bar. This is achieved by mechanical stops that confine $|\xi_{max}|$. Thus, the rate gyro is limited from below by the bearing friction torque and from above by the hard limiting of the mechanical stops.

The main parameters of a typical deflection type rate gyro are listed in Table 4.1.

Table 4.1. Data List of a Typical Deflection Rate Gyro (Sperry type RG-1000)

Functional Description	Numerical Value
Performance	
Range (±deg/sec)	150
Scale factor (volt rms/deg/sec)	0.01
Output Load (K Ohms)	51
Nonlinearity (% of F.S.)	2.0
Threshold (deg/sec)	0.015
Resolution (deg/sec)	0.015
Hysteresis (deg/sec)	0.15
Zero offset at room temperature (deg/sec)	0.50
Mass unbalance (deg/sec/g)	0.05
Natural frequency (Hz)	58 to 63
Damping coefficient with temperature	0.5 to 0.8
Spin motor	
Excitation (volt rms)	30 static; 15 run sq. wave
Phases	2
Frequency (Hz)	1250
Start power (max. watts)	56 VA
Run power (max. watts)	7.5 VA
Pickoff	
Excitation (volt rms)	9
Frequency (Hz)	3125
Power (max. watts)	0.7 (VA)
Environments	
Shock	56g 1.1 msec
Vibration (rms)	9g 50-3000 Hz
Acceleration	50g
Temperature range (deg C)	-25 - +85
Operating life (Hours)	500, 1000 starts

A cross-section of a typical rate gyro in the class of gyros discussed in this section, Sperry RG-1000, is shown in Figure 4.18.

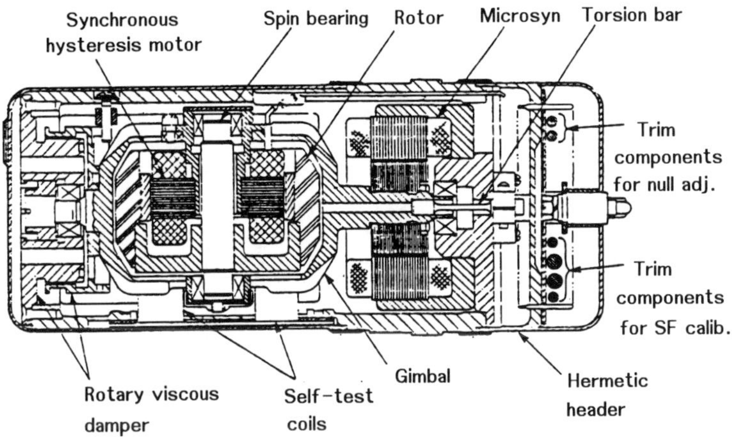

Figure 4.18. Cross section of a conventional single-axis rate gyro RG-1000 (based on Sperry Publication No. 12-1141-01).

This type of rate gyro is of the low medium quality type. Its maximum input angular rate is limited to 150 deg/sec and it is therefore incompatible with the requirement of modern fighter aircraft amounting to 400 deg/sec. The null point certainty as seen in Table 4.1, is about 0.50 deg/sec so that the dynamic range D_r is only about 300. The expected operating hours, or MTBF, is only about 500 hours, assuming 1000 starts. These performance characteristics are typical for flight control autopilot applications in moderate-performance general aviation aircraft.

4.7 The floated rate integrating gyro (RIG)

Description

The basic first-generation deflection-type rate gyro described in Section 4.8, while suitable for most conventional flight-control systems, is orders of magnitude inferior in performance to that required in high-precision guidance and navigation applications. Within the realm of electromechanical technology, the required levels of precision can only be achieved by highly specialized suspension technologies, such as the hydrostatic flotation of a rotor-carrying imbal, as will be described. A schematic description of its construction is given in Figure 4.19.

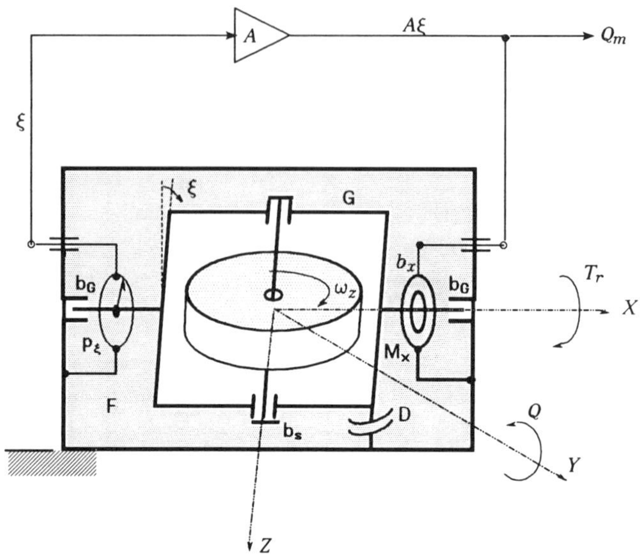

Figure 4.19. Schematic description of a floated rate integrating gyro.

The restoring spring K_s is removed and is replaced by the torque motor M_x, which has the torque constant b_x. The gimbal G is a sealed container that houses the spinner motor and the rotor R_G. Along with G and other auxiliary components, they constitute the Float. The narrow gap between C and G is filled with the fluid fluorocarbon F, which is electrically insulating and has a density of about 2 gr/cm^3. It can impart neutral buoyancy to G which is designed to match the density of F, also providing high viscous friction D between C and G supported by the needle tip of the shaft S and the diamond bearings b_G. The neutral buoyancy combined with the diamond bearings essentially remove the gimbal bearing friction and, due to the hydrostatic flotation, renders the delicate gimbaled rotor assembly insensitive to shock and vibration. Moreover, due to the floatation, the gyro assembly is completely indifferent to the effects of mass imbalance so that this source of drift rate is removed. The thermal expansion coefficient of F is larger than that of G. The excess volume is absorbed by bellows, which are an integral part of the assembly. Clearly, there is only one temperature at which the buoyancy is perfect. To reach this temperature, which must be well in excess of the ambient temperature, about 50 deg C long warm-up times, in excess of 30 min are required. This design concept, originally developed at the Charles Stark Draper Laboratory in Cambridge, Mass., in 1942, [10], has been the principal breakthrough in the realization of high precision three-axis gimbaled platforms for inertial navigation systems with an accumulating positional error of less than two nautical miles per hour. The MIT Instrumentation Laboratory 10-series, dating back to 1955, has matured as a widespread component that has been in operational use in inertial navigation systems until the present time. Its continued production for the highly specialized 0.01 deg/hr class, intended exclusively for gimbaled platforms, has drastically diminished in recent years, especially with the advent of the ring laser gyro (RLG) and the dry tuned rotor gyro (DTG), which opened the road to the widespread implementation of strapdown technology. These technologies will be addressed later.

However, in tactical applications, accuracies in the range of $0.1-1$ deg/hr are of practical interest. These can still be achieved by floated RIG technology. Therefore floated RIG's in this accuracy category have been manufactured for strapdown applications as indicated e. g., by the Lear Siegler 1903-1904 series [11]. For example $Q_{\max}$ for the 1904A Gyro is 400 deg/sec and the random drift rate is $Q_d = 0.2$ deg/hr rms. Its dimensions are 1.6 in diameter by 3.2 in length. In this design, a reasonable technical compromise has been made by discarding the temperature regulation of the floatation fluid, thus saving in cost, but giving up the precise buoyancy control, which balances the gimbal weight. A cutaway of the 1904A is shown in Figure 4.20. The principal components and parts, such as the sealed gimbal can, rotor, torquer, bellows, pickoff, and diamond bearing are clearly visible. This product indicates that the floated gyro technology is not necessarily ruled out in the era of strapdown technology for tactical applications or aided sensor systems.

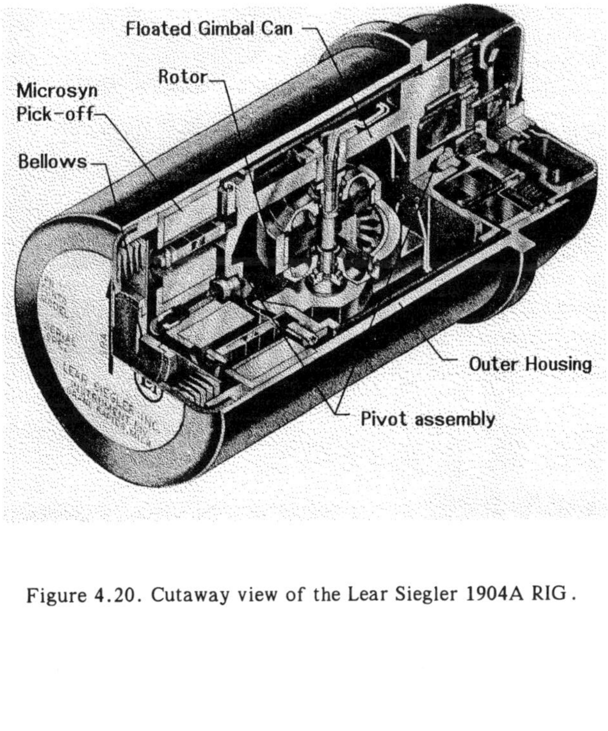

Figure 4.20. Cutaway view of the Lear Siegler 1904A RIG.

Mathematical model

We now turn to the development of the mathematical model of the floated rate integrating gyro. As indicated, it is basically similar in construction to the elementary spring-loaded rate gyro. It can be regarded as an outstanding revolutionary refinement of the technologies involved.

Recalling (4.6.10), dropping K_s and disregarding M_x for the moment, the input-output relationships are derived as follows:

$$\xi(s) = -\frac{H}{I_x s^2 + Ds} Q(s) - \frac{T_D(s)}{I_x s^2 + Ds} \tag{4.7.1}$$

In reality, for the low frequencies involved, $Ds \gg I_x s^2$ so that (4.7.1) takes the approximate form,

$$\xi(s) = -\frac{1}{s}\frac{H}{D}[Q(s) + Q_D(s)] \tag{4.7.2}$$

where $Q_D \triangleq T_D/H$. $\xi(s)$ now is proportional to the integral of $Q(s) + Q_D(s)$. This implies that angular deviation $\xi(t)$ would increase without bound, that is, it would hit the stops and render the gyro useless. Clearly, a force balance technique would confine ξ to minute levels. This can readily be accomplished by feeding the amplified signal $A\xi$ into the torque motor M_x, thus enforcing the restoring torque $T_r = Ab_x\xi$,which translates into the restoring rate

$$Q_r = \frac{T_r}{H} = \frac{Ab_x}{H}\xi \tag{4.7.3}$$

which causes the gimbal G to track the casing C. Q_r must be added to the Q terms in (4.7.2). Thus,

$$\xi(s) = -\frac{1}{s}\frac{H}{D}[Q(s) + Q_D(s) + \frac{Ab_x}{H}\xi(s)] \tag{4.7.4}$$

Solving for $\xi(s)$ we have

$$\xi(s) = -\frac{\frac{H}{D}[Q(s) + Q_D(s)]}{s + \frac{Ab_x}{D}} \tag{4.7.5}$$

In the steady state, as $s \to 0$, ξ is given by

$$\lim_{s \to 0} \xi(s) = \frac{H}{Ab_x}[Q + Q_D] \tag{4.7.6}$$

A large gain A therefore assures that only minute angular gimbal deviations ξ

are expected in force balance sensors. This, in turn, guarantees highly linear characteristics. The measurable output signal $A\xi(s)$ therefore is

$$A\xi(s) = Q_m(s) = -\frac{\frac{HA}{D}}{s + \frac{Ab_x}{D}}[Q(s) + Q_D(s)] \tag{4.7.7}$$

This first-order closed loop system has a time constant $T = D/Ab_x$, which is inversely proportional to A. In the steady state, as $s \to 0$, Q_m is given by

$$Q_m = -\frac{H}{b_x}[Q(s) + Q_D(s)] \tag{4.7.8}$$

H/b_x is the scale factor. Although the viscous damping coefficient D cancels out in (4.7.8), the temperature of the flotation fluid must be controlled within a fraction of a degree centigrade in order to guarantee exact buoyancy and make cross-axis loads on the float bearings negligible. A model of the closed-loop rate integrating gyro as a force balance sensor is readily interpreted by the block diagram shown in Figure 4.21, from which we directly write

$$Q_m = \frac{-\frac{H}{sD}A}{1 + \frac{H}{sD}\frac{Ab_x}{H}}[Q(s) + Q_D(s)] \tag{4.7.9}$$

which is identical to (4.7.5).

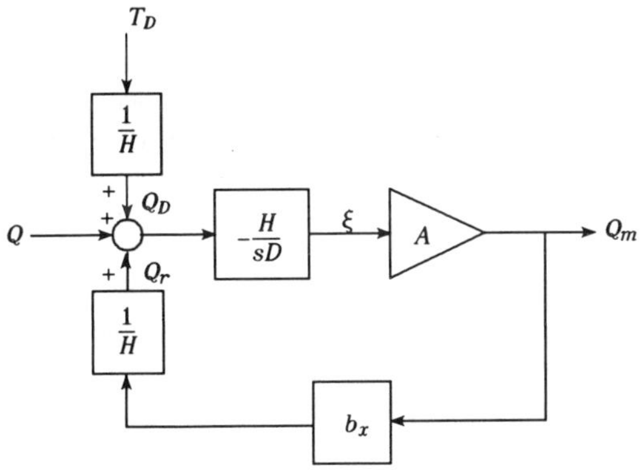

Figure 4.21. Functional block diagram of a floated RIG.

Summary of properties

1. The rate integrating gyro in the force balance configuration is a null sensor as described in Section 1.10. Two specific features are noteworthy. 1) Due to the integrating function of the gyro, there is no steady-state error due to

finite gain. 2) The noise Q_D operates at the input level and therefore is not affected by A.

2. The minute angular gimbal deviation ξ assures excellent linearity. The pickoff P_ξ, however, must be highly sensitive. Typically it is of the rotational variable differential transformer type, (RVDT).

3. Very low drift rates $Q_D(t)$, as low as 0.01 deg/hr are assured by the excellent mechanical isolation and flotation of the rotor gimbal assembly.

4. The closed-loop rate integrating gyro performs approximately like a first-order dynamical system with a time constant T that is inversely proportional to the loop gain Ab_T. Values of a few miliseconds are normally achieved.

5. The damping coefficient D, which is typically temperature sensitive, cancels out in the steady state and does not affect the scale factor.

6. To preserve precise gimbal floatation, the density of the fluid F must be carefully temperature-controlled.

7. To maintain a precise scale factor, H and therefore ω_z, must be meticulously controlled, that is, the frequency of the alternating current source of the motor must be stable according to the specifications of the gyro. The required constancy of b_z is assured by the precision temperature control of the gyro.

The principal original purpose and application of the rate integration gyro was the realization of high-precision inertially stabilized three axis platforms for the guidance of ballistic missiles. The design involves the payload of a triad of precise specific force sensors. This mechanization provides the measurement of three orthogonal components of acceleration with respect to inertial space irrespective of the attitude variations of the missile. In principle, the requirement for Q_{max} was therefore zero. In the practice of terrestrial navigation however, the requirement for Q_{max} was first of all determined by the need to maintain its alignment within the local level plane. Considering the maximum Earth rate component of Ω=15 deg/hr in addition to a possible angular rate of a high-speed aircraft traveling at, say, ~1000 km/hr, an additional ~15 deg/hr ave to be added. Thus, the required Q_{max} may amount to ~30 deg/hr. Moreover, considering a typical platform gyrocompassing alignment time of ~20 min, the total Q_{max} may be in excess of 100 deg/hr. Since the smallest detectable angular rate, Q_{min}, turned out to be in the order of 0.01 deg/hr, it follows that the required dynamic range of gimbaled rate integrating gyros for ballistic missiles or space rockets is about 10^4. With the flotation technology described earlier, this was a reachable goal. In order to provide the above values of Q_{max}, quite modest torqueing currents are adequate. This implies that the resulting temperature fluctuations are sufficiently small to permit the minute values of Q_{min}.

As described here, the floated RIG is a single-degree-of-freedom (SDF) device. This is generally the case in current inertial platform realizations. However, 2-DF floated gyros have successfully been made by Litton and American Bosh Arma for both gyrocompassing on board ships and for aircraft navigation in the 1 nmi/hr class. The design involves a spherical flotation mechanization with an additional intermediate gimbal. A brief description with further references is given in [3].

In conclusion, it is evident that floated gyro technology is primarily compatible with the concept of at least three gimbal platforms that isolate the gyros from the short-term vehicular attitude variations. The penalty is a relatively large volume and weight of the inertial measurement unit. Since warm-up time, weight and volume are not critical issues, these units have been compatible with military and civilian aircraft and with ballistic missile operational requirements.

Pendulous integrating gyro accelerometer (*PIGA*) ⊕

As indicated in Chapter 3, an unbalanced inner gimbal suspension makes the rate gyro sensitive to linear accelerations. This acceleration manifests itself as a torque and angular deflection, say, around the x axis. If free to rotate around the y axis, this torque causes a precession rate $\dot{\eta}$ around the y axis that persists as long as the acceleration does. Therefore, the accumulated angle η is proportional to the integral of the acceleration, namely, the corresponding velocity increment. This arrangement is, in effect, an electromechanical velocimeter. A schematic description of the mechanization is shown in Figure 4.22.

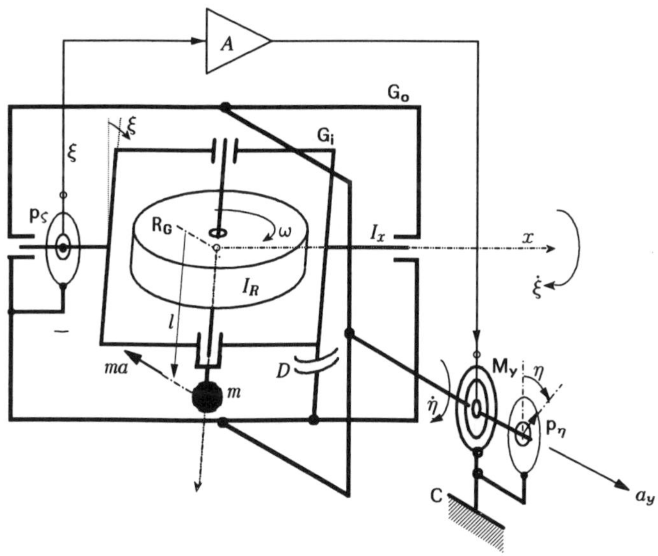

Figure 4.22. Schematic description of the pendulous integrating gyro accelerometer (PIGA).

The mass imbalance causes the pendulosity

$$P = ml \tag{4.7.10}$$

The acceleration-induced torque is Pa_y which causes the angular deflection ξ picked up by p_ξ and amplified by the factor A to produce the control signal u. This control signal drives the motor M_y at the corresponding angular rate $\dot{\eta}$. Multiplied by the gyro angular momentum H, it produces a torque $H\dot{\eta}$ around the x axis that opposes the pendulous torque, so as to drive the deflection angle ξ to zero. This device is a regular null sensor operating in a closed loop as shown. The total torque around the x axis is,

$$I_x\ddot{\xi} + D\dot{\xi} - Pa_y + H\dot{\eta} = 0 \tag{4.7.11}$$

Rearranging and Laplace transforming, we have

$$s^2\xi(s) + s\frac{D}{I_x}\xi(s) + \frac{H}{I_x}\dot{\eta} = \frac{Pa_y}{I_x} \tag{4.7.12}$$

where $H = I_R\omega$. The angular rate of the motor M_y in response to its input voltage u is given in (1.12.5)−(1.12.9). For the sake of simplicity, we disregard here the term due to friction $\ddot{\eta}_f = \dot{\omega}_f$ and the term $\alpha = (K_{\dot{\eta}} + D_a)$ due to the motor back-emf and aerodynamic drag D_a. Thus, the simplified expression for the angular rate $\dot{\eta}$ around the y axis is

$$\dot{\eta} = \frac{K_u}{I_y}u = A\frac{K_u}{I_y}\xi \tag{4.7.13}$$

Substituting (4.7.13) into (4.7.12), we have

$$\left(s^2 + \frac{D}{I_x}s + \frac{HAK_u}{I_y}\right)\xi(s) = \frac{Pa_y}{I_x} \tag{4.7.14}$$

Thus, the solution for ξ is

$$\xi(s) = \frac{P/I_x}{s^2 + \dfrac{D}{I_x}s + \dfrac{HAK_u}{I_y}}a_y \tag{4.7.15}$$

Substituting (4.7.15) into (4.7.13), we have

$$\dot{\eta}(s) = \frac{\dfrac{AK_u ml}{I_x}}{s^2 + \dfrac{D}{I_x}s + \dfrac{HAK_u}{I_y}}a_y \tag{4.7.16}$$

Finally, in the steady state, as $s \to 0$, we have

$$\dot{\eta}(s) = \frac{mlI_y}{HI_x} a_y \tag{4.7.17}$$

Performing the time integrals of both sides in (4.7.15), we have

$$\eta = \frac{mlI_y}{HI_x} V_y = \frac{PI_y}{HI_x} V_y \tag{4.7.18}$$

η is picked off by p_η. PI_y/HI_x is the scale factor. It is obvious that it is proportional to the pendulosity P and inversely proportional to H. Clearly, the high precision needed in inertial navigation applications, imposes strict requirements for low gyro drift rate and-high resolution of the pickoff p_η. The technology employed has been high-resolution shaft encoders, which are inherently linear and noise-free and are compatible with the up—down counter output electronics, which provides a natural interface with the navigation and guidance computers. Further details on implementation of scale factor and velocity resolution are to be found in, among others, in [12] and [13].

In order to assure good damping, D in (4.7.45), should be made sufficiently large, especially in view of the large value of HAK_u/I_y. This can be achieved by the gimbal flotation technique as in inertial-grade rate integrating gyros, which is required anyway in the implementation of a high-grade PIGA. Gyro drift rate $\dot{\xi}_d$ is equivalent to an unaccounted for acceleration a_{yd} manifesting itself as a drift in velocity V_{yd}.

It should be noted that since the pendulous gyro is gimbaled around the y axis, it is not sensitive to the aircraft angular rate component Q, nor to R which is along the spin axis, nor to P because the corresponding precession torque is nulled by the rigid mounting of the gyro around the z axis.

Consequently, although it is a rate gyro, and in view of its imbalanced gimbaled mechanization, it performs as an integrating accelerometer.

4.8 The dynamically tuned gyro (DTG)

Introduction

Over the last two decades, the need for small tactical missiles and unmanned aircraft known as remotely piloted vehicles or RPVs has substantially increased. For their guidance and navigation, the conventional gimbaled platforms discussed earlier have been both too bulky, costly, and unwieldy because of the lengthy preflight preparations, particularly because of the required warm-up time. The demand for smaller, more rugged, and less costly mechanizations has grown considerably since the advent of the strapdown platform design philosophy in the mid 1940s [14]. This technology is also known as the *analytical platform* [15]. This technology implies that the delicate gyros and accelerometers, which were formerly rotationally isolated by the platform from rapid vehicular attitude variations, are now fully exposed to them. As a result, the admissible Q_{max} had to be increased to $\simeq$400deg/sec in order to cope with increased vehicular roll rates. This increase is by the factor of $(400\times3600/100) = 14.400$ which implies $Q_{max} = 1,440.000$ deg/hr. Maintaining the requirement for $Q_{min} = 0.01$ deg/hr would imply a dynamic range of $D = Q_{max}/Q_{min} = 1.44\times10^8$! This is a formidable requirement that cannot be met by present day electromechanical technology. The large values of Q_{max} involve large gyro torqueing currents and consequently, frequent temperature variations. These would upset the delicate adjustment of the neutral buoyancy in floated RIGs and degrade the accuracy of the performance specification of ~ 0.01deg/hr. However, in tactical applications, accuracies in the range of 0.1 deg/hr— 1 deg/hr are of practical interest, and these can still be achieved by floated RIG technology. Therefore, floated RIGs in this accuracy category have been manufactured for strapdown applications as indicated by the Lear Siegler 1903—1904 series [11]. For example Q_{max} for the 1904A Gyro is 400 deg/sec and the random drift rate is $Q_d = 0.2$ deg/hr rms. The outer dimensions are 1.6 in diameter by 3.2 in length. Yet, for tactical applications, the cost and volume of floated RIG technology may still be too high. Its reliability in the strapdown environment may be quite good thanks to the flotation technology. However, the need for a gyro technology that is more compatible with the strapdown environment, had become evident, especially for tactical applications. Its actual development got under way in the early 1960s and emerged as, the now well established, dynamically tuned gyro or dry tuned gyro (DTG) [16] also known as the dry tuned rotor (DTR) gyro.

Description of the dry tuned rotor gyro

Unlike the common single-degree-of-freedom floated RIG, the DTG is a two-axis device suspended by an inner Cardan-like flexure joint analogous to the inner gimbal suspension of the gyroscopic seeker head described earlier. A short history of its early development, a brief description of its construction, and principle of operation can be found in [17]. There are two types of flexure suspension. The first, based on bending flexures, is the basis for the design described in some detail later. A schematic cross section of the mushroom type construction based on the design of the Kearfott-Singer CONEX SKG3940 [18], is shown in Figure 4.23. This is a low cost-version of the Kearfott Gyroflex design, which is similar to the TAMAM (Israel Precision Instrument Industries) MINITUNE gyro. A perspective drawing showing the flexures and principal parts along with notations of variables is shown in Figure 4.24. The second type is based on torsional flexing as described in [14] and reviewed in [17], is a direct extension of the classical Cardan joint. A sketch of its mechanization along with the notations of principal angular variables is shown in Figure 4.25. Both designs, if properly executed, prevent any flexing due to external accelerations. The difference in the flexure design detail has essentially no effect on their respective performances. The bending flexure design has the advantage that it allows a gimbal with a small moment of inertia. However, the advantage of the torsion flexure is more ruggedness and ease of manufacturing. The DTG using bending flexures will now be described in some detail. It is a relatively recent major technological innovation and an excellent example of a sophisticated yet simple integrated design. It provides the necessary support to the rotor but essentially allows the ideal conditions for no friction angular freedom, which is required in a conventional free gyro equipped with gyro torquers and pickoffs. In contrast to earlier designs such as the first generation rate gyro or the RIG, in which each component was a distinct entity, in the DTG some of the mechanical and magnetic components are actually inseparable integral parts.

The gyro rotor (2) is connected to the spinner shaft (4) by means of the perpendicular two-axis flexure hinges (5), (5'), which allow the rotor two degrees of freedom of bending, α and β, in response to corresponding torques, but are very stiff in response to forces perpendicular to the spin axis. It is a special two-piece rotating flexure that supports and rotationally isolates the gyro rotor, thus eliminating all frictional restraints. The design provides very low anisoelasticity. This raises the natural frequency of the suspension well above the spin speed. It is very rugged and is capable of withstanding shocks in excess of $100g$ and intense vibration environments. The shaft, is mounted in cartridged preloaded precision ball bearings (5), which provide the required stiffness. Their outer races are fitted into the cartridge sleeve (8). The shaft is connected to the inner gimbal cap (5). The rotor, driven by the stator coil (10), is bonded to the outer gimbal (5').

The motor is of the two-phase synchronous hysteresis type, which can operate from a single-phase AC supply if used with the appropriate capacitor. The frequency of the supply must be regulated to assure a precise and constant spin speed for correct dynamical rotor tuning , a crucial requirement discussed later.

The gyro rotor consists of a high-permeability steel path and a ring magnet that provides the magnetic field for the torquer (3). The torquer stator consists of a ring supporting pairs of rectangular coils. Diametrically opposed coils are connected in series and oriented in orthogonal sets coincident with the gyro sensing axes. Thus, the rotor can be torqued by the current applied to the DC torquer stator coils (3). The pickoff stator (8) consists of two pairs of poles forming a four-pole magnetic structure. Each pickoff pole consists of a soft iron laminated core on which a single bobbin wound coil is mounted. The pole faces of the stator form a plane perpendicular to the spin axis and are therefore parallel to the gyro rotor face. The pickoff magnetic circuit is completed through a return path that is an integral part of the rotor. There is a small axial gap between the pole faces and the pickoff return path. The pickoff coil pairs, on diametrically opposite poles, are connected in series opposition. The two pole pairs constitute the two orthogonal sensing axes of the rotor tilt angles θ_x, θ_y. When the air gap pairs are equal, the output from each axis is null. When the rotor is tilted with respect to the gyro casing, the respective air gaps increase and decrease so that an output is generated that is proportional to the tilt angles.

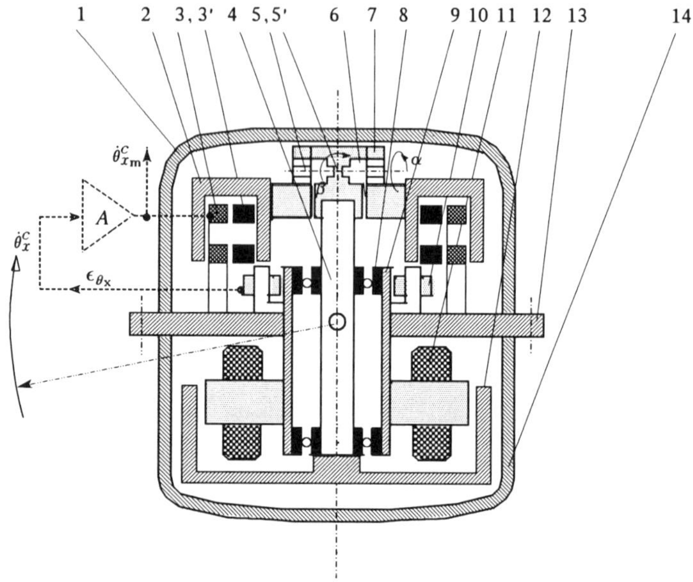

1 Top cover

2 Gyro rotor

3 Torquer coil

3' Torquer magnets

4 Spinner shaft

5 Outer flexure links gimbal to rotor

5' Inner flexure links shaft to gimbal

6 Clearance for rotor angular deflection

7 gimbal

8 Spinner bearings

9 Bearing cartridge

10 Rotor deflection pickoff

11 Motor stator

12 Motor rotor

13 Mounting flange

14 Bottom cover

Figure 4.23. Schematic cross section and main parts of a DTG.

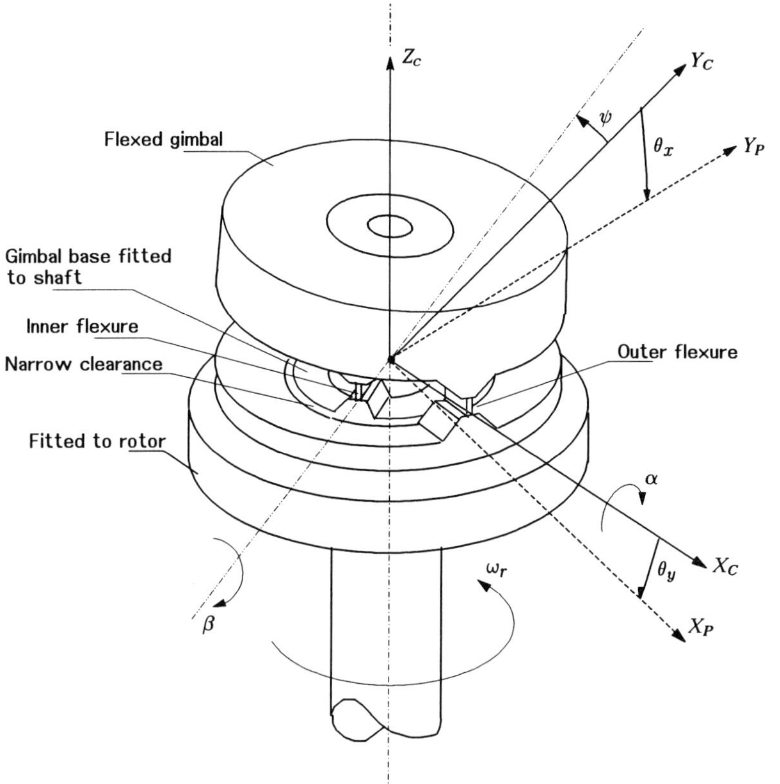

C Case axes

P In plane of rotor axes (deflected from C by X, Y rotations)

R Rotating rotor axes

G Gimbal axes

θ_x, θ_y Angular deflections of rotor plane with respect to case axes

α = Gimbal to rotor deflection

β = Gimbal to shaft deflection

Figure 4.24. Dry tuned rotor gimbals with bending flexures.

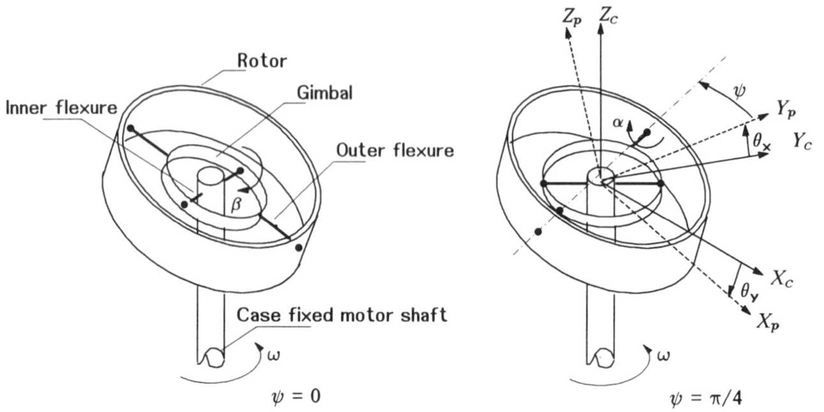

Figure 4.25. Dry tuned rotor gimbals with torsion flexures.

A cutaway view of a Teledyne SDG-5 DTG, which is of similar design, is shown in Figure 4.26.

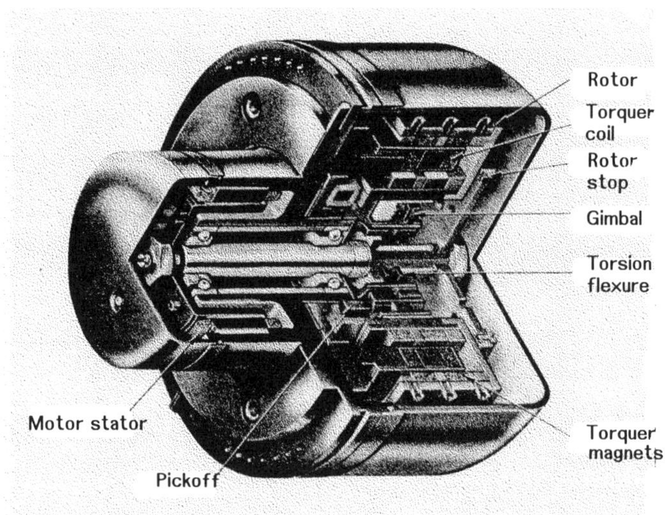

Figure 4.26. Cutaway view of a Teledyne SDG-5 DTG with bending flexures (based on Teledyne briefing, Driru II).

Principle of operation

Before going into a detailed analysis, a heuristic explanation of the principle of operation will be given. $X_cY_cZ_c$ is a gyro-case-fixed coordinate set. xyz is a coordinate set fixed to the shaft rotating with an angular velocity ω_s relative to XYZ. The angular momentum of the spinning rotor tends to maintain its spin axis fixed in inertial space whenever the gyro casing, along with the shaft, rotates with the angular rate components $\dot{\theta}_x^c$ and $\dot{\theta}_y^c$ relative to inertial space. Clearly, the gimbal flexures must flutter to permit the corresponding angular rotor deflections α and β between the rotor spin axis and the shaft while spinning. The flutter manifests itself as angular periodic motions at the frequency ω_s and angular amplitudes $\alpha_m = \theta_x$ and $\beta_m = \theta_y$. These deflections cause periodic elastic restraining torques in the hinge T_x and T_y with the peak values of $K_x\theta_x$ and $K_y\theta_y$, K_x, K_y being the respective spring constants. Once $\dot{\theta}_x^c$ and $\dot{\theta}_y^c$ return to zero, the flutter energy dissipates so that its amplitude decays to zero and the rotor spin axis realigns itself perpendicularly to the shaft axis. However, the elastic restraining torques are not the only ones. The gimbal, having a moment of inertia L_g to be defined later. While spinning at ω_s, accumulates the angular momentum $H = L_g\omega_s$. The sinusoidal flutter, around say, the x axis, reaches a peak angular rate of $\theta_x\omega_s$, which causes a periodic precession torque T_{py} around the y axis. Similarly, a corresponding precession torque T_{px} is exerted around the x axis due to $\theta_y\omega_s$. The peak precession torques T_{px} or T_{py} are therefore $\theta_x\omega_s H = \theta_x\omega_s\, L_g\omega_s = \theta_x L_g \omega_s^2$. Detailed analysis demonstrates that these precession torques are opposite in sign to the elastic restraining torques T_x, T_y. Therefore, they exhibit negative spring constants $L_g\omega_s^2$. Thus, for instance, the net torque around x is

$$T = T_x - T_{px} = K_x\theta_x - L_g\omega_s^2\theta_x = (K_x - L_g\omega_s^2)\theta_x$$

In order to achieve a zero net torque, that is, the equivalent of a free suspension, it is required that the total spring constant $K_x - L_g\omega_s^2 = 0$, namely, the spin speed must be

$$\omega_s^2 = \frac{K_x}{L_g}$$

This expression signifies the concept of *tuning*. For given values of K_x and L_g, ω_s must be precisely adjusted to fulfill the last equation. Clearly, the stability of the mechanical parameters K_x, K_y, and L_g, as well as the electronic drive signal generator frequency ω_s, are of paramount importance in order to assure close to perfect angular isolation of the rotor from the gyro casing. If the tuning is not perfect, the resulting imperfect dynamical isolation will cause leaking of torques to the rotor, consequently causing gyro drift rates. Once the tuning has been accomplished, there is no net energy dissipation in the flexure joint. The energy dissipated by the periodic flexing of the joint is compensated by the energy being pumped from the kinetic energy of the rotating structure via the negative spring constant into the flexure. It will therefore continue fluttering indefinitely and will maintain the tilt angle between rotor and shaft as long as $\dot{\theta}_x$ and $\dot{\theta}_y$ persist.

The analytical model: error sources ⊕

The analytical model presented is largely based on [14]. As shown in Figures 4.24 and 4.25, the rotor plane is tilted by the angles θ_x and θ_y with respect to the gyro case axis system. In order to maintain this angular tilt, the flex angles α and β must vary sinusoidally as $\psi = \omega t$ grows linearly with time. This relationship between α, β, and ψ is determined by the Euler transformation

$$\alpha = \theta_x \cos\psi + \theta_y \sin\psi \tag{4.8.1}$$

$$\beta = -\theta_x \sin\psi + \theta_y \cos\psi \tag{4.8.2}$$

The inverse relationship is

$$\theta_x = \alpha\cos\psi - \beta\sin\psi \tag{4.8.3}$$

$$\theta_y = \alpha\sin\psi + \beta\cos\psi \tag{4.8.4}$$

The torque equation for the gimbal and rotor about the gimbal Y axis is given by

$$T^g_{gy} = T^g_{g\alpha_y} - C_{\dot{\beta}}\dot{\beta} - K_\beta\beta + T^g_{g\epsilon_y} \tag{4.8.5}$$

or

$$T^g_{g\alpha_y} = T^g_{gy} + C_{\dot{\beta}}\dot{\beta} + K_\beta\beta - T^g_{g\epsilon_y} \tag{4.8.6}$$

in which the definitions of the notations are

T^g_{gy} — total torque on the gimbal about the gimbal Y axis;

K_β — torsional spring coefficient of the β flexure;

$C_{\dot{\beta}}$ — aerodynamic viscous damping coefficient due to angular rate $\dot{\beta}$;

$T^g_{g\alpha_y}$ — y axis reaction torque on the gimbal at the α flexure junction;

$T^g_{g\epsilon_y}$ — extraneous error torques on the gimbal about its Y-axis.

Similarly, torques on the rotor about the gimbal X, Y, Z axes can be written as follows:

$$T^g_{rx} = T^g_{rT_x} - C_{\dot{\alpha}}\dot{\alpha} - K_\alpha\alpha + T^g_{r\epsilon_x} \tag{4.8.7}$$

$$T^g_{r_y} = T^g_{rT_y} + T^g_{r\alpha_y} + T^g_{r\epsilon_y} = T^g_{rT_y} + T^g_{r\epsilon_y} - T^g_{g\alpha_y} \tag{4.8.8}$$

$$T^g_{r_z} = T^g_{rT_z} + T^g_{r\alpha_z} + T^g_{r\epsilon_z} \tag{4.8.9}$$

The above notations are defined as follows:

$T^g_{r_i}$ —total torque on the rotor about the gimbal i axis ($i = X, Y, Z$);

$C_{\dot\alpha}$ — aerodynamic viscous drag coefficient due to angular rate $\dot\alpha$;

$T^g_{r\alpha_i}$ — reaction torque on the rotor at the α flexure junction about the gimbal i axis ($i = Y$ or Z);

$T^g_{r\epsilon_i}$ — error torque on the rotor about the gimbal i-axis ($i = X, Y, Z$);

$T^g_{rT_i}$ — torque generator torque, intentionally applied to the rotor, acting about the gimbal i axis, ($i = X, Y, Z$);

The Y axis torque component $T^g_{r\alpha_y}$ in (4.8.8) acting on the rotor at the α flexure is equal to $-T^g_{g\alpha_y}$ defined in (4.8.6).

Equations (4.8.5)–(4.8.9) with (4.8.6) can now be transformed to the pickoff coordinates as follows:

$$\boldsymbol{T}^p_r = C^p_g\left\{\boldsymbol{T}^g_{rT} + \boldsymbol{T}^g_{re} + \begin{bmatrix} -C_{\dot\alpha}\dot\alpha - K_\alpha\alpha \\ -\left(T^g_{g_y} + K_\beta\beta + C_{\dot\beta}\dot\beta\right) + T^g_{g\epsilon_y} \\ T^g_{r\alpha_z} \end{bmatrix}\right\} \tag{4.8.10}$$

where

C^p_g — direction cosine matrix relating gimbal to pickoff coordinates;

$\boldsymbol{T}^g_{rT}$, $\boldsymbol{T}^g_{re}$ —torque vectors with gimbal frame components $T^g_{rT_x}$, $T^g_{rT_y}$, etc., in accordance with (4.8.5)–(4.8.9);

$\boldsymbol{T}^p_r$—net rotor torque in pickoff coordinates.

The direction cosine matrix C^p_g is determined by the transpose of the three-axis Euler transformation in accordance with Appendix A5 where small angles, are assumed for ϕ and θ. Specifically, ϕ represents α and θ represents β. In relating the gimbal coordinates to the pickoff coordinates, we note from the construction that only α comes to bear in relating p and g coordinates.

Consequently, C^p_g is given by

$$C^p_g = \begin{bmatrix} \cos\psi & -\sin\psi & -\alpha\sin\psi \\ \sin\psi & \cos\psi & \alpha\cos\psi \\ 0 & -\alpha & 1 \end{bmatrix} \tag{4.8.11}$$

Substituting (4.8.11) into (4.8.10) and expanding, we obtain the explicit set of torque equations in pickoff axes:

$$T^p_{r_x} = T^p_{rT_x} + (T^p_{r\varepsilon_x} - T^g_{g\varepsilon_y}\sin\psi) - (C_{\dot\alpha}\dot\alpha + K_\alpha\alpha)\cos\psi + (T^g_{gy} + C_{\dot\beta}\dot\beta + K_\beta\beta)\sin\psi - T^g_{R\alpha_z}\alpha\sin\psi \tag{4.8.12}$$

$$T^p_{r_y} = T^p_{rT_y} + (T^p_{r\varepsilon_y} + T^g_{g\varepsilon_y}\cos\psi) - (C_{\dot\alpha}\dot\alpha + K_\alpha\alpha)\sin\psi + (T^g_{gy} + C_{\dot\beta}\dot\beta + K_\beta\beta)\cos\psi - T^g_{R\alpha_z}\alpha\cos\psi \tag{4.8.13}$$

$$T^p_{r_z} = (T^p_{r\varepsilon_z} - \alpha T^g_{g\varepsilon_z}) + T^g_{r\alpha_z} + \alpha(T^g_{gy} + K_\beta\beta + C_{\dot\beta}\dot\beta) \tag{4.8.14}$$

In which the torque components $T^p_{r_i}$, $T^p_{rT_i}$, $T^p_{r\varepsilon_i}$ ($i = X, Y, Z$) are components of $\boldsymbol{T}^p_r$, $\boldsymbol{T}^p_{rT}$, $\boldsymbol{T}^p_{r\varepsilon}$. In (4.8.12−4.8.14) the Z component of $\boldsymbol{T}^p_{rT}$ is absent. This is so because the gyro torque is designed to apply torques about the X and Y axes only. Equation (4.8.14) can be used to solve for $T^g_{r\alpha_z}$ and substitute the solution into the X, Y equations (4.8.12)−(4.8.13). Rearranging and dropping α squared terms yields

$$T^p_{r_x} = T^p_{rT_x} - T^p_{r_z}\alpha\sin\psi - (C_{\dot\alpha}\dot\alpha + K_\alpha\alpha)\cos\psi + (T^g_{gy} + C_{\dot\beta}\dot\beta + K_\beta\beta)\sin\psi + T_{\varepsilon_x} \tag{4.8.15}$$

$$T^p_{r_y} = T^p_{rT_y} + T^p_{r_z}\alpha\cos\psi - (C_{\dot\alpha}\dot\alpha + K_\alpha\alpha)\sin\psi - (T^g_{gy} + C_{\dot\beta}\dot\beta + K_\beta\beta)\cos\psi + T_{\varepsilon_y} \tag{4.8.16}$$

where T_{ε_x}, T_{ε_y} are the composites of all the spurious torque disturbances in the X, Y equations.

The above torque equations are related to the rotational movement of the rotor by means of the angular momentum equations. The angular rate of the gimbal in gimbal axes is expressed as follows:

$$\boldsymbol{\omega}^g_g = C^g_c(\boldsymbol{\omega}^c_c + \boldsymbol{\omega}^c_s) + \dot{\boldsymbol{\beta}}^g \tag{4.8.17}$$

where

$\boldsymbol{\omega}^g_g$ —gimbal inertial angular rate vector in gimbal coordinates;

$\boldsymbol{\omega}^g_c$ — gyro case angular rate vector in case coordinates;

$\boldsymbol{\omega}^c_s$ — rotor spin rate vector in case coordinates;

$\dot{\boldsymbol{\beta}}^g$ — rate of change of the angle vector $\boldsymbol{\beta}^g$ in gimbal coordinates;

C^g_c — direction cosine matrix relating case and gimbal coordinates.

Assuming small angles for β, and considering the construction geometry, the C^g_c matrix can be shown to be

$$C_c^g = \begin{bmatrix} \cos\psi & \sin\psi & -\beta \\ -\sin\psi & \cos\psi & 0 \\ \beta\cos\psi & \beta\sin\psi & 1 \end{bmatrix} \tag{4.8.18}$$

The torque on the gimbal in gimbal coordinates is derived from the time derivative of the gimbal angular momentum vector

$$\boldsymbol{T}_g^g = I_g^g \dot{\boldsymbol{\omega}}_g^g + \boldsymbol{\omega}_g^g \times (I_g^g \boldsymbol{\omega}_g^g) \tag{4.8.19}$$

where

I_g^g is the inertia tensor of the gimbal defined in gimbal coordinates

The components of the various angular rate vectors are defined by the mechanization geometry and they are denoted by

$$\boldsymbol{\omega}_c^c = \begin{bmatrix} \omega_\mu & \omega_\nu & \omega_\lambda \end{bmatrix}^T \tag{4.8.20}$$

$$\boldsymbol{\omega}_s^c = \begin{bmatrix} 0 & 0 & \omega_s \end{bmatrix}^T \tag{4.8.21}$$

$$\dot{\boldsymbol{\beta}}^g = \begin{bmatrix} 0 & \dot{\beta} & 0 \end{bmatrix}^T \tag{4.8.22}$$

$$I_g^g = \begin{bmatrix} I_g & 0 & 0 \\ 0 & I_g & 0 \\ 0 & 0 & I_g \end{bmatrix} \tag{4.8.23}$$

where

ω_μ, ω_ν, ω_λ — X, Y, Z components of the case rate $\boldsymbol{\omega}_c$ in case coordinates.
ω_s — spin motor angular rate;
I_g — gimbal transverse moments of inertia about the X, Y axes;
J_g — gimbal polar moment of inertia about the Z axis.

We now substitute (4.8.15), (4.8.18) and (4.8.20—4.8.23) into (4.8.19). Neglecting β squared terms, recalling that $\dot{\psi}=\omega_r$, and rearranging, we obtain the explicit expression for T_{gy}^g, which appears in (4.8.15) and (4.8.16),

$$T_{gy}^g = I_g[\ddot{\beta} + (\dot{\omega}_\nu - \omega_s\omega_\mu)\cos\psi - (\dot{\omega}_\mu + \omega_s\omega_\nu)\sin\psi] + (I_g - J_g)[-\beta\omega_s^2 + \omega_s(\omega_\mu\cos\psi + \omega_\nu\sin\psi - 2\beta\omega_\lambda) + \omega_\lambda(\omega_\mu\cos\psi + \omega_\nu\sin\psi)] \tag{4.8.24}$$

The expressions for $\dot{\alpha}$, $\dot{\beta}$, $\ddot{\beta}$ for (4.8.15), (4.8.16), and (4.8.24) are derived by differentiating (4.8.1—4.8.2), and rearranging and recalling that $\dot{\psi}=\omega_s$ so that,

$$\dot{\alpha} = \dot{\theta}_x\cos\psi + \dot{\theta}_y\sin\psi + \omega_s\beta \tag{4.8.25}$$

$$\dot{\beta} = \dot{\theta}_y\cos\psi - \dot{\theta}_x\sin\psi - \omega_s\alpha \tag{4.8.26}$$

$$\ddot{\beta} = \ddot{\theta}_y\cos\psi - \ddot{\theta}_x\sin\psi - \alpha\dot{\omega}_s - 2\omega_s(\dot{\theta}_y\sin\psi + \dot{\theta}_x\cos\psi) - \omega_s^2\beta \tag{4.8.27}$$

To further develop (4.8.15–4.8.16), we need an expression that relates the torque components T^p_{rx} and T^p_{ry} in the pickoff coordinates to the rotor angular momentum in the p coordinates. The net angular rate in p coordinates is defined as the sum of the consecutive Euler angular rate vector components:

$$\boldsymbol{\omega}^p_r = C^p_c(\boldsymbol{\omega}^c_c + \boldsymbol{\omega}^c_s) + C^p_g(\dot{\alpha}^g + \dot{\beta}^g) \tag{4.8.28}$$

where

$\boldsymbol{\omega}^p_r$— total rotor angular rate in p coordinates;

$\dot{\alpha}^g$, β^g— rates of change of flexure angles in gimbal coordinates;

C^p_c, C^p_g – direction cosine matrices relating p, c, and g coordinates.

The pickoff angle θ is the angle between the case and pick-off axes. Therefore, assuming small angles, the first term in (4.8.28), θ, can be rewritten, after rearranging terms, as

$$C^p_c(\boldsymbol{\omega}^c_c + \boldsymbol{\omega}^c_r) = \boldsymbol{\omega}^c_c + \boldsymbol{\omega}^c_r - \boldsymbol{\theta}^c \times (\boldsymbol{\omega}^c_c + \boldsymbol{\omega}^c_r) \tag{4.8.29}$$

The transformation C^p_g involves only small deflection angles α and β; thus $\boldsymbol{\theta}$ is their vector sum. Therefore, the second term in (4.8.28) can be written as

$$C^p_g(\dot{\alpha}^g + \dot{\beta}^g) = C^p_g \boldsymbol{\theta}^g \cong \dot{\boldsymbol{\theta}}^c + \boldsymbol{\theta}^c \times \boldsymbol{\omega}^c_r \tag{4.8.30}$$

where

$\dot{\boldsymbol{\theta}}^c$ – angular rate of the pickoff angle vector in case or pick-off coordinates. By substituting (4.8.29) and (4.8.30) into (4.8.28), we express the total rotor angular rate in terms of known vector quantities,

$$\boldsymbol{\omega}^p_r = \boldsymbol{\omega}^c_c + \boldsymbol{\omega}^c_s - \boldsymbol{\theta}^c_c \times \boldsymbol{\omega}^c_c \tag{4.8.31}$$

The p frame rotation rate as measured in the p frame is similarly derived,

$$\boldsymbol{\omega}^p_p = C^p_c \boldsymbol{\omega}^c_c + \dot{\boldsymbol{\theta}}^c = \boldsymbol{\omega}^c_c + \dot{\boldsymbol{\theta}}^c - \boldsymbol{\theta}^c \times \boldsymbol{\omega}^c_c \tag{4.8.32}$$

The equation relating the angular momentum of the rotor to its torque in p in its p coordinates is given by the vector derivative

$$\boldsymbol{T}^p_r = I^p_r \dot{\boldsymbol{\omega}}^p_r + \boldsymbol{\omega}^p_p \times (I^p_r \boldsymbol{\omega}^p_r) \tag{4.8.33}$$

where

$\boldsymbol{T}^p_r$—net torque on the rotor assembly in pickoff coordinates;

I^p_r—inertia tensor of the rotor in p coordinates.

Specifically, the elements of the vector and matrix expressions are defined by

$$\boldsymbol{\theta}^c = \begin{bmatrix}\theta_x & \theta_y & 0\end{bmatrix}^T \tag{4.8.34}$$

$$\dot{\boldsymbol{\theta}}^c = \begin{bmatrix}\dot{\theta}_x & \dot{\theta}_y & 0\end{bmatrix}^T \tag{4.8.35}$$

$$I_r^p = \begin{bmatrix} I_r & 0 & 0 \\ 0 & I_r & 0 \\ 0 & 0 & J_r \end{bmatrix} \tag{4.8.36}$$

Using (4.8.20—4.8.23) for the rest of the vector quantities, (4.8.33) can be expanded to yield the explicit expressions for the torque–momentum relationships in scalar form in the p coordinates

$$T_{rx}^p = I_r(\dot{\omega}_\mu + \ddot{\theta}_x - \dot{\theta}_y\omega_\lambda - \theta_y\dot{\omega}_\lambda) + J_r\omega_r(\omega_\nu + \dot{\theta}_y + \theta_x\omega_\lambda)$$
$$+(J_r - I_r)\omega_\lambda(\omega_\nu + \dot{\theta}_y) \tag{4.8.37}$$

$$T_{ry}^p = I_r(\dot{\omega}_\nu + \ddot{\theta}_y - \dot{\theta}_x\omega_\lambda - \theta_x\dot{\omega}_\lambda) + J_r\omega_r(\omega_\mu + \dot{\theta}_x + \theta_y\omega_\lambda)$$
$$+(J_r - I_r)\omega_\lambda(\omega_\mu + \dot{\theta}_x) \tag{4.8.38}$$

$$T_{rz}^p \cong J_r(\dot{\omega}_r + \dot{\omega}_\lambda) \tag{4.8.39}$$

Equations (4.8.15—4.8.16), (4.8.24) and (4.8.35—4.8.39) combined with (4.8.1—4.8.2) and (4.8.25—4.8.26) result in the explicit expressions of the dynamic response of the DTG. In the process of the substitutions and rearrangements involved, the inertia terms combine to define the following terms:

$$J = J_r + \frac{1}{2}J_g \tag{4.8.40}$$

$$I = I_r + \frac{1}{2}I_g \tag{4.8.41}$$

$$L_g = I_g - \frac{1}{2}J_g \tag{4.8.42}$$

where I represents transverse inertia and J represents polar inertia.

To further define the various error sources, we express the flexure spring constants in terms of their nominal values and deviations thereof:

$$K_\alpha = K_0 + \delta K_\alpha \tag{4.8.43}$$

$$K_\beta = K_0 + \delta K_\beta \tag{4.8.44}$$

Similarly, for the spin speed

$$\omega_r = \omega_{ro} + \delta_{\omega_r} \tag{4.8.45}$$

Accordingly, the angular momentum is

$$H_0 + \delta H_0 = J\omega_{r0} + J\delta_{\omega_r} \tag{4.8.46}$$

The DTG is commanded by torque generators. Their torques T_{rT_x}, T_{rT_y} are defined in terms of the corresponding orthogonal precession command rates ω_{T_x}, ω_{T_y}. Furthermore, scale factor errors ε_x, ε_y and the effect of cross-axis coupling should be incorporated in the error model. Thus, we have

$$T_{rT_x} = \frac{H_0}{1+\varepsilon_y}\omega_{T_y} + \beta_{yx}\omega_{T_x} \tag{4.8.47}$$

$$T_{rT_y} = \frac{H_0}{1+\varepsilon_x}\omega_{T_x} + \beta_{xy}\omega_{T_y} \tag{4.8.48}$$

$$T_{e_x} = -\frac{H_0}{1+\varepsilon_y}\omega_{B_y} \tag{4.8.49}$$

$$T_{e_y} = -\frac{H_0}{1+\varepsilon_x}\omega_{B_x} \tag{4.8.50}$$

where

$\omega_{T_x}, \omega_{T_y}$ — X, Y axes torquer command rates;

β_{yx}, β_{xy} — torquer cross coupling coefficients;

$\varepsilon_x, \varepsilon_y$ — torquer X, Y scale factor errors;

$\omega_{B_x}, \omega_{B_y}$ — X, Y axis bias errors.

In addition to these error sources, there are the usual small case-axes angular misalignment errors γ_x, γ_y, γ_z with respect to the nominal X, Y, Z input angular rate components ω_x, ω_y, ω_z.

Thus, the actually sensed angular rate components are derived by means of the small angle Euler transformation:

$$\omega_\mu = \omega_x + \gamma_z\omega_y + \gamma_y\omega_z \tag{4.8.51}$$

$$\omega_\nu = \omega_y + \gamma_x\omega_z + \gamma_z\omega_x \tag{4.8.52}$$

$$\omega_\lambda = \omega_z + \gamma_y\omega_x + \gamma_x\omega_y \tag{4.8.53}$$

Finally, recall the trigonometric identities

$$\sin^2\phi = \frac{1}{2}(1-\cos 2\phi) \tag{4.8.54}$$

$$\cos^2\phi = \frac{1}{2}(1+\cos 2\phi) \tag{4.8.55}$$

$$\sin\phi\cos\phi = \frac{1}{2}\sin 2\phi \tag{4.8.56}$$

with (4.8.1—4.8.2), (4.8.15—4.8.16), (4.8.24), (4.8.37—4.8.39), and (4.8.43—4.8.44) through (4.8.54—4.8.56), can be combined to yield the complete gyro input-output dynamical equations, which include all the error sources introduced here. After rearrangement, observing that the gimbal inertia is much smaller than the rotor inertia and dropping high-order terms results in the two equations that relate the commanded torquer input angular rates to the actual rotor tilt angles $\dot{\theta}_x$ and $\dot{\theta}_y$ in the presence of all the parametric deviations, disturbance torques, biases, misalignments, and spring constant errors:

$$\omega_{T_x} = (1+\varepsilon_x)[(\omega_x+\dot{\theta}_x)+\gamma_z\omega_y-(\gamma_y+\theta_y)\omega_z+\beta_{xy}\omega_{T_y}$$
$$-\frac{1}{H_0}(\dot{\omega}_y+\ddot{\theta}_y)+\frac{(J-I)}{H_0}\omega_z(\omega_x+\dot{\theta}_x)+\frac{J}{H_0}(\omega_x+\dot{\theta}_x)\delta\omega_r]$$
$$+\omega_{B_x}-\frac{1}{H_0}[(K_0-L_g\omega_{ro}^2)-2L_g\omega_{ro}\delta\omega_r+\frac{1}{2}(\delta K_\alpha-\delta K_\beta)]\theta_y$$
$$+\frac{1}{2J}(C_{\dot{\alpha}}+C_{\dot{\beta}})\dot{\theta}_x \qquad (4.8.57)$$

$$\omega_{T_y} = (1+\varepsilon_y)[(\omega_y+\dot{\theta}_{yy})+\gamma_z\omega_x-(\gamma_x+\theta_x)\omega_z+\beta_{yx}\omega_{T_x}$$
$$-\frac{1}{H_0}(\dot{\omega}_x+\ddot{\theta}_x)+\frac{(J-I)}{H_0}\omega_z(\omega_y+\dot{\theta}_y)+\frac{J}{H_0}(\omega_y+\dot{\theta}_y)\delta\omega_r]$$
$$+\omega_{B_y}-\frac{1}{H_0}[(K_0-L_g\omega_{ro}^2)-2L_g\omega_{ro}\delta\omega_r+\frac{1}{2}(\delta K_\alpha-\delta K_\beta)]\theta_x$$
$$+\frac{1}{2J}(C_{\dot{\alpha}}+C_{\dot{\beta}})\dot{\theta}_y \qquad (4.8.58)$$

Rotor tuning

Equations (4.8.57) and (4.8.58) express the linear relationship between the torquer command rates ω_{T_x}, ω_{T_y} and the gyro input rates ω_x, ω_y. In addition to the various error sources, there is the large spring effect $(K_0 - L_g\omega_{ro}^2)\boldsymbol{\theta}$, where $\boldsymbol{\theta}$ consists of the components θ_x, θ_y. The essence of the underlying idea of the DTG is to eliminate this dominant term. This is achieved by designing the flexures so that

$$K_0 = L_g\omega_{ro}^2 \qquad (4.8.59)$$

The resulting ω_{ro} given by

$$\omega_{ro} = \sqrt{\frac{K_0}{L_g}} \tag{4.8.60}$$

is the implemented rotor spin speed is $\omega_s = \omega_{ro}$. The implementation of this equation is known as *rotor tuning*. In practice, it is achieved by careful grinding of the flexures to adjust K_0 and by adjustment screws in the gimbal to fine tune L_g. A graphical representation of the tuning condition is shown in Figure 4.27.

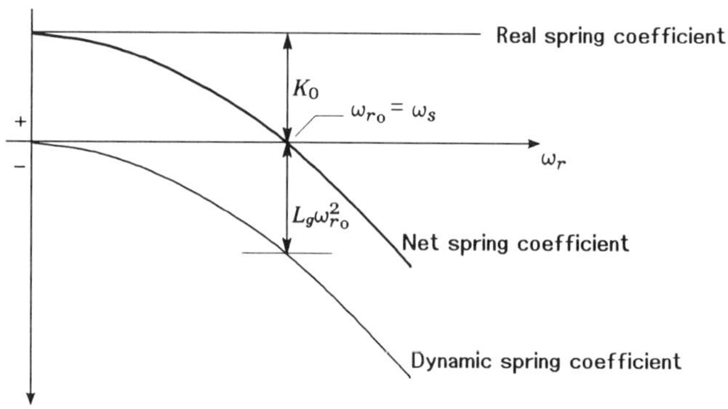

Figure 4.27. Principle of operation of a tuned rotor.

Dropping the forcing command rates on the left hand side of (4.8.58−4.8.59), reveals the contribution of each error term to the gyro drift rate. In the framework of a given choice of gyro technology, materials and manufacturing method, it is possible to predict the performance of a given design in terms of net drift rate.

Performance as a free gyro

In order to reveal the principal property of this gyro, we now assume that all the errors and command input rates in (4.8.58−4.8.59) are zero. This would leave only the terms with $\dot{\theta}_x + \omega_x = 0$ and $\dot{\theta}_y + \omega_y = 0$, namely, the ideal relationships:

$$\dot{\theta}_x = -\omega_x \tag{4.8.61}$$

$$\dot{\theta}_y = -\omega_y \tag{4.8.62}$$

Recall that $\dot{\theta}_x$, $\dot{\theta}_y$ are the rotor angular rates with respect to the case and that ω_x, ω_y are the angular rates of the case with respect to inertial space. This clearly implies that the rotor retains its orientation in inertial space regardless of the rotation of the gyro case. It should, of course, be borne in mind that this holds only for very small angular deviations θ_x, θ_y before being stopped by the narrow cylindrical gap shown in Figure 4.22. The pickoff evidently measures these rotor/case angles. Therefore, this free gyro has no practical value. If, however, it is operated in a closed loop, it can perform as a high-grade rate gyro in a fashion similar to the RIG, which, for similar reasons, is also impractical in open-loop operation. Clearly, if any or all errors are nonzero, the gyro will be subjected to a corresponding drift rate. For example, with the bias terms ω_{B_x}, ω_{B_y} present, the free gyro operation mode would yield

$$\dot{\theta}_x = -\omega_x - \omega_{B_x} \tag{4.8.63}$$

$$\dot{\theta}_y = -\omega_y - \omega_{B_y} \tag{4.8.64}$$

Similar relationships for the effects of all other error sources can be derived, including scale factor errors ε_x, ε_y, spin speed variations, misalignment errors and spring coefficient deviations from their nominal values.

Operation in closed loop

It follows from the foregoing that the DTG performs as a two-axis free gyro. As indicated earlier, its angular freedom is confined by the narrow gap (6) to about one degree. From the foregoing, shown in Figure 4.22, its usefulness as a free gyro is practically zero. Yet, its employment in the forced balance mode, renders it as a high-quality rate gyro much the same as the floated RIG in the FBS mode described in Section 4.7. Let θ_x^c, θ_y^c and θ_x, θ_y be the orientations of the gyro casing and rotor,respectively, relative to inertial space. The pickoff reading, for example, along the X axis is given by

$$\epsilon_{\theta_x} = \theta_x^c - \theta_x \tag{4.8.65}$$

The corresponding electrical signal is amplified by A and fed into the torquer (3) to precess the rotor (2). The circuit shown in dotted lines in Figure 4.22 represents the FBS loop closure. The pickoff (10) senses the deflection ϵ_{θ_x}. The loop closure shown in dotted lines in Figure 4.22 implements the following:

$$\omega_{T_x} = \dot{\theta}_x = \frac{Ab_y}{H_0}\epsilon_{\theta_x} = \frac{Ab_y}{H_0}(\theta_x^c - \theta_x) \tag{4.8.66}$$

The precession rate vector component $\dot{\theta}_x$ is normal to the torque component

Ab_y, where b_y is the torque coefficient of the y-torquer in gr cm/amp. Solving for θ_x after Laplace transforming, we have

$$\theta_x(s) = \frac{\frac{Ab_y}{H_0}}{s + \frac{Ab_y}{H_0}} \theta_x^c(s) \tag{4.8.67}$$

This result holds equally for the y axis. We will use the index xy to express the two-axis nature of the gyro. It follows that the rotor tilt θ_{xy} very closely tracks the casing angle θ_{xy}^c with a first-order time lag $\tau = H_0/Ab_{xy}$. Substituting (4.8.5) into (4.8.3) and multiplying by A, we obtain the output of A, which is denoted by $\dot{\theta}_{xym}^c$, namely, the measured angular rate, as indicated in Figure 4.22:

$$\theta_{xym}^c(s) = A\epsilon_{\theta_{xy}} = \frac{As}{s + \frac{Ab_{xy}}{H}} \theta_{xy}^c(s) = \frac{A}{s + \frac{Ab_{xy}}{H_0}} \dot{\theta}_{xy}^c(s) \tag{4.8.68}$$

In the steady state, as $s \to 0$,

$$\theta_{xym}^c(s) \cong \frac{H_0}{b_{xy}} \dot{\theta}_{xy}^c(s) \tag{4.8.69}$$

Thus, the measured angular rate of the gyro case, $\theta_{xym}^c(s)$, i.e., the vehicle body, is proportional to the actual angular rate $\dot{\theta}_{xy}^c(s)$ by the scale factor H_0/b_{xy}. The angular deflection in the closed-loop $\epsilon_{\theta_{xy}}$ is given by

$$\epsilon_{\theta_{xy}}(s) = \frac{1}{s + \frac{Ab_{xy}}{H_0}} \dot{\theta}_{xy}^c(s) \tag{4.8.70}$$

Thus, as expected, for large values of A, the deviation $\epsilon_{\theta_{xy}}$ can be made vanishingly small. This implies that the flutter amplitude in the steady state is very small implying small dynamical errors and less hinge fatigue. It is clear that large angular rates $\dot{\theta}_{xy}(s)$ require large torqueing currents. This is especially true for transients in the angular rates which may cause significant peaks and large temperature fluctuations. This factor has indeed been one of the principal drift sources in the DTG. The approximation in (4.8.6) holds in the steady state $s \to 0$. It is very similar to closed-loop dynamics of the RIG as given by (4.8.8). It demonstrates that in the closed loop, the DTG performs like a two-axis rate gyro. The scale factor precision and its firmness are determined by the ratio H_0/b_{xy}. A major issue, as in other sensors, is modeling variations of various errors as a function of environmental effects such as temperature, acceleration, and vibration. Details on this topic can be found, among others in [14]. Effects of anisoelasticity and anisoinertia are addressed in [17].

Performance characteristics

The principal performance characteristics of the TAMAM (Israel Precision Instruments Industry) moderate-performance strapdown gyro are given in Table 4.2. The figures given for the constant drift rates are without compensation.

Table 4.2. Performance Characteristics of a DTG (TAMAM Minitune)

Applications:

Stabilization: Optical line of sight, Fire control, Radar antennas
Flight control and guidance of tactical missiles
Short-term and GPS-aided Inertial Navigation

Performance Parameters:

Peak input	±200 deg/sec
Drift bias	15 deg/hr
Drift, day-to-day turn-on	0.2 deg/hr
Drift random	0.05 deg/hr (1σ)
Drift g-sensitivity	10 deg hr/g
Drift g^2-sensitivity	0.1 deg/hr/g^2 (Anisoelasticity)
Drift temp. sensitivity	0.1 deg/hr/deg C

Electrical Characteristics:

Type: Hysteresis synchronous 2ϕ
Excitation: 5.5 volts rms, 480 Hz
Running power: 3.2 watts
Starting power: 5.0 watts
Run-up time: 6 sec

Pickoff:

Excitation: 2.5 volts, 19.2 kHz 1ϕ
Phase shift: 5 deg
Sensitivity: 11.5 volt/deg ±25%

Torquer:

Sensitivity: 1100 deg/hr/mA ±10%
Linearity: 0.05% of nominal

Environmental Conditions:

Operating temperature: −40 deg C to +90 deg C
Random vibration: 12 g rms, 20–2000 Hz
Shock: 40 g for 11 msec
Acceleration: 100 g
Dimensions: 32x32x53.5 mm (Incl. electronics)
Weight: 140 gr

Summary of main properties

The DTG strapdown version was initially developed for applications in transport aircraft for the 1 nmi/hr category,with the purpose of replacing the older generation floated RIGs. The idea had its merits because the environment of these aircraft did not overly tax the requirements for the rotor torqueing rates. Besides the positional information provided by the DTG INS, additional bonuses have been high-grade flight control outputs such as attitude and heading, body angular rates, and body accelerations. However, expanding the field of applications to high-performance military aircraft and tactical missiles required peak rotor precession rates on the order of 400 deg/hr. These rates cause considerable heating of the rotor torquer coils and consequent gyro temperature fluctuations. It is hard to achieve this level of performance without giving up some of the precision. Additional development effort was required to overcome this barrier. By and large, the performance— cost characteristics of the DTG turned out to be close to earlier floated RIG's. Yet, some operational and maintenance advantages over the RIG remain. There are fewer parts, it is a two-axis gyro, there is no need for the suspension fluid with all its criticality, warm-up characteristics are more consistent and predictable, there is no need for flex leads which can introduce torque errors, and spin motor bearing design is much simpler.

These advantages are offset to some extent by the difficulty of maintaining perfect tuning of the rotor. Parasitic torques associated with the viscous coupling of the rotor with the surrounding gas which are due to residual flutter are another error source. Heating and imperfect containment of the bearing lubricant, especially in a near-vacuum environment, is another source for errors due to the gas dynamic effects.

Inertial angular rate capability, bandwidth, bias, and scale factor precision are comparable to the state-of-the-art in floated RIGs. Bandwidths of about 80% of the spin rate have been achieved. This implies bandwidth figures in the 50 to 100 Hz range.

Gyros with large angular momentum rotor wheels having low drift rates (see (4.8.58—4.8.59)) have demonstrated that calibration intervals longer than six months are adequate to maintain the specified precision level for the 1 nmi/hr range. Warm-up times of about 5 min for thermal stabilization and spin motor run-up have been demonstrated for high-precision DTG's, but with thermal modeling.

Despite all the progress made, high-precision mechanical gyros are gradually being superseded by ring laser gyros and fiber-optic gyros, which are described in the next chapter.

4.9 Very high-precision free gyroscopes ⊕

The gas-bearing gyroscope

In the gyro constructions described so far, two sets of bearings are the crucial mechanical components, namely, the rotor spin bearings b_s and the gimbal suspension bearings b_G. The shortcomings of this conventional design principle are, that b_s tends to be noisy and wear out and that the friction in b_g is the principal source of gyro drift rate. The floated RIG described in the previous section is an outstanding example of the painstaking and costly techniques that have been applied in order to minimize these error sources without circumventing them by using an entirely different design principle. The precision achieved with the floated RIG technology was, by and large, not better than 0.01 deg/hr. The extreme precision required for intercontinental ballistic missiles in the early 1950s, spurred the development of alternative gyro technologies aimed at an order of magnitude better precision. The most important development in this context is the gas-bearing gyroscope, which was originally developed by Autonetics for the Minuteman ICBM. A schematic cross-section drawing is shown in Figure 4.28.

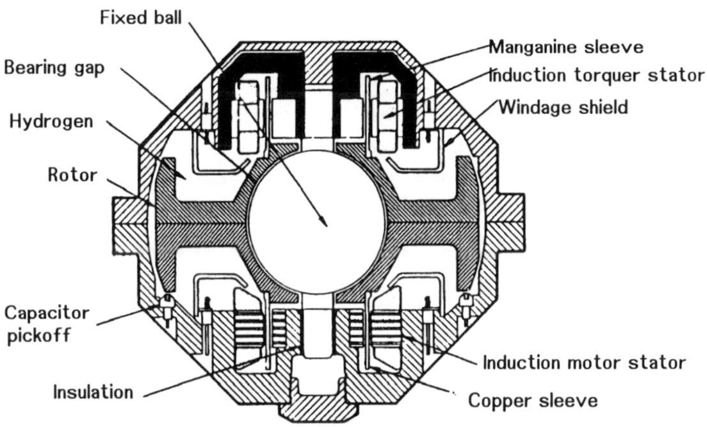

Figure 4.28. High-precision gas-bearing gyro.(based on Ref. 10).

The heart of this device is the central fixed high-precision ball, about 40 mm in diameter, which supports the spinning rotor via a 0.004 mm gap filled with pressurized hydrogen. The coils in the stator generate a rotating magnetic field, that, by inducing eddy currents in the rotor, make it into an induction motor which spins the rotor up to ~9000 rpm. Thus, the spherical joint fulfills the dual task of the rotor spin bearing b_s and the gimbal bearing b_g. It is clear that this construction has freedom of rotation about two perpendicular axes normal to the spin axis. It is therefore a two-axis free gyro. Current fed into the torquer can precess the rotor about these two perpendicular axes. The gyro housing is equipped with capacitive electrodes that measure the two components of angular tilt of the rotor with respect to the housing. From the nature of its construction, it is clear that this gyro has only very limited angular freedom. It must therefore be operated either within a stabilized platform that isolates large angular excursions of the vehicle from the gyro or in the force balance closed -loop form. In this case, it operates as a rate gyro. This option is available because the gyro is equipped with all the necessary components, namely, the torquer and the pickoff.

Since there is no mechanical contact at all between the rotor and the housing, and therefore no mechanical friction, this source of drift is eliminated. Furthermore, in view of the closed-loop operation, there is only a minute relative angular rate between the rotor spin axis and the ball. Consequently, the viscous friction due to the hydrogen between the rotor and the ball, which might cause an angular drift rate, is negligibly small. The actual drift rates achieved at Autonetics as early as 1955—1956, were as low as 0.002 deg/hr [14]. This performance level agrees with later experience with inertial platforms equipped with gas-bearing gyros, which demonstrated navigational precision of 0.1 nmi/hr. Furthermore, as a direct result of the absence of mechanical contact between the rotor and the housing, the actual mean time between failures (MTBF), has been millions of hours. In fact, gas-bearing gyros have been run continuously for years on end. This performance level is vital and compatible with the operational requirements for absolute readiness of strategic bombers and ICBMs for years on end.

Nevertheless, because of the limited hydrodynamic lift force of the gas film that supports the rotor, this kind of gyro cannot withstand accelerations higher than about 30 g. Therefore, its use has indeed been restricted to the relatively benign dynamical environment of ballistic missiles, strategic bombers, and ships. There is apparently no future for this technology with many tactical applications that do not provide the aforementioned benign environments.

The electrostatically supported gyroscope (ESG)

The ultimate in isolating the spinning rotor of a gyro from the platform which carries it, is the electrostatically supported gyro (ESG) technology. It is close to the equivalent of an artificial spinning star encapsulated in an evacuated chamber and running under its own angular momentum for many months. Since it is almost ideally isolated from its surroundings, it is as close to an ideal free driftless gyro as one can envision with state-of-the-art electromechanical technology. It was originally invented by Arnold Nordsieck in 1954 at Honeywell. Informative historical accounts of its pioneering R&D efforts are given in [10]. An exploded view of the ESG construction and its principal components is shown in Figure 4.29. In the Honeywell design [5], the rotor is made from a hollow beryllium sphere and is about 38 mm in diameter. Beryllium is a hard, stable, and low-density metal having a specific weight of about 1.8. The gap between the levitation electrodes (1,2,3,4,5,6) and the surface of the sphere is less than 10^{-2} mm. These electrodes are gold-plated on the inner spherical ceramic surface of the housing cavity. The voltage on the electrodes is nominally 150 v. This narrow gap and the high voltage develop a sufficient electrostatic force, which is inversely proportional to the square of the width of the gap and proportional to the electrode area and the square of the applied voltage. This voltage is adjusted by servo amplifiers so as to continuously position the sphere symmetrically with respect to the electrodes. Thus, it enables the stable levitation of the light beryllium sphere. It is largely due to the very high vacuum, essentially needed to minimize drag, that sparking is prevented. Because of its special features, the mathematical model of electrostatic suspension is given in Appendix B5.

In another ESG design by North American Rockwell, the rotor is a solid beryllium sphere about 1 cm in diameter. In order to provide the sphere with a well-defined principal axis, it is necessary to make the moment of inertia around one preferred axis larger than the others. This can be achieved by any number of methods, such as inserting a ring in the hollow sphere or drilling a hole in the case of the solid sphere. The angular momentum is imparted to the rotor by the spin-up coils (1,2,3,4), which generate a rotating magnetic field. These induce eddy currents in the spherical rotor and make it perform as an induction motor spinning it up to a high angular speed of $\sim$150,000 rpm. Once this intended speed has been attained, the power is disconnected from the spin-up coils and the rotor coasts on. Since there are no elements of friction or substantial drag, the rotor is slowed down only by the residual gas molecules in the gyro cavity. Hence, it has proven to run for many months on its own. These prolonged periods of time are significant in the context of permanent operational readiness in strategic missiles and bombers while the vehicle is on the ground. Therefore, it is logical that approximately every 6 months the gyro is reenergized. This operation can be viewed as part of a periodic maintenance procedure. As a result of its highly sophisticated design philosophy and refined technology, drift rates of less than 0.001 deg/hr have been achieved. The question of measuring angular shifts ϕ and θ of the housing, i.e., the vehicle with respect to the rotor, namely, inertial space, is solved as follows: marker lines, as indicated in the drawing, are etched on the surface of the sphere. These are detected by the optical pickoffs (1,2,3) and from their corresponding phase-shift variations, the required Euler angles ϕ and θ can be computed. Depending on initial conditions, the spinning motion of the sphere is also subject to nutation which would interfere with precise angular measurements of ϕ and θ and would persist indefinitely because of the lack of a damping medium. This function is accomplished by the denutation coils (1,2) which are energized so as to induce quadrature currents in the rotor that cause dissipation of energy, but do not interfere with the spin-up signals.

The ESG can be considered one of the highest summits in recent technological history. The precision required in manufacturing the component parts and their integration into a working unit has set new standards in materials engineering, metrology, and high-precision machining. Significant points of this ultimate refinement include:

- The center of the sphere must be made to coincide with its center of gravity with the utmost precision to prevent wobbling that would tax the achievable narrowness of the gap between the sphere and the levitation electrodes. This precision requires excellent homogeneity of the beryllium, which can be achieved by extruding the metal billet prior to machining it to a sphere.

- Mass imbalance is as important a factor as in gimbal bearing supported gyro rotors. The displacement of the resultant electrostatic suspension force from the rotor center of gravity will cause a precession rate proportional to the product of the acceleration field, the rotor mass, and the displacement. The position and direction of the resultant suspension force are determined by the location of the electrodes. Their positioning and forming require the utmost precision.
- The ESG cannot be equipped with a torquer. Therefore it is an open-loop pure angular position gyro. It does not enjoy the inherent linearization achieved with FBS sensors by virtue of their near null operation. Consequently, the ESG is not suitable for the strapdown mechanization, which exposes the gyro to large angular rates and excursions. Because of the lack of a torquer, use of the ESG must be restricted to gimbaled space-stabilized (not Schuler tuned, which requires torquing) mechanizations.

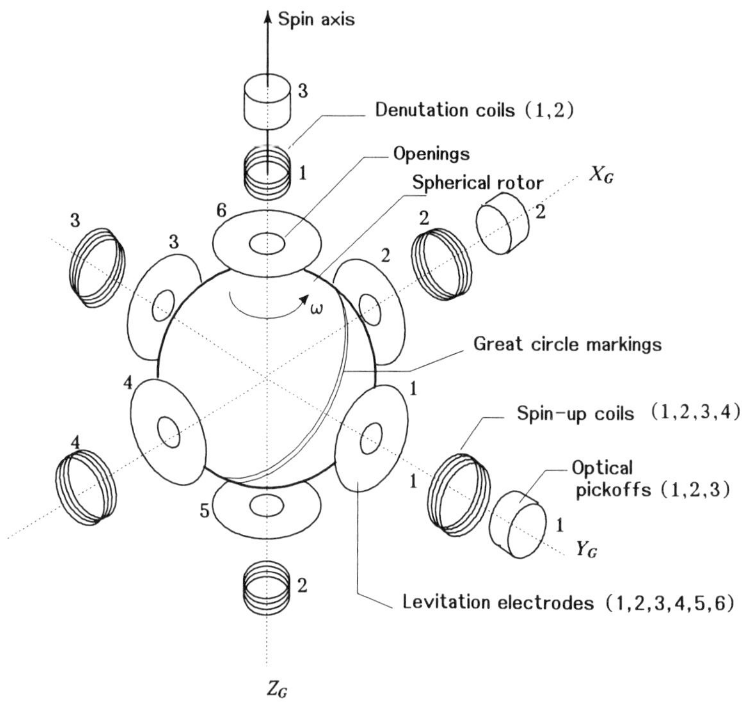

Figure 4.29. Exploded view of the Honeywell electrostatically suspended spinning sphere gyro (ESG).

- Due to its excellent mechanical and material stability, the operating characteristics of the ESG are extremely stable, repeatable, and predictable. Therefore, error compensation schemes are easily implementable.

- Being a freely suspended attitude sensor, its response time in measuring ϕ and θ is only due to possible time delays that result from the decoding of the electro-optical signals generated by the markers on the sphere.
- The ESG precision is sensitive to temperature variations. For this reason, the spin-up time is relatively long so as to minimize heating caused by the eddy currents induced by the spin-up coils.
- The extreme requirements for precision and painstaking testing make this gyro very accurate but very costly. Moreover, its very limited capability of withstanding high gs confines its use to large vehicles with moderate, long duration trajectories such as ships, submarines or heavy bombers. Drift rates achieved with the ESG have been as low as 0.0001 deg/hr, namely, about 0.01 nmi/hr if all other components in the INS are ideal.

Problems

Problem 4.1 – Gyrocompassing: Specifications and performance

A directional gyro is used as a gyrocompass. The azimuth error $\Delta\psi$ in the steady state should not exceed 1deg even in the arctic region at latitude $\lambda = 80^\circ$.

a) What is the specification for the gyro drift rate $\dot{\eta}_d$ in deg/hr?
b) In the steady-state expression that determines $\Delta\psi_d$, the angular momentum H of the gyro does not show. Yet, it is clear that H must have a strong effect on the achievable small error $\Delta\psi$.

Argue this point and resolve the apparent enigma.

Problem 4.2 – Rate gyro: Kinematically induced drift rate

A rate gyro is mounted in an aircraft in the fashion shown in Figure 4.17. The aircraft flies on a small circle with ψ=45 deg and λ=30 deg. The moment of inertia of the gyro rotor is 12 gr cm sec^2. The weight of the gimbal mount is 64.5 gr. Its center of gravity is 0.08 mm below the gimbal axis. Disregarding bearing friction and other mechanical imperfections, the gyro develops a drift rate q_d while flying along the given trajectory.

a) What would be the value of q_d if λ were changed to 45 deg?

b) What is the required spin speed ω_s of the rotor in rpm that ensures a drift rate q_d that is less than, or equal to 1 deg/hr?

Problem 4.3 – Vertical gyro: In an orbiting satellite

It is suggested to use a vertical gyro for roll control of an Earth orbiting satellite. The idea is that the centripetal acceleration pointing "upwards" counters the weight similarly to the D'Alembert force that counters the weight in an ordinary horizontal trajectory of an aircraft.

a) Is this idea applicable? Argue, and suggest whether to accept or reject.

b) Would your answer be different if the satellite were geostationary?

Problem 4.4 – Vertical gyro: Design considerations and error evaluations

The sensitivity coefficient of the bubble levels of a vertical gyro is S_β = 50 mv/deg. As a result of an alignment error, a zero point error $\Delta\beta$ = 30 mv develops.

a) Does this error cause an error $\Delta\eta$ in the steady state? If so, what is the value of $\Delta\eta$ in deg?

b) Explain in words the exact meaning of $\Delta\eta$.

c) Is it possible to eliminate $\Delta\eta$ by integral control?

Support your answers with the aid of the appropriate analytical expressions.

The angular momentum of the gyro is H = 5000 gr cm sec/rad and the erection torque coefficient b of the torque motor is 500 gr cm sec/rad.

d) What is the required gain A that provides an erection time constant of τ = 200 sec?

e) The gyro undergoes tests in a laboratory situated at latitude λ=30 deg. Disregarding all possible imperfections, will the gyro show $\Delta\eta = 0$ or $\Delta\eta \neq 0$? Support your answers with the aid of the appropriate analytical expressions.

Problem 4.5— Vertical gyro: Testing applicability as a low-cost INS

In view of the very high cost of INS, the need for low cost-platforms intended for short-duration missions, has often been considered. It has been suggested to use a vertical gyro to fulfill this need, at least in the horizontal plane, by placing two accelerometers A_x, A_y (instead of the bubble levels) on the inner gimbal.

It is assumed that the accelerometers and all the other components of the VG are completely drift-and error-free. The concept is tested with the following data.

The time constant of the erection loop is τ=5 min. The aircraft flies a straight and horizontal path. At the instant t=0, the aircraft starts accelerating by a_x=0.1 g for the duration of Δt=20 sec.

a) Write the expression for the deflection of the spin axis $\Delta\eta$ as a function of Δt and the reading of A_y assuming that $\Delta\eta$ is still a small angle.

b) Formulate an approximate linear expression for a_y based on the fact that $\Delta t \ll \tau$.

c) What is the distance error Δy derived by the double integration of a_y.

d) What is the reading of a_y as $\Delta t \rightarrow \infty$?

e) Since it has been assumed that all the components of the VG are ideal, what is the obvious remedy?

f) Discuss the rationale of trying to make use of this remedy.

Problem 4.6— Rate gyro: Design for linearity and bandwidth

The transfer function of a rate gyro $\phi(s)/q(s)$ is not truly linear because the gimbal and thus the spin axis are inclined by ϕ and are not normal to the input axis y.

a) Formulate the exact transfer function for $\phi(s)/q(s)$ that accounts for the resulting nonlinearity.

b) Determine the largest permitted deflection angle ϕ_{max} for which the nonlinearity is not greater than 1%.

The angular momentum of the rotor is $H = 10^5$ dyne cm sec.

c) What is the required spring constant K_s that assures the above requirement for ϕ_{max} with an input rate q_{max} = 1 rad/sec?

d) What is the largest permitted moment of inertia I_x that assures that the natural frequency ω_s of the gyro is not smaller than 180 rad/sec? What is the viscous friction coefficient D required to assure a damping coefficient ς that is not smaller than 0.5?

Problem 4.7— Rate gyro: Readings on a satellite

A cylindrically shaped satellite having a diameter of 10 m orbits the earth at a constant altitude of 200 km. It is equipped with a rate gyro whose input axis is parallel to the satellite axis, which is normal to the orbital plane, and an accelerometer mounted at the rim and its input axis pointing toward the satellite axis.

a) What is the reading of the rate gyro if the satellite is:

1. inertially stabilized.

2. Earth pointing.

3. moon pointing.

4. Rotating around its axis at ω = 1 rpm while orbiting.

The rate gyro has an unknown drift rate q_d.

b) How can its magnitude be identified by the on-board sensors?

c) To what precision can it be determined if the accelerometer uncertainty is 1 μg ?

Problem 4.8 – Vertical gyro: Examine equivalence to an overdamped pendulum

Note that a vertical gyro broadly behaves like a highly overdamped pendulum.

a) Formulate the dynamics of such an over damped pendulum.

b) Compare its properties in one axis to a vertical gyro in regard to its sensitivity to horizontal accelerations, and verify the similarity.

c) Does the similarity hold for all the functions of a vertical gyro? If not, point out where the similarity breaks down. State the functional significance, if any, of your answer.

Appendix A4: Euler angle transformation

The most common transformation set in flight vehicles is that between an axis system that incorporates the g vector as one of the Earth axes and the body axes x_b, y_b, z_b. The transformation is best visualized as successive rotations Ψ around Z_e, Θ round ON, and Φ around OM, as indicated in Figure 4.1. Using the usual shorthand s for sin and c for cos, these rotations are defined by:

$$[\Psi] = \begin{bmatrix} c\Psi & s\Psi & 0 \\ -s\Psi & c\psi & 0 \\ 0 & 0 & 1 \end{bmatrix} \tag{A4.1}$$

$$[\Theta] = \begin{bmatrix} c\Theta & 0 & -s\Theta \\ 0 & 1 & 0 \\ s\Theta & 0 & c\Theta \end{bmatrix} \tag{A4.2}$$

$$[\Phi] = \begin{bmatrix} 1 & 0 & 0 \\ 0 & c\Phi & s\Phi \\ 0 & -s\Phi & c\Phi \end{bmatrix} \tag{A4.3}$$

Carrying out the matrix multiplication $[E] = [\Phi][\Theta][\Psi]$, we obtain the three-axis Euler transformation:

$$[E] = \begin{bmatrix} c\Psi c\Theta & s\Psi c\Theta & -s\Theta \\ c\Psi s\Theta s\Phi - s\Psi c\Phi & s\Psi s\Theta s\Phi + c\Psi c\Phi & c\Theta s\Phi \\ c\Psi s\Theta c\Phi + s\Psi s\Phi & s\Psi s\Theta c\Phi - c\Psi s\Phi & c\Theta c\Phi \end{bmatrix} \tag{A4.4}$$

It is easily verified that (A4.4) is orthonormal, i. e., orthogonal and normal, namely,

$$[E]^T = [E]^{-1} \tag{A4.5}$$

$$\|E\| = [E]^T[E] = 1 \tag{A4.6}$$

Let $[1, m, n]^T$ and $[i, j, k]^T$ be unit vectors in Earth and aircraft axes, respectively. Then

$$[E]\begin{bmatrix} 1 \\ m \\ n \end{bmatrix} = \begin{bmatrix} i \\ j \\ k \end{bmatrix}; \qquad [E]^{-1}\begin{bmatrix} i \\ j \\ k \end{bmatrix} = \begin{bmatrix} 1 \\ m \\ n \end{bmatrix} \tag{A4.7}$$

Appendix B4: Electrostatic flotation ⊕

The basis for electrostatic suspension is the force f that acts between the two plates of a charged-up capacitor. Since the charges on the plates are of opposite signs, the force is always attraction. We assume here that the plates are planar and that the linear dimensions of the plate surface A are very large compared to the gap x. The capacitance C, under these conditions, is given by

$$C=\frac{\epsilon A}{x} \tag{B4.1}$$

where ϵ is the permittivity of the material between the plates. For vacuum or dry air, its value is $\epsilon=(36\pi\times10^{9})^{-1}=8.85\times10^{-12}$ Farad/meter. Let C be charged up to a voltage e and then open-circuited. The relationship between the electrostatic charge q in C and the voltage e across its plates is, from the basic definition of capacitance,

$$q=Ce \tag{B4.2}$$

The incremental energy stored in C for an incremental change in e is

$$\mathrm{d}W=q\mathrm{d}e=Ce\,\mathrm{d}e \tag{B4.3}$$

The total stored energy is

$$W=\int_0^e Ce'\mathrm{d}e'=\frac{Ce^2}{2}=\frac{q^2}{2C}=\frac{q^2x}{2\epsilon A} \tag{B4.4}$$

A sketch of the capacitor is shown in Figure B4..

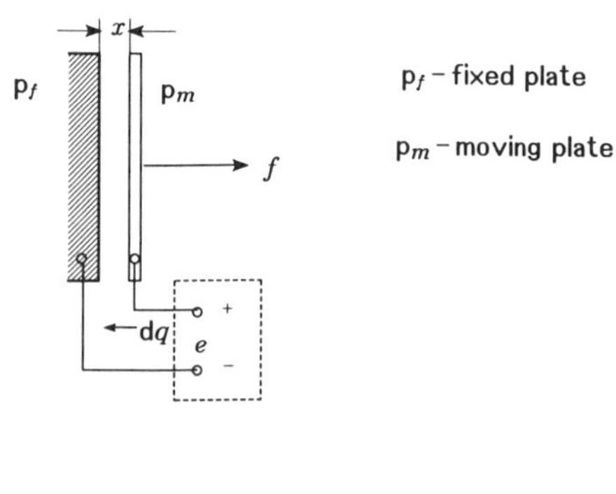

Figure B4.1. Voltage-force relationships in a plate capacitor.

The attraction force f between the capacitor plates is determined as follows: Let the nominal value of x be x_0. Assume that the moving plate is moved by $\mathrm{d}x$. Then, from (B4.4), the incremental change in energy $\mathrm{d}W$ is

$$\mathrm{d}W=\frac{q^2\mathrm{d}x}{2\epsilon A}=\frac{C^2e^2\mathrm{d}x}{2\epsilon A}=\frac{\epsilon^2A^2e^2}{x_0^2 2\epsilon A}\mathrm{d}x=\frac{\epsilon Ae^2}{2x_0^2}\mathrm{d}x=\frac{Ce^2}{2x_0}\mathrm{d}x \text{ Joule} \qquad \text{(B4.5)}$$

In order to cause this energy increment, the force f is required to act against the attraction force along the movement $\mathrm{d}x$, namely, $\mathrm{d}W=f\mathrm{d}x$. Consequently,

$$f=\frac{\epsilon Ae^2}{2x_0^2}=\frac{Ce^2}{2x_0} \text{ Joule/m} \qquad \text{(B4.6)}$$

This relationship also holds if the voltage source, shown by dotted lines in Figure B4.1, remains connected. In this case, the energy change involved in moving the incremental charge dq over the potential difference e is dqe. Thus, total energy change is

$$f\mathrm{d}x+e\,\mathrm{d}q=\mathrm{d}\left(\frac{Ce^2}{2}\right) \qquad \text{(B4.7)}$$

and f is

$$f=-e\frac{\mathrm{d}q}{\mathrm{d}x}+\frac{\mathrm{d}}{\mathrm{d}x}\left(\frac{Ce^2}{2}\right)=-e\frac{\mathrm{d}}{\mathrm{d}x}(Ce)+\frac{\mathrm{d}}{\mathrm{d}x}\left(\frac{Ce^2}{2}\right) \qquad \text{(B4.8)}$$

$$f=-e\left(C\frac{\mathrm{d}e}{\mathrm{d}x}+e\frac{\mathrm{d}C}{\mathrm{d}x}\right)+Ce\frac{\mathrm{d}e}{\mathrm{d}x}+\frac{e^2\mathrm{d}C}{2\,\mathrm{d}x}=-\frac{e^2\mathrm{d}C}{2\,\mathrm{d}x}=-\frac{e^2}{2}\left(-\frac{\epsilon A}{x^2}\right)$$

$$=\frac{\epsilon Ae^2}{2x^2} \qquad \text{(B4.9)}$$

which is the same as (B4.6) derived for the open-circuit case.

The following example demonstrates the numerical values of the attraction force f for the following constants and parameters:

$A=4\,\mathrm{cm}^2$; $x_0=2\times10^{-3}$ cm; $e=200$ v; $\epsilon=8.85\times10^{-12}$ F/m $=8.85\times10^{-14}$ F/cm;

$$f=\frac{8.85\times10^{-14}\times4\times200^2}{2\times(2\times10^{-3})^2}=15.8\times10^{-4} \text{ Joule/cm}$$

$$=15.8\times10^3 \text{ dynes}\cong 18\,\text{gr} \qquad \text{(B4.10)}$$

On the basis of the foregoing, a schematic model for an electrostatically levitated metal body B shown as a shaded rectangle, is shown in Figure B4-2. B is attracted to both condenser plates and therefore is analogous to a body attached to two stretched springs pulling in opposite directions with the corresponding forces f_1 and f_2, so that the nominal net horizontal force $f_1+f_2=0$.

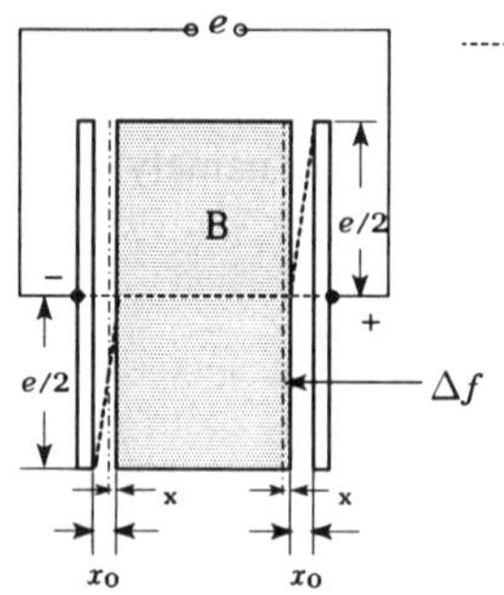

Figure B4.2. Voltage and force relationships in electrostatically suspended metal body.

A horizontal force Δf applied to B, as shown, will cause the right gap to increase by x and the left gap to decrease by x. f_1 and f_2 are given by

$$f_1=\frac{C\frac{e^2}{4}}{2(x_0+x)}; \qquad f_2=-\frac{C\frac{e^2}{4}}{2(x_0-x)} \tag{B4.11}$$

The net force is given by

$$\Delta f=f_1+f_2=\frac{Ce^2}{8}\left(\frac{1}{x_0-x}-\frac{1}{x_0+x}\right)=\frac{Ce^2}{4}\frac{x}{x_0^2-x^2} \tag{B4.12}$$

The relationship between x and Δf follows from (B4.12) as a quadratic in x,

$$x^2+\frac{Ce^2}{4\Delta f}x-x_0^2=0 \tag{B4.13}$$

which yields the solutions, of which only the one with the + sign is valid.

$$x=\frac{Ce^2}{8\Delta f}\left\{-1\pm\sqrt{1+x_0^2\left(\frac{8\Delta f}{Ce^2}\right)^2}\right\} \tag{B4.14}$$

For the foregoing numerical values preceding (B4.10), C is given from (B4.1)

$$C=\frac{\epsilon A}{x_0}=\frac{8.85\times 10^{-14}\times 4}{2\times 10^{-3}}=15.5\times 10^{-11}\ \mathrm{F}=155\ \mathrm{pF} \tag{B4.15}$$

Δf must be expressed in Joule/cm. As seen in (B4.10), $1\,\mathrm{gr}=10^{-4}$ Joule/cm. For example, for $\Delta f=20\,\mathrm{gr}=20\times 10^{-4}$ Joule/cm, which represents a solid beryllium sphere weighing 2 gr accelerated at 10 g, we have

$$x=\frac{155\times 10^{-12}\times 4\times 10^{4}}{8\times 20\times 10^{-4}}\left\{-1+\sqrt{1+\frac{4\times 10^{-6}\times 64\times 400\times 10^{-8}}{155^{2}\times 16\times 10^{8}\times 10^{-24}}}\right\}$$

$$=4.4\times 10^{-4}\{-1+\sqrt{21.4}\}\ \mathrm{cm}\cong 16\times 10^{-4}\ \mathrm{cm} \tag{B4.16}$$

This numerical result demonstrates that the deflection x is only slightly smaller than $x_0=20\times 10^{-4}$ cm indicating that the electrostatic spring is quite soft and that a high-gain control system is required to keep the sphere centered with high precision.

References

1. McRuer, D.T., Ashkenas, I., and Graham, D., *Aircraft Dynamics and Automatic Control*, Princeton University Press, Princeton, NJ, 1973.

2. Blakelock J.H. *Automatic Control of Aircraft and Missiles*, Wiley, New York, 1991.

3. Wrigley, W., Hollister, W.M., and Denhart, W.G., *Gyroscopic Theory, Design, and Instrumentation*, MIT Press, 1969.

4. Merhav, S.J. "A Nongyroscopic Inertial Measurement Unit," *AIAA Journal of Guidance, Control and Dynamics*, Vol. 5, No. 3, May—June 1982, pp. 225—235.

5. Koifman, M., and Merhav, S.J., "Autonomously Aided Strapdown Attitude Reference System," *Journal of Guidance, Control, and Dynamics*, Vol. 14, No. 6, Nov.—Dec. 1991, pp. 1164-1172.

6. Wertz, J.R., Ed., *Spacecraft Attitude Determination and Control*, Kluwer Academic Publishers, Dordrecht, Holland, 1978.

7. Kayton, W.R., and Fried, M., Ed., *Avionics Navigation Systems*, John Wiley & Sons, New York, 1969.

8. Heffley, R.K., "A Study of Key Features of Random Atmospheric Disturbance Models for the Approach Flight Phase," AIAA Atmospheric Flight Mechanics Conference, Hollywood, FL, August 1977.

9. Etkin, B., *Dynamics of Atmospheric Flight*, John Wiley & Sons, New York, 1972.

10. Mackenzie, D.A., *Inventing Accuracy, a Historical Sociology of Nuclear Missile Guidance*, MIT Press, Cambridge MA, 1990.

11. Lear Siegler Instrument Division, "Model 1903-1904 Rate Integrating Gyros,' Company publication.

12. Macomber, G.R., and Fernandez, M., *Inertial Guidance Engineering*, Prentice-Hall, Englewood Cliffs NJ, 1962.

13. Pitman, G.R., Ed. *Inertial Guidance*, John Wiley, New York, 1962.

14. Savage, P.G., "Strapdown Inertial Sensors, Theory and Applications," AGARD lecture series No. 95, UK, Italy, Norway, Germany, Denmark, June 1978.

15. Siouris, G.M., *Aerospace Avionics Systems—A Modern Synthesis*, Academic Press, New York, 1993.

16. Craig, J.G. "Theory of Operation of an Elastically Supported, Tuned

Gyroscope," *IEEE Transactions on Aerospace and Electronic Systems*, Vol. AES-8, No. 3, May 1952, pp. 280—288.

17. Lawrence, A., *Modern Inertial Technology, Navigation, Guidance, and Control*, Springer-Verlag New York, 1993.

18. Singer Kearfott Division, "The Conex Gyroscope SKG-3940," Technical Publication GTO-1815, November 1955.

5

Applications of Rate Gyros

5.0 Introduction

This chapter presents some of the foremost applications of rate gyros and other forms of angular rate sensors in flight control, guidance, and automatic navigation systems. In the following description and analysis, the effects of imperfections on overall system performance are presentedm, especially those due to dynamical errors, noise, offsets, and resolution.

The common denominator in all these applications is the following: In order to implement any form of vehicular control, the state vector of the vehicle must relate to an external reference vector. This reference vector may be a fixed or moving target, such as in guidance problems, or Earth- or space-based coordinates, such as in navigation problems. Whenever this relationship is implemented by a-body mounted sensor system, the vehicular states cannot be distinguished from vehicular attitude variations, which do not necessarily have an effect on the guidance and navigation functions. Therefore, the vehicular attitude variations must be isolated from the sensor system so that its readings, such as position, velocity and direction of the line of sight, actually represent a vehicular state. This necessary isolation can be accomplished by two principal methodologies:

1. Mechanical isolation by means of gyro-stabilized gimbaled platforms.

2. Computational isolation by the so-called analytical platform which derotates the strapdown sensor package.

A rudimentary description of such a system for inertial navigation in the reference frame of a flat and nonrotating Earth was presented in Chapter 3. Both methods have their merits and the preference of one over the other is decided by the nature of the particular application.

5.1 Two-axis platform

Description

An application of major importance using rate gyros is stabilization with respect to inertial space of a platform carrying a payload. A particularly common payload is a charge-coupled device (CCD) camera. Stabilization is commonly implemented in two axes, namely, elevation η and bearing ς, measured with respect to the body axes. Implementation is by means of a two-axis gimbaled mount. This is common practice in optical and radar seeker heads and in airborne automatic or manual terrain reconnaissance. A schematic diagram of such a platform is shown in Figure 5.1. The outer bearing b_{g_o} allows the freedom for ς and the inner bearing b_{g_i} permits the freedom for η. Any angular drift rate of the platform P, $\dot{\varsigma}$, or $\dot{\eta}$ is sensed by the corresponding rate gyros RG_ς and RG_η, which output the measurements $\dot{\varsigma}_m$ and $\dot{\eta}_m$. These feed the amplifiers A_z and A_y, which drive the torquers M_z and M_y respectively, so as to counteract the drifts $\dot{\varsigma}$ and $\dot{\eta}$.

Angular stabilization by rate gyros

With ideally driftless rate gyros and without external disturbances, P would be perfectly stabilized in space and isolated from the pitch and yaw motions of the vehicle, but not from roll motions. Roll-induced errors in the measurements of ς and η can be corrected computationally by the roll Euler transformation since the roll angle is usually known at all times. By feeding control commands $\dot{\varsigma}_c$ and $\dot{\eta}_c$ into the amplifiers A_z and A_y, the orientation of P, and thus the line of sight LOS can be changed at will. Clearly, if the platform is operated on solid ground, the stabilization function and the rate gyros are not required and only the control functions $\dot{\varsigma}_c$ and $\dot{\eta}_c$ are retained. This is the case of ground-based remotely operated devices such as cameras or other teleoperated machinery. With gyro stabilization, the attitude angles with respect to the local horizontal will change as a result of the Earth angular rate Ω_e. If its projections onto the platform axes are known, these attitude changes can be canceled by appropriate commands $\dot{\varsigma}_c$ and $\dot{\eta}_c$.

The following notations will apply:

$$\dot{\eta}_m = \dot{\eta} + \dot{\eta}_d \tag{5.1.1}$$

$$\dot{\varsigma}_m = \dot{\varsigma} + \dot{\varsigma}_d \tag{5.1.2}$$

where $\dot{\eta}_d$ and $\dot{\varsigma}_d$ are the drift rates of the rate gyros; b_x and b_y are the corresponding torque coefficients of M_x and M_y. F_D is an external force exerted on P, such as that due to vehicular acceleration, which develops torques T_{D_x} and T_{D_y}. The magnitude and sense of these torques are determined by the eccentricity of F_D from the center of gravity of the platform assembly. The moments of inertia of P are I_z and I_y. A simplified performance analysis of P, disregarding gyro dynamics, the lag of M_z, and cross-coupling in P is derived as follows. A block diagram for the ς axis only is shown in Figure 5.2. The dashed blocks indicate that the integrals are kinematic relationships and that they do not represent man-made computational operations by integrators.

The source of the input $\dot{\varsigma}_c$ is depicted as a joystick potentiometer but it can just as well be the output of an electro-optical detector in a seeker head. The external disturbance T_{Dz}/I_z, which is at the same level as the angular acceleration $\ddot{\varsigma}$, manifests itself, after the kinematic integration, as a drift rate in addition to $\dot{\varsigma}_d$. However, as shown in the following analysis, the gain A_z has no effect on the platform drift rate due to $\dot{\xi}_d$ while the drift rate due to T_{Dz} is inversely proportional to A_z.

Control loops and disturbances

The complete control loops with sensors, gimbals and payload are shown in Figures 5.1 and 5.2.

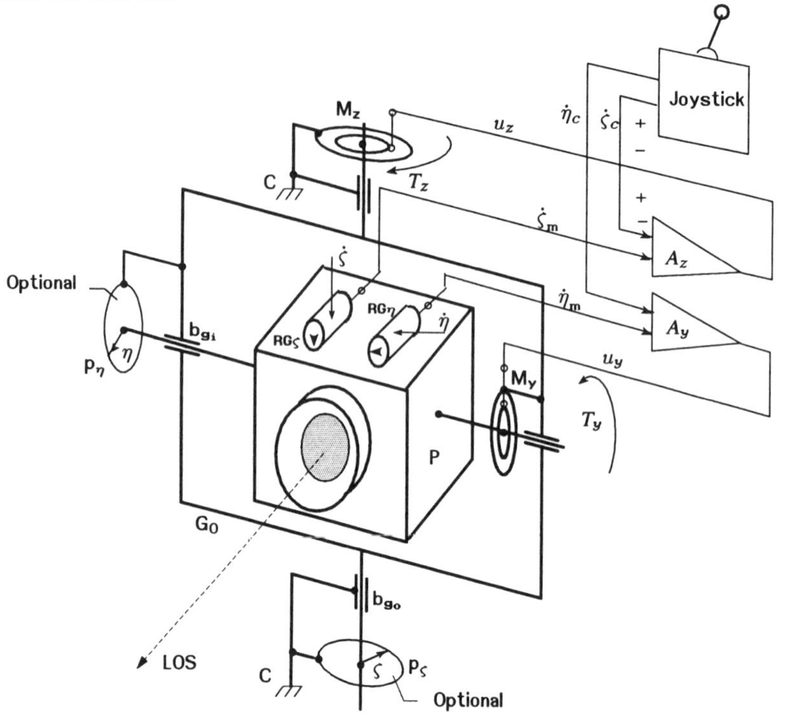

Figure 5.1. Schematic description of two-axis servoed mount for the control of an optical payload.

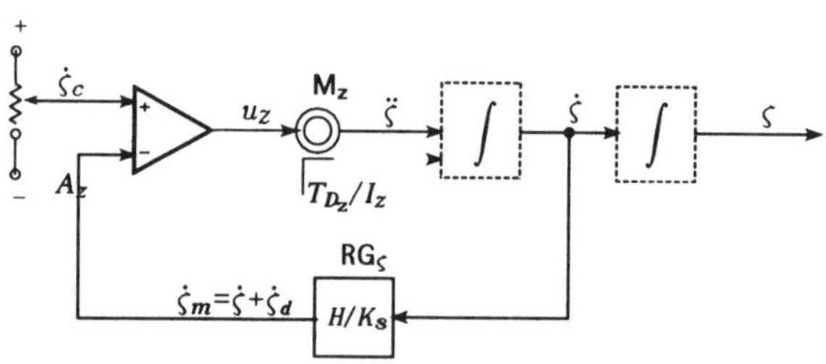

Figure 5.2. The servo control loop for one axis of the payload.

In the steady-state, as $s \to 0$,

$$[\dot{\varsigma}_c - (\dot{\varsigma} + \varsigma_d)]A_z = u_\varsigma \tag{5.1.3}$$

$$b_z u_z + T_{D_z} = I_z \ddot{\varsigma} \tag{5.1.4}$$

Thus,

$$u_z = \frac{I_z}{b_z}\ddot{\varsigma} - \frac{T_D}{b_z} \tag{5.1.5}$$

Substituting (5.1.5) into (5.1.3), we have

$$(\dot{\varsigma}_c - \dot{\varsigma}_d)A_z - \dot{\varsigma}A_z = \frac{I_z}{b_z}\dot{\varsigma} - \frac{T_D}{b_z} \tag{5.1.6}$$

Rearranging and Laplace transforming (5.1.6), the solution for $\dot{\varsigma}$ is

$$\dot{\varsigma}(s) = \frac{A_z}{s\frac{I_z}{b_z} + A_z}(\dot{\varsigma}_c - \dot{\varsigma}_d) + \frac{1}{s\frac{I_z}{b_z} + A_z}\frac{T_D}{b_z} \tag{5.1.7}$$

or

$$\dot{\varsigma}(s) = \frac{1}{s\tau + 1}(\dot{\varsigma}_c - \dot{\varsigma}_d) + \frac{1}{s\tau + 1}\frac{\tau}{I_z}T_D \tag{5.1.8}$$

where

$$\tau \triangleq \frac{I_z}{A_z b_z} \tag{5.1.9}$$

Summary of principal properties

1. The angular rate of the platform $\dot{\varsigma}$ obeys the commanded rate $\dot{\varsigma}_c$ but, with a first-order lag determined by τ and biased by the gyro drift rate $\dot{\varsigma}_d$.

2. With $\dot{\varsigma}_c = 0$ and $T_D = 0$, $\dot{\varsigma} = -\dot{\varsigma}_d$. Since the measured gyro output $\dot{\varsigma}_m = \dot{\varsigma} + \dot{\varsigma}_d$, it follows that $\dot{\varsigma}_m = 0$, i.e., it is impossible to electrically measure the gyro drift rate while it controls P in a closed loop.

3. However, the angular drift rate $\dot{\varsigma}_d$ of P can be compensated by $\dot{\varsigma}_c$ if it can be independently observed, for example, by monitoring the LOS with respect to a chosen target.

4. The effect of the external torque T_D is inversely proportional to $A_z b_z$, which is a measure of the platform stiffness. T_D adds to the drift rate $\dot{\varsigma}_d$ and persists as a bias if T_D=const. Its effect can be eliminated in the steady-state if integral control is used in the platform control loop.

5. The actual platform angle ς, being the time integral of $\dot{\varsigma}$, naturally diverges in accordance with the statistical nature of $\dot{\varsigma}(t)$ and $T_D(t)$ as determined by the expressions given in Section 4.6.

As depicted in Figure 5.2, P is stabilized and controlled with respect to free space. However, it is sometimes required to control and stabilize P with respect to the vehicle body axes. This is readily achieved by utilizing the pickoffs p_ς or p_η shown in Figure 5.2. In this mode, the rate gyros are disconnected and the outputs of p_ς and p_η are connected to A_z and A_y. This scheme implements angular position control with respect to body axes.

In the rate control mode, a given command to the joystick translates into a corresponding angular rate. This mode is favorable in manual control and it is widely used for pointing or tracking by remotely controlled cameras, laser designators, missile seeker heads, and other devices.

In the simplified expressions (5.1.7) and (5.1.8) the dynamics of M_z and RG_ς were disregarded. Thus, it appears that increasing A_z and b_z, is favorable both for achieving smaller time constants τ and higher stiffness of P for a given payload, which is characterized by its mass and moment of inertia I_z. Since A_z and b_z combine as a product, it would appear that a low torque-motor can be compensated by a high gain in A_z. This, of course, is misleading because M_z cannot merely be modeled as a linear device by its torque constant b_z except for relatively small angular accelerations $\ddot{\varsigma}$. Clearly, every actuator saturates above a certain level of torque demand. M_z must therefore be matched to the largest expected torque, which is determined by the dynamic load $I_z\ddot{\varsigma}$ and the disturbance T_D.

Other limitations on time constants and stiffness are due to gyro dynamics, as given in (4.8.11), and the electromechanical time constant of M_z. Incorporating these factors in the control model increases its relative degree from 1 to 4 at least, leading to further limitations on the overall loop gain in order to assure stability and avoid-high frequency oscillations of P. In actual designs where all these factors are accounted for, it is possible to achieve time constants τ in the range of 0.2–0.5 sec depending on the moment of inertia I_z and the required maximum angular rate $\dot{\varsigma}_{max}$.

Application to target tracking

A widely used application of the stabilized camera described earlier is in manual or automatic target tracking. An example is a target T_a on the ground being tracked by an airborne platform P. The aircraft flies at ground speed V and at altitude h. The geometry in the vertical plane only is shown in Figure 5.3(a). At any given instant, the range-to-target is R. The CCD image is projected one-to-one onto a monitor which is observed by a human operator. A fixed cursor (+) at the center of the monitor screen therefore displays the center of the camera array and thus the optical axis which coincides with the LOS, as shown in Figure 5.3(b). The human operator, or an electro-optical tracker, detects the tracking error $\{\eta_i - \eta, \varsigma_i - \varsigma\}$, which is amplified by a factor A and which outputs the control command $\dot{\eta}_c$. Tracking a target as described has numerous applications, such as in aerial reconnaissance, where target tracking freezes the image, thus diminishing its smearing and enhancing image sharpness. Another application, as stated earlier, is implementation of electro-optical and radar seeker heads, which are the basis for missile homing. Still another application is passive air-to-ground range estimation, which will be described next.

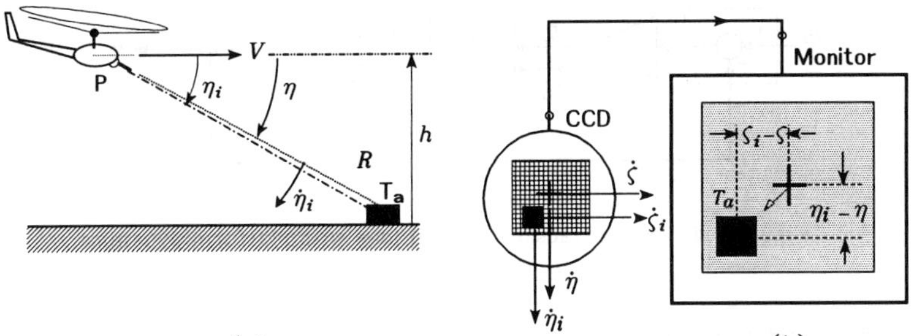

Figure 5.3. Geometry and motion in airborne target tracking.

We assume here that the human operator manages to align the cursor with T_a. This implies that $\dot{\eta}_c = \dot{\eta}$. From the geometry, we have

$$\frac{V\sin\eta_i}{R} = \dot{\eta}_i \tag{5.1.10}$$

and

$$\sin\eta_i = \frac{h}{R} \tag{5.1.11}$$

Substituting (5.1.11) into (5.1.12), we have

$$\frac{Vh}{R^2} = \dot{\eta}_i \tag{5.1.12}$$

Thus, the computed range to target R is

$$R = \sqrt{\frac{Vh}{\dot{\eta}_i}} \tag{5.1.13}$$

where V and h are assumed to be known. However, $\dot{\eta}_i$ can only be determined from its available measurements provided by P. From (5.1.8), disregarding the last term representing external disturbance, and rewriting it for η, we have,

$$\dot{\eta}(s) = \frac{1}{s\tau + 1}(\dot{\eta}_c - \dot{\eta}_d) \tag{5.1.14}$$

The complete tracking loop incorporating this dynamic model of P is shown in Figure 5.4.

Again, one should realize that the integration $1/s$ does not represent an integrator. It is simply the statement that η is the integral of $\dot{\eta}$.

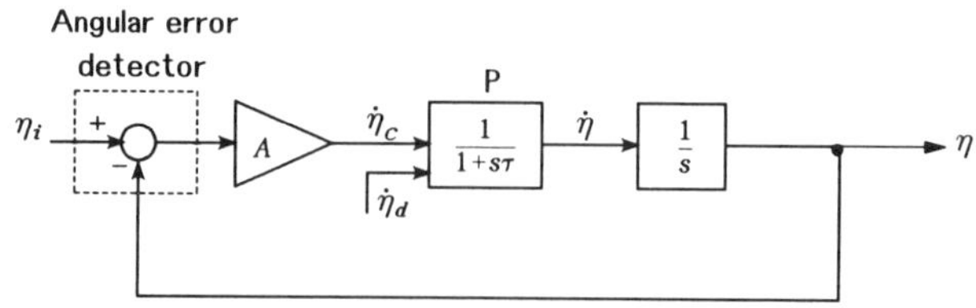

Figure 5.4. Target tracking control loop with platform dynamics.

From this model, we have

$$\frac{A}{s(1+s\tau)}(\eta_i - \eta) - \frac{1}{s(1+s\tau)}\dot{\eta}_d = \eta \tag{5.1.15}$$

From which the solution for η is

$$\eta = \frac{A}{s(1+s\tau)+A}\eta_i - \frac{1}{s(1+s\tau)+A}\dot{\eta}_d \tag{5.1.16}$$

In the steady-state, as $s \to 0$, the result is

$$\eta = \eta_i - \frac{\dot{\eta}_d}{A} \tag{5.1.17}$$

which indicates that η tracks $\dot{\eta}_i$, except for the effect of drift-rate error which is inversely proportional to A.

We now compute the measurable quantity $\dot{\eta}_c$:

$$\dot{\eta}_c = A(\eta_i - \eta) = A\eta_i(1 - \frac{A}{s(1+s\tau)+A}) - \frac{1}{s(1+s\tau)+A}\dot{\eta}_d$$

$$= A\eta_i \frac{s(1+s\tau)+A-A}{s(1+s\tau)+A} - \frac{1}{s(1+s\tau)+A}\dot{\eta}_d \tag{5.1.18}$$

Recalling that for the tracking task, which involves only low frequencies, $s\tau \ll 1$ and $s \ll A$, we have approximately

$$\dot{\eta}_c = A\eta_i \frac{s}{s+A} - \frac{1}{s+A}\dot{\eta}_d \cong \dot{\eta}_i - \frac{1}{A}\dot{\eta}_d \tag{5.1.19}$$

Conclusion: $\dot{\eta}_c$, the measurable output of A, represents the required $\dot{\eta}_i$, except for the small gyro drift-rate error $\dot{\eta}_d/A$.

Thus, subject to this small drift error, $\dot{\eta}_c$ can be substituted into (5.1.13) to determine R. The range error ΔR can be readily determined as a function of a systematic or random error.

Control by $\dot{\eta}_c$ is known as *integral control* because the resulting η is very close to the integral of $\dot{\eta}_c$. It therefore constitutes a first-order control system. It is particularly favorable in manual control because once the cursor + coincides with T_a, $\dot{\eta}_c$ can be annulled almost instantaneously, thus completing the correction without further transient overshoots, which typically occur in higher order systems. In the case of manual missile guidance by P, the human operator is always confronted with the easy task of controlling a first-order plant [3]. The missile pitch or yaw motions do not affect the control task because of the mechanical isolation of P.

5.2 Gyroscopic seeker head

Description and definition of variables

A highly specialized form of a two-axis torqued free gyro mechanization is the one known as the *gyroscopic seeker head*, also referred to as the *eye*. A schematic sketch describing the mechanization and the essential components is shown in Figure 5.5. Not all the details of the optical imaging components and assembly are shown. The seeker head consists of two primary subassemblies, the moving unit and the static unit. In order to achieve a compact mechanization and to avoid obscuring the target image, the spinning rotor, or spinner, is suspended from inner gimbals by means of the respective gimbal bearings (10), (11), which in effect constitute a Cardan joint allowing angular freedom Q around the y axis (perpendicular to the paper) and R around the z axis. The rotor incorporates a magnet with poles N, and S, which supports the primary concave spherical mirror (2), the spinner bearings (3), the support cone for the moving optical components (4), the flat secondary mirror (5), the sun baffle (7), and the optical chopper (8). This folded optics Cassegrain telescopic unit [4] provides a large focal length within a small volume and a field of view of about 1–2 degrees. The entire assembly spins around the x axis at the angular rate of Ω rad/sec. The static subassembly incorporates the spherical Irdome (6), which is transparent to the infrared incident radiation, the gimbal support holding the infrared detector (9), the cryogenic cooling unit, and the molded electromagnetic coil assembly (12). This assembly incorporates the following functional coils: Spinner motor drive, two-axis tracking torquers, phase reference, and gimbal mount caging and uncaging.

Early infrared seeker heads were designed to sense and track the jet engine flame radiating at about $0.9\,\mu$ wavelength and visible only along the jet engine tailpipe axis. This limitation posed a serious disadvantage because the missile had to close in on the target along this axis, i.e., strictly from behind the target, implying a very narrow firing zone. Consequently, the missile was able to achieve a seeker head target lock-on only while the pursuer was positioned behind the target. Therefore, the seeker head was caged prior to lock-on and uncaged just prior to firing to permit subsequent target tracking. In this wavelength region, conventional materials, primarily quartz for the Irdome and and lenses, could still be used. A lead sulfide (PbS) detector, at sufficiently close ranges in the order of a mile provided acceptable signal-to-noise ratios to facilitate proper target lock-on and subsequent tracking. In order to open up a wider firing zone, it became essential to enable the seeker head to "see" the target from much wider off target tailpipe angles extending 90 deg and eventually up to 180 deg, thus even permitting head-on attack. In technical

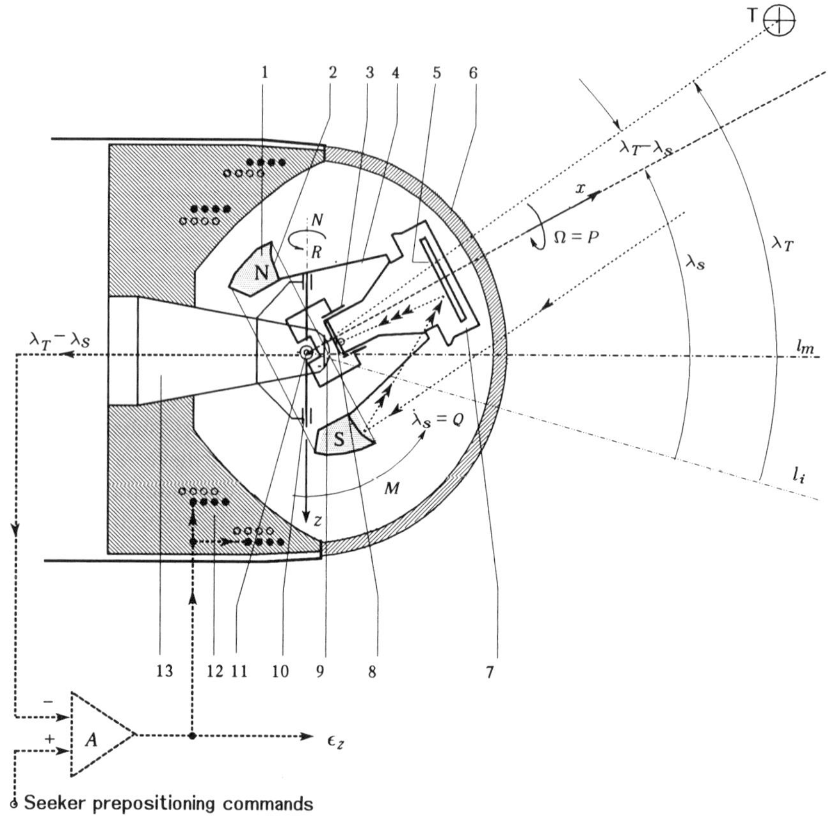

1 – Rotor magnet
2 – Spherical concave primary spinning mirror
3 – Spinner bearings
4 – Conical support of secondary mirror
5 – Secondary flat spinning mirror
6 – Spherical dome transparent to IR radiation
7 – Sun baffle
8 – Spinning optical chopper at image plane
9 – Cooled infrared detector
10 – Outer gimbal bearing
11 – Inner gimbal bearing
12 – Coil assembly for spinner, gimbal torquers, phase ref., functional control, caging, uncaging
13 – Gimbal support incorporating the cryogenic cooling unit of the detector

Figure 5.5. Schematic diagram of gyroscopic missile seeker head.

terms, this meant that the missile should see not just the flame in the jet engine tailpipe but also the relatively cool gaseous plume extending behind the target aircraft. This objective dictated the need for detecting radiation in the wavelength region of 2.2μ, which is the wavelength of a window of radiation transmission in the atmosphere. At this longer wavelength, however, the noise level is prohibitively high. In order to reduce it to an acceptable level, the detector has to be cooled down to -196 deg C by the cryogenic refrigerating unit, which may be of either the Joule Thompson expansion type or the liquid nitrogen type. At this longer wavelength, the quartz Irdome is no longer transparent, so, it is made of magnesium fluoride instead. A further advantage of the longer wavelength is that it is much easier to filter out sunlight, which has much of its energy in the 0.6μ region. The ability of the seeker head to see the target throughout a wide angular zone implies the requirement that the entire eye be steerable to any line of sight angle prior to firing. This function is known as prepositioning. It is indicated in Figure 5.5 via amplifier A. The prepositioning commands are typically obtained from the on-board radar, which tracks the target and locks onto it. Alternatively, the prepositioning commands are provided by special helmet-mounted electromagnetic, optical, or mechanical sensors which measure the target line-of-sight with respect to the body axes of the pursuing aircraft. The target line-of-sight is determined by the pilot, who points his head at the target by means of a helmet-mounted cathode ray display projected onto the inner semitransparent visor to infinity. Feeding these commands into A implements slaving of the seeker head to the helmet, namely, the pilot's line-of-sight to the target. This provides the opportunity for the seeker head to lock onto the IR target image before firing. Once this lock-on occurs, an audio signal, or a visual display is presented to the pilot who can then fire the missile. Clearly, the missile is launched while the eye may squint by a substantial angle off its longitudinal axis. The heart of the seeker head is the detector-chopper subunit, which is designed to determine the off-optical axis angular position of the target within the optical field of view. Its description is given in the next paragraph. This implies that the chopper-detector must determine both the modulus of the off-axis angle (radial deviation) along with the phase, or rather the circumference deviation from a selected phase reference point. The following are the notations pertaining to the rotor dynamics:

x, y, z —axis system of the gimbal system with respect to inertial space;
P, Q, R —respective angular rates of the rotor. $P=\Omega$;
M, N— moments around the y and z axes (also denoted by T_y, T_z);
M_y, M_z — respective torque motors of the gimbal suspension;
$b_x=b_y$ — corresponding torque constants in grcm/amp, assumed equal;
$A_x=A_y$ — amplifier gains of A_x, A_y in amp/volt;
λ_s— direction of x with respect to an arbitrary inertial reference line l_i;
λ_T— line-of-sight angle to target T with respect to l_i;
I_x, I_y, I_z — moments of inertia of rotor;
I_{xy}, I_{xz}, I_{yz} — cross moments of inertia of the rotor.

Dynamics and control

The three dynamical torque equations [1] L, M, and N are the starting point for the following analysis:

$$L = \dot{P}I_x + QR(I_z - I_y) - (PQ + \dot{R})I_{xz} \tag{5.2.1}$$

$$M = \dot{Q}I_y + PR(I_x - I_z) - R^2 I_{xz} + P^2 I_{xz} \tag{5.2.2}$$

$$N = \dot{R}I_z - \dot{P}I_{xz} + PQ(I_y - I_x) + QRI_{xz} \tag{5.2.3}$$

Since the gimbal mount is two-axes only, namely, the y and z axes, only the dynamical torque equations for M and N are relevant. Thus, we have in view of the rotational symmetry that $I_{xz} = 0$ and $I_z = I_y \triangleq I$.

Denote $I_x - I \triangleq J$; $P = \Omega$; and $J\Omega \triangleq H$. This simplifies the torque equations to

$$M = \dot{Q}I + R\Omega J = \dot{Q}I + RH \tag{5.2.4}$$

$$N = \dot{R}I - Q\Omega J = \dot{R}I - QH \tag{5.2.5}$$

Normalizing with respect to I and Laplace transforming, we have

$$sq + r\frac{H}{I} = \frac{M}{I} \triangleq \mu \tag{5.2.6}$$

$$sr - q\frac{H}{I} = \frac{N}{I} \triangleq \nu \tag{5.2.7}$$

Denoting $H/I \triangleq \omega$, the solutions of the last two equations in the time domain are

$$Q(t) = \frac{\sin\omega t}{\omega}\mu(t) + \frac{\cos\omega t}{\omega}\nu(t) - \frac{1}{\omega}\nu(t) \tag{5.2.8}$$

$$R(t) = -\frac{\cos\omega t}{\omega}\mu(t) + \frac{\sin\omega t}{\omega}\nu(t) + \frac{1}{\omega}\mu(t) \tag{5.2.9}$$

The effect of μ is to excite periodic motion of Q and R but in quadrature. The effect of ν is similar. The result is a conical motion of the gyro mount known as *coning*. The amplitude is inversely proportional to ω. For realistic values of H and I, the coning amplitude is negligibly small while ω is very large. Therefore the coning motion is disregarded for all practical purposes. However, the last terms in (5.2.8) and (5.2.9) represent the desired precession rate responses Q and R to the input torques μ and ν, which, as expected, are orthogonal to Q and R. The result, therefore, is

$$Q(t) \cong -\frac{1}{\omega}\nu(t) \tag{5.2.10}$$

$$R(t) \cong \frac{1}{\omega}\mu(t) \tag{5.2.11}$$

The photoelectric detector (9), which is coaxially aligned with the x axis senses the angular deviation $\lambda_i - \lambda_s$. The corresponding processed signal is fed into the amplifier A. The amplified signal ϵ_z is applied to the corresponding torquer coil (12), which exerts the normalized torque $\nu = N/I$,

$$\nu = \frac{N}{I} = -(\lambda_T - \lambda_s)\frac{Ab}{I} \tag{5.2.12}$$

We now substitute $Q = \dot{\lambda}_s$ and (5.2.11) into (5.2.9) so that

$$\dot{\lambda}_s = \frac{1}{\omega}(\lambda_T - \lambda_s)\frac{Ab}{I} \tag{5.2.13}$$

Rearranging, we have

$$\dot{\lambda}_s + \frac{Ab}{I\omega}\lambda_s = \frac{Ab}{I\omega}\lambda_T \tag{5.2.14}$$

Laplace transforming yields the following transfer function from λ_T to λ_s:

$$\lambda_s(s) = \frac{\frac{Ab}{I\omega}}{s + \frac{Ab}{I\omega}}\lambda_T(s) \tag{5.2.15}$$

Clearly, $\lambda_s \to \lambda_T$ as $s \to 0$. In conclusion, the gyroscopic seeker head presents a line-of-sight tracking unit with a first-order lag having the time constant

$$\tau_s = \frac{I\omega}{Ab} \tag{5.2.16}$$

With state-of-the-art technology it is possible to achieve time constants as low as $\tau_s = 0.1$ sec. The same results and observations apply to the z axis.

The principal application of this type of seeker head is in infrared missile homing, based on proportional navigation [2] which implements the law

$$\dot{\gamma} = K\dot{\lambda}_s \tag{5.2.17}$$

where $\dot{\gamma}$ is the rate of turn of the missile, K is the navigational constant, and $\dot{\lambda}_s$ is the measured rate of turn of the actual line of sight $\dot{\lambda}_T$. In order to derive $\dot{\lambda}_s$ as an available on-board signal, we return to (5.2.13), in which we substitute $\lambda_T - \lambda_s$. With the aid of (5.2.15), we have

$$\lambda_T(s) - \lambda_s(s) = \frac{s}{s + \frac{Ab}{I\omega}}\lambda_T(s) = \frac{\dot{\lambda}_T(s)}{s + \frac{Ab}{I\omega}} \tag{5.2.18}$$

Recalling that the amplifier output $\epsilon_z = (\lambda_T - \lambda_s)$ is an accessible signal, and disregarding the effect of the lag $\omega_s = Ab/I\omega$, we have

$$\epsilon_z(s) \cong \frac{A\dot{\lambda}_T(s)}{\frac{Ab}{I\omega}} = \frac{I\omega}{b}\dot{\lambda}_T(s) \tag{5.2.19}$$

Thus, the rate of turn of the missile-to-target line of sight $\dot{\lambda}_T$ is represented by the amplifier output ϵ_z scaled by the known constant $I\omega/b$. The same considerations and analysis, with small differences, hold for the y axis.

EXAMPLE 5.2–1: Seeker head tracking error due to dynamic lag

The effect of the seeker head time constant τ_s on the tracking precision has been disregarded in the approximate expression (5.2.19). The dynamical error $\epsilon_\lambda = \lambda_i - \lambda_s$ is formulated in (5.2.19). Assume $\tau_s = 1/\omega_s = 0.1$ sec and let $\dot{\lambda}_T$ be 10 deg/sec = 0.174 rad/sec. The dynamical squint error $\varepsilon_\lambda(s)$ is

$$\varepsilon_\lambda(s) = \frac{\dot{\lambda}_T(s)}{s + \omega_s} = \frac{0.174}{s + 10}$$

In the steady state, as $s \to 0$, the squint error is

$$\varepsilon_\lambda = 0.174/10 \text{ rad} = 1^0$$

This example illustrates the criticality of the seeker head dynamics. Considering the fact that a typical field of view of seeker head optics is on the order of 1–2 deg and that a λ_T of 10 ^{0}/sec is not excessive, it follows that the seeker head can easily lose the target because of the seeker head dynamical lag.

Detection of angular deviation – amplitude modulation

In accordance with Figure 5.5, the optical system focuses the radiating target T onto the spinning optical chopper (8). The resulting image is a light spot. As such, the unit constitutes a nonimaging electro-optical sensor in contrast to TV-type cameras, which are imaging sensors. The purpose of the chopper-detector is twofold: to enable the separation of the spot-like target from a warm radiating background by time modulating the light intensity of the spot image and thus enhancing the signal-to-noise ratio. and to preprocess the resulting electrical signal provided by the photoconductive detector (9) in order to facilitate the detection both of the radial deviation of the spot image from the center and its displacement with regard to a fixed circumferential reference point. The chopper, or reticle [4], is schematically shown in Figure 5.6(a). It is an optical disc divided in two halves. One half is a polar checkerboard consisting of alternating transparent and opaque trapezoidal segments. Each circular row typically consists of six segment pairs. The other half consists of a semitransparent surface that permits transmission of 50% of the light energy. Except for very close ranges, the target image falls entirely within the bounds of any trapezoidal element. Therefore, full intensity modulation is achieved. However, a larger object, such as a cloud or an Earth patch, covers a number of trapezoidal elements so that it averages out whatever brightness contrast this patch presents.

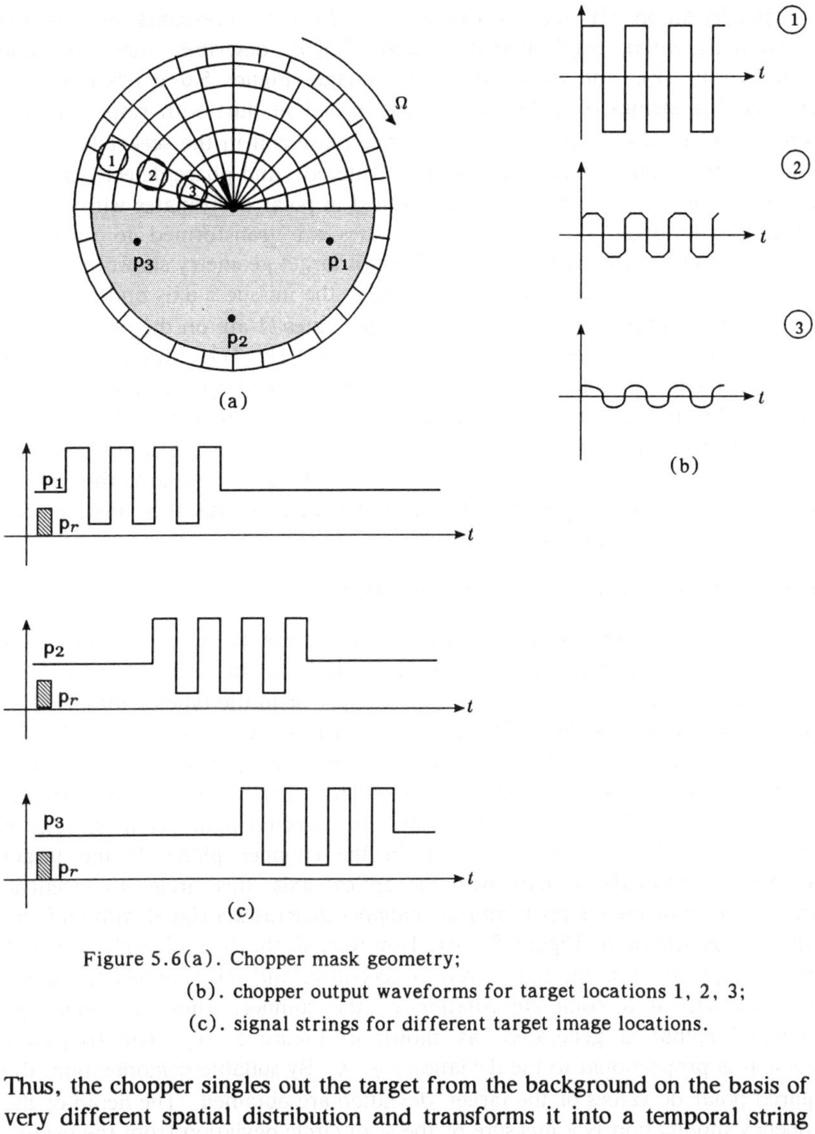

Figure 5.6(a). Chopper mask geometry;
(b). chopper output waveforms for target locations 1, 2, 3;
(c). signal strings for different target image locations.

Thus, the chopper singles out the target from the background on the basis of its very different spatial distribution and transforms it into a temporal string of pulses, as shown in Figure 5.6(b). The circles marked 1, 2, and 3 in Figure 5.6(a) describe three possible radial locations of the target image. Image 1, the farthest out, clearly generates full amplitude modulation, whereas image 3 is closer to the center so that it is not fully exposed or obscured. It therefore generates only a partial amplitude modulation as seen in Figure 5.6-b. Thus, the signal generated by the detector increases monotonically with the radial displacement. In principle, this monotonic relationship can be approximated by a linear characteristic. However, its slope, linearity, and resolution cannot be expected to be consistent. Yet, since it operates in a closed loop, these imperfections are not critical in theactual seeker head performance.

Typically an on-off characteristic is a satisfactory approximation. As for the circumferential displacement, Figure 5.6(b) describes three possible locations — P_1, P_2, and P_3 of the target image. Figure 5.6-c describes the corresponding strings of pulses which are shifted in time with respect to the reference pulse p_r provided by the reference pulse coil in the seeker head coil assembly. These pulse strings are rectified, filtered, and phase detected so that the position of P_1, for example, is determined in polar coordinates with respect to the missile body axes. These coordinates are transformed to Cartesian coordinates corresponding to the line-of-sight target geometry shown in Figure 5.5. The respective signals represent $\lambda_T - \lambda_s$ in the missile z axis and $\gamma_T - \gamma_s$ in the corresponding missile x axis. Typical spin rates Ω are on the order of 150 rps, and pulse frequencies are on the order of 1500 Hz. The entire outer rim of the chopper disk is equally divided into transparent and opaque segments that embrace different angles than in the inner part of the chopper. Whenever the target image covers this rim, the missile is very close to the target and the resulting signal, which is of a different frequency, is used to arm the missile warhead, either for the purpose of self-destruction in case of a miss, or for warhead activation in case of a near miss.

Detection of angular deviation — phase modulation

The foregoing description of the seeker head represents one of a number of different design concepts. It pertains particularly to the Sidewinder AIM-9 series. Other designs differ mainly in the chopper or in the type of torquer. For example, in the Sidewinder AIM-9L the chopper is static, as shown in Figure 5.7(a) (1), and its polar checkerboard is uniform throughout the entire circle. The secondary, flat mirror is slightly canted with respect to the spin axis, as shown in Figure 5.7(a) (3). As a result, the reflected light beam executes a conical scan (2) and traces a circle in the chopper plane. If the target line-of-sight coincides exactly with the optical axis, this circle is concentric with the center of the chopper, and the chopper generates a signal with uniform frequency as shown in Figure 5.7(b). However, if the line-of-sight deviates from the optical axis, the light circle is eccentric, and the chopped frequency varies throughout a complete rotation of the spinner. Thus, a frequency-modulated signal is generated, as shown in Figure 5.7(c). The frequency excursion is proportional to the deviation $\lambda_T - \lambda_s$. By suitable demodulation, the required polar dc values of the target deviation are obtained. The depth of the frequency modulation is a measure of the scan circle deviation from the center, while the phase of the modulated signal within the cycle is a measure of the circumferential deviation from a known reference pulse. An important advantage of this chopper concept is that its characteristic is linear for all practical purposes. This linearity contributes to the tracking precision. If the deviation $\lambda_T - \lambda_s$ is excessive, as shown in Figure 5.7(d), the traversion of the inner circle with the circular transparent reticles becomes dominant and the modulation takes on an amplitude-modulation mode, as shown in Figure 5.7-d. This can be used to indicate an onsetting miss, designed to activate a self destruct command.

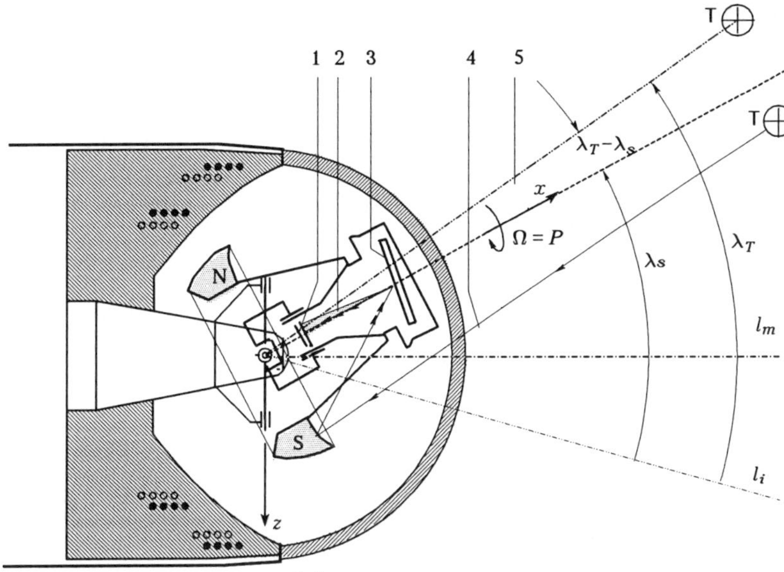

(a)

1 – Stationary chopper mounted on outer gimbal
2 – Reflected light beam tracing shifted circles in the chopper plane
3 – Spinning canted flat mirror
4 – Incident light from target
5 – Line of sight to target

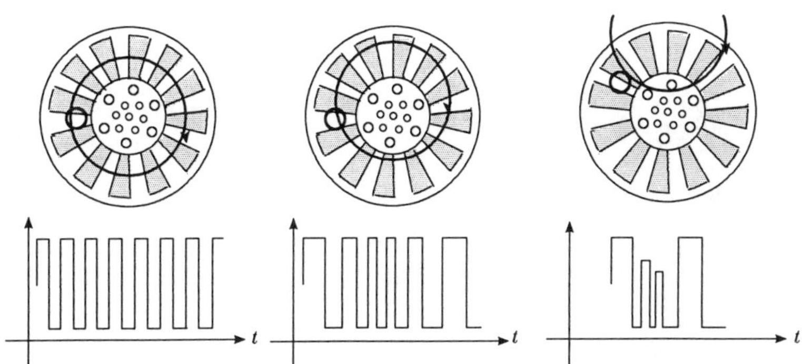

(b) (c) (d)

Figure 5.7(a). Seeker head with static chopper and canted mirror;
(b). Image trace and chopped signal when $\lambda_T-\lambda_s = 0$;
(c). $\lambda_T-\lambda_s \neq 0$ causing frequency modulation;
(d). $\lambda_T-\lambda_s$ very large, causing amplitude modulation.

Even though some monotonous dependence between the error $\lambda_T-\lambda_s$ and the detector output exists, in some designs it is only the information on the location of the target image in any of the four quadrants of the seeker head that is employed to torque the seeker head in bang-bang mode.

5.3 Application to missile homing

Homing equations

The principal application of the seeker head is in the implementation of missile homing onto a target. The seeker head is not necessarily of the gyroscopic IR type. It might just as well be of the imaging TV camera type described in Section 5.1. The case studied here is that of a stationary target on the ground. The corresponding geometry is shown in Figure 5.8.

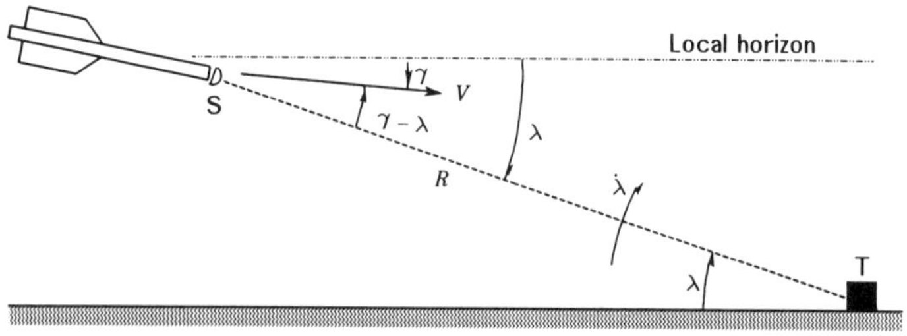

Figure 5.8. Geometry and motion in air-to-surface homing.

From the geometry, assuming small angles for λ and γ, we have

$$\dot{\lambda} = \frac{V}{R}\sin(\lambda - \gamma) \cong \frac{V}{R}(\lambda - \gamma) \tag{5.3.1}$$

where V is here assumed to be constant. In view of $R = R(t)$, (5.3.1) is a first-order time-varying differential equation. Rearranging, we have

$$\dot{\lambda} - \frac{V}{R}\lambda = -\frac{V}{R} \tag{5.3.2}$$

Disregarding the time constant τ_s of the seeker head S and possible drift rate q_d, $\lambda = \lambda_s$ and $q = \dot{\lambda}$. Proportional navigation which dictates a collision course to the target T is achieved by applying the signal $\dot{\lambda}$ to the control surfaces implementing the following law:

$$\dot{\gamma} = Kq = K\dot{\lambda} \tag{5.3.3}$$

where K is a constant that incorporates all the electronic and mechanical scale factors in the control path. K is also known as the *navigation constant.* Since $V\dot{\gamma} = a_n$, (5.3.3) implies that the intensity of the normal acceleration a_n is proportional to the rate of change of the line of sight $\dot{\lambda}$. We now rewrite (5.3.1) as

$$R\dot{\lambda} = V(\lambda - \gamma) \tag{5.3.4}$$

For small values of $\lambda - \gamma$, we have to a good approximation

$$\dot{R} = -V \tag{5.3.5}$$

Differentiating (5.3.4) we have

$$\dot{R}\dot{\lambda} + R\ddot{\lambda} = \dot{V}\dot{\lambda} - V\ddot{\gamma} \tag{5.3.6}$$

Substituting (5.3.3) and (5.3.5) into (5.3.6), leads to

$$\dot{R}\dot{\lambda} + R\ddot{\lambda} = \dot{V}\dot{\lambda} - VK\ddot{\lambda} \tag{5.3.7}$$

Denoting the total time of flight until collision by t_f, the range to T at any instant is

$$R = V(t_f - t) \tag{5.3.8}$$

Substituting (5.3.8) into (5.3.7), leads to

$$-V\dot{\lambda} + V(t_f - t)\ddot{\lambda} = \dot{V}\dot{\lambda} - VK\ddot{\lambda} \tag{5.3.9}$$

Rearranging we have the homogeneous time-varying differential equation,

$$\ddot{\lambda} + \frac{K-2}{t_f - t}\dot{\lambda} = 0 \tag{5.3.10}$$

Denoting the initial condition for $\dot{\lambda}$ by

$$\dot{\lambda}(t=0) = \dot{\lambda}_0 \tag{5.3.11}$$

the solution for $\dot{\lambda}$ is

$$\dot{\lambda} = \dot{\lambda}_0 \left\{ \frac{t_f - t}{t_f} \right\}^{K-2} \tag{5.3.12}$$

Conclusions:

1. To facilitate the flight path that approaches a straight line i.e., $\dot{\gamma} \to 0$, in view of (5.3.3), it is necessary, that $\dot{\lambda} \to 0$.

2. In order to assure this, it is necessary that $K > 2$. If this condition is fulfilled, the flight path always terminates with zero maneuver, i.e., $\dot{\lambda} = \dot{\gamma} = 0$.

3. If indeed $\dot{\lambda} = 0$, it follows from (5.3.1) that $\gamma = \lambda$, which implies a hit. This result clearly holds only for a stationary target.

The effect of gyro drift rate — stationary target

Assume now that the seeker head is disturbed by a gyro drift rate q_d. Equation (5.3.3) now takes the form

$$\dot{\gamma} = K(q + q_d) = K\dot{\lambda} + Kq_d \tag{5.3.13}$$

Substituting (5.3.13) instead of (5.3.3) into (5.3.10) yields the equation

$$\ddot{\lambda}+\frac{K-2}{t_f-t}\dot{\lambda}=-\frac{K}{t_f-t}q_d \tag{5.3.14}$$

Thus, the drift-rate q_d imposes a forcing function, the intensity of which increases as $t \to t_f$. The general solution of (5.3.14) is of the form

$$\dot{\lambda}=\dot{\lambda}_p+C\left\{\frac{t_f-t}{t_f}\right\}^{K-2} \tag{5.3.15}$$

where $\dot{\lambda}_p$ is the particular solution and C is an unknown constant. We now assume that $\dot{\lambda}_p = k_0$, where k_0 is a constant so that $\ddot{\lambda}_p=0$ and (5.3.14) reduces to

$$(K-2)k_0=-Kq_d \tag{5.3.16}$$

namely,

$$k_0=-\frac{K}{K-2}q_d \tag{5.3.17}$$

Substituting (5.3.17) into (5.3.15), we have

$$\dot{\lambda}=-\frac{K}{K-2}q_d+C\left\{\frac{t_f-t}{t_f}\right\}^{K-2} \tag{5.3.18}$$

It now remains to determine the constant C. Since for $t=0$, $\dot{\lambda}=\dot{\lambda}_0$, we have from (5.3.18)

$$C=\frac{K}{K-2}q_d+\dot{\lambda}_0 \tag{5.3.19}$$

Substituting (5.3.19) into (5.3.18) and rearranging, the solution takes the form

$$\dot{\lambda}=\frac{K}{K-2}q_d\left[\left(\frac{t_f-t}{t_f}\right)^{K-2}-1\right]+\dot{\lambda}_0\left(\frac{t_f-t}{t_f}\right)^{K-2} \tag{5.3.20}$$

Conclusions:

1. For $t=0$, $\dot{\lambda}=\dot{\lambda}_0$ as before, implying that initially, the effect of q_d is zero.
2. For $t=t_f$, it follows that $\dot{\lambda}=-K/(K-2)q_d$ and $\dot{\gamma}=-K^2/(K-2)q_d$. Therefore, the missile maneuver persists until the end and sharply grows as $t \to t_f$. As demonstrated in the next paragraph, the drift rate q_d is equivalent to a maneuvering target but opposite in sense.
3. If K is only slightly larger than 2, the effect of q_d on $\dot{\lambda}$ and $\dot{\gamma}$ can assume very large values to the extent of disrupting the homing process. Therefore, in practice, K is set to 3 or more as a result of the effect of drift-rate bias or maneuvering targets.

Effect of gyro drift – moving target

We now review the above for the more general case when the target T is moving at a constant evader speed V_e as indicated in Figure 5.9. The missile pursuer speed is now denoted by V_p.

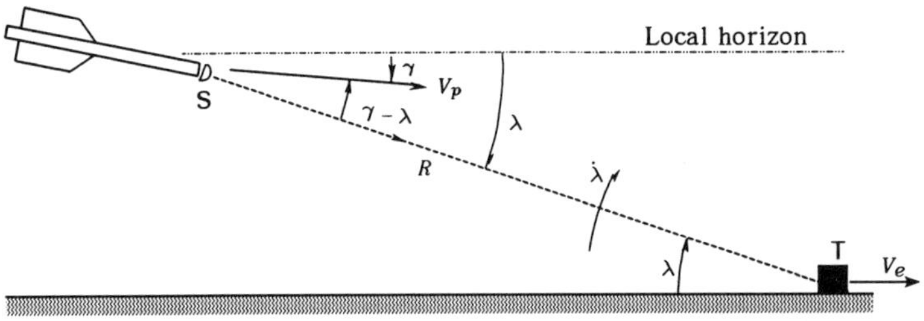

Figure 5.9. Air–to–surface homing – moving target.

From the small angle geometry we have,

$$\dot{\lambda} \cong \frac{V_p}{R}(\lambda - \gamma) - \frac{V_e}{R}\lambda \tag{5.3.21}$$

or

$$R\dot{\lambda} \cong (V_p - V_e)\lambda - V_p\gamma \tag{5.3.22}$$

The instantaneous range for small angles is

$$R \cong (V_p - V_e)(t_f - t) \tag{5.3.23}$$

We denote the closing speed by V_c, where

$$V_c \cong V_p - V_e \tag{5.3.24}$$

and

$$\dot{R} = -V_c \tag{5.3.25}$$

Differentiating (5.3.22),

$$\dot{R}\dot{\lambda} + R\ddot{\lambda} = V_c\dot{\lambda} - V_p\dot{\gamma} \tag{5.3.26}$$

Substituting (5.3.3) for $\dot{\gamma}$, and (5.3.25) into (5.3.26), we have

$$-V_c\dot{\lambda} + V_c(t_f - t)\ddot{\lambda} = V_c\dot{\lambda} - V_pK\dot{\lambda} \tag{5.3.27}$$

or

$$V_c(t_f - t)\ddot{\lambda} + (V_pK - 2V_c)\dot{\lambda} = 0 \tag{5.3.28}$$

which can be rewritten as

$$\ddot{\lambda} + \frac{(K\frac{V_p}{V_c} - 2)}{t_f - t}\dot{\lambda} = 0 \tag{5.3.29}$$

Denoting

$$K' \triangleq K\frac{V_p}{V_c} \tag{5.3.30}$$

Equation(5.3.30) is equivalent to (5.3.10), except that K is replaced by K' which is the effective navigation constant. Since for a receding target, as shown here, $V_p > V_c$, it follows, in view of (5.3.12) and replacing K by K', that the condition for proper homing now is

$$K > \frac{2V_c}{V_p} \tag{5.3.31}$$

which is smaller than 2. On the other hand, for an approaching target, K must clearly be greater than 2. This is an additional argument for setting K to be greater than 2 in order to cope with the variety of situations encompassing moving targets.

Accelerating target

We now review the homing process for the case of an accelerating or maneuvering target T. Referring again to (5.3.22), but allowing a time varying $V_e = V_e(t)$, (5.3.25) is modified to

$$\dot{R}\dot{\lambda} + R\ddot{\lambda} = V_c\dot{\lambda} - V_p\dot{\gamma} - \dot{V}_e\lambda \tag{5.3.32}$$

Substituting $\dot{\gamma} = K\dot{\lambda}$, $\dot{R} = -V_c$, and recalling (5.3.8), we have

$$-V_c\dot{\lambda} + V_c(t_f - t)\ddot{\lambda} = V_c\dot{\lambda} - V_pK\dot{\lambda} - \dot{V}_e\lambda \tag{5.3.33}$$

Rearranging and dividing through by $t_f - t$,

$$\ddot{\lambda} + \frac{K\frac{V_p}{V_c} - 2}{t_f - t}\dot{\lambda} = -\frac{\dot{V}_e}{V_c(t_f - t)}\lambda \tag{5.3.34}$$

which can be rewritten as

$$\ddot{\lambda} + \frac{K' - 2}{t_f - t}\dot{\lambda} = -\frac{\dot{V}_e}{R}\lambda \tag{5.3.35}$$

Since $\dot{V}\lambda$ is the component of linear acceleration normal to the line of sight, we define $(\dot{V}\lambda)/R \triangleq \ddot{\lambda}_e$. Thus,

$$\ddot{\lambda} + \frac{K' - 2}{t_f - t}\dot{\lambda} = -\ddot{\lambda}_e \tag{5.3.36}$$

Equation (5.3.36) is equivalent to (5.3.14). The right-hand side of the line-of-sight angular acceleration forcing function $\ddot{\lambda}_e$ due to the acceleration of the target is equivalent to the forcing function due to q_d. This may cause very large missile accelerations toward the termination of the flight path.

5.4 Beam riding guidance

Geometry and motion

The seeker head described earlier can be applied to other guidance algorithms. In a highly simplified version, the entire gimbal and electro-optical unit is caged to the casing. In doing so, the seeker head performs as a simple two-axis goniometer. By pointing its optical axis in any desired direction with the aid of a coaligned sight indicated in Figure 5.10, a reference line is set up in space. If a luminous target is within the field of view, the chopper detector unit can measure both component deviations, λ_x and λ_y , of the target line of sight from the optical axis. This measurement provides the basis for beam riding guidance.

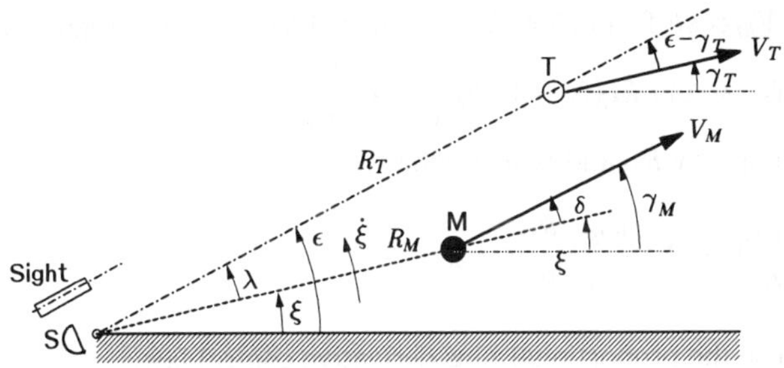

T	Target
M	Missile
R_T	Range to target
R_M	Range to missile
ϵ	Line of sight (LOS) angle to target
ξ	Line of sight angle to missile
γ_T	Target flight path angle
γ_M	Missile flight path angle
λ	Angular deviation of missile LOS from target LOS detected by S.
δ	Missile flight path angle relative to line of sight

Figure 5.10. Beam-riding geometry and motion.

The underlying principle of operation and the technique outlined here have been utilized in numerous antitank missile systems such as the U.S. TOW.

Guidance equations and effect of target maneuver

From Figure 5.10, we have

$$\dot{R}_M = V_M \cos\delta \tag{5.4.1}$$

$$R_M\dot{\xi} = R_M(\dot{\epsilon} - \dot{\lambda}) = V_M \sin\delta \tag{5.4.2}$$

The missile acceleration a_N to V_M is

$$a_N = V_M\dot{\gamma}_M = V_M(\dot{\xi} + \dot{\delta}) = V_M(\dot{\epsilon} - \dot{\lambda} + \dot{\delta}) \tag{5.4.3}$$

The guidance task is to annul λ, caused by the control forces exerted on **M**. Define the control variable u causing a_N proportional to it, namely,

$$a_N = Ku \tag{5.4.4}$$

From (5.4.3) and (5.4.4), we have

$$\dot{\epsilon} - \dot{\lambda} + \dot{\delta} = \frac{Ku}{V_M} \tag{5.4.5}$$

Differentiating (5.4.2), we have

$$R_M(\ddot{\epsilon} - \ddot{\lambda}) + \dot{R}_M(\dot{\epsilon} - \dot{\lambda}) = V_M \cos\delta\, \dot{\delta} \tag{5.4.6}$$

Substituting $V_M \cos\delta$ from (5.4.1), $\dot{\delta}$ from (5.4.5), and rearranging we have

$$R_M(\ddot{\lambda} - \ddot{\epsilon}) + \dot{R}_M(\dot{\lambda} - \dot{\epsilon}) = \dot{R}_M\left(\dot{\epsilon} - \dot{\lambda} - \frac{Ku}{V_M}\right) \tag{5.4.7}$$

Dividing through by $\dot{R}_M$ and rearranging,

$$\frac{R_M}{\dot{R}_M}\ddot{\lambda} + 2\dot{\lambda} + \frac{Ku}{V_M} = \frac{R_M}{\dot{R}_M}\ddot{\epsilon} + 2\dot{\epsilon} \tag{5.4.8}$$

We now choose the proportional plus derivative control law,

$$u = \lambda R_M + T\frac{\mathrm{d}}{\mathrm{d}t}(\lambda R_M) \tag{5.4.9}$$

Substituting (5.4.9) into (5.4.8) and rearranging, we have

$$\frac{R_M}{\dot{R}_M}\ddot{\lambda} + 2\left(\dot{\lambda} + \frac{K}{V_M}TR_M\right)\dot{\lambda} + \left(\frac{KR_M}{V_M} + \frac{K}{V_M}T\dot{R}_M\right)\lambda = \frac{R_M}{\dot{R}_M}\ddot{\epsilon} + 2\dot{\epsilon} \tag{5.4.10}$$

We now substitute

$$\frac{R_M}{\dot{R}_M} = \frac{R_M}{V_M} = t \tag{5.4.11}$$

where t is the time elapsed from the instant of launch, $t=0$. Substituting (5.4.11) into (5.4.10), and dividing through by t, we obtain

$$\ddot{\lambda} + \left(\frac{2}{t} + KT\right)\dot{\lambda} + \left(K + \frac{KT}{t}\right)\lambda = \ddot{\epsilon} + \frac{2}{t}\dot{\epsilon} \tag{5.4.12}$$

This is a time-varying linear second-order differential equation in λ. As t

grows the terms with $1/t$ can be neglected and (5.4.12) reduces to the ordinary second-order differential equation,

$$\ddot{\lambda} + KT\dot{\lambda} + K\lambda = \ddot{\epsilon} \tag{5.4.13}$$

On the other hand, for $t \cong 0$, the $1/t$ terms become dominant and (5.4.12) reduces to the first order differential equation,

$$\dot{\lambda} + \frac{KT}{2}\lambda = \dot{\epsilon} \tag{5.4.14}$$

For large values of t, the forcing function $\ddot{\epsilon}$ implies that for $\dot{\epsilon} = \text{const.}$, $\ddot{\epsilon} = 0$ and λ converges to zero in the steady state. This means that the beam rider can perfectly track a target that presents a constant rate of turn of the target line of sight to the seeker. On the other hand, if $\ddot{\epsilon} \neq 0$, an error in λ develops which will be sustained for the duration of $\ddot{\epsilon}$. Since $\ddot{\epsilon}$ can be caused by a target maneuver, the foregoing implies that if **T** represents an evader, and if the timing of the onset of $\ddot{\epsilon}$ is correct, the magnitude and the timing of λ can cause a miss. If $\ddot{\epsilon}$ is modeled as a random stationary process characterized by the autocorrelation function $\phi_{\ddot{\epsilon}\ddot{\epsilon}}(\tau)$, it is possible to evaluate the rms value of λ as shown in Chapter 4. Specifically, expressing the self-spectrum $\Phi_{\ddot{\epsilon}\ddot{\epsilon}}(s)$ of $\ddot{\epsilon}$ and expressing the transfer function from $\ddot{\epsilon}$ to λ by

$$H_\lambda(s) = \frac{1}{s^2 + KTs + K} \tag{5.4.15}$$

the variance of λ which settles to a stationary value, can be determined by

$$\sigma_\lambda^2 = \frac{1}{2\pi j}\int_{-j\infty}^{j\infty} H_\lambda(s)H_\lambda(-s)\Phi_{\ddot{\epsilon}\ddot{\epsilon}}(s)\,ds \tag{5.4.16}$$

The variance of the deviation from the missile line of sight therefore is

$$\sigma_d^2 = R_M^2\sigma_\lambda^2 \tag{5.4.17}$$

Thus, σ_d grows linearly with the missile range R_M.

The effect of seeker head noise

The line-of-sight deviation λ detected by the seeker head **S** is accompanied by an error denoted by ε_λ, which may consist of a constant offset and wideband noise. Incorporating ε_λ into the control law u leads to

$$u = (\lambda + \varepsilon_\lambda)R_M + T\frac{d}{dt}(\lambda + \varepsilon_\lambda)R_M \tag{5.4.18}$$

$$= \lambda R_M + T\dot{\lambda}R_M + T\lambda\dot{R}_M + \varepsilon_\lambda R_M + T\dot{\varepsilon}_\lambda R_M + T\varepsilon_\lambda\dot{R}_M \tag{5.4.19}$$

The last three terms, after normalizing by $\dot{R}_M$ and multiplying by K, add to (5.4.12) to yield

$$\ddot{\lambda} + \left(\frac{2}{t} + KT\right)\dot{\lambda} + \left(K + \frac{KT}{t}\right)\lambda = \ddot{\epsilon} + \frac{2}{t}\dot{\epsilon} - KT\dot{\varepsilon}_\lambda - K\left(1 + \frac{T}{t}\right)\varepsilon_\lambda \qquad (5.4.20)$$

The first two terms on the right-hand side of (5.4.18) are due to target maneuvers, while the last two are due to seeker head noise. Disregarding the first two and the $1/t$ terms, we have

$$\ddot{\lambda} + KT\dot{\lambda} + K\lambda = -KT\dot{\varepsilon}_\lambda - K\varepsilon_\lambda \qquad (5.4.21)$$

The transfer function from ε_λ to λ is determined by,

$$\lambda(s) = -\frac{K(1 + Ts)}{s^2 + KTs + K}\varepsilon_\lambda \qquad (5.4.22)$$

$$H_{\varepsilon_\lambda}(s) = \frac{K(1 + Ts)}{s^2 + KTs + K} \qquad (5.4.23)$$

A special case is $\epsilon_\lambda =$ const. i. e., a bias in the LOS measurement. It is evident that in the steady state, $\lambda_{ss} \rightarrow \epsilon_\lambda$. If $\dot{\epsilon}_\lambda =$const., the steady state-error is $\lambda_{ss} = T\dot{\epsilon}_\lambda$. If the spectrum of the seeker head noise $\Phi_{\varepsilon_\lambda \varepsilon_\lambda}(s)$ is known, its contribution to σ_λ can be determined from

$$\sigma_\lambda^2 = \frac{1}{2\pi j}\int_{-j\infty}^{j\infty} H_{\varepsilon_\lambda}(s) H_{\varepsilon_\lambda}(-s) \Phi_{\varepsilon_\lambda \varepsilon_\lambda}(s)\,ds \qquad (5.4.24)$$

Broadly, $\Phi_{\varepsilon_\lambda \varepsilon_\lambda}(s)$ can be modeled as white noise. Its origin is basically the shot noise in the detector of the seeker head. However, the signal that it accompanies undergoes synchronous detection at the frequency determined by the spin speed of the seeker rotor. Thus, the noise is also modulated, and it typically has a bandwidth of a few hundred Hz, which can still be modeled as white noise.

It should be noted that $\dot{\epsilon}$ is not the primary variable that dictates the target maneuver. The appropriate variable is actually γ_T, which determines $\dot{\epsilon}$ by the following relationship, as can be gathered from the geometry of Figure 5.10:

$$\dot{\epsilon} = \frac{V_T}{R_T}\sin(\epsilon - \gamma_T) \qquad (5.4.25)$$

If one assumes small angles for $(\epsilon - \gamma_T)$, we have, after rearranging,

$$\dot{\epsilon} + \frac{V_T}{R_T}\epsilon = \frac{V_T}{R_T}\gamma_T \qquad (5.4.26)$$

This, again, is a time-varying differential equation the solution of which yields ϵ and $\dot{\epsilon}$ in terms of the actual target maneuver γ_T. The substitution of its derivatives into (5.4.12) establishes an explicit relationship between the target maneuvers due to γ_T and the guidance error λ.

5.5 Three-axis platform for inertial navigation

Reference axes

In outer space, the platform stabilization is with respect to an arbitrary reference line. In atmospheric flight close to Earth, where the vehicle is supported by a vertical lift force, the stabilization needs to be with respect to the local vertical or horizon and with respect to local north. Consequently, the platform must be rotated in accordance with the diurnal rotation of Earth and the motion of the vehicle relative to Earth. The geometry and motion are shown in Figure 5.11. The total angular rate of the vehicle with respect to inertial space is the vector sum of $\boldsymbol{\Omega}_e$ and $\boldsymbol{V}/R$. In Figure 5.12 they are shown as a scalar sum. The vehicle flies at the altitude h and at an eastern ground speed V_x. These variables are not necessarily constant. It is assumed that $h \ll R_e$. At any instant, the elevation angle of the platform's longitudinal axis with respect to the arbitrary inertial reference line is η. The direction of the local horizon with respect to that line is η_H. The geometry and the definition of the reference lines in the vertical plane are described in Figure 5.11. Since we are dealing with a nominally geocentric flight path, the natural coordinate system is polar (R,θ), here $R = R_e + h$, with the origin at Earth center C_e. θ is the subtended angle due to the aircraft motion $\dot{\theta} = V_x/R$ starting from an arbitrary initial value θ_0. The lowercase θ should not be confused with the uppercase Θ, which represents the aircraft pitch angle between the longitudinal body axis x_b and the local level. The kinematic relationships that follow from the geocentric motion are not at all related to the aircraft attitude angles. As such, the aircraft is represented by a point. In the following analysis, for the sake of simplicity, we assume that $h \ll R_e$.

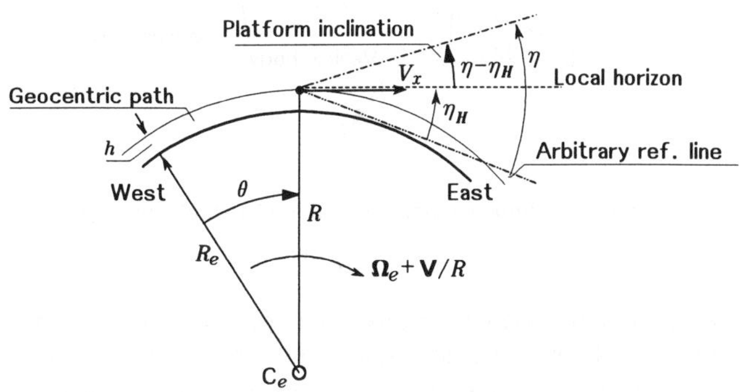

Figure 5.11. Geometry and motion of an airborne level platform in geocentric flight.

Gimbaled mechanization

The three-axis inertially stabilized platform provides an on-board inertial reference axis system that is the basis for inertial navigation. It is a direct implementation of Newton's second law. The orientation of the platform is controlled by a triad of orthogonal RIGs and a triad of specific force sensors mounted on this platform. The accelerometer triad therefore measures the Cartesian components of the vehicle's acceleration vector with respect to inertial space. A description of its three-axis mechanization is given e.g., in [5]. A single-axis schematic of the platform control system around its y axis is shown in Figure 5.12. By means of the platform-mounted accelerometers, e.g., A_x, it implements the local level, which is continually monitored by the rate gyros sensing the angular rates of the platform. At the same time, the system provides the required inertial velocity component V_x and the covered distance $X = R_e\theta$. as shown in Figure 5.12. The schematic also indicates the appropriate pickoff p_Θ, which measures the pitch angle Θ between the inertial platform attitude angle η and the longitudinal body axis x_b. Θ would be a perfect measurement of the aircraft pitch angle if the error $\epsilon_\eta = \eta - \eta_H$ were zero. Thus, pitch angle can be provided only to the precision of the platform leveling. This is not a significant shortcoming however, because ϵ_η is generally orders of magnitude smaller than the permissible error in Θ. The measured and computed angular rates $\dot{\eta}_m$, $\dot{\eta}_c$ are compared and amplified to provide the platform control command u_η which slaves it to track $\dot{\eta}_c$.

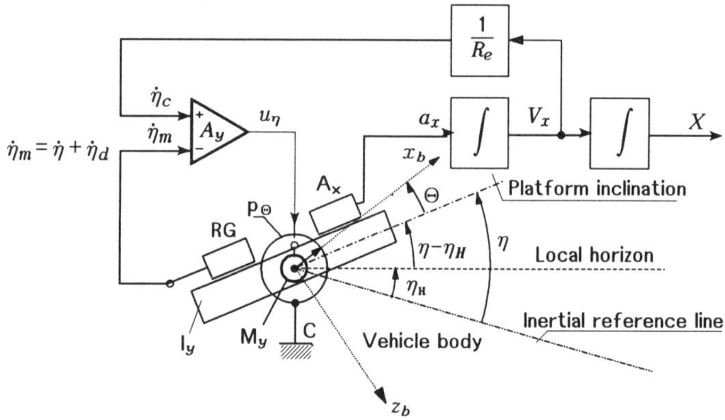

Figure 5.12. Gimbaled mechanization of a stable platform with velocity Schuler tuning.

A schematic isometric view of a three-axis platform supporting the gyro and accelerometer triads is shown in Figure 5.13. The outer and inner gimbals G_o and G_i, respectively, are shown along with their respective pickoffs and torque motors. The three axis mechanical isolation of the stable platform from the vehicle body is self-evident.

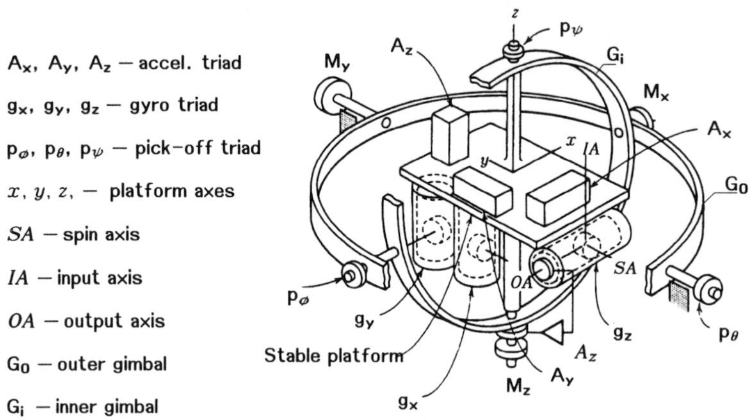

Figure 5.13. Perspective view of a three-axis gimbaled platform.

As indicated earlier, the basic inertial sensor package A_x, A_y, A_y, g_x, g_y, g_z has no relation to the vehicle attitude angles ϕ, θ, ψ that need to be known for the roll, pitch and yaw control functions. There is no direct method to determine the angular rates *P*, *Q*, *R*, which are often required for stability augmentation.

The mathematical model of the velocity Schuler-tuned platform

The platform angular control loop is essentially the same as the one described in Section 5.1, with the following exceptions: 1) in INS applications, as discussed here, the platform is driven to align itself with the local level by means of its accelerometer payload, and 2) very high-precision rate gyros are required. The reading a_x of the accelerometer A_x is

$$a_x = g(\eta - \eta_H) + \dot{V}_x \tag{5.5.1}$$

where η_H varies in accordance with

$$\eta_H(t) = \int_0^t -\frac{V_x}{R_e} d\tau \tag{5.5.2}$$

or

$$\eta_H(s) = -\frac{1}{s}\frac{V_x}{R_e} \tag{5.5.3}$$

The minus sign is in accordance with the conventions of a right-hand axis system, considering that the y axis points out of the paper plane. The division of V_x by R_e provides the signal $\dot{\eta}_c$, which is given by

$$\dot{\eta}_c(s) = -\frac{1}{sR_e}[g(\eta - \eta_H) + \dot{V}_x] \tag{5.5.4}$$

The platform torqueing dynamics are equivalent to (5.1.8). The corresponding platform time constant now denoted by $\tau_p \triangleq I_y/A_y b_y$ is assumed to be negligibly small so that (5.1.8) simplifies to

$$\dot{\eta} \cong \dot{\eta}_c - \dot{\eta}_d \tag{5.5.5}$$

The second term in (5.1.8) is considered as incorporated in $\dot{\eta}_d$. The rate integrating gyro is assumed to operate in a closed loop mode in accordance with (4.10.8) so that it performs like a high-precision rate gyro. Substituting (5.5.5) into (5.5.4), and since $\eta = \dot{\eta}/s$,

$$\dot{\eta} + \dot{\eta}_d = -\frac{1}{sR_e}\left[g\left(\frac{\dot{\eta}}{s} + \frac{1}{s}\frac{V_x}{R_e}\right) + sV_x\right] \tag{5.5.6}$$

Recalling from (3.2.10—3.2.11) that $\omega_s^2 = g/R_e$,

$$\dot{\eta} + \dot{\eta}_d = -\frac{\omega_s^2}{s^2}\dot{\eta} - \frac{g}{R_e^2}\frac{V_x}{s^2} - \frac{V_x}{R_e} = -\frac{\omega_s^2}{s^2}\dot{\eta} - \frac{V_x}{R_e}\frac{\omega_s^2}{s^2} - \frac{V_x}{R_e} \tag{5.5.7}$$

Rearranging, we have

$$\dot{\eta}\left(1 + \frac{\omega_s^2}{s^2}\right) = -\frac{V_x}{R_e}\left(1 + \frac{\omega_s^2}{s^2}\right) - \dot{\eta}_d \tag{5.5.8}$$

$\therefore$

$$\dot{\eta} = -\frac{V_x}{R_e} - \frac{s^2}{s^2 + \omega_s^2}\dot{\eta}_d \tag{5.5.9}$$

Summary of principal properties

1. The mechanization shown here complies exactly with the Schuler tuned mathematical pendulum described in Section 3.2. Whereas an Earth-radius-long mathematical pendulum or a nanometer-long physical pendulum is a technical absurdity, the present solution, based on inertial sensors and servo technology, has proven to be totally practical.

2. The platform tracks the direction of the local horizon with respect to inertial space at the angular rate $-V_x/R_e$. The first integration in Figure 5.12 yields the velocity V_x and $\dot{\eta}_c$ which facilitates the Schuler tuning. The second integration yields the accumulated distance X covered.

3. $\dot{V}_x$ is not present in (5.5.9), therefore, the platform does not respond to horizontal acceleration components.

4. The analysis, presented here in the vertical west-east plane, is equally valid for vertical planes in any other direction. The actual platform design is at least a three-axis configuration, so, one may think of the Schuler tuning in the context of the global sphere, not just a circle.

Positional Schuler tuning

The platform control mechanization shown in Figure 5.12 has the advantage of simplicity because the loop closure around the RIG moves its zero frequency pole out to more than 2000 rad/sec and renders it almost as a pure gain coefficient. In other words, the loop dynamics practically presents a first order system. Its disadvantage is that it does not cancel out steady-state errors due to T_D, as pointed out in connection with (4.10.8). To achieve this, integral control is needed. This can be readily accomplished by two modifications: Eliminating the loop closure around the RIG and thus thus re establishing the pole at zero frequency. and driving the platform by $\eta(s)$ instead of $\dot{\eta}(s)$. This can be accomplished by using $\eta_c = X/R_e$, as shown in Figure 5.14. Clearly, the order of the control loop is raised by one and stability considerations require a lead network in order to provide the necessary damping.

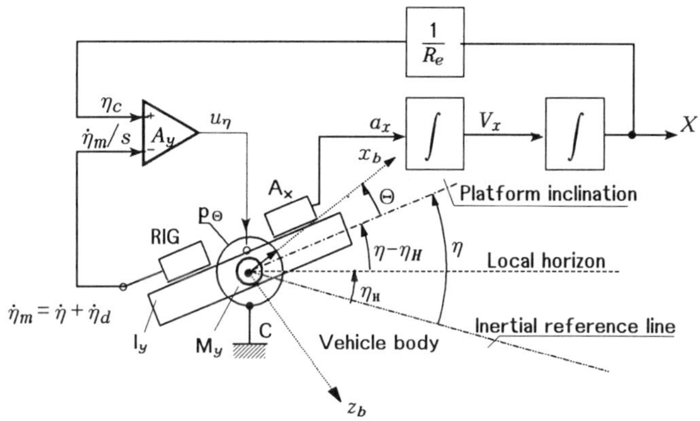

Figure 5.14. Gimbaled stable platform with position Schuler tuning.

The platform responds to the drift process $\dot{\eta}_d$, but in accordance with the Schuler bandpass filtering operation:

$$\eta(s) = -\frac{V_x}{R_e s} - \frac{s}{s^2 + \omega_s^2}\dot{\eta}_d(s) = \eta_H - \frac{s}{s^2 + \omega_s^2}\dot{\eta}_d(s) \tag{5.5.10}$$

The acceleration error Δa_x due to platform drift rate is

$$\Delta a_x = g(\eta - \eta_H) = g\frac{s}{s^2 + \omega_s^2}\dot{\eta}_d(s) \tag{5.5.11}$$

The ground-speed error $\Delta V_x = \Delta a_x / s$ is

$$\Delta V_x = g\frac{1}{s^2+\omega_s^2}\dot{\eta}_d(s) = H_v(s)\dot{\eta}_d(s) \tag{5.5.12}$$

The corresponding position error $\Delta X = \Delta V_x / s$ is

$$\Delta X = g\frac{1}{s(s^2+\omega_s^2)}\dot{\eta}_d(s) = H_x(s)\dot{\eta}_d(s) \tag{5.5.13}$$

where

$$H_v(s) \triangleq g\frac{1}{s^2+\omega_s^2} \tag{5.5.14}$$

$$H_x(s) \triangleq g\frac{1}{s(s^2+\omega_s^2)} \tag{5.5.15}$$

The impulse response corresponding to $H_v(s)$ is

$$h_v(t) = g\frac{\sin\omega_s t}{\omega_s} \tag{5.5.16}$$

and the one corresponding to $H_x(s)$ is

$$h_x(t) = \frac{g}{\omega_s^2}(1-\cos\omega_s t) \tag{5.5.17}$$

Velocity and positional error propagation

The conventional model for the drift rate $\dot{\eta}_d$ is a first order Markov process with the following autocorrelation function:

$$\phi_{\dot{\eta}_d\dot{\eta}_d}(\tau) = \sigma_{\dot{\eta}_d}^2 e^{-a|\tau|} = \frac{\sigma_{\dot{\eta}_d}^2}{a} a e^{-a|\tau|} \tag{5.5.18}$$

It is easily verified that the $ae^{-a|\tau|}$ assumes the properties of a Dirac function $\delta(\tau)$, for $a\to\infty$ and $\tau\to 0$. $\sigma_{\dot{\eta}_d}^2/a$ is the power spectral density of the drift process when modeled as a white noise process. The result that $H_v(s)$ and $H_x(s)$ have free poles on the imaginary axis in the s plane implies that on integrating the corresponding power spectra, the resulting power grows to infinity. The nature of this growth, namely, the divergence in time, was presented in Section 2.7. The specific time divergence of the Schuler-tuned platform for various drift statistics will be presented later. In general, the velocity and positional error due to the gyro drift rate, as shown in Section 2.7, are given by the double integrals

$$\sigma^2_{\Delta V}=\sigma^2_{\dot{\eta}_d} g^2 \int_0^t \int_0^t h_v(\sigma) h_v(\sigma') \rho_{\dot{\eta}_d\dot{\eta}_d}(\sigma-\sigma') d\sigma d\sigma' \tag{5.5.19}$$

and

$$\sigma^2_{\Delta x}=\sigma^2_{\dot{\eta}_d} g^2 \int_0^t \int_0^t h_x(\sigma) h_x(\sigma') \rho_{\dot{\eta}_d\dot{\eta}_d}(\sigma-\sigma') d\sigma d\sigma' \tag{5.5.20}$$

With these assumptions for a narrow normalized autocorrelation $\rho_{\dot{\eta}_d\dot{\eta}_d}(\tau)$, i.e., white noise, and in view of Section 2.7, (4.12.18) and (4.12.19) specialize to

$$\sigma^2_{\Delta V}=\frac{\sigma^2_{\dot{\eta}_d}}{a}\int_0^t h_v^2(\sigma) d\sigma = \frac{\sigma^2_{\dot{\eta}_d}}{a}\frac{g^2}{\omega_s^2}\int_0^t \sin^2(\omega_s\sigma) d\sigma \tag{5.5.21}$$

$$\sigma^2_{\Delta X}=\frac{\sigma^2_{\dot{\eta}_d}}{a}\int_0^t h_x^2(\sigma) d\sigma = \frac{\sigma^2_{\dot{\eta}_d}}{a}\frac{g^2}{\omega_s^4}\int_0^t (1-\cos\omega_s\sigma)^2 d\sigma \tag{5.5.22}$$

Since $\omega_s^2 = g/R_e$, we have $g^2/\omega_s^4 = R_e^2$ so that

$$\sigma^2_{\Delta V}=\frac{\sigma^2_{\dot{\eta}_d}}{a/\omega_s} R_e^2 \int_0^\varsigma \sin^2\xi d\xi \tag{5.5.23}$$

$$\sigma^2_{\Delta X}=\frac{\sigma^2_{\dot{\eta}_d}}{a\omega_s} R_e^2 \int_0^\varsigma (1-\cos\xi)^2 d\xi \tag{5.5.24}$$

Denoting $a/\omega_s \Delta \alpha$, we have $\omega_s^2\alpha = a\omega_s$. Substituting into (5.5.21) and (5.5.22) and carrying out the integrations, we have

$$\sigma^2_{\Delta V}=\sigma^2_{\dot{\eta}_d} R_e^2\left[\frac{1}{\alpha}\left(\varsigma-\frac{\sin 2\varsigma}{2}\right)\right] \tag{5.5.25}$$

$$\sigma^2_{\Delta X}=\frac{\sigma^2_{\dot{\eta}_d}}{\omega_s^2} R_e^2\left[\frac{1}{\alpha}\left(3\varsigma-4\sin\varsigma+\frac{\sin 2\varsigma}{2}\right)\right] \tag{5.5.26}$$

Defining

$$v \triangleq \left[\frac{1}{\alpha}\left(\varsigma-\frac{\sin 2\varsigma}{2}\right)\right]^{1/2} \tag{5.5.27}$$

$$u \triangleq \left[\frac{1}{\alpha}\left(3\varsigma-4\sin\varsigma+\frac{\sin 2\varsigma}{2}\right)\right]^{1/2} \tag{5.5.28}$$

the relationships between the rms velocity and position error $\sigma_{\Delta V}$, $\sigma_{\Delta X}$ and the drift rate $\sigma_{\dot{\eta}_d}$ are, respectively,

$$\sigma_{\Delta V} = \sigma_{\dot{\eta}_d} R_e v \tag{5.5.29}$$

$$\sigma_{\Delta X} = \frac{\sigma_{\dot{\eta}_d}}{\omega_s} R_e u \tag{5.5.30}$$

Conclusion: The nature of the divergence of the velocity and positional error with time, for drift rates which can be approximated by a white noise model is proportional to the rms drift rate, and it depends on the bandwidth ratio $\alpha = a/\omega_s$.

The foregoing development is a valid approximation if the bandwidth a of the drift rate process is relatively large in comparison with ω_s. If this assumption does not apply, and if it can be assumed that $a \to 0$, so that $\dot{\eta}_d$ = const., we have $\phi_{\dot{\eta}_d \dot{\eta}_d}(\tau) = C^2$ for $\forall\ \tau$. The expression for the positional error variance is

$$\sigma^2_{\Delta V} = C^2 \int_0^t \int_0^t h_v(\sigma) h_v(\sigma') \mathrm{d}\sigma \mathrm{d}\sigma' = \sigma^2_{\dot{\eta}_d} g^2 \left[\int_0^t h_v(\sigma) \mathrm{d}\sigma \right]^2 \tag{5.5.31}$$

$$\sigma^2_{\Delta X} = C^2 \int_0^t \int_0^t h_x(\sigma) h_x(\sigma') \mathrm{d}\sigma \mathrm{d}\sigma' = \sigma^2_{\dot{\eta}_d} g^2 \left[\int_0^t h_x(\sigma) \mathrm{d}\sigma \right]^2 \tag{5.5.32}$$

Substituting $h_v(t)$ and $h_x(t)$ from (5.5.16) and (5.5.17) into (5.5.31) and (5.5.32) and integrating, we obtain for v and u

$$v = 1 - \cos\varsigma \tag{5.5.33}$$

$$u = \varsigma - \sin\varsigma \tag{5.5.34}$$

which, when substituted into (5.5.31) and (5.5.32), yield the corresponding $\sigma_{\Delta V}$ and $\sigma_{\Delta X}$ for $a \to 0$, i. e., a drift rate with random bias.

For the intermediate, realistic, case when a is of the same order of magnitude as ω_s, the approximations (5.5.21), (5.5.22), (5.5.32) and (5.5.33) are not valid, and exact integrations of (5.5.19) and (5.5.20) must be carried out. Substituting (5.5.16) and (5.5.17) into (5.5.19) and (5.5.20), the double integrals take the following respective specific forms:

$$I_v = \frac{1}{\omega_s^2} \int_0^t \int_0^t \sin\omega_s\sigma \sin\omega_s\sigma' e^{-a|\sigma - \sigma'|} \mathrm{d}\sigma \mathrm{d}\sigma'$$

$$=\frac{1}{\omega_s^4}\int_0^\varsigma\int_0^\varsigma \mathrm{s}\xi\mathrm{s}\xi' e^{-\alpha|\xi-\xi'|}\mathrm{d}\xi\mathrm{d}\xi' \tag{5.5.35}$$

$$I_x=\frac{1}{\omega_s^4}\int_0^t\int_0^t (1-\mathrm{c}\omega_s\sigma)(1-\mathrm{c}\omega_s\sigma')e^{-a|\sigma-\sigma'|}\mathrm{d}\sigma\mathrm{d}\sigma'$$

$$=\frac{1}{\omega_s^6}\int_0^\varsigma\int_0^\varsigma (1-\mathrm{c}\xi)(1-\mathrm{c}\xi')e^{-\alpha|\xi-\xi'|}\mathrm{d}\xi\mathrm{d}\xi' \tag{5.5.36}$$

Setting $\alpha \triangleq a/\omega_s$; $\xi \triangleq \omega_s\sigma$; $\varsigma \triangleq \omega_s t$; $\mathrm{c}(\cdot) \triangleq \cos(\cdot)$, and, $\mathrm{s}(\cdot) \triangleq \sin(\cdot)$; changing variables, $\lambda = \xi - \xi'$ and separating the integrals and the negative and positive branches of $e^{-\alpha|\xi-\xi'|}$, (5.5.35) and (5.5.36) transform into

$$I_v=\frac{1}{\omega_s^4}\int_{\xi=0}^{\varsigma}\mathrm{s}\xi\left\{\int_{\xi}^{\xi-\varsigma}\mathrm{s}(\xi-\lambda)e^{-|\lambda|}(-\mathrm{d}\lambda)\right\}\mathrm{d}\xi$$

$$=\frac{1}{\omega_s^4}\int_0^{\varsigma}\mathrm{s}\xi\left\{\int_{\xi-\varsigma}^{0}\mathrm{s}(\xi-\varsigma)e^{\alpha\lambda}\mathrm{d}\lambda+\int_0^{\xi}\mathrm{s}(\xi-\lambda)e^{-\alpha\lambda}\mathrm{d}\lambda\right\}\mathrm{d}\xi \tag{5.5.37}$$

$$I_x=\frac{1}{\omega_s^6}\int_{\xi=0}^{\varsigma}(1-\mathrm{c}\xi)\left\{\int_{\xi}^{\xi-\varsigma}[1-\mathrm{c}(\xi-\lambda)]\mathrm{e}^{-\alpha|\lambda|}(-\mathrm{d}\lambda)\right\}\mathrm{d}\xi$$

$$=\frac{1}{\omega_s^6}\int_{\xi=0}^{\varsigma}(1-\mathrm{c}\xi)\left\{\int_{\xi-\varsigma}^{0}[1-\mathrm{c}(\xi-\lambda)]\mathrm{e}^{\alpha\lambda}\mathrm{d}\lambda+\int_0^{\xi}[1-\mathrm{c}(\xi-\lambda)]e^{-\alpha\lambda}\mathrm{d}\lambda\right\}\mathrm{d}\xi \tag{5.5.38}$$

The results of these integrations are respectively,

$$I_v=\frac{1}{(\alpha^2+1)^2}\left\{\alpha(\alpha^2+1)\left(\varsigma-\frac{\sin 2\varsigma}{2}\right)-(\alpha^2\mathrm{s}^2\varsigma-\mathrm{c}^2\varsigma)-2(\alpha\,\mathrm{s}\varsigma+\mathrm{c}\varsigma)e^{-\alpha\varsigma}+1\right\} \tag{5.5.39}$$

$$I_x=\frac{1}{\alpha^2(\alpha^2+1)^2}\{\alpha\varsigma(3\alpha^2+2)(\alpha^2+1)+2(1-\mathrm{e}^{-\alpha\varsigma})(\alpha^2+1)[\alpha(\mathrm{c}\varsigma-\mathrm{s}\varsigma)-1]$$

$$+2\alpha^3e^{-\alpha\varsigma}(\alpha\mathrm{c}\varsigma-\mathrm{s}\varsigma)-\alpha^3(\alpha^2+1)4(\mathrm{s}\varsigma-\mathrm{s}(2\varsigma)/8)-\alpha^2(\alpha^2+\alpha^2\mathrm{c}^2\varsigma-\mathrm{s}^2\varsigma)\} \tag{5.5.40}$$

With (5.5.39) and (5.5.40), we define,

$$v(\varsigma,\alpha) \triangleq \sqrt{I_v} \tag{5.5.41}$$

$$u(\varsigma,\alpha) \triangleq \sqrt{I_x} \tag{5.5.42}$$

The last two expressions indeed degenerate into (5.5.27) and (5.5.28) for $\alpha \to \infty$, and into (5.5.33) and (5.5.34) for $\alpha \to 0$. $v \triangleq \sqrt{I_v}$ and $u \triangleq \sqrt{I_x}$ as functions of ς and for $\alpha = .01-20$ are plotted in Figures 5.15(a) and Figure 5.15(b), respectively.

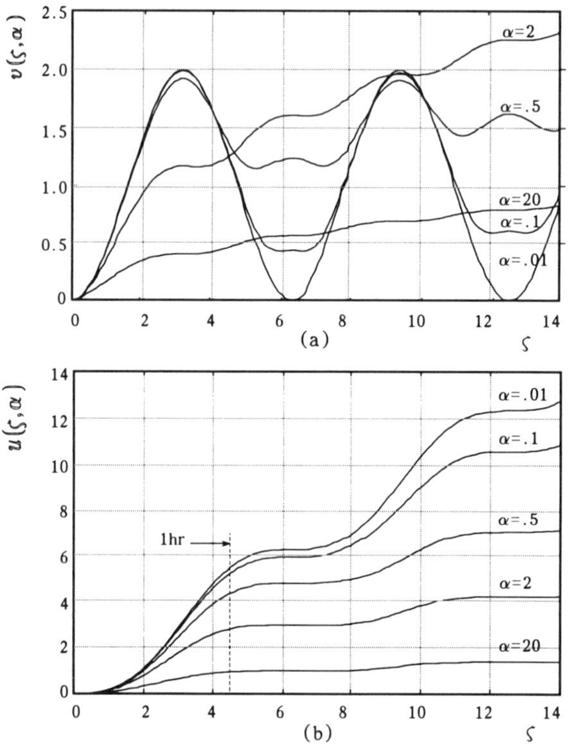

Figure 5.15(a). Normalized velocity error propagation as a function of ς for different α;

(b) normalized position error propagation as a function of ς for different α.

In view of (5.5.29) and (5.5.30), the rms velocity and positional errors along x

are

$$\sigma_{\Delta V} = \sigma_{\dot{\eta}d} R_e v(\varsigma, \alpha) \tag{5.5.43}$$

$$\sigma_{\Delta x} = \sigma_{\dot{\eta}d} \frac{R_e}{\omega_s} u(\varsigma, \alpha) \tag{5.5.44}$$

In view of Section 2.7, it should be kept in mind that (5.5.43) and (5.5.44) do not represent the divergence histories of any particular test. They are the expected rms values of entire ensembles of tests.

Additional results for INS error propagation due to constant gyro scale factor errors, accelerometer bias, initial position and velocity errors, as a function of time, can be found in [5].

The foregoing analysis removes the restriction to constant error sources and provides the tools for evaluating navigation errors due to Markov error processes spanning models from white noise to DC.

EXAMPLE 5.5–1: Random position error propagation in an INS system

The rate integrating gyro in a gimbaled inertial platform has a rms random drift rate $\sigma_{\dot{\eta}_d} = 0.01$ deg/hr. The time constant of its autocorrelation function is $T = 30$ min. What is the rms value of the positional error after 1 hr of flight?

The drift rate in natural units rad/hr is: $\sigma_{\dot{h}_d} = 0.01/57.3 = 1.745 \times 10^{-4}$ rad/hr.
T is 0.5 hr, and the bandwidth of the random drift process is $a = 1/T = 2$ hr^{-1}.
The parameter $\alpha = a/\omega_s = 2/4.46 = 0.448 \cong 0.5$. From Figure 5.15(b), the value of $u(\varsigma, \alpha)$ for $\varsigma = 4.46$ and $\alpha \cong 5$ is $u(4.46, 0.5) = 4.3$; $R_e = 6348$ km, the positional 4 rms error is

$$\sigma_{\Delta x} = 1.745 \times 10^{-4} \times \frac{6348}{4.46} \times 4.3 = 1.066 \text{ km}$$

For the less favorable case, namely $\alpha = 0.01 \cong 0$, $\sigma_{\Delta x}$ would be 1.363 km. This example demonstrates the well known state-of-the-art figure, that to achieve a positional error within 1 n mi, a gyro with a drift rate no greater than 0.01 deg/hr is required.

Conclusions:

1. The processes of velocity and positional divergence are determined by the dimensionless functions $v(\varsigma, \alpha)$ and $u(\varsigma, \alpha)$ respectively. The latter is a nondecreasing function of the normalized time ς and the normalized noise bandwidth α. The effect of the Schuler frequency is clearly visible. It moderates the process of positional divergence. The numerical parameters are the platform angular drift rate $\sigma_{\dot{\eta}_d}$ and the geophysical constants R_e and ω_s.

2. It is interesting to note that as a result of the oscillatory nature of the Schuler-tuned platform, there is no significant difference in positional error between missions of one or two hours of flight.

3. The worst value for α is $\alpha = 0$ which corresponds to a$=0$. That is, the drift rate is a random bias that should be reckoned with because it is almost impossible to completely and indefinitely cancel out a bias.

4. The larger α, the smaller the positional divergence. This implies that high-frequency noise is not a crucial contributor to positional divergence.

Strapdown mechanization and the analytical platform

Gimbaled platforms which emerged as an established technology in the 1940s were reasonably compatible with large vehicles, such as bombers, ballistic missiles, submarines, ships, and land vehicles. Two major factors caused the gradual decline of the gimbaled platform in most applications:

1) Their mechanical complexity and inherent limited reliability required large maintenance expenditures to ensure operational readiness at all times.

2) The need for a much smaller platform in view of the limited free space on board the instrument crowded military airplanes, missiles, and remotely piloted vehicles.

3) The success in the development of new generations of inertial and optical sensors which was instrumental in realizing the long-awaited concept of the analytical platform.

4) The dramatic and revolutionary progress in powerful minicomputers, which provide the means for on-line precise computation of the complex relationships in the so-called analytical platform.

In contrast to the gimbaled platform, which isolates the inertial sensors from the vehicular rotational motions, in the strapdown mechanization, all six inertial sensors are rigidly strapped to the vehicle body. Thus, all the inertial measurements of $\mathbf{F}$ and $\mathbf{\Omega}$ are with respect to body axes and not with respect to inertial space as they should be. The triad of specific forces $\mathbf{a}_b = [a_{xb}, a_{yb}, a_{zb}]^T$ must be transformed into geographical or space reference axes. This can be implemented by a computational algorithm, that implements the Euler transformation. The Euler angles can be computed from the nine parameter direction cosine algorithm as initially indicated in Chapter 3. It is easily recognized that this transformation process is the analytical equivalent of the mechanical gimbal system. This is why this methodology is known as the analytical platform. The computational algorithm that realizes the analytical platform is shown in Appendix A5.

Despite its simplicity, this concept has given rise to three major challenges:

1. The stringent requirements of the gyros that are now exposed to vehicular attitude variations, which may be two orders of magnitude larger than those in the gimbaled mechanization, require, nevertheless, the same null-point stability and resolution. This implies a dynamical range $D_r \simeq 10^8$ for the class of 1 nmi/hr INS.

2. The sensors are no longer protected from the linear and rotational vehicular vibrations, as is the case for the shock-mounted mechanical gimbaled platform. This imposes tougher specifications with regard to the ruggedness of the inertial components, motivating initiatives for new design concepts.

3. The concept poses very stringent requirements on the digital computational algorithm, which must provide the platform orientation with a precision on the order of a microradian for long periods of time.

The primary error sources resulting from the strapdown computational shortcomings are due to the following factors:

a) *Finite word length*: Instead of analog quantities, numerical values consisting of a finite number of bits are used.

b) *Truncation error*: The solution of the first-order differential equation (A5.13) can be expressed as an infinite series consisting of weighted powers of the argument. Since only a finite number of elements in this series can be afforded, a truncation error results.

c) *Noncommutativity error*: The solution of (A5.13) which assumes the small angle approximation, also implies commutativity. For small time increments dt, this is mostly justified. However, additional nonlinear terms that exist violate the commutativity. Their disregard may cause a substantial error.

The preceding three points cause the following errors:

1. In the direct computation of the DCM by (A5.13):

a) A deviation from unity of the rows and columns in the DCM constitute a scale error.

b) A skewness error results if the rows and columns in the DCM are not orthogonal.

c) A computational drift error, which is a quantity that determines the deviation of the actual computed platform orientation from the intended one.

2. In the computation of the DCM via the computed quaternions:

a) Scale error as in 1.a).

b) Drift error as in 1.c).

There is no skewness error because the columns and rows computed from quaternions are always orthogonal.

The following comments are noteworthy:

- By orthogonalizing the DCM, the scale and the skew errors are eliminated, but the drift error, that is, the scale and skew, translate into drift.

- When orthogonalizing a quaternion-derived DCM, only the drift error remains. In this case, it is not necessary to eliminate the skew, only the scale error.
- If the quaternion is normalized before its transformation to the DCM, an orthogonal matrix is obtained. That is, the scale error that exists in the quaternion derived DCM is eliminated by the prior quaternion normalization.

The foregoing factors have posed limitations on strapdown technology for at least three decades, during which the gimbaled platform continued to be considered superior to the analytical platform. However, in recent years digital computers have become so powerful and the algorithms have become refined to such a degree that the aforementioned error sources have become negligible. Thus, strapdown technology has become at least comparable to the most elaborate gimbaled mechanizations. The overall platform precision is limited by the sensors in both technologies.

An extensive organized survey of the state of the art of strapdown technology is given in [6].

5.6 Stability augmentation–effect of gyro bandwidth

Background

As a general rule, every mechanical aerospace system functioning in 3-D space under automatic or manual control has a tendency toward instability which causes the system to be lightly damped causing an oscillatory time response. Examples range from an entire aircraft which, under the pilot's manual control, often develops pilot-induced oscillations (PIO) to the automatic control of a gimbaled instrumented platform. From a control systems aspect, this problem and its traditional remedy, known as *stability augmentation* (SA), have been widely studied and established as several standard design procedures. In the case of aircraft control, the system which implements it, is known as the stability augmentation system (SAS), that is viewed as an integral part of the autopilot. The principle involved the incorporation of a means in the feedback path, which provides inertial angular rate. This is almost always implemented by a conventional rate gyro, as described in Chapter 4.

It has been common practice, as indicated e.g., in [7], to represent the rate gyro solely by its sensitivity coefficient H/K_s, as defined in Chapter 4. Its dynamics, determined by its natural frequency ω_s and its damping coefficient ς are presently ignored. The argument for disregarding the gyro dynamics is practically valid since the natural frequency of the aircraft is, as a rule, much lower than that of the gyro. It is, however, important to examine how the system design as a whole is affected if the gyro dynamics is taken into account. Such an examination clarifies why the requirement of a relatively large gyro bandwidth is important. This problem is studied in some detail in [8] with regard to the roll autopilot of an aircraft. The relevance of gyro dynamics becomes more pronounced if additional modes due to actuator dynamics and flexible modes are taken into account.

Effect on roll rate control

In the following example, the effect of the roll rate gyro dynamics in conjunction with an aircraft rigid body model is illustrated by means of a simple root locus study. Models for aircraft roll dynamics, besides the examples in [8], can be found in [1], [3], and [7].

EXAMPLE 5.6—1: Effect of rate gyro dynamics on roll control loop stability

The roll rate–rigid body dynamics $\dot{\phi}$ of an aircraft is described by the approximate relationship

$$\dot{\phi}(s) = \frac{L_{\delta_a}}{(s - L_p)}\delta_a$$

where δ_a is the actual aileron deflection angle commanded by δ_{a_c}. This is executed via a servo mechanism traditionally modeled as a first-order lag $\tau_s = \omega_s^{-1}$ such that

$$\delta_a = \frac{\omega_s}{s + \omega_s}\delta_{a_c}$$

L_{δ_a} is the aileron roll control derivative and L_p is the roll rate stability derivative also known as the *roll subsidence mode*. The aerodynamic coefficients and flight condition parameters that determine L_{δ_a} and L_p are given, along with numerical examples in [1] and [3]. The time constant of the response of $\dot{\phi}(t)$ to a step function is $\tau = 1/L_p$. L_p is typically in the range 1–3 $\sec^{-1}$, namely, τ=0.3–1 sec. The overall flight control system design to stability may require a significant reduction of this time constant. This is traditionally accomplished by means of a rate gyro through which a control loop is closed as shown in Figure 5.16. As stated earlier, the gyro dynamics is often disregarded so that the rate gyro is represented solely by the coefficient H/K_s. If ω_s is assumed to be very large, the servo lag can be disregarded. This situation thus poses a trivially easy problem because only a single open-loop pole at $s=-L_p$ is involved. This implies that the control loop is absolutely stable and that the loop gain K can be increased at will, pulling the single closed-loop pole to the left as desired thus achieving an arbitrarily small roll rate time constant. If the servo lag is not disregarded, the open loop system has two poles at $s_1 = -L_p$ and $s_2 = -\omega_s$. The corresponding root locus has two asymptotes at ± 90 deg, and the system is still asymptotically stable. However, if the gyro dynamics are accounted for, the system is no longer absolutely stable and, as expected, it will go unstable for a sufficiently large loop closure gain.

Disregarding drift rate, the rate gyro dynamics is derived from (4.8.11). Replacing the pitch rate Q, which was chosen there as an example, by the roll rate P discussed here, and writing $P = \dot{\phi}$, we have

$$\xi(s) = \frac{H/I_x}{s^2 + 2\omega_n\zeta s + \omega_n^2}\dot{\phi} = \frac{(H/K_s)\omega_n^2}{s^2 + 2\omega_n\zeta s + \omega_n^2}\dot{\phi}$$

The numerical values are taken from Example 4.8—1, namely,

H = 10 gr-f cm sec;

I_x = 0.02 gr-f sec^2;

ω_n = 200 sec^{-1};

K_s = 800 gr-f cm/rad.

The roll rate closed control loop is described in the following diagram, in which K is the loop closure gain. The numerical values for the parameters of the aircraft dynamics are given as

$L_{\delta_a} = 30$ rad/sec^2;

$L_p = 2$ rad/sec;

$\varsigma = 0.5$.

The servo lag is assumed to be $\omega_s = 10$ rad/sec.

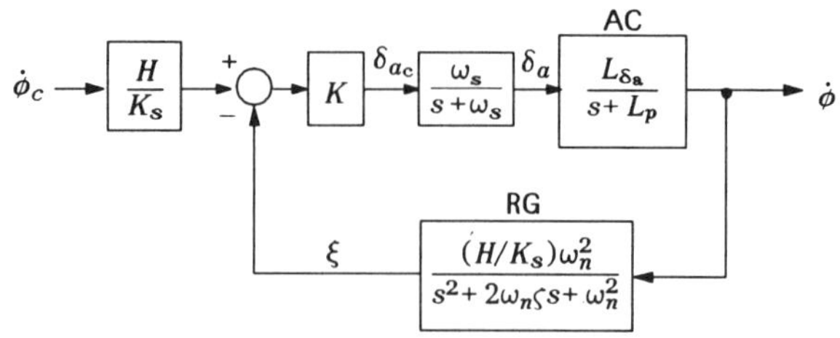

Figure 5.16. Aircraft roll rate control loop including rate servo and rate gyro dynamics.

The input block H/K_s is incorporated to set the scaling. Substitution of the preceding numerical values yields the following open-loop transfer function $G(s)$:

$$G(s) = \frac{30 \times 10 \times (10/800) \times 40000}{(s+2)(s+10)(s^2+200s+40000)}$$

$$= \frac{150000}{s^4 + 212s^3 + 4220s^2 + 48200 + 800000}$$

The root locus of the closed-loop system of Figure 5.16 is shown in Figure 5.17. It demonstrates the potential destabilization of the closed-loop system due to the dynamics of the rate gyro.

The consequence is that in the design of the SAS, the loop gain must be more conservative in order to account for the gyro dynamics. This becomes more pronounced if the roll angle ϕ itself, not $\dot{\phi}$, is the controlled variable. This is because an additional pole due to the kinematic integration of $\dot{\phi}$ is introduced at the origin $s=0$.

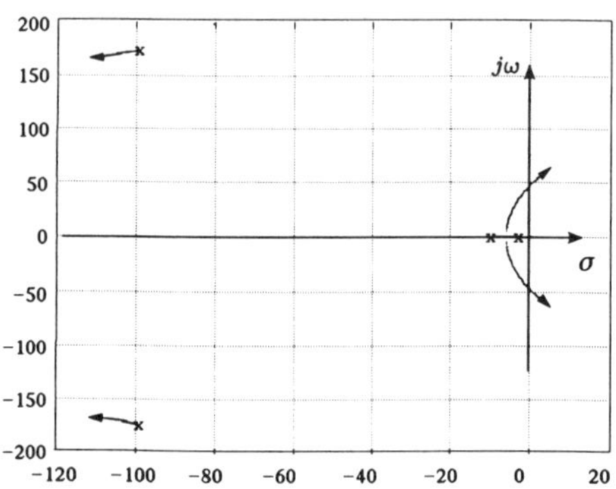

Figure 5.17. Root locus of aircraft roll rate control system servo and rate gyro dynamics.

Problems

Problem 5.1– Gyrocompassing: Specifications and performance

A directional gyro is used as a gyrocompass. It is required that the azimuth error $\Delta\psi$ in the steady state should not exceed 1deg, even in the arctic region at latitude $\lambda = 80$ deg.

a) What is the specification for the gyro drift rate $\dot{\eta}_d$ in deg/hr?
b) In the steady-state expression that determines $\Delta\psi_d$, the angular momentum H of the gyro does not show. Yet, it is clear that H must have a strong effect on the achievable low error $\Delta\psi$. Argue this point and resolve the apparent enigma.

Problem 5.2– Rate gyro: Kinematically induced drift rate

A rate gyro is mounted in an aircraft so that it measures pitch rate q. The aircraft flies on a small circle with $\psi=45^0$ and $\lambda=30^0$. The moment of inertia of the gyro rotor is 12 gr cm sec^2. The weight of the gimbal mount is 64.5 gr. Its center of gravity is 0.08 mm below the gimbal axis. Disregarding bearing friction and other mechanical imperfections, the gyro develops a drift rate q_d while flying along the given trajectory.

a) What would be the value of q_d if λ were changed to 45 deg?

b) What is the required spin speed ω_s of the rotor in rpm so that the drift rate q_d is less than or equal to, 1 deg/hr?

Problem 5.3–Vertical gyro: In an orbiting satellite

It is suggested to use a vertical gyro for the roll control of an Earth orbiting satellite. The idea is that the centripetal acceleration pointing upwards counters the weight similarly to the D'Alembert force that counters the weight in an ordinary horizontal trajectory of an aircraft.

a) Is this idea applicable? Argue, and suggest whether to accept or reject.

b) Would your answer be different if the satellite were geostationary?

Problem 5.4–Vertical gyro: Roll control of an aircraft

An aircraft, roll controlled by a vertical gyro, enters a station keeping horizontal circular trajectory and executes a perfect coordinated turn. The roll angle is $\phi = 30^0$. The aircraft ground speed is v = 200/sec. The data of the gyro are:

Rotor angular momentum	$H = 4\times10^6$ dyne cm sec
Sensitivity of the electrolytic bubble levels	$S_\beta =$ 1.0 volt/rad
Torque constant of the torque motors	$b = 10^5$ dyne cm/amp
Amplification in the erection loop	$A = 0.02$ amp/volt

a) Determine the time constant τ of the erection loop and the angular deviation $\xi = \xi(t)$ of the gyro spin axis from the vertical as a function of time from the onset of the roll angle ϕ at t=0.

b) What is the value of ξ as $t \to \infty$?

c) Show by means of a sketch the inclination of the horizon as seen by the pilot on the panel roll display and through the window: at t=0 and at $t \to \infty$.

d) Discuss your findings in c) and suggest a remedy if necessary.

Problem 5.5— Schuler tuned platform: Navigation and recalibration on the moon

A land vehicle on the moon is equipped with a Schuler-tuned INS platform. The moon mass is $M_m = 0.0123\ M_e$, where M_e is the Earth mass, and $R_m = 0.3R_e$.

a) What are the Schuler frequency and period on the moon?

b) What can you deduce concerning the overall similarity of the matter of the moon to that of Earth?

Starting from a known point, after two hours of travel, the driver stops the vehicle and compares his position, as displayed by the INS, with a position fix based on a known landmark and he finds a position error $\Delta x = 2650$ m. Assuming that the sole error source is a platform gyro drift rate q_d, the driver computes q_d and compensates for it for the rest of the mission.

c) Use the INS block diagram, to show this compensation can be implemented.

d) What is the value of q_d he has determined?

Problem 5.6— Vertical gyro: Design considerations and error evaluations

The sensitivity coefficient of the bubble levels of a vertical gyro is $S_\beta = 50$ mv/deg. As a result of an alignment error, a zero-point error of $\Delta\beta = 30$ mv develops.

a) Does this error cause an error $\Delta\eta$ in the steady state? If so, what is the value of $\Delta\eta$ in deg?

b) Explain in words the meaning of $\Delta\eta$.

c) Is it possible to eliminate $\Delta\eta$ by integral control?

Support your answers with the appropriate analytical expressions.

The angular momentum of the gyro is $H = 5000$ grcm sec/rad and the torque coefficient of the erection torque motor is $b = 500$ gr cm sec/rad.

d) What is the required gain A that provides an erection time constant of $\tau = 200$ sec?

e) The gyro undergoes tests in a laboratory situated at latitude $\lambda = 30^0$. Disregarding all possible imperfections, will the gyro show a $\Delta\eta = 0$, or $\Delta\eta \neq 0$? Support your answers with the aid of the appropriate analytical expressions.

Problem 5.7— Vertical gyro: Testing applicability as a low-cost INS

In view of the very high cost of INS, the need for low-cost platforms, intended for short-duration missions, has often been considered. It has been suggested to use a vertical gyro to fulfill this need, at least in the horizontal plane, by placing two accelerometers A_x, A_y (instead of the bubble levels) on the inner gimbal.

It is assumed that the accelerometers, and the other components of the VG are completely drift- and error-free. The concept is tested with the following data.

The time constant of the erection loop is $\tau = 5$ min. The aircraft flies a straight and horizontal path. At the instant $t = 0$, the aircraft starts accelerating by $a_x = 0.1$ g for the duration of $\Delta t = 20$ sec.

a) Write the expression for the deflection of the spin axis $\Delta\eta$ as a function of Δt and the reading of A_y assuming that $\Delta\eta$ is still a small angle.

b) Formulate an approximate linear expression for a_y based on the fact that $\Delta t \ll \tau$.

c) What is the distance error ΔY derived by the double integration of a_y?

d) What is the reading of a_y as $\Delta t \to \infty$?

e) Since it has been assumed that all the components of the VG were assumed to be ideal, what is the obvious remedy?

f) Discuss the rationale of trying to make use of the above remedy?

Problem 5.8— Rate gyro: Design for linearity and bandwidth

The transfer function of a rate gyro $\phi(s)/q(s)$ is not truly linear because the gimbal, and thus the spin axis, is inclined by ϕ and is not normal to the input axis y.

a) Formulate the exact transfer function for $\phi(s)/q(s)$ that accounts for the resulting nonlinearity.

b) Determine the largest permitted deflection angle ϕ_{max} for which the nonlinearity is not greater than 1%.

The angular momentum of the rotor is $H = 10^5$ dyne cm sec.

c) What is the required spring constant K_s *that* assures the above requirement for ϕ_{max} with an input rate $q_{max} = 1$ rad/sec?

d) What is the largest permitted moment of inertia I_x that assures that the natural frequency ω_s of the gyro is not smaller than 180 rad/sec, and what is the required damping coefficient D to assure a damping coefficient ς that is not smaller than 0.5?

Problem 5.9— Rate gyro: Readings on a satellite

A cylindrically shaped satellite having a diameter of 10 m orbits Earth at a constant altitude of 200 km. It is equipped with a rate gyro whose input axis is parallel to the satellite axis, which is normal to the orbital plane, and an accelerometer mounted at the rim with its input axis pointing inward toward the satellite axis.

a) What is the reading of the rate gyro if the satellite is:

1. inertially stabilized?
2. Earth pointing?
3. moon pointing?
4. rotating around its axis at $\omega = 1$ rpm while orbiting?

The rate gyro has an unknown drift rate q_d.

b) How can its magnitude be identified by the on-board sensors?

c) To what precision can it be determined if the accelerometer uncertainty is 1 μg?

Problem 5.10— Rate gyro: Air-to-surface range estimation

A remotely piloted aircraft is equipped with a two-axis gyro stabilized platform carrying an electro-optical camera. The camera is pointed at a target on the ground and is electro-optically locked on it. The aircraft flies at a constant sea level altitude h and at a constant ground speed V. The rate gyro output is q deg/sec.

a) What is the slant range R to the target?

b) Determine the sensitivity $\partial R/\partial q$ and state the requirement for the largest permitted gyro drift rate q_d that assures $\Delta R/R \ll 1\%$.

c) What is the range error at h = 2000 m and V = 60 m/sec if the rate gyro has a drift rate of 2 deg/min.

Problem 5.11 — Vertical gyro: Examine equivalence to an overdamped pendulum

Note that a vertical gyro, by and large, behaves like a highly overdamped pendulum.

a) Formulate the dynamics of such an overdamped pendulum.

b) Compare its properties in one axis to a vertical gyro in regard to its sensitivity to horizontal accelerations and verify the similarity.

c) Does the similarity hold for all the functions of a vertical gyro? If not, point out where the similarity breaks down. State the functional significance, if any, of your answer.

Problem 5.12 — Gimbaled airborne payload: error analysis

A single-axis gyro stabilized platform, mounted in an RPV flying at the ground speed V = 40 m/sec, carries a down-looking TV camera. The technical data are:

1. angular rate scale factor SF=50 mv/(deg/sec);
2. torque coefficient of the gimbal torque motor b = 4 gr-f cm/amp;
3. total weight of the mount W = 100 gr-f;
4. displacement of center of gravity below the y axis of the mount ϵ = 0.012 cm;
5. ,oment of inertia of the mount $I_y = 2\times10^3$ dyne cm sec^2.

a) Describe schematic of the stabilized mount indicating the design parameters and the gyro axis IA, SA, OA.

b) Determine the amplifier gain that provides a time constant $\tau \leq 0.25$ sec.

c) Indicate in the schematic where to apply an angular rate command q_c. What is the command voltage for a q_c = 10 deg/sec?

At a certain instant, the RPV starts decelerating for 1 min losing half of its initial speed.

d) What is the drift rate of the camera mount, around which axis, and what is the accumulated angular deviation by the end of the deceleration period?

e) Can this drift rate be manually compensated? Show where to apply it. What is this manual compensating signal in volts?

Appendix A5: Direction cosines and quaternions

Direction cosines

The 3×3 Euler angle transformation matrix (A4.4), whose elements are determined by the choice of the order of rotation Φ, Θ, Ψ, exemplifies the more general 3×3 direction cosine matrix (DCM) $[C]$ formulated in (A5.4). It uniquely defines the transformation of a vector from one coordinate frame to another. The DCM is determined by computing the cosines of the angles between each axis of one frame and every axis of the other frame. The Euler angles, which determine these direction cosines, can be computed from the strapped-down rate gyro measurements P, Q, R by means of the Euler differential equations:

$$\dot{\Theta} = Q\cos\Phi - R\sin\Phi \tag{A5.1}$$

$$\dot{\Phi} = P + Q\sin\Phi\tan\Theta + R\cos\Phi\tan\Theta \tag{A5.2}$$

$$\dot{\Psi} = Q\frac{\sin\Phi}{\cos\Theta} + R\frac{\cos\Phi}{\cos\Theta} \tag{A5.3}$$

These equations provide the three Euler angular rates $\dot{\Theta}$, $\dot{\Phi}$, and $\dot{\Psi}$ as long as nonsingularity is retained. This is assured as long as the flight vehicle performs shallow trajectories, namely, moderate vertical maneuvers that do not involve pitch angles Θ that approach $\pm 90^0$. If this occurs, $\dot{\Phi}$ and $\dot{\Psi}$ become undefined, a situation known from the mechanical terminology of stabilized platforms as Gimbal Lock. It can be visualized from Figure 4.2 if the outer gimbal $\mathbf{G_0}$ becomes parallel to inner gimbal $\mathbf{G_I}$. This is a statement of the fact that for nose-up or nose-down attitudes, roll and heading cannot be distinguished from one another. The reason is that heading Ψ is measured about the vertical axis and roll Φ is measured about the vehicle longitudinal axis, which also becomes vertical for $\Theta = \pm 90^0$. A simple compromise remedy could be to hold Φ and Ψ at their last computed values at some maximum acceptable pitch angle Θ_{max} until Θ again resumes values sufficiently removed from $\pm 90^0$. Another remedy that solves the problem entirely is to incorporate an additional gimbal in the mechanization or to introduce an additional degree of freedom in the computational algorithm. Further details on the DCM and Euler angle transformation can be found in Chapter 2 of [10]. Equations (A5.1)–(A5.3), in conjunction with (A4.4) and (4.07)–(4.09), constitute a complete strapdown attitude-determination algorithm. The 3×3 DCM is formulated by

$$[C] \triangleq \begin{bmatrix} C_{11} & C_{12} & C_{13} \\ C_{21} & C_{22} & C_{23} \\ C_{31} & C_{32} & C_{33} \end{bmatrix} \tag{A5.4}$$

for the Euler angle transformation matrix given in (A4.4), the elements c_{ij} are

$$C_{11} = c\theta\, c\psi$$

$$C_{12} = -c\phi\, s\psi + s\phi s\theta c\psi$$

$$C_{13} = \ s\phi\, s\psi + c\phi\, s\theta\, c\psi$$

$$C_{21} = c\theta s\psi$$

$$C_{22} = c\phi c\psi + c\phi s\theta s\psi$$

$$C_{23} = -s\phi c\psi + c\phi s\theta s\psi$$

$$C_{31} = -s\theta$$

$$C_{32} = s\phi c\theta$$

$$C_{33} = c\phi c\theta$$

The Euler angles themselves can be extracted from the precedinge as follows:

$$c\theta = \sqrt{1 - C_{31}^2} \tag{A5.5}$$

$$\phi = \tan^{-1}\frac{C_{32}}{C_{33}} \tag{A5.6}$$

$$\theta = \tan^{-1}\frac{-C_{31}}{\sqrt{1 - C_{31}^2}} \tag{A5.7}$$

$$\psi = \tan^{-1}\frac{C_{21}}{C_{11}} \tag{A5.8}$$

The angles themselves ϕ, θ, ψ are often needed for the realization of flight control systems or other payloads.

Equation (A4.4) together with (A5.1)–(A5.3), in principle constitutes a complete algorithm for the determination of the Euler angle transformation matrix using the measurements of P, Q, R. This algorithm, however is not efficient because of its nonlinearities and the required three-stage computations (A5.1)–A5.3), (4.0.7)–(4.0.9), and (A4.4).

Incremental Euler transformations

A much more efficient algorithm is based on the following reasoning. The rotations ϕ, θ, ψ are considered to be the accumulation of infinitesimal rotation elements $d\theta_1$, $d\theta_2$, $d\theta_3$ which are components of the rotation vector $\boldsymbol{d\theta}$. Its direction is along the instantaneous rotation axis given by

$$\boldsymbol{d\theta} \triangleq \boldsymbol{\omega_A}\, dt \tag{A5.9}$$

where $\boldsymbol{\omega_A}$ is an aircraft angular rate vector and dt is a time increment. This formulation allows linearization of the DCM and its efficient computation by means of a linear vector differential equation, which can be easily solved by using standard computational routines. We define three coordinate frames: 1) The inertial frame I; the aircraft axes at time t; and the aircraft axes at time t +dt. The transformation from the first to the second frame is denoted by $\mathbf{T}_{A|I}$. The transformation from the second to the third, that is, the transition matrix $\mathbf{T}_{(t+d)|t}$, is obtained by combining the infinitesimal Euler rotations $[d\theta_1]_x$, $[d\theta_2]_y$, $[d\theta_3]_z$, defined by the assumptions $\sin(d\theta) \cong d\theta$; $\cos(d\theta) \cong 1$ and disregarding products of infinitesimals. These transformations commute, that is, the order of their rotational sequence is immaterial. With these assumptions, we have

$$\mathbf{T}_{(t+dt)|dt} = [d\theta_1]_x\,[d\theta_2]_y\,[d\theta_3]_z = \begin{bmatrix} 1 & d\theta_3 & -d\theta_2 \\ -d\theta_3 & 1 & d\theta_1 \\ d\theta_2 & -d\theta_1 & 1 \end{bmatrix}$$

$$= I_{3\times 3} + \begin{bmatrix} 0 & d\theta_3 & -d\theta_2 \\ -d\theta_3 & 0 & d\theta_1 \\ d\theta_2 & -d\theta_1 & 0 \end{bmatrix}$$

$$= I_{3\times 3} + \begin{bmatrix} 0 & \omega_{A3} & -\omega_{A2} \\ -\omega_{A3} & 0 & \omega_{A1} \\ \omega_{A2} & \omega_{A1} & 0 \end{bmatrix} dt = \mathbf{\Omega}_A dt \tag{A5.10}$$

where $\mathbf{\Omega}_A$ is defined by the skew symmetric matrix

$$\mathbf{\Omega}_A \triangleq \begin{bmatrix} 0 & \omega_{A3} & -\omega_{A2} \\ -\omega_{A3} & 0 & \omega_{A1} \\ \omega_{A2} & \omega_{A1} & 0 \end{bmatrix} \tag{A5.11}$$

The complete transformation from the first to the third frame is determined by the product $\mathbf{T}_{(t+dt)|dt}\mathbf{T}_{A|I}(t)$, where $\mathbf{T}_{A|I}(t)$ is the matrix to be solved. Thus, multiplying $\mathbf{T}_{A|I}$ by (A5.10) we have

$$\mathbf{T}_{A|I}(t+dt) = \mathbf{T}_{(t+dt)|dt}\mathbf{T}_{A|I}(t) = \mathbf{T}_{A|I}(t) + \mathbf{\Omega}_A \mathbf{T}_{A|I}(t)\,dt \tag{A5.12}$$

from which

$$\frac{\mathbf{T}_{A|I}(t+dt) - \mathbf{T}_{A|I}(t)}{dt} \triangleq \dot{\mathbf{T}}_{A|I}(t) = \Omega_A \mathbf{T}_{A|I}(t) \tag{A5.13}$$

This is a homogeneous differential equation in $\mathbf{T}_{A|I}(t)$ with $\mathbf{\Omega}_A$ as the system matrix. The elements ω_{Ai}, i=1, 2, 3, being incremental Euler angles, are apparently unknown. However, in view of the small incremental angle assumptions and observing (A5.1—A5.3), ω_{Ai} can be considered, within each time increment dt, as equal to the measurable body axes angular rates P, Q, R provided by the rate gyros. The solution of (A5.13) which, given the initial conditions $\mathbf{T}_{A|I}(t=0)$, yields the required time-varying Euler angle transformation from which the Euler angles can be extracted in accordance with (A5.5)–(A5.8).

Two shortcomings still prevail in the foregoing algorithm:

1) Nine parameters are required in the matrix differential equation (A5.13) involving the simultaneous solution of nine differential equations.
2) The problem of possible singularity remains unresolved.

Further relationships of the DCM, relating the aircraft frame to the geographical frame, can be found in [9]. These transformations are particularly important in terrestrial navigation.

Quaternions

So far, the coordinate transformation matrix has been represented by three Euler angles. The same orientation can be uniquely defined by nine direction cosines as shown earlier. An alternative representation is by the four parameter quaternions.

The use of quaternions resolves these shortcomings as follows:
1) The quaternion leads to a 4×4 matrix differential equation.
2) Quaternions introduce an additional degree of freedom and thus resolve the problem of potential singularity.

The quaternion q is the basis for all four-parameter transformations. Essentially, it is a method for representing angular orientation with four parameters only, which is the smallest redundancy that removes the singularities of a three-parameter (Euler angles) representation. Thus, a rotation of one coordinate frame relative to another can be expressed uniquely by four parameters, three to define the direction of the axis of rotation and the fourth to specify the amount of rotation. Quaternion algebra, due to W. R. Hamilton in 1843, is an extension of three-dimensional vector algebra. q is a quadruple of real numbers arranged in a definite order similarly to a three-dimensional vector

$$q=(q_0, q_1, q_2, q_3)^T \tag{A5.14}$$

It can also be defined as a hypercomplex number of the form

$$\bar{q} = q_0 + q_1\mathbf{i} + q_2\mathbf{j} + q_3\mathbf{k} = (q_0, \mathbf{q}) \tag{A5.15}$$

where q_0, q_1, q_2, q_3 are real numbers and the set $\{1, \mathbf{i}, \mathbf{j}, \mathbf{k}\}$ is the basis of the quaternion vector space of real and imaginary numbers. Since $\mathbf{i}$, $\mathbf{j}$, $\mathbf{k}$ can represent an orthogonal basis, quaternion algebra incorporates scalar and vector algebra. The basic mathematical properties and operations on quaternions such as addition, subtraction, multiplication, cyclic symmetry, inverse, norm, identity and equality uphold the rules of basic vector algebra. These relationships are presented in detail, e.g., in [10]. As in vectors, we can define the conjugate $\bar{q}^*$ of $\bar{q}$, that is,

$$\bar{q}^* = q_0 - q_1\mathbf{i} - q_2\mathbf{j} - q_3\mathbf{k} \tag{A5.16}$$

The norm of a quaternion $N(Q)$ is a scalar defined as

$$N(Q) = \bar{q}\bar{q}^* = \bar{q}^*\bar{q} = q_0^2 + q_1^2 + q_2^2 + q_3^2 \tag{A5.17}$$

A zero quaternion is one having a zero scalar and a zero vector, that is,

$$\bar{q}_0 = 0 + 0\mathbf{i} + 0\mathbf{j} + 0\mathbf{k} \tag{A5.18}$$

A unit quaternion is one whose norm equals 1, namely,

$$\bar{q}_1 = 1 + 0\mathbf{i} + 0\mathbf{j} + 0\mathbf{k} \tag{A5.19}$$

From (A5.17) and (A5.19) the constraint equation on a unit quaternion is

$$q_0^2 + q_1^2 + q_2^2 + q_3^2 = 1 \tag{A5.20}$$

We now discuss how a quaternion is used to express a coordinate frame transformation. Let $\mathbf{p} = (X, Y, Z)^T$ and $\mathbf{p}' = (X', Y', Z')^T$ represent two arbitrary coordinate frames. A quaternion represents a transformation from $\mathbf{p}$ to $\mathbf{p}'$ by

$$\bar{q}^*\mathbf{p}\bar{q} = \mathbf{p}' \tag{A5.21}$$

The explicit expression of (A5.21) is

$$(q_0 - q_1\mathbf{i} - q_2\mathbf{j} - q_3\mathbf{k})(X\mathbf{i} + Y\mathbf{j} + Z\mathbf{k})(q_0 + q_1\mathbf{i} + q_2\mathbf{j} + q_3\mathbf{k}) = \mathbf{p}' \tag{A5.22}$$

By using the multiplication rules of unit vectors,

$$\mathbf{ij} = -\mathbf{ji} = \mathbf{k}$$

$$\mathbf{jk} = -\mathbf{kj} = \mathbf{i}$$

$$\mathbf{ki} = -\mathbf{ik} = \mathbf{j}$$

(A5.22) can be expanded as follows:

$$\begin{aligned}
\mathbf{p}' = {} & X[q_0^2 + q_1^2 - q_2^2 - q_3^2] + Y[2q_3q_0 + 2q_1q_2] + Z[2q_1q_3 - 2q_0q_2]\} \\
& + \mathbf{j}\{X[2q_1q_2 - 2q_0q_3] + Y[q_0^2 - q_1^2 + q_2^2 - q_3^2] + Z[2q_1q_0 + 2q_3q_2]\} \\
& + \mathbf{k}\{X[2q_0q_2 + 2q_1q_3] + Y[2q_2q_3 - 2q_2q_3] + Z[q_0^2 - q_1^2 - q_2^2 + q_3^2]\}
\end{aligned} \tag{A5.23}$$

In matrix notation, we have

$$\begin{bmatrix} X' \\ Y' \\ Z' \end{bmatrix} = \begin{bmatrix} [q_0^2 + q_1^2 - q_2^2 - q_3^2] & [2q_3q_0 + 2q_1q_2] & [2q_1q_3 - 2q_0q_2] \\ [2q_1q_2 - 2q_0q_3] & [q_0^2 - q_1^2 + q_2^2 - q_3^2] & [2q_1q_0 + 2q_3q_2] \\ [2q_0q_2 + 2q_1q_3] & [2q_2q_3 - 2q_2q_3] & [q_0^2 - q_1^2 - q_2^2 + q_3^2] \end{bmatrix} \begin{bmatrix} X \\ Y \\ Z \end{bmatrix} \tag{A5.24}$$

This 3×3 matrix relating **p** to **p'** can be regarded as equivalent to the 3×3 DCM (A5.4). Equating it term by term with (A5.24), we obtain

$$C_{11} = \mathrm{c}\theta\,\mathrm{c}\psi = q_0^2 + q_1^2 - q_2^2 - q_3^2 \tag{A5.25}$$

$$C_{12} = \mathrm{c}\theta\,\mathrm{s}\psi = 2(q_1q_2 - q_0q_3) \tag{A5.26}$$

$$C_{13} = -\mathrm{s}\theta = 2(q_1q_3 - q_0q_2) \tag{A5.27}$$

$$C_{21} = \mathrm{s}\phi\,\mathrm{s}\theta\,\mathrm{c}\psi - \mathrm{c}\phi\mathrm{s}\psi = 2(q_1q_2 - q_0q_3) \tag{A5.28}$$

$$C_{22} = \mathrm{s}\phi\,\mathrm{s}\theta\,\mathrm{s}\psi + \mathrm{c}\phi\mathrm{c}\psi = q_0^2 - q_1^2 + q_2^2 - q_3^2 \tag{A5.29}$$

$$C_{23} = \mathrm{s}\phi\,\mathrm{c}\theta = 2(q_1q_2 - q_1q_0) \tag{A5.30}$$

$$C_{31} = \mathrm{c}\phi\,\mathrm{s}\theta\,\mathrm{c}\psi + \mathrm{s}\phi\mathrm{s}\psi = 2(q_1q_3 - q_0q_1) \tag{A5.31}$$

$$C_{32} = \mathrm{c}\phi\,\mathrm{s}\theta\,\mathrm{s}\psi - \mathrm{s}\phi\mathrm{c}\psi = 2(q_2q_3 + q_0q_1) \tag{A5.32}$$

$$C_{22} = \mathrm{s}\phi\,\mathrm{s}\theta\,\mathrm{s}\psi + \mathrm{c}\phi\mathrm{c}\psi = q_0^2 - q_1^2 + q_2^2 - q_3^2 \tag{A5.33}$$

$$C_{33} = \mathrm{c}\phi\,\mathrm{c}\theta = q_0^2 - q_1^2 - q_2^2 + q_3^2 \tag{A5.34}$$

Assume that q_0, q_1, q_2, q_3 have been previously computed. Equations (A5.25)–(A5.34) translate the quaternion into the familiar format of the DCM and by the procedure given in (A5.5)–(A5.8), the Euler angles themselves.

The procedure for the computation of the quaternion (q_0, q_1, q_2, q_3) itself is similar to (A5.13) for the computation of the DCM matrix. It can be shown that

$$\frac{\mathrm{d}}{\mathrm{d}t}q = \frac{1}{2}[Bq] \tag{A5.35}$$

where

$$B = \begin{bmatrix} 0 & -\omega_x & -\omega_y & -\omega_z \\ \omega_x & 0 & \omega_z & -\omega_y \\ \omega_y & -\omega_z & 0 & \omega_x \\ \omega_z & \omega_y & -\omega_x & 0 \end{bmatrix} \tag{A5.36}$$

It can be shown that in terms of the quaternion scalar components, the set of linear differential equations is

$$\dot{q}_0 = -\frac{1}{2}[q_1\omega_x + q_2\omega_y + q_3\omega_z] \tag{A5.37}$$

$$\dot{q}_1 = \frac{1}{2}[q_0\omega_x + q_2\omega_z - q_3\omega_y] \tag{A5.38}$$

$$\dot{q}_2 = \frac{1}{2}[q_0\omega_y - q_1\omega_z + q_3\omega_x] \tag{A5.39}$$

$$\dot{q}_3 = \frac{1}{2}[q_0\omega_z + q_1\omega_y - q_2\omega_x] \tag{A5.40}$$

The integration of these linear differential equations is straightforward, given the initial conditions $Q(0) = Q(t_0)$ and the components ω_x, ω_y, ω_z of the relative angular rate between the two coordinate frames. These are also the readings of the body-mounted rate gyros. Further details on the computational procedure of the quaternion and its application to analytical platform computation are given in [9] and [10].

References

1. Blakelock, J.H., *Automatic Control of Aircraft and Missiles*, John Wiley & Sons, New York, 1991.

2. Murtaugh, S.A., and Criel, H.E., "Fundamentals of Proportional Navigation," *IEEE Spectrum*, Vol. 3, 1966, pp. 75–85.

3. McRuer, D.T., Graham, D., and Krendel, E., "Manual Control of Single-Loop Systems," Pts. I and II, *Journal of the Franklin Institute*, Vol. 238, No. 1, January, 1967, and No. 2, February 1967.

4. Hudson, R.D. Jr., *Infrared System Engineering*, John Wiley & Sons, New York, 1969.

5. Leondes, C.T., Editor, *Guidance and Control of Aerospace Vehicles*, McGraw-Hill, New York, 1963.

6. Garg, S.C., Morrow, L.D., and Mamen R., "Strapdown Navigation Technology: A Literature Survey," *AIAA Journal of Guidance and Control*, Vol. 1, No. 3, May–June 1978, pp. 161–172.

7. McLean, D., "*Automatic Flight Control Systems*", Prentice-Hall, Englewood Cliffs, 1990.

8. Nesline, F.W., and Zarchan, P., "Why Modern Controllers Can Go Unstable," *AIAA Journal of Guidance, Control, and Dynamics*, Vol. 7, No. 4, July-August, 1984, pp. 495–500.

9. Farrell, J.L., *Integrated Aircraft Navigation*, Academic Press, New York, 1976.

10. Siouris G.M., *Aerospace Avionics Systems—A Modern Synthesis*, Academic Press, New York, 1993

6

Coriolis Angular Rate Sensors

6.0 Introduction

This chapter presents the principle of operation of a class of nongyroscopic electromechanical sensors based on the Coriolis effect. The significant ongoing technological development effort in a number of leading industries in the United States now yielded prototypes that have demonstrated promising performance. These efforts may eventually lead to the replacement of the relatively cumbersome gyroscopic spinning mass technology.

The growing need for highly rugged miniature angular rate sensors has initiated a number of studies and prototype product development programs. These products are potentially suitable for medium-to low-accuracy applications typical in tactical flight-vehicles, in particular, those involving strapdown mechanization. One principal approach is the Coriolis angular rate sensor. The underlying idea is to put an accelerometer in motion that is relative to the rotating vehicle body. The twice cross-product of the linear relative velocity V and the vehicular inertial angular rate Ω is the Coriolis specific force $a_c = 2\Omega \times V$, which is perpendicular to both V and Ω in accordance with the definition of a right-hand axis system. The development of the basic concept is given in [1], where the realization and analysis are provided, particularly, for rotating accelerometers. The rotation causes a sinusoidal velocity of the accelerometer and assures that the relative motion of the accelerometers, one for each vehicle axis, is confined to a small space. An alternative mechanization is through vibrating accelerometers, also indicated in [1]. The leading idea is that accelerometers are potentially much cheaper, smaller, and more rugged than gyroscopic devices. Furthermore, the moving accelerometer senses both the inertial angular rate Ω and the specific force F and outputs a signal that is the sum of their readings. Fortunately, the two readings have entirely different characteristics. Ω is an amplitude-modulated periodic signal whose frequency is precisely known, while F is an unmodulated accelerometer reading. This facilitates the separation of the Ω and F signals by synchronous detection to provide the corresponding readings on two output channels.

6.1 Rotating Coriolis angular rate sensors

Description

A schematic description of the rotating mechanization is given in Figure 6.1.

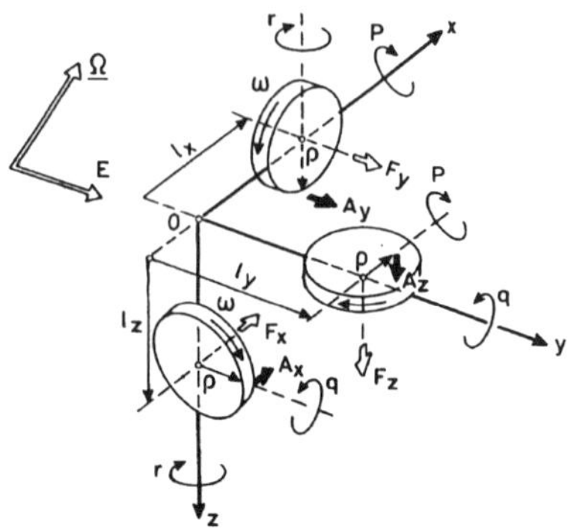

Figure 6.1. Mechanization of a triad of rotating accelerometers performing as angular rate and specific force sensors.

Three orthogonally oriented accelerometers A_x, A_y, A_z are rotated at the identical angular rate ω as shown. The vehicle rotates at the inertial angular rate $\mathbf{\Omega}$ and is subjected to the force $\mathbf{F}$. The accelerometers sense both the Newtonian force components F_x, F_y, F_z and the corresponding Coriolis force components F_{cx}, F_{cy}, F_{cz}, which are proportional to Ω_x, Ω_y, Ω_z and to $v = \omega\rho$. The accelerometer readings therefore incorporate information on $\mathbf{\Omega}$ along with $\mathbf{F}$. Fortunately, since the components of $\mathbf{F}_c$ are periodic at the frequency ω, it is relatively easy to separate them from $\mathbf{F}$.

General equations for specific force

The general equation for the total specific force measurable at a point mass moving within a rotating system is [2],

$$a = F + \dot{\Omega} \times r + 2\Omega \times \left.\frac{dr}{dt}\right|_b + \Omega \times (\Omega \times r) + \left.\frac{d^2 r}{dt^2}\right|_b \tag{6.1.1}$$

where Ω is the inertial angular rate of the system and r is the instantaneous distance of the point mass from the center of rotation of the vehicle. The subscript b signifies that the differentiation is with respect to the body axes. In particular, referring to Figure 6.1, $r = \rho + l$, where ρ is the instantaneous vector distance of the point mass from its center of revolution, and l is the fixed distance vector of the element's center of revolution to center of rotation of the vehicle. Equation (6.1.1) can now be rewritten as follows:

$$a = F + \dot{\Omega} \times r + 2\Omega \times \left.\frac{d\rho}{dt}\right|_b + \Omega \times [\Omega \times (\rho + l)] + \left.\frac{d^2\rho}{dt^2}\right|_b \tag{6.1.2}$$

Accelerometer triad readings

Substituting $\rho = [\rho_x \sin\omega t,\ \rho_y \sin\omega t,\ \rho_z \sin\omega t]^T$, incorporating the accelerometer noise components n_x, n_y, n_z, and resolving a into a_x, a_y, a_z, respectively while setting $\rho_x = \rho_y = \rho_z = \rho$ and rearranging terms, we have

$$\begin{aligned} a_x &= F_x + l_z(\dot{q} + pr) + 2\omega\rho\cos\omega t\left(q - \frac{\dot{r}}{2\omega}\right) + 2\omega\rho\sin\omega t\left(r + \frac{\dot{q}}{2\omega}\right) \\ &\quad + p\rho(q\cos\omega t + r\sin\omega t) + n_x \end{aligned} \tag{6.1.3}$$

$$\begin{aligned} a_y &= F_y + l_x(\dot{r} + qp) + 2\omega\rho\cos\omega t\left(r - \frac{\dot{p}}{2\omega}\right) + 2\omega\rho\sin\omega t\left(p + \frac{\dot{r}}{2\omega}\right) \\ &\quad + q\rho(r\cos\omega t + p\sin\omega t) + n_y \end{aligned} \tag{6.1.4}$$

$$\begin{aligned} a_z &= F_z + l_y(\dot{p} + rq) + 2\omega\rho\cos\omega t\left(p - \frac{\dot{q}}{2\omega}\right) + 2\omega\rho\sin\omega t\left(q + \frac{\dot{p}}{2\omega}\right) \\ &\quad + r\rho(p\cos\omega t + q\sin\omega t) + n_z \end{aligned} \tag{6.1.5}$$

The accelerometer outputs therefore are the sums of the unmodulated specific force components of F along with the unmodulated accelerations due to $\dot{\Omega} \times l$, and the $\cos\omega t$ and $\sin\omega t$ modulated angular rate components.

To provide an inertial measurement unit, the F components must be separated from the Ω components. Each of the noise components n_x, n_y, n_z, is assumed to consist of three components, $n = n_d, n_v, n_r$, where n_d is slowly varying drift, n_v is sinusoidal vibration interference, and n_r is random zero mean wide-band noise. In order to directly measure F, the accelerometers are assumed to respond from dc out to a few hundred Hz. Equations (6.1.3)–(6.1.5) indicate that the accelerometer outputs, a_x, a_y, a_z, consist of the F terms representing the actual specific force components acting on the vehicle and the third terms, which essentially are the Coriolis forces, proportional to p, q, r. The second and fourth terms are due to dynamical coupling, which can be compensated to a large extent, as shown in [1].

6.2 Combined angular rate and acceleration sensing

Signal separation by demodulation

A major issue is to separate the signals representing F from those representing Ω as precisely as possible. Substituting a_x, a_y, a_z, from (6.1.3)–(6.1.5) and denoting the computed values of p, q, r by $\hat{p}$, $\hat{q}$, $\hat{r}$, they can be determined by the following demodulation algorithm, which establishes the Ω channel, namely,

$$\hat{p} = \frac{1}{8\rho}\int_0^T a_z \operatorname{sgn}\cos\omega t dt \tag{6.2.1}$$

$$\hat{q} = \frac{1}{8\rho}\int_0^T a_x \operatorname{sgn}\cos\omega t dt \tag{6.2.2}$$

$$\hat{r} = \frac{1}{8\rho}\int_0^T a_y \operatorname{sgn}\cos\omega t dt \tag{6.2.3}$$

or, $\hat{q}$, $\hat{r}$, $\hat{p}$ can be determined using the demodulation function $\operatorname{sgn}\sin\omega t$. All the dc components, including accelerometer biases, and terms incoherent with $\cos\omega t$, drop out over T. Actually, if $\operatorname{sgn}\sin\omega t$ is applied in parallel with $\operatorname{sgn}\cos\omega t$ the signals representing p,q, q,r, and p,r can be separated simultaneously so that each rotating accelerometer can perform as a two-axis rate gyro. The factor $1/8\rho$ is required for proper scaling. $T = 2\pi/\omega$ is the period of one revolution. For p, for example, it results from the relationship $a_z = 2\omega\rho p = 2\times 2\pi f \times (2/\pi)\rho = 8f\rho$, where $2/\pi$ is the full wave rectification factor and $f = 1/T$. The specific force terms can be determined by simple integration over the period T, which establishes the F channel, namely,

$$\hat{F}_x = \frac{1}{T}\int_0^T a_x dt = F_x + l_z pr + n_{d_x} \tag{6.2.4}$$

$$\hat{F}_y = \frac{1}{T}\int_0^T a_y dt = F_y + l_x qp + n_{d_y} \tag{6.2.5}$$

$$\hat{F}_z = \frac{1}{T}\int_0^T a_z dt = F_z + l_y rq + n_{d_z} \tag{6.2.6}$$

The wideband noise terms n_v and n_r are initially assumed to be zero. All the periodic terms in (6.1.3)−(6.1.5) drop out upon integration over the cycle time. The coupling terms in (6.2.4)−(6.2.6) can be compensated by using the computed terms $\hat{p}$, $\hat{q}$, $\hat{r}$. The bias terms n_d of the accelerometers are retained. Equations (6.2.1−6.2.3), in conjunction with (6.2.4)−(6.2.6), constitute the basic algorithm for separating the angular rate vector $\mathbf{\Omega}$ from the specific force vector $\mathbf{F}$. Its operation is illustrated in Figure 6.2(a)−6.2(g) for F_z and p. It demonstrates that only the signal in plot (e) is the extracted signal required for p. All the other terms average out over the period T. Note that any bias in the accelerometers that is included in F_z also cancels out, as seen in (g). The effects of cross-coupling terms, the presence of terms proportional to $\dot{p}$, $\dot{q}$, and $\dot{r}$, and the effect of accelerometer noise are discussed in [1]. Details of the signal processor design are also given in [1] and in [3].

Imperfections

In its basic implementation, as described above, the accelerometers respond to high frequency vibration along their respective input axes. These extraneous signals cannot be entirely removed by the signal processor. This is especially true for frequency components which are close to ω. It is easily verified that they can leak into the Ω readings and that they may substantially degrade precision.

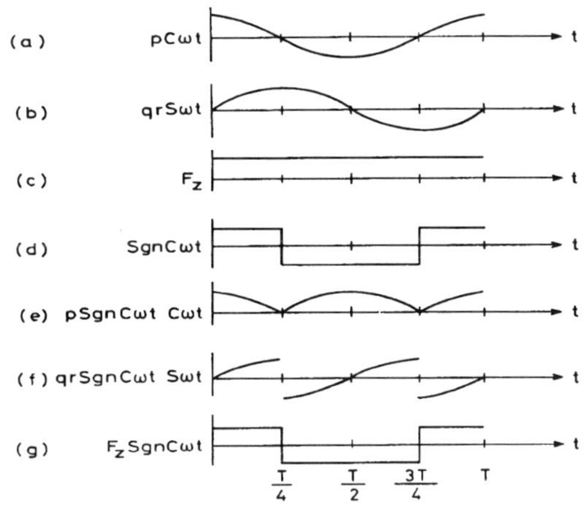

Figure 6.2. Signal processing for the separation of angular rate Ω from specific force F.

This shortcoming can be largely overcome by arranging pairs of matched accelerometers. Plot (e) in Figure 6.2 indicates that, due to the periodic zero-mean motion, $V=\omega\rho\cos\omega t$ periodically crosses zero. At these instances the corresponding Coriolis acceleration a_x^c, through which p is determined, vanishes. If at such instances p is significant, a fragment of information on p is lost. This loss is equivalent to zero-mean noise and, upon integration of p to obtain ϕ, an error will develop over time.

6.3 Rockwell-Collins rotating Coriolis sensor

Schematic description

The concept of rotating accelerometers was independently and simultaneously developed in the early 1980s by companies such as Kearfott, Rockwell-Collins, and others. In view of size, weight and low-cost requirements, but at the penalty of lower precision, the accelerometers that could be used were small, prestressed piezoelectric crystals that respond only to time-varying forces. The pre-stressing is required to provide sensitivity to the direction of the applied forces. Therefore, the mechanization and the signal extraction are somewhat different from the one described earlier. The rotating mechanization [4], which inevitably incorporates relatively noisy ball bearings and slip rings, both of which affect reliability, poses additional limitations which rule out the justification for the use of high-precision costly accelerometers. Therefore, these developments have been confined to medium- and low-accuracy tactical missile and conventional aircraft flight control applications. In the Kearfott and Collins mechanizations two quadruple sets of accelerometers are required for each axis. Each consists of two pairs of small piezoelectric slabs displaced by 45^0, one quadruple for the Ω channel and the other for the F channel. A sketch of the crystal arrangement, here referred to as *benders*, is shown in Figure 6.3.

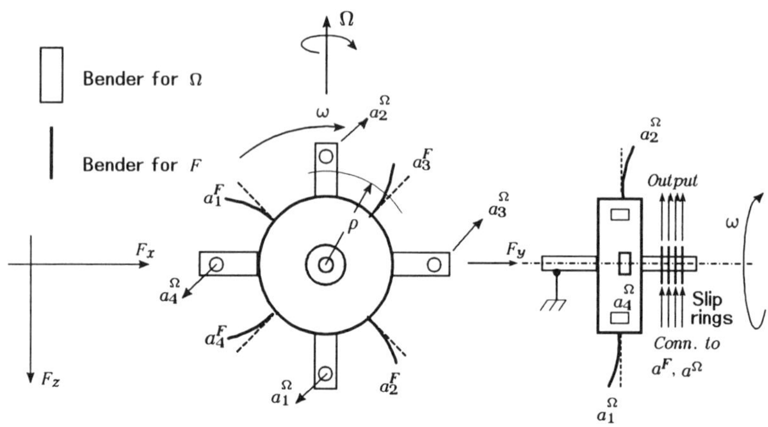

Figure 6.3. Rotational specific force and Coriolis sensors using eight piezoelectric benders.

Principle of operation and signal processing

As shown in Figure 6.3, the F benders are subjected to the force field F so that a_1^F bends to the left while a_2^F bends to the right with respect to the rotor structure. Thus, their corresponding signals are in opposite phase. The same holds for a_3^F and a_4^F. Subtracting the outputs of each bender in the pair, yields the peak outputs during the revolution of $1/2(a_1^F - a_2^F) = a_{12}^F$ and $1/2(a_3^F - a_4^F) = a_{34}^F$. Throughout a revolution, each of the F benders flexes back and forth with respect to the rotor. Thus, the 1-2 pair outputs the sinusoidal signal $a_{12}^F \sin\omega t$, while the 3-4 pair, being in quadrature, outputs $a_{34}^F \cos\omega t$. Modulating the first output pair by $\sin\omega t$ and the second by $\cos\omega t$, assuming for simplicity of presentation that F=const., $a_{12}^F = a_{34}^F$, and adding the two results in

$$a^F(\omega t) = a_{12}^F \sin^2\omega t + a_{34}^F \cos^2\omega t = a^F = \text{const.} \tag{6.3.1}$$

The Ω benders e. g., a_1^Ω, a_2^Ω on the other hand, deform in the same direction with respect to the rotor structure, as shown in Figure 6.3. Therefore their respective outputs are in phase, and they must be added. Following a similar procedure as for F, we have

$$a^\Omega(\omega t) = a_{12}^\Omega \sin^2\omega t + a_{34}^\Omega \cos^2\omega t = a^\Omega = \text{const.} \tag{6.3.2}$$

An important issue in Coriolis angular rate sensors is their potential sensitivity to mechanical vibrations. This is particularly true because they are based on accelerometers. Therefore the design must provide for the elimination, or at least the minimization, of this vulnerability in the Ω channel. The device described in Figure 6.3 inherently provides this property. The force, e.g., F_y, causes both a_1^Ω, a_2^Ω to bend to the left, i. e., in opposition with respect to the rotor structure. But since their outputs are added, the response of the pair to F_y, if their matching is perfect, cancels out. The effect of F_x and F_z is negligible because of the great stiffness of the Ω benders in the x, z plane. The effect of vibrations on the F channel is not an issue because the vibration force components are legitimate inputs into the F benders.

Technical data

A perspective drawing of the Collins Attitude-Heading AHS-85 device, briefly described in [4], is shown in Figure 6.4. Each device, as stated earlier, can measure angular rate and, of course, acceleration about two axes. Therefore, two such sensors can constitute a full three-axis attitude-heading reference while one axis remains redundant and can be used for extra reliability.

The rotating structure has a spin rate of about 3120 rpm corresponding to 52 Hz, or 326 rad/sec. This is much lower than the spin rate of conventional gyros which operate in the region of 24000 rpm. The sinusoidal signals from the rotating transducers are transmitted by inductive coupling to eliminate slip rings. To evaluate the feasibility with state-of-the-art sensors, consider the following example.

EXAMPLE 6.3–1: Required accleration measurement threshold in Coriolis rate sensor

It is required to determine the desired acceleration threshold of the piezo benders in a Collins-type angular rate sensing device. Assume the following parameters.

The effective radius is $\rho = 1.0$ cm; in accordance with the foregoing, $\omega = 326$ rad/sec. The required smallest detectable angular rate is $p_{min} = 2$ deg/hr. What is the required threshold of the accelerometer?

The required relationship is:

$a_{th} = 2\omega\rho p_{min} = 2 \times 326 \times 1 \times 10^{-2} \times 2/(57.3 \times 3600) \times 10^6 = 63\ \mu g$. Thus, a piezo bender with a threshold of $\sim 60\ \mu g$, given the foregoing device parameters, can detect angular rates as low as 2 deg/hr.

According to ref [4], the performance with prototype models demonstrated ngular drift rates of less than 2 deg/hr, scale factor error of less than 0.05% and cross-axis coupling of less than 0.16%. Acceleration bias drift is quoted to be less than 50 μg over one hour and less than 100 μg over 10 hours. Flight tests for more than 1000 hours have reportedly been performed on conventional aircraft such as the Gulfstream 1. Devices of this kind have become known in the technical literature as *mltisensor systems*.

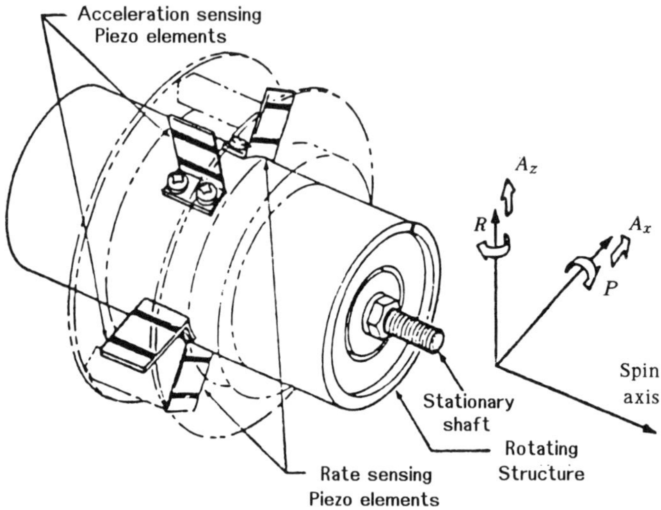

Figure 6.4. Perspective view of the Collins attitude-heading AHS-5 multisensor device (source, [4]).

A similar multisensor device has been under development at Kearfott Singer. Its effective radius ρ was 7 mm but its spin rate was 12000 rpm or 1260 rad/sec. To reduce slip ring contact noise, the rings were deep seated. The reported angular-rate null-point uncertainty of the device was about 1 deg/hr.

6.4 Dithered accelerometers

Geometry

An alternative mechanization for providing the relative velocity V to the accelerometers is to vibrate them sinusoidally at a constant amplitude ρ and frequency ω, as described in Figure 6.5.

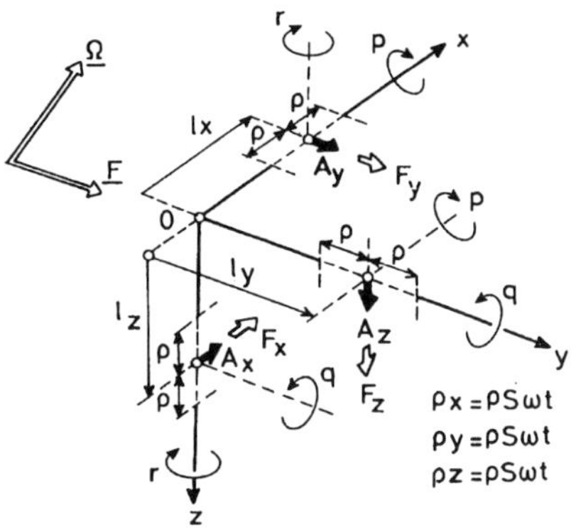

Figure 6.5. Coriolis angular rate and specific force sensor implemented by dithered accelerometers.

The instantaneous distances from the center of rotation of the vehicle are:

$$\begin{bmatrix} r_{A_x} \\ r_{A_y} \\ r_{A_z} \end{bmatrix} = \begin{bmatrix} 0 & 0 & l_z+\rho\sin\omega t \\ l_x+\rho\sin\omega t & 0 & 0 \\ 0 & l_y+\rho\sin\omega t & 0 \end{bmatrix} \begin{bmatrix} 1_x \\ 1_y \\ 1_z \end{bmatrix} \tag{6.4.1}$$

Resolving the resulting corresponding Coriolis forces along with the Newtonian specific force components of F_x, F_y, F_z results in

$$a_x = F_x + l_z(pr + \dot{q}) + 2\omega\rho\cos\omega t\, q + \rho\sin\omega t(pr + \dot{q}) \tag{6.4.2}$$

$$a_y = F_y + l_x(pq + \dot{r}) + 2\omega\rho\cos\omega t\, r + \rho\sin\omega t(pq + \dot{r}) \tag{6.4.3}$$

$$a_z = F_z + l_y(qr + \dot{q}) + 2\omega\rho\cos\omega t\, p + \rho\sin\omega t(qr + \dot{p}) \tag{6.4.4}$$

In comparison with (6.1.3)–(6.1.5) the last three equations are more favorable mainly because all the parasitic terms due to cross-coupling of angular rates are not coherent with $\cos\omega t$. In principle, they cancel out upon demodulation by $\operatorname{sgn}\cos\omega t$.

Signal separation by demodulation

To underline this property, assume for the moment that F and Ω are constant throughout T. Then we have

$$a_x = F_x + l_z pr + 2\omega\rho\cos\omega t\, q + \rho\sin\omega t\, pr \tag{6.4.5}$$

$$a_y = F_y + l_x pq + 2\omega\rho\cos\omega t\, r + \rho\sin\omega t\, pq \tag{6.4.6}$$

$$a_z = F_z + l_y qr + 2\omega\rho\cos\omega t\, p + \rho\sin\omega t\, qr \tag{6.4.7}$$

Substituting into (6.2.1)–(6.2.3) and solving, we have, with the noise terms disregarded,

$$\hat{p} = p \tag{6.4.8}$$

$$\hat{q} = q \tag{6.4.9}$$

$$\hat{r} = r \tag{6.4.10}$$

While the components of $\hat{\mathbf{F}}$ are the same as in (6.2.4)–(6.2.11), the following observations can be made:

1. The device operates as a single-axis rate gyro.

2. The Ω channel is free of constant drift rate components.

3. The F channel is the same as in any conventional inertial mechanization.

4. The error terms due to cross-coupling are modulated by $\sin\omega t$; therefore their effect is minute.

5. In computing Ω, the scale factor remains constant.

6. There is no need for slip rings.

7. If ρ is confined to a small value, ω has to be increased accordingly to provide a sufficiently large value for $V = \omega\rho$.

8. As in the rotatory mechanization, high-frequency noise and vibration components tend to contaminate the Ω readings.

EXAMPLE 6.4–1: Acceleration threshold for specified minimum angular rate

Let the vibration amplitude be $\rho = 0.3$ mm and let the vibration frequency be $f = 600$ Hz, that is, $\omega = 2\pi f = 3770\ \text{sec}^{-1}$. The Coriolis device must be able to measure roll rate down to 5 deg/hr. What is the requirement for the smallest detectable acceleration?

Recalling (6.4.4) and referring to the second term, which is the principal and coherent Coriolis term representing p, we have, in view of the full-wave rectification illustrated in Figure 6.2,

$$a_z = \frac{2}{\pi} 2\omega\rho\, p = 0.637 \times 2 \times 3770 \times 0.3 \times 10^{-3} \times 5 \times 0.0174 \times (1/3600)$$
$$= 310 \times 60^{-7}\ \text{m sec}^{-2} = 310 \times 10^{-8}\ \text{g} = 3.1\ \mu\text{g}$$

This relatively low threshold can be achieved with the vibrating beam accelerometers (VBA) by interpolation between successive time intervals, as will be shown in (6.7.1).

6.5 Dithered accelerometer pairs

Geometry

As stated earlier, a mechanization that uses a single accelerometer for each axis is susceptible to interferences due to vibration accelerations. These interferences, often being in the range of $1g$ or more, can introduce intolerable errors in determining p, q, and r. This shortcoming can be largely overcome by a paired mechanization as described in [5] and illustrated in Figure 6.6. For each axis x, y, z, pairs of accelerometers are arranged back to back, with the sensitive elements separated by the distances $2L$. The respective input axes are in opposition. The amplitude of the parallel sinusoidal counter motion is $\rho/2$, and its frequency is ω.

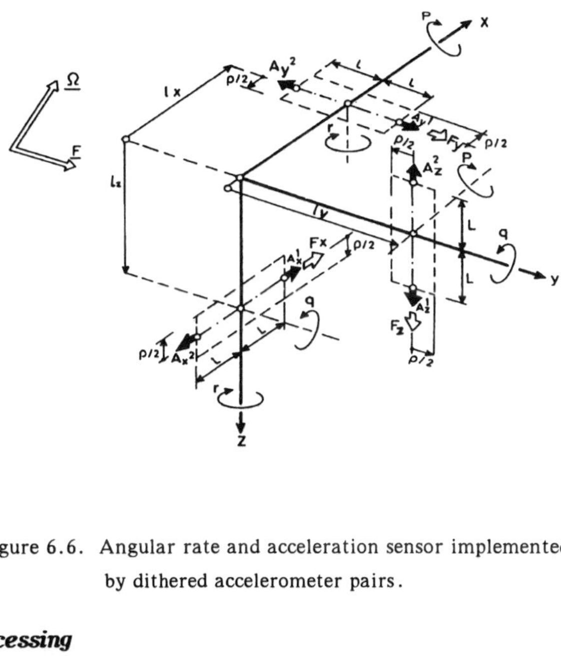

Figure 6.6. Angular rate and acceleration sensor implemented by dithered accelerometer pairs.

Signal preprocessing

The principal feature of this mechanization is that vibration parallel, for example, to the z axis produces a 180deg out-of-phase signal in A_z^1 and A_z^2 while the Coriolis signals due to p are in phase. Thus, based, for example, on (6.4.4), two separate signal channels, one for Ω and one for F, can be set up as follows:

$$a_z^p = a_z^1 + a_z^2 + v_z^1 - v_z^2 = 2\omega\rho\cos\omega t\, p + \tilde{v}_z \quad \text{(addition)} \tag{6.5.1}$$

$$a_z^{F_z} = \frac{1}{2}[a_z^1 - (-a_z^2)] + v_z^1 - (-v_z^2) = F_z + v_z \,\text{(subtraction)} \tag{6.5.2}$$

$\tilde{v}_z = v_z^1 - v_z^2$ is the residual vibration-induced noise. The better the matching between the accelerometer pair, the smaller is $\tilde{v}_z$. In the Ω channel, the F and v readings, which are opposite in sign, almost cancel. Residual phase shifts in the two F measurements must be compensated to be as equal as possible. The Ω terms in the F channel are in phase, and, on subtraction, they cancel.

Figure 6.7 shows the basic computational flow diagram that summarizes (6.5.1) and (6.5.2) along with the sgn$(\cos\omega t)$ demodulation function.

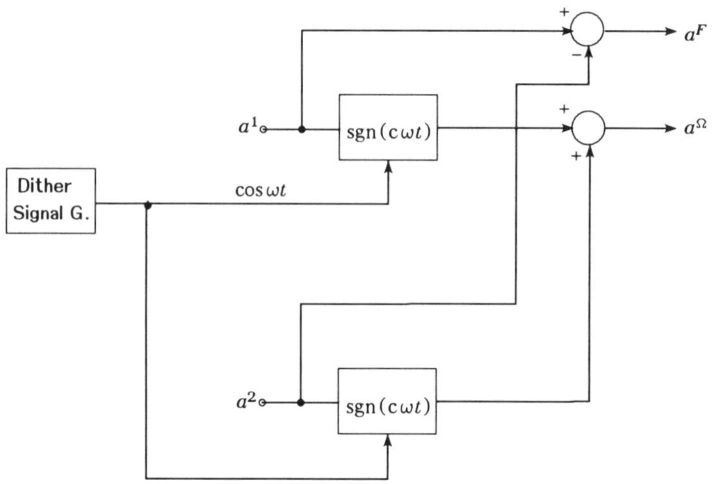

Figure 6.7. Signal processor for separating the Coriolis angular rate signals from acceleration signals.

a^1 and a^2 are the raw accelerometer outputs, and a^F and a^Ω are the preprocessed outputs for any axis x, y, z. This representation is in analog format whose output is voltage or current, as described in Chapter 3. In reality, analog accelerometers of the FBS class, such as the Q-flex, although they have the required precision, are too large and heavy to permit a realistic technical design. Feasible design relies on miniature lightweight vibrating beam accelerometers (VBA) whose output format is a deviation from a constant carrier frequency of about 35 kHz. Decoding these frequency variations requires suitable counter techniques especially designed to meet the stringent requirements of high resolution. These techniques, based on interpolation, will be described in more detail, below.

Readings of accelerometer pairs

The foregoing preprocessing operations in (6.5.1) and (6.5.2) for the z axis only are now formulated in vector format for all three axes, as follows:

$$\begin{bmatrix} a_x^q \\ a_y^r \\ a_z^p \\ a_x^{F_x} \\ a_y^{F_y} \\ a_z^{F_z} \end{bmatrix} = \begin{bmatrix} 1 & 1 & 0 & 0 & 0 & 0 \\ 0 & 0 & 1 & 1 & 0 & 0 \\ 0 & 0 & 0 & 0 & 1 & 1 \\ \frac{1}{2} & -\frac{1}{2} & 0 & 0 & 0 & 0 \\ 0 & 0 & \frac{1}{2} & -\frac{1}{2} & 0 & 0 \\ 0 & 0 & 0 & 0 & \frac{1}{2} & -\frac{1}{2} \end{bmatrix} \begin{bmatrix} a_x^1 \\ a_x^2 \\ a_y^1 \\ a_y^2 \\ a_z^1 \\ a_z^2 \end{bmatrix} \tag{6.5.3}$$

In accordance with Figure 6.6, the instantaneous distances of the accelerometer sensitive elements from center of rotation of the vehicle are given by:

$$\begin{bmatrix} r_{A_x}^1 \\ r_{A_x}^2 \\ r_{A_y}^1 \\ r_{A_y}^2 \\ r_{A_z}^1 \\ r_{A_z}^2 \end{bmatrix} = \begin{bmatrix} L & 0 & l_z+\frac{1}{2}\sin\omega t \\ -L & 0 & l_z-\frac{\rho}{2}\sin\omega t \\ l_x+\frac{\rho}{2}\sin\omega t & L & 0 \\ l_x-\frac{\rho}{2}\sin\omega t & -L & 0 \\ 0 & l_y+\frac{\rho}{2}\sin\omega t & L \\ 0 & l_y-\frac{\rho}{2}\sin\omega t & -L \end{bmatrix} \begin{bmatrix} 1_x \\ 1_y \\ 1_z \end{bmatrix} \tag{6.5.4}$$

Substituting (6.5.4) into (6.1.1), expanding and rearranging, we have for the readings of the accelerometer pairs,

$$a_x^1 = F_x + l_z(pr+\dot{q}) + \frac{\rho}{2}s\omega t(pr+\dot{q}) + 2\omega\frac{\rho}{2}c\omega t\,q - L(q^2+r^2) \tag{6.5.5}$$

$$a_x^2 = -F_x - l_z(pr+\dot{q}) + \frac{\rho}{2}s\omega t(pr+\dot{q}) + 2\omega\frac{\rho}{2}c\omega t\,q - L(q^2+r^2) \tag{6.5.6}$$

$$a_y^1 = F_y + l_x(pq+\dot{r}) + \frac{\rho}{2}s\omega t(pq+\dot{r}) + 2\omega\frac{\rho}{2}c\omega t\,r - L(p^2+r^2) \tag{6.5.7}$$

$$a_y^2 = -F_y - l_x(pq+\dot{r}) + \frac{\rho}{2}s\omega t(pq+\dot{r}) + 2\omega\frac{\rho}{2}c\omega t\,r - L(p^2+r^2) \tag{6.5.8}$$

$$a_z^1 = F_z + l_y(qr+\dot{p}) + \frac{\rho}{2}s\omega t(qr+\dot{p}) + 2\omega\frac{\rho}{2}c\omega t\,p - L(p^2+q^2) \tag{6.5.9}$$

$$a_z^2 = -F_z - l_y(qr+\dot{p}) + \frac{\rho}{2}s\omega t(qr+\dot{p}) + 2\omega\frac{\rho}{2}c\omega t\,p - L(p^2+q^2) \tag{6.5.10}$$

Force and angular rate components

Substituting (6.5.5)–(6.5.10) into (6.5.3) and assuming that F and Ω are constant throughout the small time interval T, the angular rate components are

$$a_x^q = 2\omega\rho q \cos\omega t + \rho \sin\omega t\, pr \; - 2L(q^2 + r^2) \qquad (6.5.11)$$

$$a_y^r = 2\omega\rho r \cos\omega t + \rho \sin\omega t\, pq \; - 2L(p^2 + r^2) \qquad (6.5.12)$$

$$a_z^p = 2\omega\rho p \cos\omega t + \rho \sin\omega t\, qr \; - 2L(p^2 + q^2) \qquad (6.5.13)$$

and the force components are

$$a_x^{Fx} = F_x + l_z pr \qquad (6.5.14)$$

$$a_y^{Fy} = F_y + l_x pq \qquad (6.5.15)$$

$$a_z^{Fz} = F_z + l_y qr \qquad (6.5.16)$$

Finally, substituting (6.5.11)–(6.5.13) into (6.2.1)–(6.2.3) and (6.5.14)–(6.5.16) into (6.2.4)–(6.2.6), the results are as before for the single accelerometer mechanization, namely,

$$\hat{p} = p \qquad (6.5.17)$$

$$\hat{q} = q \qquad (6.5.18)$$

$$\hat{r} = r \qquad (6.5.19)$$

and

$$\hat{F}_x = F_x + l_z pr \qquad (6.5.20)$$

$$\hat{F}_y = F_y + l_x pq \qquad (6.5.21)$$

$$\hat{F}_z = F_z + l_y qr \qquad (6.5.22)$$

Comments:

1) The estimates $\hat{p}$, $\hat{q}$, and $\hat{r}$ are unbiased and, subject to residual noise terms, they are accurate representations of p, q, r.

2) The additional terms $l_z pr$, $l_x pq$, and $l_y qr$ can be compensated using the estimates $\hat{p}$, $\hat{q}$, and $\hat{r}$ and the known values of l_x, l_y, and l_z.

6.6 Silicon mechanization of dither

The shuttle mechanism

The vibrating accelerometer pair concept is the basis of a recent development, the heart of which is the mechanization of a counter moving shuttle with periodic counter motion of the accelerometer pair. The implementation is schematically described in Figure 6.8. The accelerometer pair A_1, A_2 executes a linear counter motion by means of the closed-loop servo drive, which enforces the displacement ρ at the dither frequency ω. The accelerometer input axes IA are in opposition and are perpendicular to the paper plane. The angular rate Ω is perpendicular to the direction of ρ and IA. The flexure permits the small parallel motion ρ, as indicated. Being a flexible parallelogram, it also executes a small motion perpendicular to ρ in the paper plane, but this clearly has no effect on the performance of the device. The flexure is attached to the vehicle's body as indicated in Figure 6.8. The counter motion of A_1, A_2 perfectly balances the dither forces if all the mechanical parts of the pair are identical in mass and shape. Consequently, no vibration, which could otherwise cause interference with the other two devices of the triad, is transmitted to the vehicle body. A number of prototype vibrating Coriolis devices, documented in U.S. patent [6] have been developed and tested and are described in [7]. These tests have demonstrated encouraging results. The lessons learned from these developments indicated that a major obstacle in achieving high precision, repeatability, and robustness to environmental influences have been temperature gradients and stresses within the device. These are primarily caused by the different materials used in the electrical and mechanical components that are joined to each other.

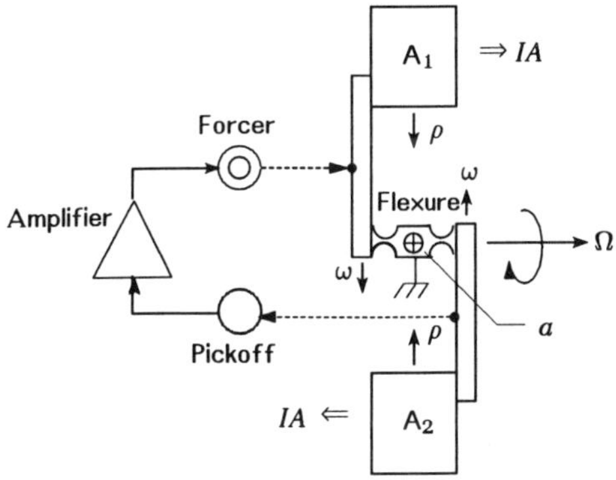

Figure 6.8. The shuttle mechanism and drive implementing dithered accelerometer pair.

Silicon micromechanization

In view of this experience, and with the advent of micromachining technology involving integrated microelectronic circuitry, a new device, photolithically micro machined from a monolithic silicon wafer, has recently been developed at SDC. It is known as μSCIRAS (Micro Sundstrand Coriolis Inertial Rate and Acceleration Sensor). A sketch of its mechanism depicted for the angular rate component r and the acceleration component a_x is shown in Figure 6.9 [7].

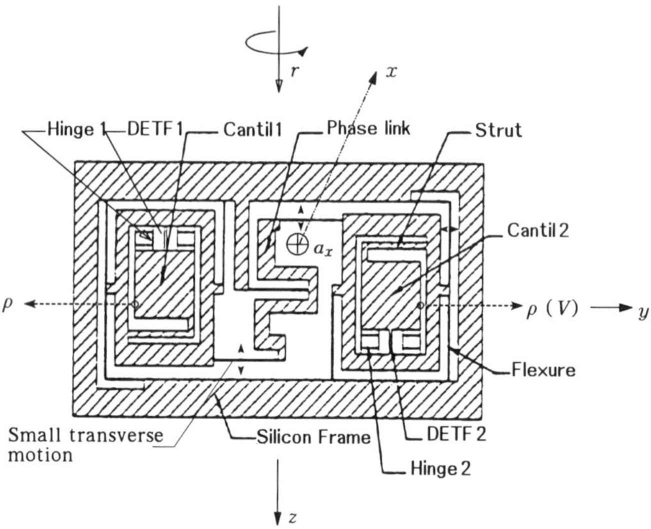

Figure 6.9. Silicon chip realization of the dithered accelerometer pair.

The two accelerometers are of the vibrating beam accelerometer (VBA) type described in Chapter 3. They are arranged in antiparallel fashion to implement the motion described in Figure 6.8, and they are connected by the unique arrangements of flexures etched in the Silicon. These establish the dither natural frequency and provide phase linking of the dither counter motion. The flexures and phase links are etched during the same multistep process which forms the DETF tines and hinges. The vibrating tines, namely, the DETF's, (Double-ended tuning forks) are on opposite sides of the silicon wafer. The proof mass hinges are on the opposite sides of the wafer for each cantilever CANTIL 1 and CANTIL 2. The electronic drives utilize reaction forces of electric current flowing in the presence of a uniform magnetic field that provides the motive force for maintaining the individual oscillations of the tines and the dither. Small gold traces are deposited on the tines and the flexures during fabrication. An oscillator circuit for the tines and the dither drive generates an electrical current through the gold traces at the resonant frequency of the dither mechanism. This implies that only minute power is required to enforce the dither. A magnetic field is applied perpendicular to the surface plane of the silicon wafer. The resultant forces cause the tines and the flexures to oscillate orthogonally to the magnetic field and the direction of the current at their corresponding oscillation and dither frequencies of 35 kHz and 600 – 1200 Hz, respectively. The dither amplitude ρ is on the order of 0.025 mm. The mechanism illustrated in Figure 6.9 is approximately 12×8 mm. By the very nature of its design, technology, and fabrication method, it has the best possible matching characteristics of the accelerometers, low temperature gradients, and excellent mechanical long-term stability and ruggedness under the severe conditions of shocks up to 20,000 g, which are typical in artillery. It is the innovative silicon micromechanical technology that has reduced the weight and size of the moving parts by an order of magnitude, thus allowing the realization of the alternating velocity amplitude $v = \omega\rho = 7536 \times 0.025 \times 10^{-3} = 0.188$ m sec^{-1} a practical reality with minute power requirements and physical dimensions.

6.7 Sensor output signal processing

As indicated in Section 6.2, after the first stage of signal preprocessing separating the linear acceleration F signals from the Coriolis Ω signals, a second stage of signal processing is required. The task of this stage is to decode the frequency-modulated accelerometer readings and present them in digital format which is compatible with those commonly used in flight control and navigation systems.

Conversion of frequency variations to pulse counts

A counter-ASIC (application specific integrated circuit), one for each sensor mechanism, counts the DETF cycles over half of a dither period. This is implemented by sampling the sinusoidal output signal at each of its zero crossings. For the dither frequency of $f_d = 640$ Hz, the available data rates are therefore 1280 Hz. Thus, for $f_n = 35$ kHz, the maximum number of counts per sample is $N_{\max} = 27.343$. Clearly, only the integer 27 can be counted, and, accordingly, the resolution would only be 1/27. To enhance the resolution of the signal frequency counts, it is necessary to provide a between-samples interpolation that can provide the fractional counts. This is accomplished by a high-frequency clock running at, e.g., $f_{\text{clock}} = 50$ MHz which acts like a vernier. This high-speed clock subdivides the period of each tine cycle, thus providing for fractional counting of each tine cycle. The sample period synchronization is obtained from a synchronizing ASIC running at the dither frequency f_d. The set of signals that makes the interpolation possible is shown in Figure 6.10.

Signal description

Signal 1 represents the 50 MHz clock; Signal 2, which is synchronized with signal 1, is the sampling signal f_s running at 1280 Hz; Signal 3 represents the switched tine oscillation at the variable frequency f_n around 35 kHz; Signal 4, whose switch-on at point i-1 is synchronized with Signal 2 and whose switch-off is synchronized with the instant of the completion of the tine oscillation period $T_n = 1/f_n$ in signal 3, provides the partial leading-edge count P_{i-1}. Similarly, at the trailing edge point i of the sample period in signal 2, the clock sampling in Signal 4 is switched on, and switched off again by the completion of the tine oscillation period T_n. This provides the next partial count P_i. Signal 5, which is synchronized with signal 3, provides a full count F_i over the entire period T_n near the leading edge $i-1$, and again, the full count $F_i = F_{i-1}$ near the trailing edge i. The required fractional counts are given by the ratios P_{i-1}/F_{i-1} and P_i/F_i. These partial counts are added to the integer tine count N_i to obtain the total precise count N_c as follows:

$$N_c = N_i + \frac{P_{i-1}}{F_{i-1}} - \frac{P_i}{F_i} \tag{6.7.1}$$

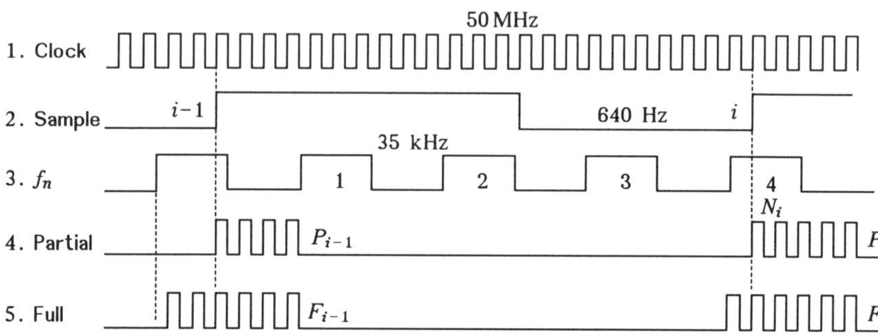

Figure 6.10. Principle of operation of the signal processor, including high-resolution vernier interpolation counting (based on Ref. 7).

The foregoing scheme basically provides the resolution of the fractional counts to half a period of the 50 MHz clock frequency at the leading and trailing edges. In combination, according to (6.7.1), the resolution is one quarter of T_{clock}, which is $1/(4f_{\text{clock}}) = 1/(4\times 50\times 10^6) = 5\times 10^{-9}$ sec.

The actual resolution must be referred to the maximum frequency deviation $\Delta f = 0.1 f_n = 0.1\times 35000 = 3500$ kHz for which the cycle period is $T_{\Delta f} = 1/3500 = 2.8\times 10^{-4}$. The dynamic range of the angular rate measurement per sample, therefore, is

$$D_r = \frac{2.8\times 10^{-4}}{5\times 10^{-9}} = 5.6\times 10^4 \tag{6.7.2}$$

EXAMPLE 6.7–1: Smallest resolvable angular rate for a given dynamic range

The maximum roll rate of a missile is $p_{\max} = 100$ deg/sec. If the dynamic range of a Coriolis sensor is $D_r = 5\times 10^4$, what is the smallest resolvable angular rate p_{th} in deg/hr?

$$p_{max} = 100 \times 3600 = 3.6 \times 10^5 \text{ deg/hr}$$

$$p_{th} = 3.6 \times 10^5 / (5 \times 10^4) = 7.2 \text{ deg/hr}$$

This result demonstrates a performance level potentially suitable in numerous tactical guidance and medium-class navigation applications.

It should be borne in mind that the dynamic range derived in (6.7.2) is for a single sample. Averaging for example, over one second, i.e., $N = 1280$ samples, can decrease p_{th} by $1/N$, as indicated in Chapter 3.

6.8 Projected performance characteristics

The μSCIRAS is presently in an advanced stage of development. Projected performance, based on the detailed knowledge of the potential of the technology involved, is given in Table 6.1 [7].

Table 6.1. Projected performance characteristics of a Coriolis angular rate/force sensor (Sundstrand μSCIRAS)

		Near Term	Ultimate Goal
ACCELEROMETER (3 axes)	Bias Scale Factor Alignment Capture Range	500 μg 25 ppm 50 μradians 40 g to 120 g	100 μg 25 ppm 50 μradians 20 g to 50 g
GYRO (3 axes)	Bias Scale Factor Alignment Maximum Rate Random Walk	10 to 30 deg/hr 50 ppm 75 μradians 10 rps 0.2 deg/rt-hr	1 to 5 deg/hr 50 ppm 75 μradians 10 rps 0.1 deg/rt-hr
PACKAGE	Size	1" Dia. × 1" (Single)	2" Dia. × 1" (Triax)
ENVIRONMENTS	Shock (survival) Vibration Temp. Range Dormancy	10,000 g's 7 g's RMS -55° to 95°C 10 years	20,000 g's 10 g's RMS -55° to 95°C 20 years
OUTPUT	Format/Rate	Digital/ 1200 Hz	Digital/ 2400 Hz

Concluding comments

The principal properties and potential advantages of the integrated silicon technology as manifested in the μSCIRAS development and summarized in Table 6.1 are

1. Extremely large dynamical range from the low input rate limited by the bias uncertainty of, say 2 deg/hr to the maximum input rate of 3600 deg/sec $=13\times10^6$ deg/hr. This amounts to a dynamic range of $D_r=6.5\times10^6$.

2. Extremely good long-term stability, implying insignificant maintenance and repeated testing.

3. Very high shock and vibration resistance, which renders the device suitable for artillery-fired guided projectiles.

4. The very wide operating temperature range mainly due to stability of the silicon technology.

Silicon sensors provide many advantages over other materials, such as quartz or metal, for microsized rate sensor development. These advantages include excellent scale factor matching and stability, long life, bias stability, virtually no degradation, and the ability to handle larger stress levels. This permits large values for ρ, which is beneficial for better output signal-to-noise ratios, thus reducing the random error component in the angular rate. The inherent stability of silicon implies a predictable response, permitting reliable error compensation by calibration and relaxing the need for ultraprecise and costly machining. Further advantages of silicon sensors include selecting the geometry of the tuning fork tines for precise control of operating characteristics. For the multisensor application with two accelerometers, fabricated out of the same silicon wafer and placed in close proximity, the inherent matching characteristics should reduce the VBA's error sources to the order of a few parts per million.

Since this class of sensors outputs frequency changes in response to the respective inertial and kinematic inputs $\mathbf{F}$ and $\mathbf{\Omega}$, they are, as shown earlier, natural for operation in strapdown INS or flight control systems [8].

An excellent survey on the state of the art of rotation and force microsensors and their potential applications is provided in [9] along with the better known mechanical and optical gyros. Figures 6.11 and 6.12 describe performance contours of current sensor technologies for gyros and accelerometers. The closed contours are mapped in terms of bias and scale factor uncertainties. The various applications are indicated within these contours. Thus, one can see the association between current technologies and their operational utilization at a glance.

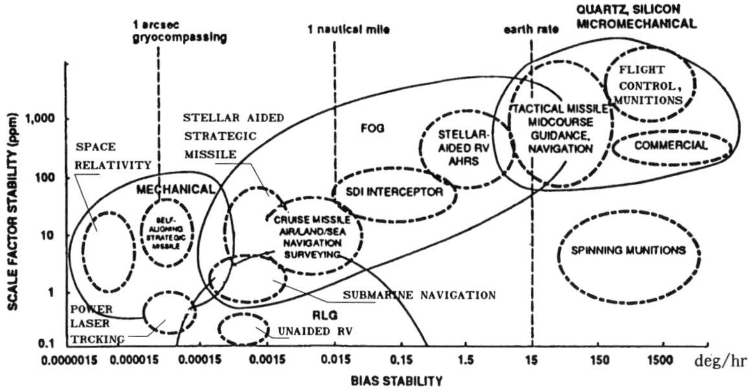

Figure 6.11. Current gyro technology applications (based on [9]).

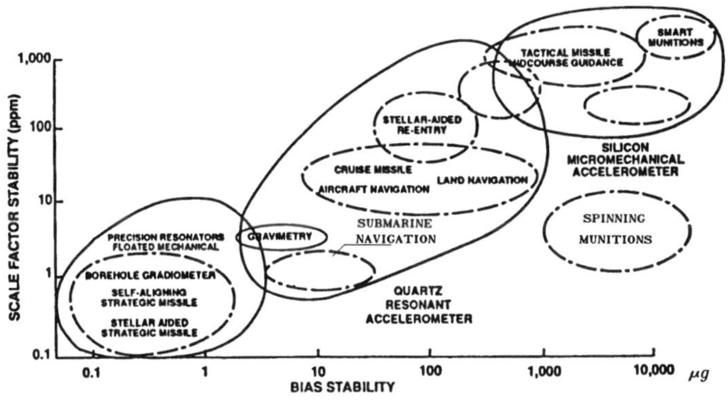

Figure 6.12. Current accelerometer technology applications (based on [9]).

Problems

Problem 6.1—Principle of operation of the Coriolis sensor: Equivalence to the spinning wheel gyro

As presented in Chapters 4 and 6, the gyroscope and vibrating Coriolis angular rate sensor appear to be based on entirely different principles, conservation of angular momentum on one hand and Coriolis force on the other. Yet, they measure exactly the same angular rate with respect to free space.

a) Explain the operation of the spinning wheel gyro by means of Coriolis forces. (Hint: Refer to two arbitrary mass points diametrically situated on the rotor.)

b) Explain the operation of the rotational and dithered Coriolis sensors by the principle of conservation momentum. (Hint: Recall the Foucault pendulum.)

Problem 6.2—Dynamical tradeoffs in the vibrating Coriolis angular rate sensor

The Coriolis specific force was shown to be $a_c = 2\Omega \times v$, where v derives from the vibration $\rho = \rho_0 \sin \omega t$. Clearly, for high sensitivity, v must be made as large as possible. The design constraint is on the peak vibration acceleration a_{v_m}.

a) For different values of a_{v_m}, show the tradeoff relationship between v and ω.

b) What is preferable, high or low frequency ω?

c) What limits the choice of very low ω for a specified value of v_m?

Due to the periodic nature of v, a_{c_m}, the measured value of a_c crosses zero at the frequency 2ω. This implies that Ω_c, the computed value of Ω, is periodically lost at the frequency 2ω, because $\Omega_c = a_{c_m}/(2v)$.

d) With this effect in mind, what is preferable, a low or high value of ω?

e) What is the nature of the error in the measured Ω? Is it an additive or a scale factor error?

f) Does this effect exist in the rotational Coriolis (Collins type) sensor?

Problem 6.3—Elimination of zero crossing error in vibrating Coriolis sensors

In view of the Collins type rotational Coriolis sensor, propose a scheme for eliminating the zero crossing error.

a) Show the functional block diagram and formulate the required demodulating signals.

b) Discuss the effect on overall performance if one of the Coriolis sensors fails.

c) If the frequency of an extraneous vibration is at the dither frequency ω, what type of error will be induced?

Problem 6.4—Separation of angular rate from specific force: Underlying assumptions

The signal processing algorithm that permits the separation of Ω from F is expressed in (6.2.1)–(6.2.3) and (6.2.4)–(6.2.6). These derive from (6.1.3)–(6.1.5).

a) Does the validity of these relationships depend on some underlying assumptions as to the statistical nature of Ω or F? If so, what are they?

b) In view of your answer to a), is the signal processing algorithm related to the notion of sensor bandwidth in analogy to measurement bandwidth in general?

c) If such a relationship exists, what role does the dither frequency ω play in this issue?

Problem 6.5— Operating parameters: Criticality and required precision

Conventional strapdown IMUs require g compensation. Does the same apply to the Coriolis sensors described in this Chapter?

a) Justify your yes or no explicitly on the basis of the sensor arrangement and relevant analytical expressions.

b) What is the requirement for the precision of matching accelerometer pairs in order to assure that the navigation error will not exceed 5 nmi/hr?

c) In view of the relationship e.g., $a_z = 2\omega\rho p$, and the signal processing algorithm leading to (6.2.1)–(6.2.8), what is the critical operating parameter: ω, ρ, or both?

References

1. Merhav, S., "A Nongyroscopic Inertial Measurement Unit," *AIAA Journal of Guidance, Control and Dynamics*, Vol. 5, No. 3, pp. 227–235, May–June, 1982.

2. Goldstein, H., *Classical Mechanics*, Addison-Wesley, Reading, 1959.

3. Merhav, S., "Apparatus and Method for measuring Specific Force and Angular rate," United States Patent No. 4,445,376, May, 1984.

4. Klass, P.J., "Inertial Sensor Utilizes Coriolis Effect," *Aviation Week and space Technology*, pp. 89–91, October 10, 1983.

5. Merhav, S., "Apparatus for Measuring Inertial Specific Force and Angular Rate of a Moving Body," United States Patent No. 4,590,801, May 1986.

6. Hulsing, R., "Multi Axis Angular Rate Sensor Having a Single Dither Axis," United States Patent No. 4,821,572, April 1989.

7. Hulsing, R., and MacGugan, D., "Miniature IMU Based on Micromachined Coriolis Sensor," ION paper, January 21, 1993.

8. Savage, P.G., "Strapdown Inertial Sensors, Theory and Applications," AGARD lecture Series No. 95, June 1978.

9. Barbour, N.M., Elwell, J.M., and Setterlund R.H., "Inertial Instruments—Where to Now", *Proc. AIAA Guidance, Navigation and Control Conference*, paper AIAA-92-4414-CP, Hilton Head, SC, August 1992, pp. 566-574.

7

The Interferometric Fiber–Optic Gyro

7.0 Introduction

This chapter describes the principle of operation and the main properties of the interferometric fiber optic gyro (IFOG) which, along with the ring laser gyro (RLG) introduced in the next chapter, has been a principal departure from the well-paved road of spinning mass gyros. Both are based on a fundamental law of physics, namely, the universal constancy of the speed of light in vacuum underlying relativity. The laser gyro is not really a gyro in the classical Newtonian sense of conservation of momentum, but, it takes advantage of the 'inertia' of light as will be explained below. The technology has evolved along two principal paths: The RLG, being a closed-path optical oscillator, is *active*, while the IFOG, which draws its light energy from an external source, is a *passive* interferometer.

The principle involved in both is the Sagnac effect [1]. The first successful mechanization of the Sagnac effect is the RLG described in Chapter 8, in which a closed optical path is provided to sense inertial angular rate by means of light interference of two counterrotating beams. The principle was first demonstrated by Macek and Davis [2] in 1963. Since that time, governments and industrial organizations have vigorously pursued the development of the RLG technology. In the United States alone, hundreds of millions of dollars have been spent to demonstrate its feasibility, to evaluate and imorove its performance and reliability, and construct the required but very costly laboratory and manufacturing facilities.

The leading companies in the U.S. engaged in this effort have been Honeywell, Litton, Northrop, Raytheon, Singer-Kearfott, and Sperry. Most of these organizations have built prototypes that demonstrated performance and reliability compatible with military and commercial applications. Some of these companies have also been engaged in the development and production of RLG-based INS systems. Honeywell, the leading company in this endeavor, has produced hundreds of Model GG1342 INSs per month, mainly intended for the Boeing 757/767 INS, which has been successfully operational for more than two decades with MTBF levels in excess of 5000 hrs. Some European companies engaged in the development and production of RLGs have been Ferranti, Thompson CSF, Sperry, and British Aerospace.

The more recent emerging technology discussed in this chapter is the interferometric fiber-optic gyro (IFOG) [3], also based on the Sagnac effect. It provides the closed optical path by a multiturn optical fiber coil wound on a bobbin. It is more compact and potentially of lower cost than the RLG.

7.1 The Sagnac interferometer

Mechanization by discrete optical components

Both the RLG and the FOG are optical angular rate sensors that sense changes in the transition time of light along the clockwise (cw) and counterclockwise (ccw) closed optical paths. The difference in the two transition times is proportional to the inertial angular rate of the device. In the RLG, which is of the active resonant type, the difference in transition time is inherently converted into detectable variations in the laser oscillation frequency. FOGs, on the other hand, can be mechanized in two fashions: In the first, similar to the RLG, the optical fiber path acts like the resonator cavity; it is therefore known as the *resonator optical fiber gyro* (RFOG). The second is a direct measurement of the phase difference between the two interfering light paths. Hence, it is known as the *interferometric FOG* (IFOG). The latter has reached an advanced stage of development and has been exposed to flight tests and integrated in tactical strapdown IMUs. The following presentation will therefore principally relate to the IFOG.

The roots of the IFOG can be found in the experimental methodology and techniques used by Michelson and Morely in their early "ether drag" experiments. A brief review of the Michelson-Morely interferometer is given in [4]. The classical optical Sagnac interferometer, which is an extension of the Michelson-Morely interferometer, is shown in Figure 7.1. It is designed to set up two optical beams in the clockwise (cw) and counterclockwise (ccw) senses. A light source S emits a beam which is split by the beam splitter 0 along l_1 (cw) and l_2 (ccw). l_2 is reflected from the mirrors 2, 3, and 1 and passes through 0, while l_1 is reflected from the mirrors 1, 3, 2, and 0 where it combines with l_2. Thus, the beam splitter 0 also fulfills the function of an optical coupler. The combined cw and ccw beams are picked up by the detector D_e, which can sense possible interference patterns if phase differences exist between l_1 and l_2.

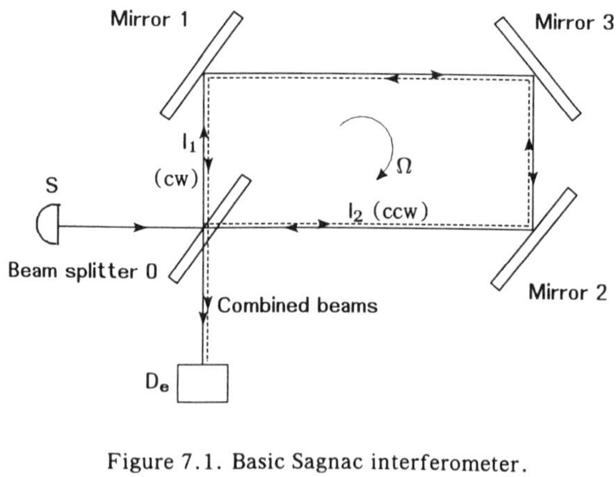

Figure 7.1. Basic Sagnac interferometer.

Fiber-optic mechanization

In a practical realization, the two counterrotating light paths are not implemented by discrete optical components such as mirrors and beam splitters shown earlier, but by optical fibers, which act as wave guides that bend the light as desired. Integrated optics that incorporate beam splitters, couplers and phase modulators complete the 'optical network'. A schematic description of the fiber version of the Sagnac interferometer is shown in Figure 7.2.

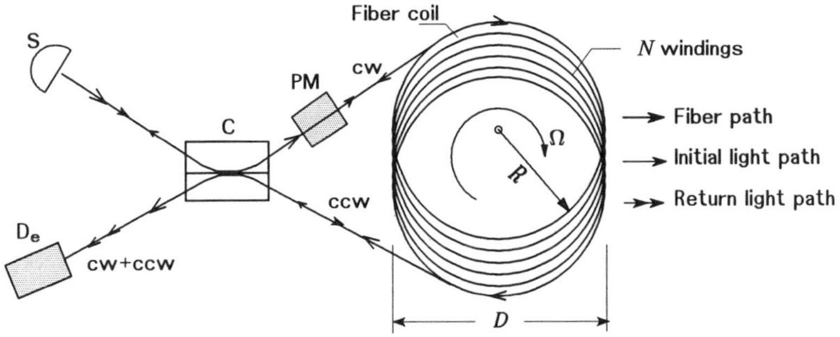

Figure 7.2. Optic fiber realization of the Sagnac interferometer.

Light from S is fed into the 50/50 coupler C, where it is split into two paths passing through the coil in two opposite directions. A phase modulator PM, placed, for example, in the cw sense, should introduce a $\pi/2$ phase shift, the purpose of which will be explained below. On their return to the coupler, the light paths recombine and resplit. One half returns to S to no avail and the other half, namely the sum of the cw and the ccw interfering waves, goes to the detector D_e.

7.2 Effect of angular rate on Sagnac phase shift

Effect in vacuum

The operation of the basic fiber-optic Sagnac interferometer is derived by the following considerations. The length of a single fiber loop is $L=\pi D$; under rotation Ω, the tangential velocity of any particle of the fiber is $V=\Omega D/2$; the transition time at rest around the loop with N windings is $T=N2\pi R/c_m$; c_m is the speed of light in a given medium, $c_m=c_v/n$. c_v is the speed of light in vacuum and n is the index of refraction of the medium. However, we initially assume that the material of the fiber has a refraction index $n=1$ so that $c_m=c_v$, implying that the fiber is equivalent to vacuum. This assumption permits an elementary kinematic approach in developing the IFOG transfer function. Since the exit point of the photon has receded, a photon traveling cw travels the distance $L^+>L$ to complete the loop, while a photon traveling ccw covers the distance $L^-<L$. The corresponding transition times are $\tau^+=L^+/c_v$ and $\tau^-=L^-/c_v$, respectively. For N windings, we have

$$L^{+} = 2N\pi R + V\tau^{+} = 2N\pi R + \Omega R\frac{L^{+}}{c_{v}} \tag{7.2.1}$$

$$L^{-} = 2N\pi R - V\tau^{-} = 2N\pi R - \Omega R\frac{L^{-}}{c_{v}} \tag{7.2.2}$$

The solutions for L^{+} and L^{-} are

$$L^{+} = \frac{2N\pi R}{1 - \dfrac{\Omega R}{c_{v}}} \tag{7.2.3}$$

$$L^{-} = \frac{2N\pi D}{1 + \dfrac{\Omega R}{c_{v}}} \tag{7.2.4}$$

The difference in the path lengths is

$$L^{+} - L^{-} \triangleq \Delta L = \frac{\dfrac{4N\pi R^{2}}{c_{v}}\Omega}{1 - \dfrac{\Omega^{2}R^{2}}{c_{v}^{2}}} \tag{7.2.5}$$

Since in reality $c_{v}^{2} \gg \Omega^{2}R^{2}$, the transit time difference in vacuum $\Delta\tau_{v}$ is

$$\Delta\tau_{v} = \tau^{+} - \tau^{-} = = \frac{\Delta L}{c_{v}} = \frac{\dfrac{4N\pi R^{2}}{c_{v}^{2}}\Omega}{1 - \dfrac{\Omega^{2}R^{2}}{c_{v}^{2}}} \cong \frac{4N\pi R^{2}}{c_{v}^{2}}\Omega = \frac{4NA}{c_{v}^{2}}\Omega \tag{7.2.6}$$

where $A = \pi R^{2}$ is the area enclosed by the coil. Let the wavelength of the light be λ. The corresponding wave period is $\tau_{w} = \lambda/c_{v}$, so that

$$c_{v} = \frac{\lambda}{\tau_{w}} \tag{7.2.7}$$

Substituting (7.2.7) into (7.2.6) and rearranging, we have

$$\Delta\tau_{v} = \frac{4N\pi R^{2}}{c_{v}}\frac{1}{c_{v}}\Omega = \frac{4N\pi R^{2}}{c_{v}}\frac{\tau_{w}}{\lambda}\Omega \tag{7.2.8}$$

The fractional fringe interference is defined by $\Delta\tau_{v}/\tau_{w}$. Denoting the optical path length by $L_{p} = 2N\pi R$,

$$\frac{\Delta\tau_v}{\tau_w} = \frac{(2N\pi R)2R}{c_v\lambda}\Omega = \frac{L_p D}{c_v\lambda}\Omega \tag{7.2.9}$$

The corresponding Sagnac phase shift is

$$\Psi_s = 2\pi\frac{\Delta\tau_v}{\tau} = \frac{2\pi L_p D}{c_v\lambda}\Omega = \frac{8\pi NA}{c_v\lambda}\Omega = K_s\Omega \tag{7.2.10}$$

where $A = \pi D^2/4$ is the area enclosed by the coil, and

$$K_s \triangleq \frac{2\pi L_p D}{c_v\lambda} = \frac{8\pi NA}{c_v\lambda} \tag{7.2.11}$$

is the Sagnac scale factor. The fringe shift x_R, in rad, due to the rotation of the IFOG, corresponding to (7.2.10), is $\Psi_s/2\pi$, namely,

$$x_R = \frac{\Psi_s}{2\pi} = \frac{8\pi NA}{2\pi c_v\lambda}\Omega = \frac{4NA}{c_v\lambda}\Omega = \frac{N\pi D^2}{c_v\lambda}\Omega = \frac{L_p D}{c_v\lambda}\Omega \tag{7.2.12}$$

Effect in optical medium

In order to adapt (7.2.12) to the real situation of a physical fiber where $n>1$ and the velocity of light is c_m, one might be tempted to simply substitute $c_m = c_v/n$ so that x_R would appear to be given by $(L_p Dn)/(c_v\lambda)\Omega$. This, however, is not correct because moving from the basic situation of vacuum to a physical medium creates a different kinematic situation. The light now travels at the speed $c_m = c_v/n$, comoving or countermoving with respect to the fiber substance which travels at the tangential velocity $R\Omega$. In this case, both the Fizeau and Doppler effects occur and they compensate the effect of the index of refraction n. Consequently, (7.2.12) holds, irrespective of the medium. This issue is discussed in [5]. When the IFOG is at rest, light travels at the speed c_m both in the cw and ccw paths. Both waves are still in phase after the propagation time $\tau_m = 2\pi R/c_m = n\tau_v$, and no fringe effect will occur. However, when the IFOG rotates, the beam splitter will have moved through a length $\Delta L_m = R\Omega\tau_m = R\Omega\tau n = n\Delta L_v$ during the time τ_m. In this situation, the speeds of the cw and ccw light are no longer the same when observed in the motionless inertial frame of reference. A Fresnel-Fizeau drag effect occurs that depends on the relative directions of the light and the medium motion. The total speeds c^+ and c^- of the corotating and counterrotating waves are determined by the relativistic additions of the velocities c_0/n and $R\Omega$ as follows [6]:

$$c^+ = \frac{c_v/n + R\Omega}{1 + \dfrac{(c_v/n)\,R\Omega}{c_v^2}} = \frac{c_v/n + R\Omega}{1 + \dfrac{R\Omega}{nc_v}} \cong \frac{c_v}{n}\left(1 - \frac{R\Omega}{nc_v}\right) + R\Omega\left(1 - \frac{R\Omega}{nc_v}\right)$$

$$= \frac{c_v}{n} - \frac{R\Omega}{n^2} + R\Omega - \frac{R^2\Omega^2}{nc_v} \cong \frac{c_v}{n} + R\Omega\left(1 - \frac{1}{n^2}\right) \tag{7.2.13}$$

Similarly, for the counterrotating beam, we have

$$c^- = \frac{c_v/n - R\Omega}{1 - \frac{(c_v/n)\,R\Omega}{c_v^2}} = \frac{c_v/n - R\Omega}{1 - \frac{R\Omega}{nc_v}} \cong \frac{c_v}{n}\left(1+\frac{R\Omega}{nc_v}\right) - R\Omega\left(1+\frac{R\Omega}{nc_v}\right)$$

$$= \frac{c_v}{n} + \frac{R\Omega}{n^2} - R\Omega - \frac{R^2\Omega^2}{nc_v} \cong \frac{c_v}{n} - R\Omega\left(1-\frac{1}{n^2}\right) \tag{7.2.14}$$

where $R^2\Omega^2/nc_v$ is neglected in the presence of the other terms.

Substituting c^+ and c^- for c_v in (7.2.1) and (7.2.2), respectively, recalling that $\tau^+ = L^+/c^+$ and $\tau^- = L^-/c^-$, and solving for τ^+ and τ^-, we get

$$\tau^+ = \frac{2N\pi R}{c^+ - R\Omega} \tag{7.2.15}$$

$$\tau^- = \frac{2N\pi R}{c^- + R\Omega} \tag{7.2.16}$$

The difference in transition times is

$$\Delta\tau_m \triangleq \tau^+ - \tau^- \tag{7.2.17}$$

so that,

$$\Delta\tau_m = 2N\pi R\left[\frac{c^- + R\Omega - c^+ + R\Omega}{(c^+ - R\Omega)(c^- + R\Omega)}\right] \cong 2N\pi R\,\frac{2R\Omega - (c^+ - c^-)}{c^+c^-} \tag{7.2.18}$$

Substituting c^+ and c^- from (7.2.13) and (7.2.14), we have

$$\Delta\tau_m = 2N\pi R\,\frac{2R\Omega - 2R\Omega\left(1-\frac{1}{n^2}\right)}{\left[\frac{c_v}{n} + R\Omega\left(1-\frac{1}{n^2}\right)\right]\left[\frac{c_v}{n} - R\Omega\left(1-\frac{1}{n^2}\right)\right]}$$

$$\cong 2N\pi R\,\frac{2R\Omega - 2R\Omega\left(1-\frac{1}{n^2}\right)}{c_v^2/n^2} = \frac{4\pi NR^2n^2\left(1-1+\frac{1}{n^2}\right)}{c_v^2}\Omega$$

$$= \frac{4\pi NR^2}{c_v^2}\Omega = \frac{\pi ND^2}{c_v^2}\Omega \tag{7.2.19}$$

Comparing (7.2.19) with (7.2.6), we have the important result

$$\Delta\tau_m = \Delta\tau_v \tag{7.2.20}$$

Thus, $\Delta\tau_v$ is valid for a fiber made of any appropriate optical substance.

EXAMPLE 7.2–1: Required precision for Sagnac phase-shift measurement

The following example demonstrates the precision required in the determination of Ψ_S. For an IFOG intended for inertial navigation, the following numerical values are assumed:

L_p = 1000 m

D = 4 in = 10.16 cm

λ = 820 nanometer

n = 1.5

$c_m = c_v/n = 2\times10^8$ m/sec

In view of (7.2.20), n is irrelevant, so that the resulting Sagnac scale factor K_S is

$$K_S = 6.28\times1000\times0.1016/(3\times10^8\times820\times10^{-9}) = 2.706 \text{ sec}$$

To detect,e.g., an angular rate of Ω = 0.01 deg/hr = (0.01/57.3)/3600 rad/sec = 4.847 $\times10^{-8}$ rad/sec, which corresponds to an error of 1n_mi/hr, it is required to detect a Sagnac phase shift as small as

$$\Psi_S = 2.706\times4.847\times10^{-8} \text{ rad} = 0.131\ \mu\text{rad} = 0.0257 \text{ arc sec}$$

This result demonstrates that for high-grade INS requirements, a phase detector of extreme precision is required.

7.3 Relationship between output power and phase shift

Phase characteristics—biasing and modulation

The light amplitude at the detector D_e is the sum of the cw and ccw waves which are shifted by the phase angle Ψ_s. Thus,

$$E_D = E^+ + E^- e^{-j\Psi_s} \tag{7.3.1}$$

Since the beam splitter is assumed to be precisely 50/50, we have $E^+ = E^- = E$, and

$$E_D = E + Ee^{-j\Psi_s} = E(1+e^{-j\Psi_s}) = E(1+\cos\Psi_s - j\sin\Psi_s) \tag{7.3.2}$$

The light power, or intensity, is

$$I_D = E_D^2 = E^2[(1+\cos\Psi_s)^2+\sin^2\Psi_s] = E^2(2+2\cos\Psi_s)$$
$$= 2I(1+\cos\Psi_s) = 2I[1+\cos(K_s\Omega)] \tag{7.3.3}$$

where $I \triangleq E^2$. Since by (7.3.3), I_D, is even in Ψ_s and in Ω, it is impossible to determine whether its sense is positive or negative. Moreover, for small values of Ψ_s, namely, Ω, the sensitivity of I_D to variations in Ψ_s is close to zero, as can be seen by

$$\frac{dI_D}{d\Psi_s} = -2I\sin\Psi_s \tag{7.3.4}$$

or, in terms of Ω,

$$\frac{dI_D}{d\Omega} = -2IK_s \sin(K_s\Omega) \tag{7.3.5}$$

The need for a 90 deg phase shift

The sensitivity reaches its peak at $\Psi_s = \pi/2$ and drops to zero as $\Psi_s \to \pi$. A bias of $\pi/2$ must be introduced. Thus, the intensity at the detector of the biased interferometer is

$$I_D = 2I[1 + \cos(\Psi_s - \pi/2)] = 2I(1 + \sin\psi_s) \tag{7.3.6}$$

The one holds no phase information and is disregarded. Equation (7.3.6), shown in Figure 7.3(a), is odd in Ψ_s, so that the sense of Ω can be determined from it. An illustration of $\sin\Psi_s$ in the region $0-\pi$ is shown in Figure 7.3(a). The bias to $\pi/2$ is also indicated. It serves as the new origin of Ψ_s. We wish to determine the sensitivity at the points $\psi^+ = \Psi_s - \pi/2$ and $\psi^- = \pi/2 - \Psi_s$. The sensitivity of the $\pi/2$-shifted interferometer for any positive or negative value of Ψ_s, is

$$\left.\frac{dI_D}{d\Psi_s}\right|_{\Psi_s=\psi^+} = 2I\cos(\Psi_s - \pi/2)(+1) = 2I\sin\Psi_s \tag{7.3.7}$$

$$\left.\frac{dI_D}{d\Psi_s}\right|_{\Psi_s=\psi^-} = 2I\cos(\pi/2 - \Psi_s)(-1) = -2I\sin\Psi_s \tag{7.3.8}$$

which are consistent with (7.3.4) as expected. The (+) or (−) signs are determined by inspecting the trigonometric circle diagrams for ψ^+ and ψ^-.

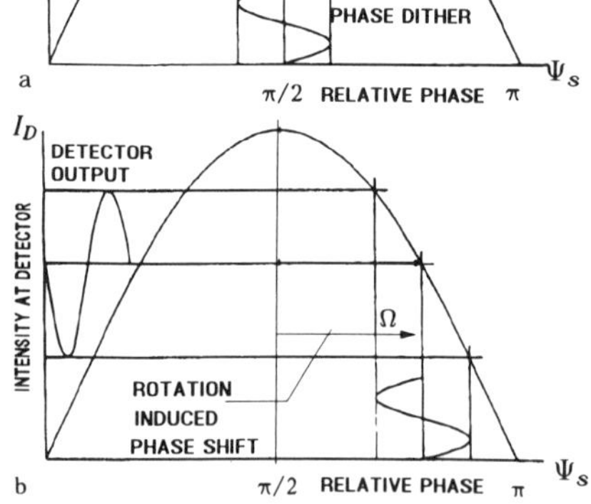

Figure 7.3. (a) 90 deg phase bias; (b) phase dither in Sagnac interferometer (based on [9]).

The last two expressions determine that the derivatives at points ψ^+ and ψ^- are positive and negative, respectively. They also show that these derivatives disclose the respective signs of Ψ_s, namely, the positive or negative sense of Ω. Moreover, the respective magnitudes $\pm\sin\Psi_s$ disclose the magnitude of Ω. The foregoing relationships provide the prescription for implementing the actual algorithm to derive Ω in magnitude and sign.

Extraction of angular rate by phase dither

In order to determine the derivative, a sinusoidal phase dither of constant amplitude I must be superimposed on the constant phase shift of $\pi/2$, as indicated in Figure 7.3(a). It demonstrates that the response of the non-rotating IFOG consists of second harmonics only, thus resulting in zero output. Figure 7.3(b) demonstrates the response of a rotating IFOG, which consists largely of the first harmonic. As shown, the phase dither is shifted by $+\Psi_s = K_s(+\Omega)$. The phase dither is provided by an optical phase modulator (PM), to be explained below. If the dither amplitude is made vanishingly small, the response r provided by the detector almost truly represents the derivative. The actual extraction of the derivative is provided by the synchronous detection of r. Since the dither amplitude a is known, the derivative is also known. However, making I too small may cause a large noise to signal ratio. A large value of I will introduce distortions, as indicated Figure 7.3(b). These distortions, however, have no real influence because the higher harmonics caused by them are rejected by the synchronous detector.

Since the characteristic provided by (7.3.7) and (7.3.8) is $\sin\Psi_s$, the IFOG is reasonably linear only over a limited input range of $\simeq \pm 30$ deg. Therefore, as an open loop device, the IFOG would not be satisfactory in high precision applications such as INS.

The $\pi/2$ bias can be implemented by a number of methods such as an additional transit delay in one of the circulating beams. Some of these methods are surveyed in [7] and other references. The effective measurable range of Ψ_s would, however, still be limited to less than $\pm\pi/2$. The angular rate Ω, by which the IFOG must be biased, is determined by $K_s\Omega_{\pi/2} = \pi/2$. Substituting K_s from (7.2.10), we have

$$\Omega_{\pi/2} = \frac{\pi}{2K_s} = \frac{c_v\lambda}{16NA} \tag{7.3.9}$$

This bias must be sufficiently large to allow the actual angular rate range typical in aircraft or missiles, which may be in excess of ± 400 deg/sec. The compatibility of the design parameters given in Example 7.2–1 with the corresponding values determined by (7.3.9) is examined in Example 7.3–1.

EXAMPLE 7.3–1: Examination of the numerical value of $\Omega_{\pi/2}$

In order to examine the value of $\Omega_{\pi/2}$, we use the parameters of Example 7.2–1 that resulted from the inertial-grade requirement of $\Omega_{min} = 0.01$ deg/hr. For the numerical values assumed there, we have

$$N = L_p/\pi D = 1000/(3.14\times 0.107) = 3134 \text{ windings}$$

$$\lambda = 820 \text{ nanometer}$$

$$c_v = 3\times 10^8 \text{ m/sec}$$

$$A = \pi D^2/4 = 0.785\times 0.1016^2 = 8.1\times 10^{-3} \text{ m}^2$$

so that

$$\Omega_{\pi/2} = \frac{c_v\lambda}{16NA} = \frac{3\text{x}10^8\times 820\times 10^{-9}}{16\times 3134\times 8.1\times 10^{-3}} = 0.606 \text{ rad/sec} = 34.65 \text{ deg/sec}$$

This result demonstrates that the IFOG, under the inertial-grade requirement, would not be suitable for a strapdown mechanization on-board aircraft or tactical missiles. It would, however, present no difficulty if mechanized with a gimbaled platform or if used in a strapdown mechanization on-board slowly rotating vehicles.

Implementing nonreciprocity by phase modulation

The foregoing analysis is based on the condition that the length of the cw path must appear to be longer (or shorter) by $\lambda/2$ than the ccw light path. As indicated earlier in Figure 7.3, this can be accomplished by a suitable phase modulator. Common phase modulators, however, are *reciprocal*, i.e., they cause the same phase shift, regardless of the direction of travel of the light. Therefore, the returning ccw light, which, after the first transition, simultaneously passes through the phase modulator in the opposite direction, will undergo an identical phase shift so that the net relative phase shift between the two paths will cancel out. This deficiency pertains to all kinds of phase modulation, not necessarily to the $\pi/2$ shift. The implication clearly is that the phase modulator PM must be operated so that it *appears* to be nonreciprocable. This is an issue of foremost importance. The phase shift ϕ along the fiber length L_p is given by

$$\phi = 2\pi f\tau = \frac{2\pi f n L_p}{c_v} \tag{7.3.10}$$

The magnitude of ϕ does not depend on the direction of the light propagation, so that $\phi_{cw} = \phi_{ccw} = \phi$. Therefore, if we vary just L_p, n, or f, no relative phase shift between the cw and the ccw beams can be achieved. In order to generate a nonreciprocal phase shift, we must do one of the following:

1) $L_{p\,cw} \neq L_{p\,ccw}$;
2) $n_{cw} \neq n_{ccw}$, or
3) $f_{cw} \neq f_{ccw}$.

These three methods have been attempted and implemented in the course of IFOG technology development, and they are described and discussed in [3] and [7]. The one that has survived and become an established technique is method 2), in which a phase modulator, through which the fiber passes, locally changes the refraction index n, thus causing a change in the phase shift in accordance with (7.3.10). This method will be described later.

Implementation of a 90 deg phase shift

The most common method for implementing nonreciprocity is by suitably time modulating the PM. For a fiber coil of length L_p and refractive index n, the transition time is

$$\tau = \frac{L_p n}{c_v} \tag{7.3.11}$$

The PM is connected to one end of the fiber. For its driving signal $V(t)$, we generate a square wave of amplitude $V_{\pi/2}$ at a modulation frequency f_m that ideally must be set to $f_m = 1/2\tau$. This signal is converted by the PM into a phase shift $\pi/2$ which can make L_p appear to be $\lambda/4$ longer. This square wave is generated, for example, by means of

$$V(t) = \frac{V_{\pi/2}}{2}\left[1 + \text{sign}\left(\sin(\frac{2\pi}{2\tau}t)\right)\right] = \frac{V_{\pi/2}}{2}\left[1 + \text{sign}\left(\sin(2\pi f_m t)\right)\right] \tag{7.3.12}$$

which is described in Figure 7.4.

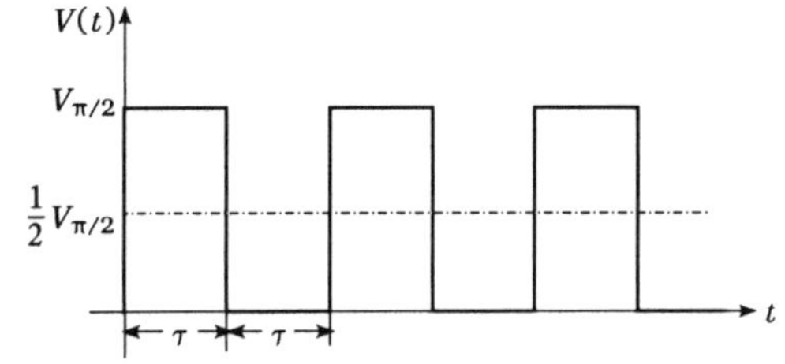

Figure 7.4. Implementation of 90 deg phase shift in the Sagnac interferometer.

The cw beam enters the PM immediately after emerging from the beam splitter, while the ccw beam travels through the entire coil length L_p and thus enters the PM after τ sec. During this transition period τ, $V_{\pi/2}$ is fully applied to the PM to enforce the required $\pi/2$ shift. At the beginning of the next τ seconds $V_{\pi/2}$ is annulled so that the returning ccw beam passes the PM without the shift, thus sustaining the relative $\pi/2$ shift between the two beams.

For a typical fiber coil with L_p = 1000 m and an index of refraction n = 1.5, $\tau = L_p n/c_v = 1000\times1.5/(3\times10^8) = 0.5\times10^{-5}$ sec. The required f_m therefore is

$$f_m = \frac{1}{2\tau} = \frac{1}{2\times0.5\times10^{-5}} = 100 \text{ kHz}$$

If f_m is not exactly set to $1/2\tau$ or if the scale factor of the modulator does not precisely convert $V_{\pi/2}$ into $\pi/2$, the required nonreciprocity is not assured, and the resulting error would manifest itself as an uncontrolled bias.

From the preceding, we conclude that for the implementation of a practical IFOG, two phase modulators are required. One for the constant $\pi/2$ phase shift, and the other for the serrodyne shifter.

The foregoing implies that the biased IFOG, if operated as an open-loop device, is unacceptably nonlinear, especially if a substantial dynamic range is required. To overcome this shortcoming, a scheme is required for implementing a loop closure that causes the IFOG to operate as a null sensor. As a rule, this technique obeys the properties of null sensors presented in Chapters 1, 4, and 6. It therefore can assure both excellent linearity and a wide dynamic range. Since the IFOG response is in terms of phase variations Ψ_s, the loop closure implies phase nulling. This technique has become known as the *phase-nulling fiber-optic laser gyro* (PNLG) [4].

7.4 Implementing the IFOG in a closed loop

Phase nulling laser gyro (PNLG)

The basis for phase nulling is the relationship between the Sagnac phase shift and a corresponding frequency shift of the light. The fringe shift x_R rad, induced by the rotation of the IFOG, as given by (7.2.12), is

$$x_R = \frac{L_p D}{c_v \lambda}\Omega = H_s \Omega \tag{7.4.1}$$

which is proportional to the frequency difference Δf between the cw and ccw waves. $H_s = L_p D / c_v \lambda$ is the scale factor of a 'sensor' in which the input is Ω rad/sec and the output is x_R which is eventually represented in volts. In order to facilitate the loop closure, we need a 'transducer' which can translate phase information into a linear representation of the measured Ω. This can be established by the following reasoning.

The transition time delay Δt_d around the fiber is determined by the fiber length L_p and the speed of light in the fiber $c_m = c_v / n$,

$$\Delta t_d = \frac{L_p}{c_m} = \frac{L_p}{c_v} n \tag{7.4.2}$$

Assume that a frequency shift Δf is enforced by a suitable modulator. This shift, multiplied by Δt_d, causes a fringe shift x_F:

$$x_F = \Delta t_d \Delta f = \frac{L_p n}{c_v} \Delta f \tag{7.4.3}$$

The difference $x_R - x_F \triangleq x_\varepsilon$ is the shift error,

$$x_\varepsilon = \frac{L_p D}{c_v \lambda}\Omega - \frac{L_p n}{c_v}\Delta f \tag{7.4.4}$$

which must be driven to zero by a suitable servo loop. The only nonconstant parameter in the second term of (7.4.4) is Δf. If it can be variedin order to achieve the nulling of x_ε, we obtain the solution

$$\Delta f = \frac{D}{\lambda n}\Omega \tag{7.4.5}$$

which establishes a linear relationship between the value of Ω and the measurable quantity Δf. It is interesting to note that the induced frequency shift Δf does not depend on the length L_p of the optical fiber. In view of (7.2.12), x_ε is equivalent to the phase error Ψ_ε by virtue of $2\pi x_\varepsilon = \Psi_\varepsilon$. A number of methods exist to implement this Δf modulation. Some of these, such as the acoustic-optic modulator or the piezostretching of the fiber, are reviewed in [7].

Implementation by Serrodyne shifter

Here, we will refer only to the *Serrodyne frequency shifter* [8], which is a phase modulator driven by a sawtooth voltage, as will be explained below. It is implemented by integrated optics technology which provides practical low-power solutions to the modulation issue. The phase modulator is an element placed in the optical path that changes its index of refraction when a voltage is applied to it in accordance with the following linear relationship:

$$\phi(t) = K_\phi V(t) \tag{7.4.6}$$

The instantaneous level of light at any given point is described by

$$a = a_m \sin(2\pi f t + \phi) \tag{7.4.7}$$

where

a_m — amplitude of light wave;
f — frequency of light wave;
ϕ — phase shift.

The sawtooth phase-modulating signal manifests itself as follows. Making $\phi = \phi(t) = 2\pi t/T$ and substituting into (7.4.7) yields

$$a = a_m \sin(2\pi f t + 2\pi t/T) = a_m \sin[2\pi(f + 1/T)t] \tag{7.4.8}$$

$\phi(t) = 2\pi t/T$ is an infinite ramp function. As such, it cannot be implemented by analog means because this would require unbounded voltages. However, if implemented as a sawtooth, as shown in Figure 7.5, though bounded, it is equivalent to a linear ramp since at any given instant t, $\phi(t)$ obeys the ramp $2\pi t/T$.

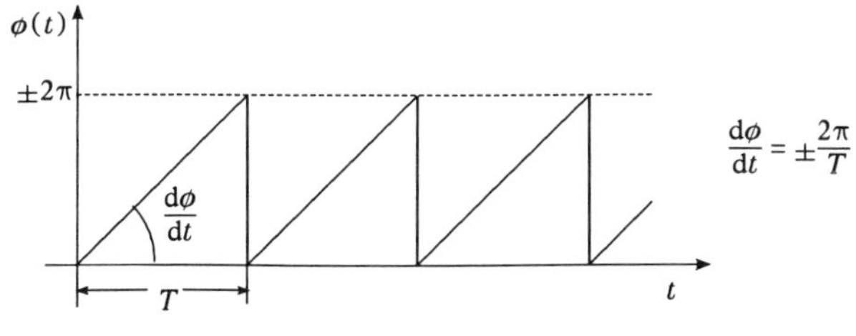

Figure 7.5. Sawtooth phase modulation in a Sagnac interferometer.

The sawtooth signal $2\pi t/T$ must be applied to the phase modulator to cause the frequency modulation of the light in accordance with the right-hand side of (7.4.8), which can be rewritten as

$$a = a_m \sin[2\pi(f + \Delta f)t] \tag{7.4.9}$$

where $\Delta f = 1/T$. Thus, by the changes in the slope $1/T$ both in magnitude and sign, the frequency modulation is controlled. The periodic reset, or rollover, must be implemented with great precision to avoid sidebands.

The operational block diagram of the closed loop phase nulling-system in analog terms is schematically shown in Figure 7.6. K is a gain factor large enough to assure a tight loop closure. The purpose of the integrator is to provide a first-order control loop. u, which can be positive or negative, is applied to the ramp generator, which consists of another integrator equipped with reset and restart switching. Whenever its output reaches the value of $+2\pi$ or -2π, as shown in Figure 7.5, it resets to zero and restarts with the current input u. The larger the absolute value of u, the smaller is T and the larger is the absolute value of $\Delta f = 1/T$. This establishes the linear relationship to x_ε.

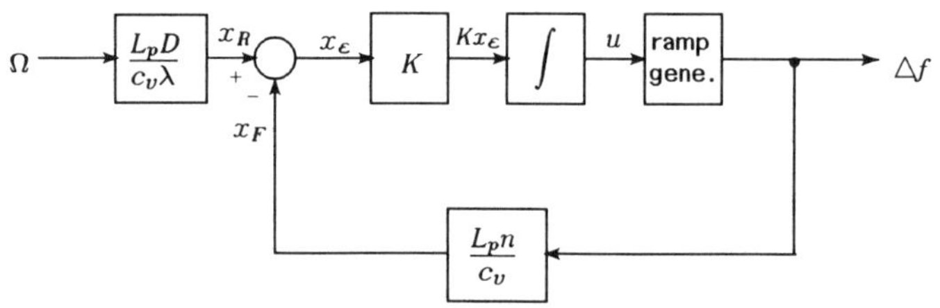

Figure 7.6. Operational block diagram of the PNLG showing translation from Ω to Δf.

As such, the ramp generator is better known as a voltage controlled-oscillator (VCO). The closed-loop transfer function is derived from

$$\left(\frac{L_pD}{c_v\lambda}\Omega - \frac{L_pn}{c_v}\Delta f\right)\frac{K}{s} = \Delta f \tag{7.4.10}$$

from which the solution for Δf is

$$\Delta f = \frac{\dfrac{L_pD}{c_v\lambda}K}{s + \dfrac{L_pn}{c_v}K}\Omega \tag{7.4.11}$$

In the steady state, as $s \to 0$, (7.4.11) reduces to

$$\Delta f = \frac{D}{\lambda n}\Omega \tag{7.4.12}$$

which is identical to (7.4.5). Thus, ideally, the nulling system yields the correct linear relationship between Δf and Ω in which the firm constants λ and D are precisely known. n, on the other hand, is susceptible to change by external influences, such as pressure, and electric and magnetic fields. Therefore, the closed-loop scale factor is not quite immune to such disturbances. The parameters in the input block $L_p D/(c_v \lambda)$ of Figure 7.6 are all firm and, as such, do not cause scale factor errors. However, bias errors may be associated with this block resulting from nonreciprocal phase shifts between the cw and the ccw beams as will be explained later. Variations in the scale factor of the ramp generator, and the gain K cancel out in the closed loop. The system operates around $x_\varepsilon \cong 0$, hence, excellent dynamic range and linearity are assured. Most importantly, the relationship (7.4.12) does not depend on the light intensity, detector sensitivity, or open-loop gain which, in principle, would show in the detailed analysis that would relate to Figure 7.7. This property is common to all high-gain nulling sensors.

We now turn to the description of the technical realization of the PNLG.

Technical realization of the PNLG ⊕

A basic form of the PNLG, based on [9], is described in Figure 7.7. Light from a single-frequency laser (1) is split into cw and ccw waves by the beam splitter (2).

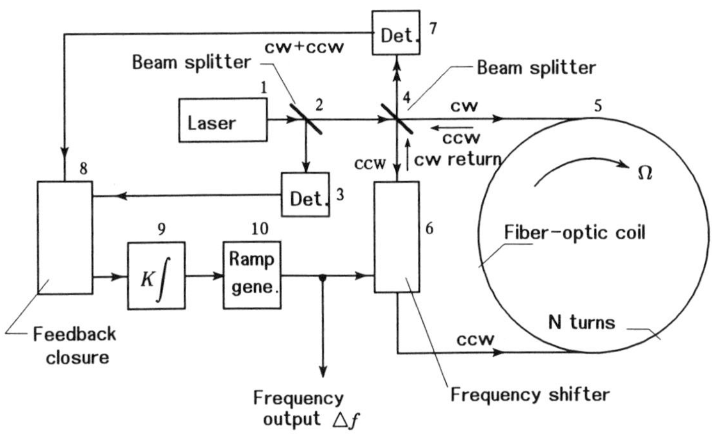

Figure 7.7. Schematic of the realization of a PNLG (based on [9]).

One beam is reflected onto the detector (3), the role of which is to monitor extraneous laser light intensity fluctuations and to regulate them by a gain control amplifier incorporated in (8). The second beam is intercepted by the beam splitter (4) and fed into each end of the fiber coil as two counterpropagating beams. When the coil rotates at inertial angular rate Ω in the paper plane, the two beams exhibit a nonreciprocal phase shift, as explained earlier. Before entering the fiber coil (5), the ccw beam passes through the frequency shifter (6), while the cw beam enters the coil and then passes through (6), which has the following functions. 1) By changing the frequency of one counterpropagating beam with respect to the other, it generates a phase shift $2\pi x_F$ that can be used to null $2\pi x_R$, as explained earlier. 2) It serves to introduce the constant $\pi/2$ phase shift as explained in conjunction with (7.3.9). 3) It can be used to provide the phase dither necessary to determine the Sagnac phase shift as explained in conjunction with (7.3.7) and (7.3.8). For any value Ω, a corresponding signal is detected, the first harmonic of which is proportional to Ω. This signal is fed into a phase-sensitive synchronous demodulator that is also incorporated in (8). There, it is rectified and low-pass filtered, and is available as an analog dc signal. This signal is fed into the integrator (9), the output of which drives the Serrodyne ramp generator (10), whose function is explained in the previous section. The cw+ccw waves, recombined in the beam splitter combiner, are detected by the detector (7) whose output is fed into (8), in which the PNLG loop is closed. The output Δf can readily be represented in digital format by means of suitable clock driven counters similar to those presented in Section 6.7 in connection with the μSCIRAS sensor.

The beam splitters in Figure 7.7 are illustrated as discrete components. In reality, however, they are made as part of an integrated optics component thath also comprises additional functions such as phase modulation, coupling, and polarization. The components are fused to the fiber. This subassembly is referred to as the *multifunction integrated optics chip* (MIOC) [9]. Its primary functions and its integration in the PNLG path are illustrated in Figure

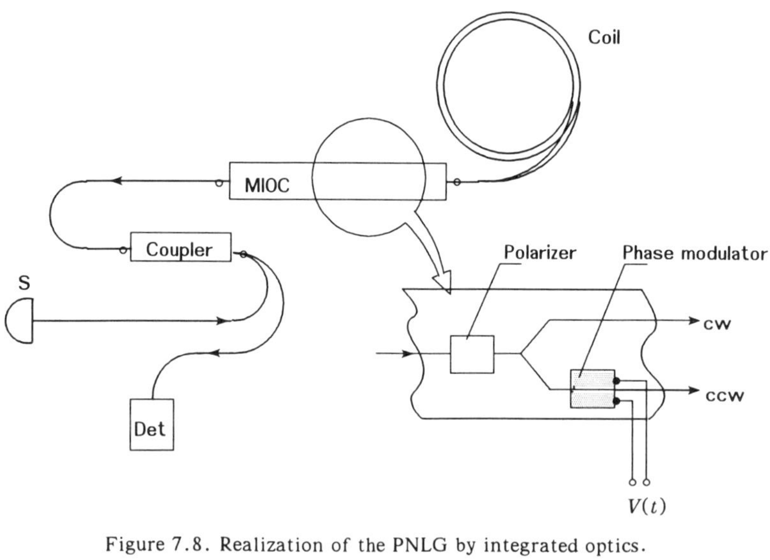

Figure 7.8. Realization of the PNLG by integrated optics.

The key function of the device is the modulation of light. For this purpose, an active electro-optic crystal, most commonly lithium niobate, is employed. By applying an electric field transversely across the fiber, its refractive index is locally varied in proportion to the applied voltage. This technology is representative of the novel field of integrated optics which has evolved during the past decade. It involves the direct processing of light by familiar functions such as modulators, couplers, amplifiers and switches. In essence, it has been the translation of electronics and microwave technology into crystal and fiber-optics technology [10]. The field of applications reaches far beyond the IFOG discussed here. It has evolved as a major technology in modern computer and telecommunication networking.

Sensitivity and resolution

Equation (7.4.44) shows that the frequency shift required to nullify a rotation rate Ω does not depend on the fiber length L_p but rather on the fiber material, namely, its refraction index n, its coil diameter D and the wavelength of the light λ. For example for a glass fiber with $n=1.46$, a wavelength $\lambda=0.84\ \mu$m, and a diameter $D=12.7$ cm, the numerical value of the sensitivity coefficient $\Delta f/\Omega$ is determined by

$$\Delta f=\frac{0.127}{0.84\times10^{-6}\times1.46}\Omega=1.03\times10^{5}\Omega \tag{7.4.13}$$

where Δf is in Hz and Ω is in rad/sec. If, for instance, the PNLG is designed to measure vehicular angular rates Ω up to ±400 deg/sec, the corresponding frequency sweep range $-\Delta f\rightarrow+\Delta f$ required of the ramp generator would be $1.03\times10^{5}\times800$ deg/57.3 $=1.43\times10^{6}$ Hz $=1.43$ MHz. This is an acceptable figure even for standard electronic circuitry.

The precision of the PNLG throughout the range of angular rates $-\Omega\rightarrow+\Omega$ is primarily determined by its resolution, namely, its ability to detect and nullify small fringe shifts x_R. As evident from (7.2.12), the larger the fiber length L_p the larger the corresponding x_R for a given Ω. Thus, L_p is the principal factor in determining the angular rate resolution. According to [9], for a wavelength $\lambda=0.63\ \mu$m, $L_p=4000$ m, and $D=0.127$ m, a resolution of 4.7×10^{-6} fringe has been achieved. In accordance with (7.4.1), the smallest Ω that can be resolved is determined by

$$\Omega_{\min}=\frac{c_v\lambda}{L_pD}x_R=\frac{3\times10^{8}\times0.63\times10^{-6}}{4000\times0.127}\times4.7\times10^{-6}=17.5\times10^{-7}\ \text{rad/sec}$$

$$=17.5\times10^{-7}\times57.3\times3600=0.36\ \text{deg/hr} \tag{7.4.14}$$

More precise experiments, as quoted in [9], with $\lambda=0.84\ \mu$m and the same values for L_p and D, have resolved 2.0×10^{-8} fringe and have yielded

$\Omega_{min} = 0.002$ deg/hr with an averaging time of 1sec. The expected fringe shifts x_R for various values of Ω and the frequency shifts Δf required for their nulling are shown in Figure 7.9 which concisely relates the design parameters L_p, D, n, λ to the PNLG performance parameters Δf and x_R. It is a useful tool in testing the effect of varying a design parameter on performance and in evaluating and validating performance of prototypes and products.

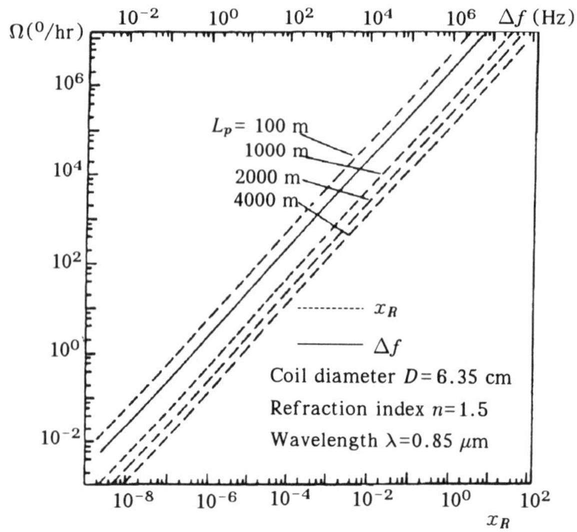

Figure 7.9. Chart relating principal design parameters and variables of a PNLG. (based on [9])

Discrete phase nulling

In order to provide a natural discrete output to represent frequency variations, and in order to facilitate digital interfacing, the continuous sawtooth ramp is replaced by a repeated staircase signal as illustrated in Figure 7.10. The following description is based on [4].

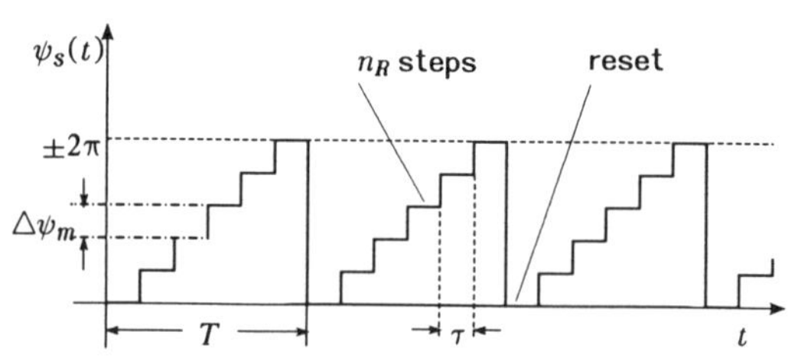

Figure 7.10. Serrodyne realization by a periodic staircase.

T is the reset or 'rollover' period, implying that once ψ has reached 2π, the angle ψ is abruptly rolled back to zero in a ratchet-like manner. τ is the transit time around the coil. $n_R = T/\tau$ is the number of transit times within the reset period T. Let $\Delta\psi_m$ be the required modulator phase shift. Then, at reset we have

$$n_R \Delta\psi_m = 2\pi \tag{7.4.15}$$

For the inertial angular rate Ω, to the angular increment $\Delta\theta$ over the reset period T is

$$\Delta\theta = \Omega T = \Omega\, n_R \tau \tag{7.4.16}$$

In accordance with (7.4.10), which relates Ω to the required Sagnac phase nulling shift,

$$K_s\Omega - \Delta\psi_m = 0 \tag{7.4.17}$$

Solving Ω and substituting into (7.4.16),

$$\Delta\theta = \frac{\Delta\psi_m}{K_s} n_R \tau = \frac{2\pi}{K_s}\tau \tag{7.4.18}$$

where $\Delta\theta$ determines the resolution of the discrete ramp implementation.

For example, for the IFOG of Example 7.2.-1, with a coil length L_p=1000 m, the transit time is $\tau = L_p/c_m = 1000/(2\times10^8) = 5\ \mu$sec. The Sagnac scale factor was found to be K_s=2.706 sec. In accordance with (7.4.18), the resolution of this coil is

$$\Delta\theta = 2\pi\times5\times10^{-6}/2.706 = 11.6\ \mu\text{rad} = 2.38 \text{ arc sec} \tag{7.4.19}$$

Recalling (7.2.10) and that $\tau = L_p/c_m$, we derive the following simpler expression, which naturally yields the same result for $\Delta\theta$,

$$\Delta\theta = \frac{2\pi c_v \lambda L_p}{2\pi L_p D c_m} = \frac{\lambda n}{D} \text{ rad} = \frac{0.82\times10^{-6}\times1.5}{0.106} = 11.6\ \mu\text{rad} \tag{7.4.20}$$

Conclusions:

1) The discrete serrodyne modulator yields, as its natural output, angular increments of the measured attitude angle as is common in the design of digitally implemented INS systems.

2) The resolution of the discrete PNLG, as described, is solely determined by the ratio of the optical wave length λ to the coil diameter D.

Limitations and imperfections

In the process of modeling the IFOG error sources, we have to distinguish between their fundamental physical causes and their technological imperfections. For example, the limited resolution presented in the previous section, or the bounds on precision imposed by the detector shot noise discussed below, are dictated by the laws of physics and are not related tothe quality of the workmanship and technique. On the other hand, delays in the fiber due to extraneous mechanical stresses, electrical fields, or uncontrolled variations in the index of refraction come under the category of technical imperfections, which can be mitigated though, as a rule, at considerable cost. This is often the origin of the cost against quality tradeoffs. In the following section, we review some of the most important error sources.

7.5 Effect of photon shot noise

Effect on Sagnac phase uncertainty

The corpuscular nature of light is the origin of the random nature of the detector current. The impact of photons on the detector manifests itself as an average electrical current i_D on which a randomly fluctuating current i_s, known as *photon shot noise*, is superimposed. The number of discharged electrons per photon is determined by the quantum efficiency η_D of the detector. i_s is naturally interpreted as a random fluctuation in the intensity of light which, in turn is interpreted by the IFOG as a phase angle fluctuation $d\Psi_s$ or, by virtue of (7.2.10), an angular rate fluctuation $d\Omega$. It is the intensity of i_s and its bandwidth b that determine the smallest resolvable change $d\Omega$ and thus the precision of the IFOG. The variance of the shot noise i_s is given by [11]

$$\sigma_{i_s}^2 = 2ei_D b \tag{7.5.1}$$

where

e — electron charge;
b — bandwidth of noise measurement;
i_D — average detector current.

i_s is sampled at the time intervals t_s. According to the sampling theorem [14], the bandwidth of the sampled process is given by

$$b = \frac{1}{2t_s} \tag{7.5.2}$$

Substituting b into (7.5.1), we have

$$\sigma_{i_s}^2 = \frac{ei_D}{t_s} \tag{7.5.3}$$

The electronic noise of the detector and amplifier, namely the dark noise, is disregarded here. In accordance with (2.6.17), ei_D in (7.5.3) is the power spectral density S_w of a white noise process over a bandwidth $b=1/2t_s$. In accordance with (7.3.7), the fringe slope is $dI_D/d\Psi_s|_{\psi_S=\psi^+} = 2I\sin\Psi_s$, which varies with Ψ_s, more specifically with $\Delta\Psi_s = \Psi_s - \pi/2$ or $\pi/2 - \Psi_s$. The transformation from the shot noise i_s to the resulting phase uncertainty $\tilde{\Delta}\Psi_s$

is determined by the fringe slope $dI_D/d\Psi_s$, as shown in Figure 7.11. However, since it varies widely in the range 0 to π, we choose as an equivalent slope the average fringe slope i_D/π shown by the dash-dot line.

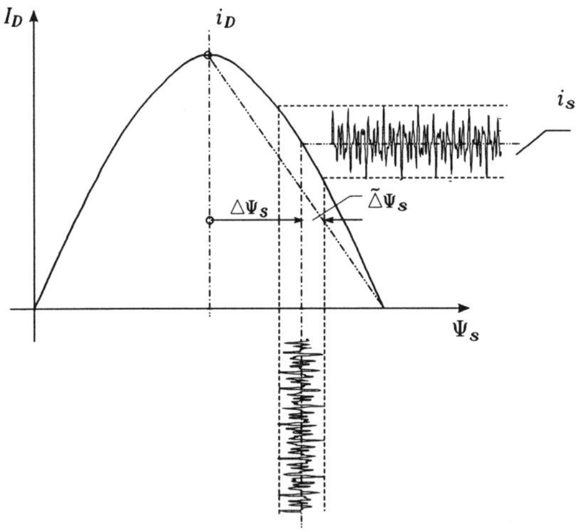

Figure 7.11. Effect of shot noise on phase uncertainty in an IFOG.

The phase uncertainty therefore is

$$\tilde{\Delta}\Psi_s = \frac{\sigma_{i_s}}{i_D/\pi} = \frac{\left(\frac{e i_D}{t_s}\right)^{\frac{1}{2}}}{i_D/\pi} = \frac{(e i_D t_s)^{\frac{1}{2}}}{i_D t_s/\pi} = e^{\frac{1}{2}}(i_D t_s)^{-\frac{1}{2}}\pi \tag{7.5.4}$$

In accordance with (7.2.10), the angular rate perturbation $\tilde{\Omega}$ due to noise is given by

$$\tilde{\Omega} = \frac{\lambda_v c_v}{2\pi L_p D}\tilde{\Delta}\Psi_s = \frac{\lambda_v c_v}{2\pi L_p D} e^{\frac{1}{2}}(i_D t_s)^{-\frac{1}{2}}\pi = \frac{\lambda_v c_v}{2 L_p D}\frac{1}{(i_D t_s)^{1/2} e^{-1/2}} \tag{7.5.5}$$

(7.5.5) can also be expressed more explicitly as follows: The current i_D in the detector is given by

$$i_D = n_{ph}\eta_D e \tag{7.5.6}$$

where,

n_{ph} — number of photons/sec impinging on the detector sensitive surface.

η_D — quantum efficiency of the detector, the average number of electrons emitted for every impinging photon.

Substituting (7.5.6) into (7.5.5), we have

$$\tilde{\Omega}=\frac{\lambda_v c_v}{2L_pD}\frac{1}{(n_{ph}e\eta_D t_s)^{1/2}e^{-1/2}}=\frac{\lambda_v c_v}{2L_pD}\frac{1}{(n_{ph}\eta_D t_s)^{1/2}} \tag{7.5.7}$$

Clearly, t_s must be restricted to reasonably small values to ensure a sufficient bandwidth to cope with the measurement requirements. Accordingly, in aerospace systems, t_s should not exceed the order of $\sim$1sec. $\tilde{\Omega}$ is white noise having a PSD $S_{\tilde{\Omega}}$ and a bandwidth b in accordance with the notations used in (2.6.17).

$$S_{\tilde{\Omega}}=\frac{\sigma^2_{\tilde{\Omega}}}{2b}=\left(\frac{\lambda_v c_v}{2L_pD}\right)^2\frac{e}{i_Dt_s}t_s=\left(\frac{\lambda_v c_v}{2L_pD}\right)^2\frac{e}{i_D}\ (\text{rad/sec})^2\ \text{sec} \tag{7.5.8}$$

or, in view of (7.5.6) and recalling that $K_s=2\pi L_pD/c_v\lambda_v$,

$$S_{\tilde{\Omega}}=\frac{\sigma^2_{\tilde{\Omega}}}{2b}=\frac{\tilde{\Omega}^2}{2b}=\left(\frac{\lambda_v c_v}{2L_pD}\right)^2\frac{1}{(n_{ph}\eta_D t_s)}t_s=\left(\frac{\pi}{K_s}\right)^2\frac{1}{(n_{ph}\eta_D)} \tag{7.5.9}$$

The sampling interval t_s cancels out in the PSD of $\tilde{\Omega}$ because from (7.5.7), $\tilde{\Omega}^2$ is inversely proportional to t_s while its PSD is proportional to it.

The resulting angular error denoted by $\tilde{\theta}$ is given, in accordance with (2.7.36), by

$$\sigma^2_{\tilde{\theta}}=S_{\tilde{\Omega}}t==\left(\frac{\pi}{K_s}\right)^2\frac{1}{(n_{ph}\eta_D)}t\ \text{rad}^2 \tag{7.5.10}$$

so that

$$\sigma_{\tilde{\theta}}=\frac{\pi}{K_s}\frac{1}{\sqrt{n_{ph}\eta_D}}\sqrt{t}\ \text{rad} \tag{7.5.11}$$

EXAMPLE 7.5–1: Angular error propagation due to shot noise in an IFOG

Given the same coil and light data as in Example 7.1–1, we have $K_s=2.706$ sec. With e.g., $n_{ph}=3\times10^{15}$ photons/sec, and $\eta_D=0.3$ and $t_s=$ 1sec, [3]. From (7.5.9),

$$\sigma_{\tilde{\Omega}}=\sqrt{\tilde{\Omega}^2}=\frac{\pi}{K_s}\frac{1}{(n_{ph}\eta_D t_s)^{1/2}}=\frac{3.14}{2.706}\frac{1}{(30\times10^{14}\times0.3\times1)^{1/2}}=1.4\times10^{-8}\ \text{rad/sec}$$

$$=2.88\times10^{-3}\ \text{deg/hr}$$

The resulting angular divergence with time is, in accordance with (7.5.11),

$$\sigma_{\tilde{\theta}}=\frac{\pi}{K_s}\frac{1}{(n_{ph}\eta_D)^{1/2}}\sqrt{t}\ \text{rad}=4.32\times10^{-3}\ \text{deg/sqrt(hr)}\sqrt{t}$$

Consequently, the expected divergence of $\sigma_{\tilde{\theta}}$ after e.g., 17 hrs, would be

$$\sigma_{\tilde{\theta}}=4.32\times10^{-3}\sqrt{17}=1.78\times10^{-2}\ \text{deg}.$$

This result demonstrates that shot noise for the high-quality IFOG of Example 7.2–1 is not a major error source. However, for a lower-grade IFOG, say with L_p = 100 m and $D=0.05$ m, $\sigma_{\tilde{\theta}}$ would amount to 0.178 deg. Such an IFOG could still be used in conventional attitude heading reference systems (AHRS).

Sources of bias errors

We have seen earlier that a phase modulator is required to turn the basic Sagnac interferometer into a practical rotation sensor. For the $\pi/2$ phase shift, a minute increase of only a quarter wavelength $\lambda/4$ in the difference between the effective path lengths of the cw and ccw beams is required. Any unaccounted-for error $\tilde{\Psi}_s$ in this $\lambda/4$ shift manifests itself as a bias error that translates into $\tilde{\Omega} = \tilde{\Psi}_s / K_s$. Such phase errors, which are the result of differences in the cw and the ccw beams, are termed *nonreciprocal* phase errors. Common sources of such errors are reviewed in Chapter 12 of [7] and in Chapter 5 of [3] and are briefly summarized here. All these errors, including the one due to shot noise, are *additive*. Therefore, in view of the properties of forced balance systems (FBS) described in Chapter 1, closing the phase loop has only a minor effect on the magnitude of the bias error of the IFOG.

1. Power losses: In (7.5.14), it was assumed that $E^+ = E^- = E$. In the actual realization, the bonding between fiber sections and electro-optical components may involve partial reflections in either of the optical paths so that $E^+ \neq E^-$. This can induce phase error and thus a bias error in Ω. This error may be temperature-sensitive and thus not easily compensable.

2. Kerr effect: As mentioned earlier, the refractive index n is affected by an electric field. If applied purposely, it serves a means for the desired phase modulation. However, if it is the result of a stray time-varying electric field, it may cause an extraneous bias in ϕ and in Ω as a result of nonreciprocal phase shift, which may be quite difficult to cancel.

3. Strain in the fiber: One of the methods referred to in the literature [7] for achieving phase modulation, is by purposely stretching and relaxing the fiber as a function of time. The strain in the fiber, which causes a change in the refractive index n and thus in ϕ, is generally temperature-dependent. Thus, if a time-dependent temperature gradient across the fiber exists, nonreciprocity may occur, resulting in a temperature dependent bias error.

4. Acousto-optic effect: Sound waves can affect the performance of a Sagnac interferometer by the pressure-induced changes of the index of refraction n. As in the previously mentioned induced error sources, the effect can be beneficially exploited as in the case of the Bragg cell which changes the frequency of a light wave and, in general, can be applied in acousto-optic phase modulation. For high-precision applications, the acousto-optic interference can be substantial so that acoustic isolation is required.

5. Magnetic effect: An IFOG is influenced by external magnetic fields, causing an increased bias instability with increasing magnetic field strength. In an experimental laboratory set-up, it has been demonstrated [3] (p. 337) that a bias uncertainty in Ω of about 10 deg/hr can occur if the fiber coil is exposed to Earth's magnetic field. If the coil is magnetically shielded, this bias error can be decreased by a factor of ~ 10. This still leaves a substantial bias uncertainty of ~ 1 deg/hr which is hard to compensate because the projection of the magnetic field in body axes varies with time, position, and the vehicle orientation.

Conclusion: All the foregoing phase errors denoted by $\tilde{\Psi}$ are additive and they relate to the corresponding angular rate error $\tilde{\Omega}$ via $\tilde{\Omega} = [c_v\lambda/(2\pi L_p D)]\tilde{\Psi}$ (7.5.10). Therefore, for a given $c_v\lambda$, $\tilde{\Omega}$ is inversely proportional to the subtended IFOG area $A = L_p D$. This relationship is somewhat analogous to the property of mechanical gyros, in which additive angular rate errors induced by extraneous torques are inversely proportional to the angular momentum $H = I\omega$.

Recent IFOG test results

IFOG technology has reached a stage of maturity that by now makes it available for very high-grade strapdown rotation sensors in a class better than 0.01 deg/hr and a dynamic range better than 10^8. A recent publication by Litton industries [12] summarizes extensive and meticulous tests whose results attest to the high degree of perfection that has been achieved. Some of these are shown below. Navigation grade LN-250 IFOGs incorporating a fiber of $L_p = 1000$ m packaged into a diameter less than 3 in were tested on a high-precision rate table mounted on a seismic isolation pier. A computer-controlled oven enclosed the rotating rate able platform. All data were acquired at intervals from 1 to 60 sec under computer control. Advanced data analysis techniques were applied to allow the separation of different error sources and to assess residual errors persisting after compensating the IFOG output. Thermal models were established and applied to bias, scale factor, and input axis alignment data to assess the residual errors after compensation. The residual bias uncertainty versus temperature variation is shown in Figure 7.12.

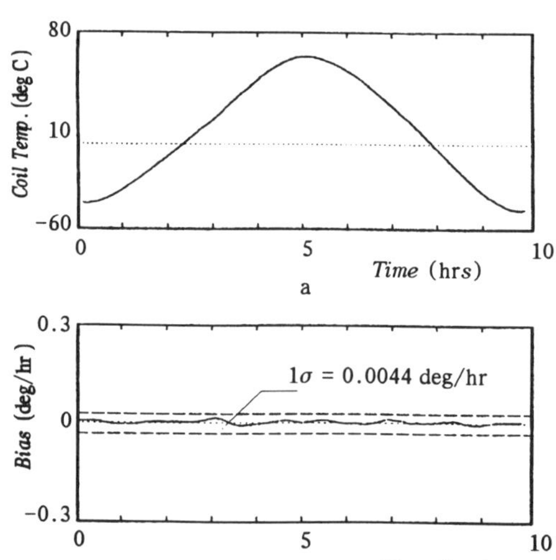

Figure 7.12(a) Temperature profile at gyro coil over 10 hours; (b) residual bias uncertainty (based on [12]).

Bias data were collected in 10 sec samples over the full military temperature range of −55 to +71 deg C. After stabilizing at −55 deg C the oven temperature was ramped up at +0.5 deg C/min to +71 deg C, stabilized, and then ramped back down to −55 deg C. The bias residual data shown in Figure 7.12− b were generated by applying a thermal model to the raw bias data and after applying a 30 min filter to reduce the angle random walk (ARW) effects. The resulting bias standard deviation was found to be 0.0044 deg/hr.

The scale factor data were taken once every minute with the IFOG rotating at 100 deg/sec and reversing the rotation direction every 15 revolutions. During the eighteen hour test, the temperature was ramped from +71deg C at rates of ±0.5 deg C per minute, as shown in Figure 7.13(a). The residual scale factor variation versuss temperature after modeling is shown in Figure 7.13(b). The compensated scale factor was found to have a standard deviation of 9.1 ppm after applying a 5 min filter to the data.

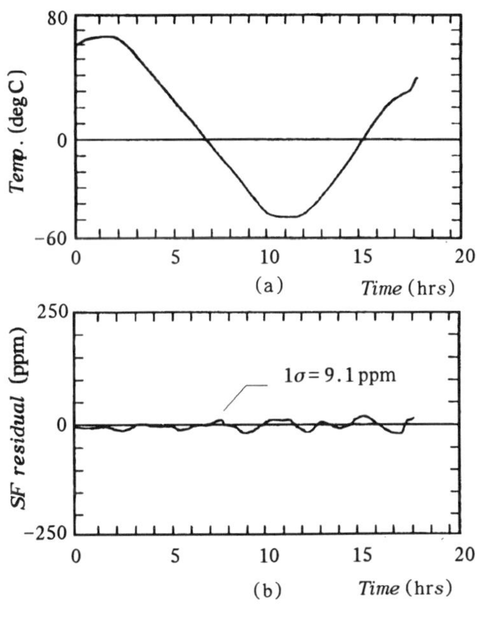

Figure 7.13(a) Temperature profile at gyro coil over 20 hrs;
(b) scale factor error residual.

In addition to the temperature tests, nine IFOGs were subjected to vibration tests at 25 deg C. The average performance did not exceed 80% of the 0.01 deg/hr specification. The above results reported in [12] indicate that INS-grade IFOG technology is apparently very close to production.

The key to achieving the required high degree of precision and repeatability is the stability of the materials that go into the IFOG and its manufacturing process. It is the firmness of the characteristics that permit reliable calibration and compensation.

Another recent publication by Honeywell industries [13] addresses the stochastic modeling of IFOG errors due to noise. This work provides insights into the relationships between key gyro parameters and residual noise and bias. These noise models and the IFOG response to random inputs are important in evaluating the overall INS performance for various missions.

Problems

Problem 7.1— Optimal fiber length

It has was shown in section 7.2 that the fiber index of refraction n has no on the effect IFOG sensitivity K_S, which is proportional to the fiber length L. Another property of the fiber material is its exponential attenuation of the light intensity along the fiber length expressed by $I(l) = I_0 e^{-\alpha l}$. $I_0 = I(0)$, α is the fiber attenuation coefficient, and l is the distance along the fiber.

a) Is the performance of the IFOG affected by this property of the material? If so, how does it manifest itself in the quality of the IFOG?

b) Does the photon shot noise come to bear in your considerations? Derive an expression that determines the appropriate performance parameter. (Define the signal-to-noise ratio.)

c) In view of your answers to the foregoing, is there an optimal fiber length?

d) If the answer is yes, derive an expression for this optimal length, and demonstrate the importance of achieving the smallest possible value for α.

Problem 7.2— Equality of light intensity in the cw and ccw light paths

In Section 7.3, an expression is derived for the sum of the cw and ccw light intensities. It has been assumed there that the corresponding light intensities E^+ and E^- at the optical combiner are equal.

a) Is the assumption $E^+ = E^-$ always justifiable?

b) Assume now that they are different by a fraction δ, namely, $E^+ = E^-(1+\delta)$. Derive the expression for I in accordance with (7.3.3), in which δ is a parameter.

c) How, if at all, does δ affect the fringe effect as expressed in (7.3.5) and (7.3.6)?

Problem 7.3— Resolution of a discrete PNLG IFOG

It is required to design a discrete PNLG IFOG with an angular resolution $\Delta\theta = 2$ arcsec. The index of refraction of the fiber is $n = 1.6$ and the light wave length λ is 820 nanometers.

a) What is the diameter D of the fiber coil?

b) Does D dictate the smallest resolvable angular rate Ω_{min}? Support your answer by the relevant relationships in this chapter.

c) With this value of D and a resolution of 10^{-6} fringe, what is the fiber length L_p required to resolve $\Omega_{min} = 0.2$ deg/hr?

Problem 7.4— Dimensional design tradeoffs

In the design of an IFOG, D and L_p are principal parameters.

a) Show how they decide the scale factor coefficient relating the input Ω and the output in terms of the fringe Δf.

b) It appears that it is just the product L_pD that counts. Bring arguments in favor of a large L_p and a small D and vice versa.

c) What determines Ω_{max}? Develop an expression for the IFOG's dynamic range D_r.

References

1. Post, E.J., "Sagnac Effect", *Review of Modern Physics*, Vol. 39, April, 1967, pp. 475–493.,

2. Macek, W.M., and Davis, D. T. M., "Rotation Rate Sensing with Traveling Wave Ring Lasers," *Applied Physics Letters*, 2, February 1, 1963.

3. Ezekiel, S. and Arditty, H. J., *Fiber Optic Rotation Sensors*, Springer Series in Optical Sciences, Springer-Verlag, New York, 1982, p. 7.

4. Matthews, A., "Utilization of Fiber Optic Gyros in Inertial Measurement Units," *Journal of the Institute of Navigation*, Vol. 37, No. 1, Spring 1990, pp. 17–38.

5. Arditty, H.J. and Lefèvre, H.C., "Sagnac Effect in Fiber Gyroscopes," Optics letters, Vol. 6, No. 8, April 1981 pp. 401-403.

6. Ezekiel, S., Smith, S.P., and Zarintetchi, F., "*Basic Principles of Fiber-Optic Gyroscopes*", Chapter 1, Academic Press, New York, 1994.

7. Lawrence, A., *Modern Inertial Technology, Navigation, Guidance, and Control*, Springer-Verlag New York, 1993.

8. Cummings, R.C., "The Serrodyne Frequency Modulator," *Proceedings IRE*, Vol. 3, 1957.

9. Chalil, R.F. and Udd, E. "Phase Nulling Fiber-Optic," *Optics letters*, Vol. 4, No. 3, March 1979.

10. Yariv, A., "Guided Waves Optics," *Scientific American*, Vol 240, No. 1, January 1979, pp. 64–72.

11. Davenport, W.B., and Root, W.L., *An Introduction to Random Signals and Noise*, McGraw-Hill New York, 1958

12. Pattterson, R.A., Rozelle, A.C., and Goldner, E.L., "Inertial Navigation IFOG Performance over Static and Dynamic Environments," AIAA Conference on Guidance, Navigation, and Control, Phoenix, AZ, August, 1994.

13. Bielas, M.S., "Stochastic and Dynamic Modeling for Fiber Gyros," *SPIE*, Vol. 229, July 1994.

14. Franklin, G.F., Powell, J.D., and Workman, M.L., *Digital Control of Dynamic Systems*, Addison-Wesley, Reading, MA, 1990.

8

The Ring Laser Gyro

8.0 Introduction

This chapter introduces the ring laser gyro, which is considered one of the most revolutionary twentieth century developments in the aerospace sensor field. It appeared on the scene in the early 1960s and since then has matured as a dominant product in the aerospace industry, having accumulated millions of successful operational flight hours in military and commercial aircraft.

The often quoted paper by Killpatrick [1], is a comprehensive tutorial that explains the principle of operation and the basic properties of this device. As in many later publications on the subject, the author explains that the RLG is really not a gyro in the sense of the conventional concept of a spinning mass. However, as pointed out in [1], the original meaning of the Greek term "$\gamma\upsilon\rho o s$" is *ring*, which, linguistically, may be more appropriate than the conventional connotation of a spinning mass. The outstanding property of ring laser gyros (RLG) is that there are no moving and sliding components. The sensitive element consists of a pair of colocated, closed path clockwise and counterclockwise light beams. Their interaction causes interference, which manifests itself as fringes resulting from the corresponding changes of the wavelengths and oscillating frequencies. These changes are proportional to the inertial angular rate. The intrinsic readout is by counting the pulses due to the fringes. This constitutes a natural interface with the digital processor and computer. This development has significantly contributed to the practical implementation of guidance and inertial navigation systems, especially of strapdown mechanization.

8.1 Operating principle

Basically, the RLG is a Sagnac interferometer. Thus, it has much in common with the IFOG and RFOG. The common denominator is the interferometry. The principal difference is that unlike the passive IFOG, it is an active device essentially operating like an electro-optical oscillator. The device is a laser that incorporates three or more reflectors, arranged so that the light beams form a polygon enclosing an area A. The trianglular shape is lowest cost, but it encloses the least area A for a given size and is therefore the least sensitive. A comparative geometrical analysis of different polygonal shapes is given in [2]. The reflecting mirrors, together with the light-amplifying medium in the light path, constitute an oscillator. There are two such oscillators, each propagating light energy along the same path but in opposite directions. The frequencies at which these oscillators operate are determined by the length of the optical path. For the oscillation to be sustained, as in any oscillator, two conditions must be fulfilled: 1) the overall gain around the loop must be unity at a power level determined by the amplifying medium, and 2) the phase shift around the loop must be zero. This condition implies that the number of light wavelengths must be an exact integer, namely, the wavelength must be an exact integer fraction of the path length. This condition is actually the one that determines the oscillation frequency of the laser. Further details can be found in [3] and [4]. If the ring laser is stationary in inertial space, the cw and ccw beams oscillate at exactly the same frequency. However, if put into rotation, a slight difference in the cw and the ccw path lengths occurs in the same manner as in the passive IFOG. Consequently, the corresponding wavelengths and thus the oscillating frequency will differ. This difference is proportional to the inertial rotation rate Ω of the RLG. The measurement of this difference, which is the essence of the operation of the RLG, is by monitoring this frequency difference.

Description

A schematic description of the basic operating elements of a triangular RLG is given in Figure 8.1. A closed cavity in a polygonally shaped Zerodur block contains the two counterrotating single-frequency light beams that are generated from the lasing action of the Helium–Neon gas discharge within the optical cavity. Zerodur is a ceramic glass that has an extremely low thermal expansion coefficient. Thus, the dimensions of the RLG are very firm constants which are vital to maintain a constant lasing frequency. The gas is typically a mixture of one part of neon to ten parts of helium at a total pressure of 1–10 mm of mercury. The reflecting surfaces are dielectric mirrors designed to selectively reflect the oscillation frequency of the Helium Neon gas mixture. The mirrors require high-quality optical seals to prevent any contaminants from leaking into the gas mixture, thus degrading performance and shortening life-time. Since the apparent path lengths of the cw and ccw waves differ slightly, and since the corresponding light wavelengths λ_{cw} and λ_{ccw} must be N integer fractions of the path lengths L_{cw} and L_{ccw}, it follows that the corresponding oscillating frequencies f_{cw} and f_{ccw} must differ too. N is typically in the range of 10^5–10^6. The path length difference ΔL is related to Ω by the Sagnac effect, given by

$$\Delta L = \frac{4A}{c_m}\Omega \tag{8.1.1}$$

which is in close correspondence with (7.2.6). As before, A is the enclosed area and c_m is the velocity of light in the medium. With the RLG at rest, i.e., $\Omega = 0$, we have

$$\lambda = \frac{L}{N} \tag{8.1.2}$$

The increment in wavelength is

$$\triangle\lambda = \frac{\triangle L}{N} \tag{8.1.3}$$

and in view of the change $\triangle\lambda$ caused by $\triangle L$, the corresponding relative change in frequency is

$$\frac{\triangle f}{f} = \frac{\triangle L}{L} \tag{8.1.4}$$

Substituting $\triangle L$ from (8.1.1), we have

$$\triangle f = \frac{\triangle L}{L} f = \frac{4A\Omega}{c_m L} f = \frac{4A}{\lambda L}\Omega \tag{8.1.5}$$

The cw and ccw beams at their corresponding frequencies f and $f+\triangle f$ are combined, as indicated in Figure 8.1 by the readout optics.

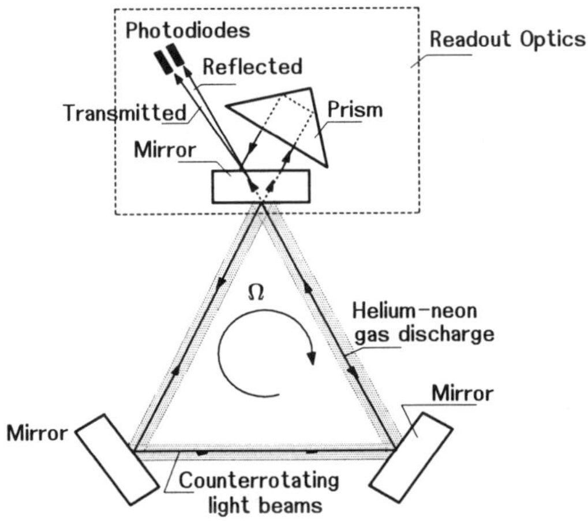

Figure 8.1. Schematic description of the ring laser gyro.

$\triangle f$ is measured by allowing a small fraction of the light to escape through the mirror, which is part of the readout optics. The prism is designed to reflect one of the beams so that it partially overlaps the other by making it inclined by a very small angle, referred to as the *wedge angle*. Due to the finite width of the beams, the wedge angle permits partial merging of the two beams.

The interference fringe pattern

This merging, which is just a summation, creates a fringe pattern in the readout zone. The spatial light signals are $E_{cw} = E_1 \sin 2\pi f t$ and $E_{ccw} = E_2 \sin 2\pi(f+\triangle f)t$.

Assuming that $E_1 = E_2 = E$, and since $\triangle f \ll f$, their sum x is

$$x = E[\sin 2\pi ft + \sin 2\pi(f+\Delta f)t] = 2E\sin\frac{1}{2}2\pi(2f+\Delta f)t\cos\frac{1}{2}(2\pi\Delta ft)$$

$$\cong 2E\sin(2\pi ft)\cos(\pi\Delta ft) = 2E\sin(2\pi ft)\cos\left\{\pi\frac{4A}{\lambda L}\Omega t\right\} \qquad (8.1.6)$$

which consists of the carrier signal at frequency f modulated by $\cos(\pi\Delta ft)$. It is required to express the combined readout in terms of its intensity, namely, $I = x^2$. We therefore have

$$I = 4E^2\sin^2(2\pi ft)\cos^2\left\{\pi\frac{4A}{\lambda L}\Omega t\right\} = 4E^2\frac{1-\cos 4\pi ft}{2}\;\frac{1+\cos\left\{\frac{2\pi 4A}{\lambda L}\Omega t\right\}}{2}$$

$$= E^2[1-\cos(4\pi ft)][1+\cos(\frac{8\pi A}{\lambda L}\Omega t)]$$

$$= E^2[1-\cos(4\pi ft)][1+\cos(2\pi\gamma\theta)] \qquad (8.1.7)$$

where

$$\gamma = \frac{4A}{\lambda L} \qquad (8.1.8)$$

is the spatial angular frequency and $\theta = \Omega t$ is the angle through which the gyro has rotated. γ is exactly the constant that translates Ω to Δf in accordance with (8.1.5). The so-called period is

$$\Pi = \gamma^{-1} = \frac{\lambda L}{4A} \qquad (8.1.9)$$

If the RLG were annular in shape, (8.1.9) would be given by

$$\Pi = \frac{\lambda}{D} \qquad (8.1.10)$$

which is exactly the expression for resolution of the circular IFOG, as given in (7.4.20). This result is indeed to be expected because the angular period Π, namely, the length of the standing wave, is exactly the natural unit of the resolution.

In (8.1.7), the first term in brackets is non-negative and represents the average light intensity at frequency f. The second non-negative term in brackets expresses the 100% modulation caused by the interference between the two combined counterrotating beams and thus represents the fringes. They appear only if $\Omega \neq 0$ and their angular frequency is proportional to Ω. This means that at a given point in the rotating RLG axis system, say the reflection point from the readout mirror seen by the photodiodes, the periodic light intensity variation behaves as if an inertially fixed annular standing wave outside the RLG is moving across an open window as the RLG rotating at the inertial angular rate Ω. This interpretation is depicted in Figure 8.2 as an inertially fixed ring on which a periodic light intensity wave having the angular period Π is superimposed and that is sensed by the RLG which rotates at the angular rate Ω. Digital integrated angular rate increment pulses are generated by the pair of diodes depicted in Figure 8.1. The operation of the RLG from the point of view of the user is similar to a digital rotational encoder in which the optical bar-code is fixed in inertial space. The function of the electro-optical pickoff is essentially the same. Any drift of the RLG manifests itself as a slow rotation of the annular standing wave with respect to inertial space.

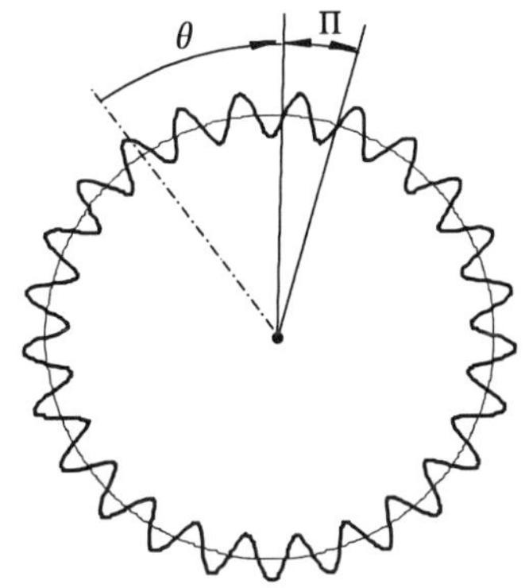

Figure 8.2. Annular standing wave interpretation of the RLG.

As the intensity fringes cross a diode, a corresponding sinusoidal signal is generated. Two diodes are required in order to determine whether Ω is clockwise or counter-clockwise. The diodes are placed at 90^0 with respect to each other in the coordinates of the fringe angle γ. With this arrangement, it is possible to determine which of the two diodes precedes the other. Digital pulse conditioning and logic convert the sinusoidal outputs into a train of pulses whose frequency is directly proportional to Ω and whose accumulated count represents the angle θ. The pulse size depends on the wavelength λ and the path length between mirrors. Thus, the RLG, in conjunction with its output logic, performs as a digital rate integrating gyro. As such, its null point is always dictated by the point at which the θ counter is initialized. This is precisely what is needed to implement a strapdown digital attitude reference system, commonly used in modern inertial navigation units.

EXAMPLE 8.1–1: Distance error propagation to limited resolution

The dimensions of a ring laser gyro are as follows: $L = 30$ cm; $A = 87$ cm^2; $\lambda = 0.63\ \mu$. In accordance with (8.1.9), the period Π or the resolution is

$$\Pi = \frac{0.63\times10^{-6}\times30\times10^{-1}}{4\times87\times10^{-4}} = 54.3\ \mu\text{rad} = 11.2 \text{ arc sec}$$

This zero-mean resolution error, which may be either positive or negative, can be modeled as a white noise process having an rms value of $\Pi\sqrt{2}$ and a bandwidth determined by the sampling interval t_s. The corresponding PSD is therefore given by

$$S_{\Pi\Pi} = \frac{\Pi^2}{2} t_s$$

The corresponding acceleration PSD is given by

$$S_{xx} = \frac{g^2\Pi^2}{2} t_s$$

The accumulated positional variance is

$$V_{xx} = S_{xx}\frac{t^3}{3}$$

and the resulting positional rms value for t_s= 1 sec, and throughout 1hr of operation is determined in accordance with (4.7.37) by

$$\sigma_x = \sqrt{\frac{S_{xx}}{3}}\, t^{\frac{3}{2}} = \frac{g\Pi}{\sqrt{6}} t_s^{\frac{1}{2}} t^{\frac{3}{2}} = \frac{9.81\times 54.3\times 10^{-6}}{\sqrt{6}} \times 1^{\frac{1}{2}} \times 3600^{\frac{3}{2}} = 47 \text{ m}$$

The conclusion is that for this level of resolution this minor additional error is negligible in comparison with the total errors of 1 n mi typical in high-performance INS.

8.2 Technical description

Spontaneous oscillation

A technical sketch of an RLG is shown in Figure 8.3. The Ne–He gas-filled cavity in the Zerodur block is equipped with a cathode and two anodes and supplied with a high voltage of more than 600 volts which causes ionization of the gas. The ionized gas provides the conditions for spontaneous emission of photons in an avalanchelike process, thus constituting a source of energy. This source provides the required amplification, which more than compensates for the energy losses in the optical system. This is not unlike the amplification of a vacuum tube which can burst into oscillation by virtue of the energy drawn from the power supply. If this gain indeed provides energy in excess of the energy losses to the cavity walls, the mirrors, and unexcited atoms in the medium, light waves will be generated traveling in both directions. These waves generate a standing wave whose length, as stated earlier, must be an integer fraction of the cavity length. The precision of the RLG critically depends on the effect of possible variations in the laser cavity geometry on the lasing beams.

Path length control

An essential requirement in this respect is that the average path length around the lasing polygon, in this case a triangle, be constant to ensure lasing at peak average power. This condition of peak power corresponds to the lasing frequency f, which is centered at the peak of the He–Ne gas discharge Doppler gain curve [5]. Because of this stringent requirement for path length precision, in addition to the inherent stability of the Zerodur material, a closed-loop control of the path length is incorporated in the laser system. As indicated in Figure 8.3, a power photodetector is located near the right hand side mirror. It measures a small fraction of the radiation of both the cw and ccw beams. The signal from a discriminating function derived from the power detector drives the path length transducer at the left-hand side of the laser block. Thus, the laser is controlled to operate at its peak power, assuring the constancy of the lasing frequency at rest.

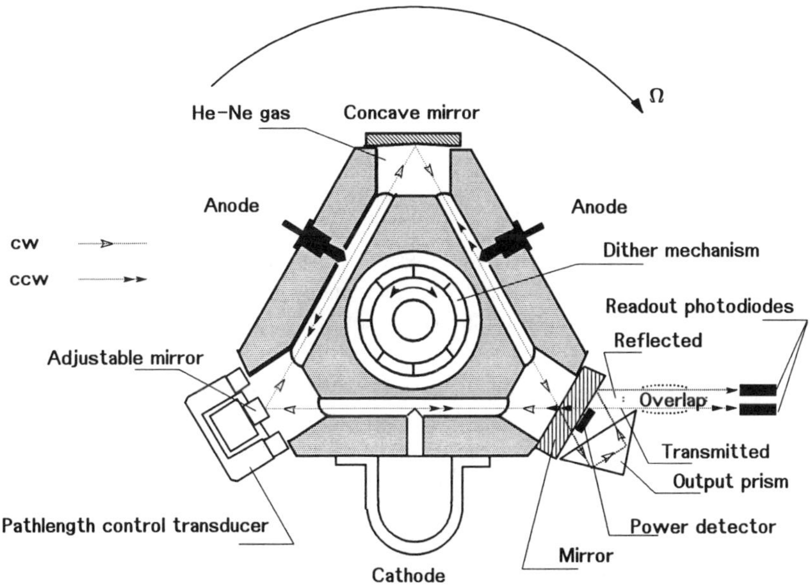

Figure 8.3. Technical sketch of a triangular RLG including functional, control, and dither components.

8.3 The Lock–in phenomenon

The effect of lock–in on the RLG characteristic

Unfortunately, at very low angular rates Ω, the frequencies of the cw and ccw beams pull together so that $\Delta f = 0$. This phenomenon, known as *lock–in*, results from imperfections in the lasing cavity, in particular the mirrors, which produce backscattering of light energy from the cw to the ccw beam, causing unwanted coupling between them. Lock–in is a familiar phenomenon also encountered in mechanical or electronic oscillators if closely located and when operating at slightly different frequencies. Clearly, if $\Delta f = 0$, the RLG output is zero. At a certain critical angular rate Ω_l the two oscillations pull apart so that the RLG response restarts.

The effect is a significant dead zone which can be as wide as $\cong 0.1$deg/sec. The first simple remedy that comes to mind to overcome lock in, is to rotate the RLG by some large constant biasing Ω_b so that the output would be $\Omega+\Omega_b$ thus moving the operating point way out of the lock-in zone. Ω_b needs to be removed from the output to retain the symmetry of the RLG characteristic. With angular rates in the order of Ω=400 deg/sec common in military aircraft, and a null-point precision of 0.01 deg/hr required in INS, this implies a precision of 10^{-8} in Ω_b. This difficulty can be overcome by mechanical rotational dither which periodically switches the sign of Ω_b so that its average is not affected. The solid line in Figure 8.4 is a typical characteristic of a dithered RLG relating a steady angular rate Ω to Δf as a function of the dither amplitude α_d. In the neighborhood of the dead zone Δf can expressed as function of Ω by

$$\Delta f = 0 \qquad \text{if } \Omega^2 \leq \Omega_l^2 \tag{8.3.1a}$$

$$\Delta f = (4A/L_p\lambda)\Omega\sqrt{1-\frac{\Omega_l^2}{\Omega^2}} \qquad \text{if } \Omega^2 > \Omega_l^2 \tag{8.3.1b}$$

Equation (8.3.1) does not reflect the physics of the lock-in phenomenon. It is a mere description of this observed phenomenon. However, it is useful for examining the RLG response in the region of the lock-in zone formulated by (8.3.1a) and dfor estimating scale factor errors for small values of Ω. For $\alpha_d \rightarrow \infty$, Δf re assumes the linear RLG relationship (8.1.5) indicated by the dotted line in Figure 8.4, corresponding to (8.3.1b).

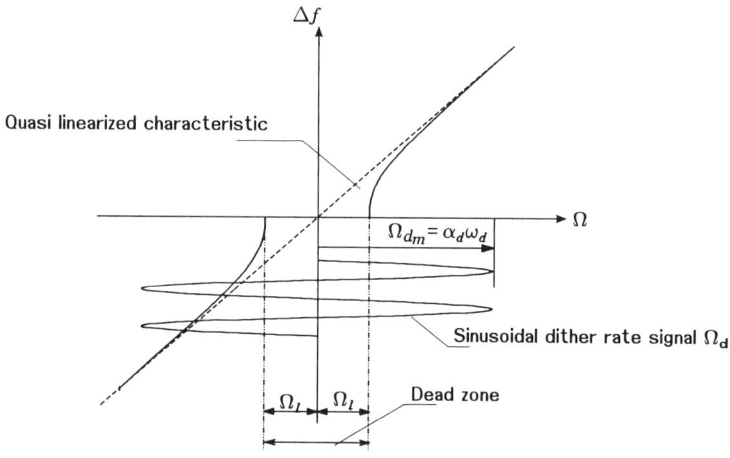

Figure 8.4. RLG response characteristic showing dead zone due to the lock-in phenomenon.

The lock-in phenomenon, which was observed at the beginning of the RLG development as early as 1963 [6] would have rendered it a useless device. The lock-in rate Ω_l is found to depend on the wave length λ, the coupling factor r, and the enclosed area A. In order to reduce Ω_l, it is desirable to reduce λ and r, while increasing A. In other words, the bigger the RLG, the shorter the wavelength, the better the cavity and the mirrors, the smaller is Ω_l. However, even with the most meticulous technology and the largest acceptable size, it is unfeasible to reduce Ω_l below 0.01 deg/sec = 36 deg/hr.

This is unacceptable in high grade INS technology. Typically, $\Omega_l = 0.1$ deg/sec. As mentioned above, the obvious remedy is by sustained periodic rotational dither. This has been, and still is, common practice.

Prevention of lock-in by mechanical dither

In theory, any shape for the dither wave form, including a square wave, is admissible. However, because of mechanical limitations, a sinusoidal wave form is always used. Thus, a sinusoidal signal drives the dither mechanism which is shown in Figure 8.3. Its spoked wheel-like structure, which acts like a torsional spring, is driven by a piezoelectric transducer. It causes the entire RLG to execute a small rotational periodic motion at the mechanical resonance frequency of the RLG mount, typically at a few hundred Hz. Thus, it is put in motion for almost all the time. For example, if the dither amplitude $\alpha_d = 0.1$ deg and the dither frequency $\omega_d = 400$ Hz, the peak dither angular rate is $\Omega_{dm} = 800\pi \times (0.1\,\text{deg}/180\,\text{deg}) \times \pi = 14$ rad/sec = 800 deg/sec, which is three orders of magnitude larger than Ω_l. Under such conditions, the output periodically swings out of the dead zone. If, indeed, $\Omega_{dm} \gg \Omega_l$, this will be the case for most of the time, so that the dominant portion of the characteristic is virtually linear. Linearization by dither has been known and used by control system engineers since the early fifties [7]. The input angular rate Ω adds to Ω_d and causes the dither to be biased. The resulting asymmetry in the net positive and negative pulse counts represents Ω down to very low values in the order of 0.01 deg/hr and less.

The sinusoidal dither motion, however, involves a periodic nulling of Ω_d at twice ω_d whenever the dither amplitude reaches its peak α_d. This apparently permits lock-in instances, causing some loss of Ω information. However, since Ω_d is quite high, the dwell time in the Ω_l zone is so short that lock-in actually has no chance to develop. Thus, the RLG has uninterrupted constant resolution over its entire operating rate range. However, there is a small residual effect due to the lock-in phenomenon. This effect is a small scale factor nonlinearity, usually negligible, resulting from the averaging of the input rate across the lock-in dead zone. Furthermore, each time the RLG block passes through the lock-in threshold, a small random rate noise pulse is generated at twice ω_d. This noise source is integrated by the output pulse counter and causes random walk which, in accordance with Chapter 4, progresses as $t^{1/2}$. So far, the discussion has referred to the averaged angular rates but disregarding the substantial dither signal in the RLG output, which in principle, has no effect on the overall INS performance. Yet, for practical reasons relating to the gyro signal processor and electronics, the dither signal should be removed from the RLG output. Since ω_d is precisely known, this can be accomplished by a digital filter that suppresses the unwanted dither component from the output while somewhat affecting the overall sensor bandwidth.

Removing the dither signal from the output

An alternative method is to cancel the dither signal altogether at its source. This can be accomplished as follows. Normally, the pair of photodiodes and the gyro optics block along with the output prism constitute a common solid unit. Thus, if this unit rotates, say, clockwise, the combined fringe motion including the dither, appears to the photodiodes as moving in the opposite sense, as if the fringes were a stationary pattern with respect to inertial space. However, if the photodiodes are mounted on the nondithered gyro base, the photodiodes appear to perform exactly the same dithered motion as the fringes with respect to the output prism. This clearly implies complete cancellation of the dither signal at its origin. The detection of the desired input angular rate Ω is obviously unaffected because the diodes rotate with respect to inertial space at the rate Ω while the fringes, which result from a standing wave fixed in inertial space, do not.

In spite of the generally successful performance achieved by mechanical dither, it has never been a completely satisfactory solution. On the level of sound design principles, it contradicts the concept of a motionless gyro and the notion of elegant engineering. On the practical level it introduces an additional noisy device which may adversely affect reliability. Therefore, since the 1970s, continuing efforts have been devoted to remove the lock-in problem without resorting to any form of mechanical dither.

Alternative methods for lock-in compensation

A review of some principal approaches is given in [8]. These include the use of a Faraday cell placed within the laser cavity. By applying a magnetic field across the cell, a differential shift of the index of refraction is created between the cw and ccw laser beams, thus producing a differential change in the corresponding optical path lengths of the two beams. Consequently, corresponding changes in the oscillation frequencies take place in accordance with the frequency and amplitude of the applied magnetic field. Clearly, this is equivalent to the mechanical dither. An important advantage is that the dither signal must not necessarily be sinusoidal. With a square wave dither, there is only a negligible dwelling time in the lock-in zone. Thus, the noise associated with crossing the dead zone threshold is substantially reduced. The dither at the output can easily be removed by electronic means.

Another approach is the magnetic mirror which uses the transverse Kerr effect. One of the laser mirrors is coated with a magnetosensitive layer. Applying a magnetic field perpendicularly to this mirror, a differential phase shift between the cw and the ccw beams is created. This phase shift, as pointed out in Section 7.1 in regard to the IFOG, manifests itself as a differential path length change of the two beams. Clearly, this also translates as a differential frequency shift. Again, for a magnetic dither of suitable amplitude Ω_{d_m} and frequency ω_d, a motionless lock-in compensation device is provided. The magnetic mirror can serve additional purposes, such as bias compensation. As in the Faraday cell, much more freedom is offered for the waveform of the dither signal. For example, there are hardly any limitations in using a square wave dither signal. Its advantage is that the dwell time in the dead zone $\pm\Omega_l$ is practically reduced to zero so that residual scale factor errors are still further reduced.

The differential laser gyro — performance data

A third development effort, known as the *multioscillator* or the *differential laser gyro* (DILAG), has been under way since the mid 1970s at Litton and Honeywell. An important feature is the elimination of any kind of periodic dither altogether. Not many details on the subject have been disclosed so far. This method also involves the use of a Faraday cell in a way that can cancel the unwanted causes of bias shift resulting from the insertion of the cell inside the cavity. The idea of the multioscillator is to use a polarizing crystal inside the laser cavity so that two pairs of counterrotating beams are created, where each pair is polarized in opposition to the other. In effect, two separate ring lasers are thus created in the same cavity. Each one can be separated by using a polaroid filter at the output. The opposite polarization has the effect that each beam responds to the Faraday bias Ω_{bF} in the opposite sense with respect to the other. Therefore, one gyro output becomes biased oppositely to the other. These biases are made to pull the lasers out of their respective dead zones. By summing the two output signals, the two Faraday biases, in principle, are canceled at the output level along with the extraneous biases due to uncertainties. The sensed input rotation rate Ω at the output is doubled. The composite characteristic is shown in Figure 8.5.

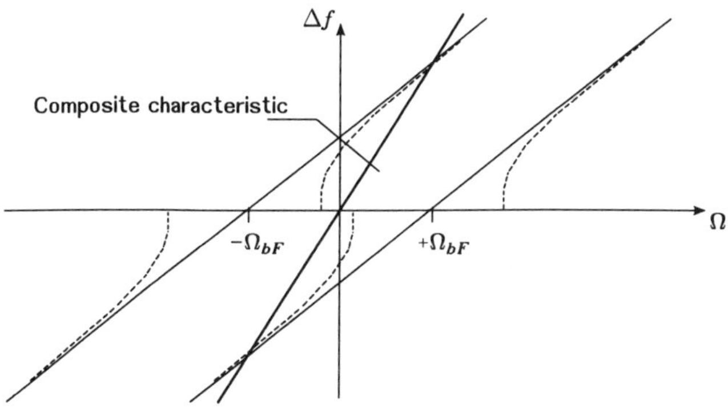

Figure 8.5. Composite response characteristic of a zero-lock RLG showing elimination of the dead zone.

Clearly, the need for the periodic dither is eliminated along with the need to remove the dither signal from the output. If care is taken to choose $|\Omega_b|>|\Omega_{max}|$, where Ω_{max} is the largest expected Ω, net angular rates can be held above $|\Omega_l|$ so that lock-in will never occur.

Typical performance data

A recent publication [8] provides up to date technical and performance data achieved with the fourth-generation Litton S18-4 ZLG Zero Lock Gyro, which has been under development since 1975. This is an 18 cm path length gyro currently in large series production. It is intended for 1 nmi/hr INS class for military fixed-wing and rotor craft. Specific performance data quoted in [8] disclose the following overall excellent results, throughout the entire input rate range of 0 – 400 deg/sec, obtained in series production:

1. Scale factor (ppm): Thermal residual, thermal sensitivity (ppm/deg C), linearity, asymmetry, all on order of a fraction of 1 ppm.
2. Random walk: Worst case 15×10^{-6} deg/$\sqrt{\text{hr}}$ in a prototype series; average value 5×10^{-6} deg/$\sqrt{\text{hr}}$).
3. Turn-on-to turn-on bias: The majority of a series of 20 gyros demonstrated less than 0.004 deg/hr bias.

Eleven flight tests of about one and a half hours duration carried out during 1990 demonstrated an average radial error of 0.42 nmi/hr. Table 8.1 lists the principal error related properties of a typical RLG for the 1 n mi/hr class.

Table 8.1. Principal error data and properties of 1nmi RLG

Property	Numerical value
Bias error	
Repeatability	0.001 deg/hr
Stability	0.005 deg/hr (1σ) 0.05 deg/hr, z axis]
Random walk	0.004 deg/$\sqrt{\text{hr}}$ (1σ) 0.04 deg/$\sqrt{\text{hr}}$, z axis]
Scale factor	
Stability	< 5 ppm (1σ)
Nonlinearity	< 5 ppm (1σ)
Asymmetry	< 1 ppm (1σ)
Input axes alignment	
Error	$\pm$1 mrad
Stability	10 micro rad
Axis-to-axis orthogonality	< 2 arc sec
Input angular rate range	0 – $\pm$400 deg/sec
Temperature range	-54 deg C – +71 deg C
Shock (half sine, 10 ms)	20 g
Vibration	5 g
Lifetime	
Operating	>50.000 hrs
Storage	>5 years
Pulse weight (resolution)	$\leq$ 2 arc sec
Input power	5 watts

The precision of RLGs is strongly related to their physical dimensions. Scaling laws for laser gyros are different for different configurations. Honeywell's, and Litton's experience with mechanically dithered laser gyros has been that the lock-in zone and bias uncertainty vary inversely between the square and the cube of the gyro path length, while the scale factor uncertainty varies inversely with path length. A similar trend holds for the overall precision of RLG's in general. Based on technical publications of both Litton and Honeywell, performance precision of typical rectangular RLG's, which relate optical path lengths with the corresponding nmi/hr divergence figures, is listed in Table 8.2.

Table 8.2. Typical drift rate values for different RLG path lengths

Path length (cm)	Angular drift (^{o}/hr)	Application
28	0.01	1 nmi/hr navigation
17	0.1	AHRS & midcourse guidance
12	1.0	Tactical missile guidance
6	10	Autopilot & attitude control

Figures such as those presented in Table 8.1 can serve to evaluate the total RLG error in accordance with the following error model referred to the x axis:

$$\Omega_{out} = (1+\epsilon)(\Omega_x + \gamma_z\Omega_y - \gamma_y\Omega_z) + \Omega_b \tag{8.3.2}$$

where the scale factor error ϵ is defined by

$$\epsilon = \epsilon_0 + f(|\Omega_x|) \quad (\text{or} + g(\Omega_x)) \tag{8.3.3}$$

$$\Omega_b = \Omega_{bo} + \Omega_{n_1} + \Omega_{n2} \tag{8.3.4}$$

where

- Ω_o — total gyro output signal;
- Ω_x — gyro angular input rate referred to x axis;
- ϵ — gyro scale factor error;
- Ω_b — gyro bias error;
- γ_y, γ_z — misalignment of the lasing plane with respect to the input axis;
- Ω_y, Ω_z — rotation rates normal to the input axis;
- ϵ_0 — constant scale factor offset;
- $f(|\Omega_x|)$ — symmetrical linearity error;
- $g(\Omega_x)$ — linearity error with symmetrical and asymmetrical components;
- Ω_{bo} — constant bias error;
- Ω_{n_1} — random bias error with unbounded integral value;
- Ω_{n2} — random bias error with bounded integral value.

ϵ_0 in (8.3.3) is primarily the result of gain-loss changes in the cavity, factors that may affect the path length, such as manufacturing tolerances, imperfections in the path length control loop and residual thermal expansions of the Zerodur block. Recall (8.1.5) to verify that the effective path length A/L determines the sensitivity, i.e., the scale factor.

The term $f(|\Omega_x|)$ is a residual effect of lock-in for gyros that employ mechanical dither.

The term $g(\Omega_x)$ is a residual effect due to lock-in for RLGs employing nonmechanical lock-in compensation. The scale factor linearity error $g(\Omega_x)$ for a given input rate Ω is proportional to the extent to which the biased gyro input is shifted from the lock-in region Ω_l, normalized by the sensed input rate Ω, that is, the linearity error, as pointed out in Chapter 1, is defined as a fraction of the input. The higher the Faraday bias Ω_{b_F} in comparison with Ω_l, the smaller $g(\Omega_x)$.

The constant bias error Ω_{b_0} is caused by the following factors: 1. Ion (Langmuir) flow phenomena in the cavity that bring about variations in the effective path lengths of the cw and the ccw beams; 2. scattering caused by cavity interference with the lasing function, such as, beam interactions with imperfect mirror surfaces, which introduce differential phase shifts between the cw and ccw beams; 3. residual errors caused by the nonmechanical lock-in compensation device.

The n_1 noise component is attributed to random instabilities in the bias producing mechanisms taking place in the lasing cavity. It is associated with the dither function. In the case of mechanically dithered RLGs, it manifests itself as a random angular error that occurs each time the dithered angular rate crosses the lock-in threshold. Being associated with bias, it is an additive noise. It is described as a zero-mean white or colored stationary process in accordance with the Markov model presented in Chapter 4. In general, unless there is reason to assume otherwise, the power spectral density of this noise source is nonzero as $s \to 0$. Therefore, its integral builds up as the square root of time. It is expressed in terms of deg/$\sqrt{t}$. This error divergence is not related to the positional progression due to the vehicle motion. It would build up with time even if it is at rest.

The bounded n_2 term is associated with scale factor errors originating in the mechanisms intended to eliminate lock-in. A reason for SF errors in RLG's with a base mounted output prism can be due to an off-nominal location of the center of the RLG mechanical dither oscillation [4]. In this case, the apparent oscillatory motion of the fringes is not precisely canceled. Recall that a scale factor error, by definition, is proportional to the input, in our case, Ω. Thus, if e.g., $\Omega = 0$, there are no fringes. As such, a scale factor error does not cause divergence with time as a bias error. Whether moving or at rest, the scale factor error remains the same. Clearly, in navigation applications an angular SF error causes a positional error proportional to the distance traveled. Note, however, that if the flight path is closed, that is, the vehicle returns to its point of departure, the SF-induced error is zero.

Scale and drift error maps of angular rate sensors

The descriptions in the last two chapters of many of the mechanical and optical rotational and specific force sensors have shown that bias and scale factor errors are the principal parameters that characterize the performance of these sensors. This characterization is valid from the crudest to the most precise and sophisticated devices. For this reason, the principal classes of angular rate sensors are described here in bias-scale factor error coordinates, on logarithmic scales, as shown in Figure 8.6. The closed contours signify the regions within which various sensor products within their class are to be found. As a general trend, one can expect that within all the closed regions and in general within the entire bias-SF coordinate plane, the closer one is to the axes' origin, the higher the cost of the product that should be expected. All the classes addressed here range from those that have matured as vital products in strategic or tactical weapon systems, such as 3 or 4 which may no longer be in production, to those that are still in early or advanced R&D stages, such as 7 or 8.

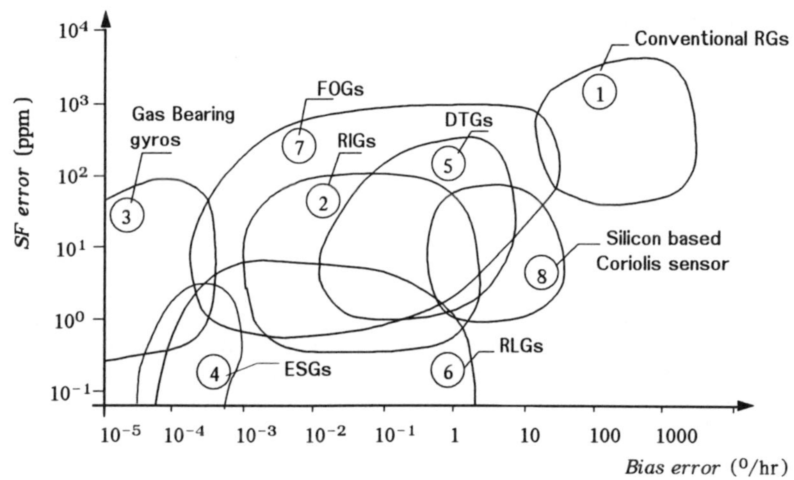

Class	Mission
1 Conventional rate gyros	Autopilot, attitude estimation, vehicular damping
2 Rate integrating gyros	Midcourse guidance, 1 nmi/hr class INS
3 Gas bearing gyros	<0.1 nmi/hr INS for Intercontinental missions
4 Electrostatic gyros	<0.1 nmi/hr INS for ICBM missions
5 Dry tuned rotor gyros	≥1 nmi/hr mainly for tactical INS and guidance
6 Ring laser gyros	≥0.1nmi/hr missions military and civil missions
7 Fiber optic gyros	≥1 nmi/hr for midcourse guidance and INS
8 Silicon-based Coriolis gyros	Ultimate, 1nmi/hr INS; intermediate, tactical missiles

Figure 8.6. Scale factor performance maps and applicatons of the most important mechanical and optical rate gyros.

Concluding remarks

1) The RLG, in contrast to the IFOG, is inherently a linear open loop device having an extremely stable input-output characteristic. It therefore does not require any form of feedback for linearization or bandwidth enhance- ment.

2) During the past 20 years, the RLG has progressively proven itself as the best technology in terms of performance and reliability. Thousands of failure-free flight operating hours have become common practice. Its relatively high cost of acquisition is more than offset by its very low cost of ownership.

3) Operationally, the RLG has the important advantage that warmup time is practically zero. After powering up, it is instantly ready for operation. Similarly, the RLG hardly requires recalibrations. Once calibrated, it maintains operational readiness for thousands of hours.

4) It is still a relatively bulky component, typically occupying some 80 cubic inches, or 1500 cc, with outer dimensions of about $7 \times 6 \times 2$ in, or $18 \times 15 \times 5$ cm. Since three RLGs are required, the total volume for such gyros alone amounts to some 4.5 liters. This implies that it is suitable only in fairly large aircraft. With the advent of ZLG, a substantial reduction in physical dimensions is achieved.

5) Since the output is represented by pulse rates proportional to the input angular rate, a natural interface with the digital pulse counter and navigation computer is provided.

6) As compared to electromechanical gyros, the RLG, by its very nature, is free of a host of inevitable error sources, such as mass unbalance, anisoinertia, anisoelasticity, vibration rectification sensitivity, residual friction torques, pickoff nonlinearities, and offsets. All these are nonexistent in the RLG.

7) Presently, while being well established in aerosensor technology, the most serious contender of the RLG is the IFOG, as described in Section 7.1. For the 1 nmi/hr class, the IFOG is much smaller and potentially less costly, and it has become comparable in terms of reliability.

Problems

Problem 8.1— Enclosed area of different polygonal shapes

Derive an expression for the enclosed areas $A(n)$ of n-sided polygons. The constraint is the largest distance D_{max} between two polygon corners.

a) Compare the values of $A(n)$ for $n = 3,4,5$ and a given D_{max} and conclude as to the relative advantage of having $n>3$.

b) Are there diminishing returns in increasing n beyond, e.g., $n=5$? If so, what are they?

An RLG with a small A is known to be less accurate than one with a large A.

c) Explain the reasons for this. Are they due tobecause

a higher noise level?
larger values of Ω_{max} the RLG can measure?
a higher scale factor coefficient?

Support your answers with the appropriate relationships.

Problem 8.2— Characteristic of the RLG with dither

The lock-in zone of an RLG is $\pm\Omega_l$, as shown in Figure 7.12. The sinusoidal dither signal causes the RLG block to oscillate at $\alpha(t) = \alpha_d \sin\omega_d t$ where α_d is the angular dither amplitude and ω_d is the frequency.

a) Derive an algorithm for the computation of the describing function $\Delta f = F(\Omega,\Omega_{dm})$, where $\Omega_{dm} = \alpha_d\omega_d$ is the dither parameter qualitatively shown in Figure 7.12. For simplicity, set $4A/L_p\lambda = 1$.

Hint: For the computation of $F(\Omega,\Omega_{dm})$, determine the area under the dither sine wave that exceeds Ω_l for different angular rate amplitudes $\alpha_d\omega_d$. The sine wave must be added to the input Ω. Normalize the result by the total half sine area. (For large α_d's $F(\Omega,\Omega_{dm}) \rightarrow \Omega$).

b) Define the error $\epsilon_d = \Omega - F(\Omega,\Omega_{dm})$ with the corresponding values obtained from (8.3.1) and conclude its validity for different values of Ω_{dm}.

c) For $\Omega_{dm}=100\Omega_l$, how large must Ω/Ω_l be to assure that $\epsilon_d \leq 10^{-4}$.

Problem 8.3— General features of the RLG

Recall and list the principal relationships underlying the characteristics of the RLG.

a) In view of its characteristic features, can you support an argument for operating an RLG in a closed-loop force-balance mode? Is such an idea conceivable for an RLG?

The wavelength of a triangular RLG is $\lambda = 0.63\ \mu$. Its path length is $L = 28$ cm.

b) What is its angular resolution and its scale factor (Sagnac) coefficient?

This RLG is intended for a strapdown INS to be used in a fighter aircraft which may develop roll rates up to 400 deg/sec.

c) What is the corresponding pulse train frequency provided by the output device?

d) Is there a limit to the largest measurable input rate Ω_{max} and to $D_r = \Omega_{max}/\Omega_{min}$? What factors determine the value of Ω_{min}?

References

1. Killpatrick, J., "The Laser Gyro," *IEEE Spectrum*, October 1967, pp. 44–55.

2. Siouris, G.M., *Aerospace Avionics Systems, a Modern Synthesis*, Book, Academic Press, New York, 1993.

3. Lawrence A., *Modern Inertial Technology, Navigation, Guidance, and Control*, Springer-Verlag, New York, 1993.

4. Savage, P.G., "Strapdown Inertial Sensors, Theory and Applications," AGARD Lecture Series No. 95, UK, Italy, Norway, Germany, Denmark, June 1978.

5. Garret, C.G.B., *Gas lasers*, McGraw-Hill, New York, 1967.

6. Macek, W.M. and Davis, D.T.M. "Rotation Rate Sensing with Traveling Wave Ring Lasers," *Applied Physics Letters*, 2, February 1, 1963.

7. Kochenburger R.J., "Frequency Response Method for Analyzing and Synthesizing Contactor Servomechanisms," *Transactions AIEE*, Vol. 60 (communication and electronics) 1950, pp. 270–284.

8. Volk C.H., Canfield J.M., and Gillespie S.C., "Zero Lock", AIAA Guidance, Navigation and Control Conference, August 1991, AIAA Paper 91-2764.

9

Filtering, Estimation, and Aiding

9.0 Introduction

In this chapter, the principle of operation and the general requirements of sensors operating as a measurement system are presented. The idea of mutual aiding of sensors by filtering and estimation is realized by complementary, or Kalman, filtering. A variety of application examples underlines the power of utilizing the complementary properties of the sensors within the system.

In the pursuit of high precision guidance and navigation, the obvious approach was initially to employ the best available sensors. In some cases, such as in strategic ballistic missiles, this was done with little regard to cost. Often, even the best of such choices does not meet the operational requirements. Efforts for further improvement of the sensors per se within a given technological state of the art often turn out to be futile in terms of cost-effectiveness and reliability. The outstanding example is inertial navigation technology. Even the best mechanical gyros, in the class of 0.001 deg/hr, cause a position error of about 150 m after one hour of flight. The error sources broadly fall into two principal categories, namely, wideband errors due to noise, and narrowband errors due to long-term drift or uncompensated biases. Fortunately, this distinction holds the key to mutual aiding or *measurement fusion*, by which the readings of at least two types of sensors, each belonging to one of the above categories, are blended to aid each other mutually in the frequency regions where any one of them performs poorly. This mutual aiding has been achieved by a simple method known as *complementary filtering* [1], initially implemented by analog component technology, and later, with the advent of high-speed large-memory digital computers, by *Kalman Filtering* [2]. In order to achieve an optimal blending by the Kalman filter, the statistics of the measurements must be known at least approximately. In the simple case of high-frequency noise suppression in a single sensor, the classical Wiener filtering approach [3] is adequate, provided that the signal and noise spectra are reasonably well known and substantially separated. In situations where these assumptions on the two types of sensors are not valid, and whenever the frequencies of the noise and the signal are in the same region, neither of the above methods applies. In such situations, the noise can be significantly suppressed by *adaptive noise cancellation* [4] provided that an auxiliary measurement is available.

9.1 Complementary Filtering

Basic formulation

Assume two linear time-invariant filters $F_1(s)$ and $F_2(s)$ which have the following properties:

$$\lim_{s \to 0} F_1(s) = 0; \quad \lim_{s \to \infty} F_2(s) = 0 \tag{9.1.1}$$

$$F_1(s) + F_2(s) = 1; \quad F_1(s) = 1 - F_2(s) \tag{9.1.2}$$

Thus, $F_1(s)$ is high-pass, $F_2(s)$ is low-pass and $F_1(s) + F_2(s)$ is all-pass. Now consider the filter arrangement in parallel shown in Figure 9.1.

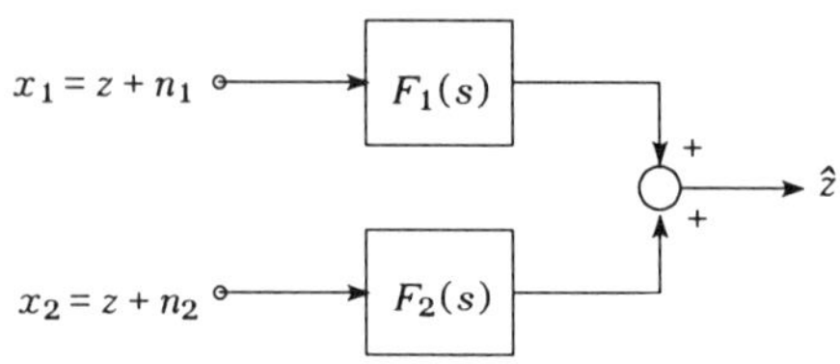

Figure 9.1. Conceptual diagram of complementary filtering.

x_1 and x_2 are the two respective sensor readings of the same variable z. Their filtered sum is

$$\hat{z}(s) = [F_1(s) + F_2(s)]\, z(s) + F_1(s)n_1(s) + F_2(s)n_2(s) \tag{9.1.3}$$

In view of (9.1.1) and (9.1.2), we have

$$\hat{z}(s) = z(s) + [1 - F_2(s)]n_1(s) + F_2(s)n_2(s) \tag{9.1.4}$$

Assume that n_1 is narrowband and n_2 is wideband, then n_1 is blocked, and n_2 is low-pass filtered. Thus, the desired signal z is unaffected and both noise sources are attenuated.

Compass-aided heading gyro

This principle is demonstrated by the fusion of the measurements of a magnetometer and a directional gyroscope by means of $F_2(s) = k/(k+s)$ from which, by (9.1.2), $F_1(s) = s/(k+s)$, as shown in Figure 9.2.

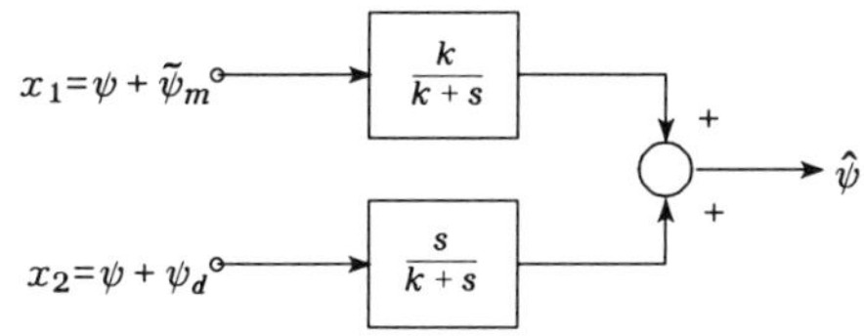

Figure 9.2. Example of complementary filtering for magnetic aiding of a directional gyroscope.

The estimated value of ψ is therefore

$$\hat{\psi} = \frac{s}{k+s}\psi + \frac{k}{k+s}\psi + \frac{s}{k+s}\psi_d + \frac{k}{k+s}\tilde{\psi}_m$$

$$= \psi + \frac{\dot{\psi}_d}{k+s} + \frac{k}{k+s}\tilde{\psi}_m \tag{9.1.5}$$

As expected, the azimuth angle ψ is perfectly represented in the estimate $\hat{\psi}$. The drift rate $\dot{\psi}_d$ is bounded and, in the steady state, it settles to the value $\dot{\psi}_d/k$, and the high-frequency noise $\tilde{\psi}_m$ is low-pass filtered and attenuated. This result is identical with the result in (4.6.19), which deals with the actual slaving of a directional gyroscope to a compass. The slaving is therefore an implementation of complementary filtering CF. The actual realization of CF is not as shown in Figure 9.1 but rather as shown in Figure 9.3. The estimation error $\psi_m - \hat{\psi} = \tilde{\psi}$ is fed back into the integrator via k. The measurements shown here are by a yaw rate gyro and a magnetometer. Just integrating $\dot{\psi}$ to yield ψ is impractical because of the gyro drift $\dot{\psi}_d$, which would also be integrated ,thus accumulating a large heading angle error. As indicated earlier, the magnetometer, though relatively noisy, bounds this error. This function of sensor fusion is also known in the current literature as *aiding*. Thus, the magnetometer in the present example aids the rate gyro.

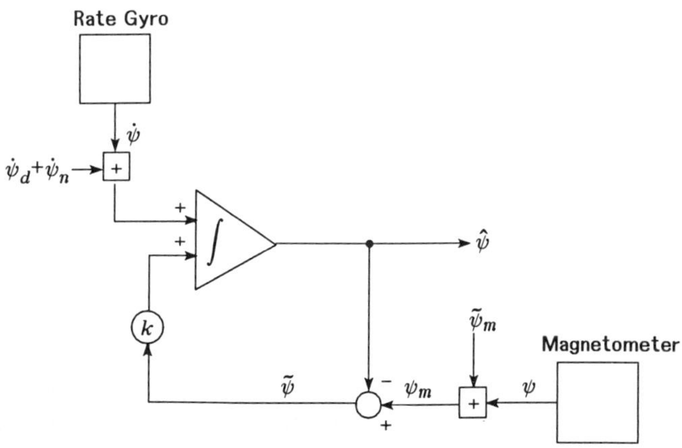

Figure 9.3. Example of closed-loop implementation of CF.

The following relationship holds:

$$[\dot{\psi}+\dot{\psi}_d+\dot{\psi}_n+k(\psi_m-\hat{\psi})]\frac{1}{s}=\hat{\psi} \tag{9.1.6}$$

from which

$$\dot{\psi}+\dot{\psi}_d+\dot{\psi}_n+k\psi+k\tilde{\psi}_m-k\hat{\psi}=s\hat{\psi} \tag{9.1.7}$$

Laplace transforming and solving for $\hat{\psi}$, we have

$$\hat{\psi}=\psi+\frac{k}{k+s}\tilde{\psi}_m+\frac{\dot{\psi}_d}{k+s}+\frac{\dot{\psi}_n}{k+s} \tag{9.1.8}$$

This result is identical to (9.1.5), and, as stated earlier, to (4.6.19). In the choice of k, two conflicting factors come to bear. On the one hand, the attenuation of $\tilde{\psi}_m$ requires a small value of k. On the other hand, too small a value will cause a large bias $\dot{\psi}_d/k$ and an often ignored, large rate gyro noise component that diverges, in accordance with Chapter 4, as a result of the almost pure integration that prevails for small k. If the statistics of $\tilde{\psi}_m$ and $\dot{\psi}_d$ are known, it is a straightforward matter to compute the optimal value of a by minimizing the variance $\sigma^2_{\tilde{\psi}}$. The optimal value of k is determined by the ratio $\sigma^2_{\tilde{\psi}_m}/\sigma^2_{\dot{\psi}d}$. This approach to optimization, however, becomes unwieldy for higher order filters. In reality, the optimization is carried out by Kalman filtering KF [2], by which the optimal gain k is directly determined on the basis of a priori assumptions on the statistics of $\tilde{\psi}_m$ and $\dot{\psi}_d$. Actually, it has been pointed out [5] that the CF and the fixed gain KF are equivalent. The apparent conflict in the choice of the gain k is significant if it is a constant. However, if it starts out as a large number and gradually decreases, the conflict is essentially eliminated. This becomes clear by expressing the error $\tilde{\psi}=(\psi_m-\psi)$, weighted by k, which is at the input of the integrator but with opposite sign:

$$(\psi_m-\hat{\psi})k=\left(\psi_m-\psi-\frac{k}{k+s}\tilde{\psi}_m-\frac{1}{k+s}\dot{\psi}_d-\frac{1}{k+s}\dot{\psi}_n\right)k$$

$$=\left(\psi+\tilde{\psi}_m-\psi-\frac{k}{k+s}\tilde{\psi}_m-\frac{1}{k+s}\dot{\psi}_d-\frac{1}{k+s}\dot{\psi}_n\right)k$$

$$=\frac{k}{k+s}\tilde{\psi}_m-\frac{k}{k+s}\dot{\psi}_d-\frac{k}{k+s}\dot{\psi}_n \tag{9.1.9}$$

The second term in (9.1.9) is the estimate of the drift rate term $\dot{\psi}_d$. If k is initially large, this estimate rapidly converges, but it is still contaminated by the first and third noise terms. If, subsequently, k is gradually decreased, the initially noisy estimate of the drift rate settles down because the noise terms are subject to substantially greater low-pass filtering. This process underlies the principle of operation of the variable-gain KF in particular in the context of estimating unknown constant biases.

Doppler–inertial ground speed estimation

Much of the aforementioned conflict can be resolved by adding an additional integration stage to the CF as shown in Figure 9.4, which describes the fusion of the inertial measurement of acceleration, integrated to yield inertial velocity, with the noisy readings of a Doppler radar [6].

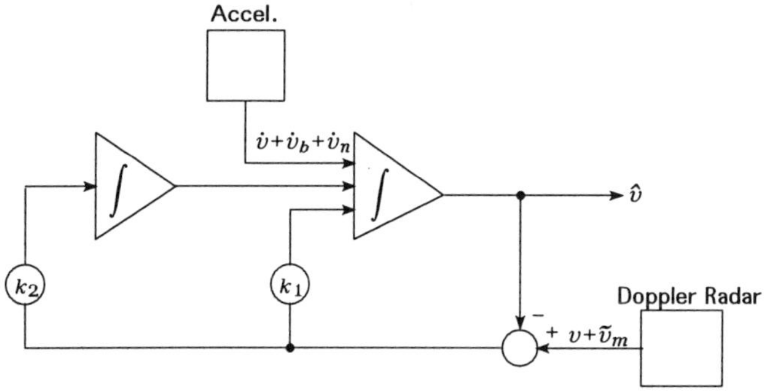

Figure 9.4. Doppler aided inertial ground speed estimation with removal of accelerometer bias effect.

The acceleration measurement description shown here is highly oversimplified. In reality, the accelerometer is either mounted on a gimbaled platform to isolate it from the vehicular angular motions, as discussed in Chapter 3 or its readings are processed by a computational algorithm known as the *analytical platform* [7]. The accelerometer is generally in error due to the presence of an unknown bias $\dot{v}_b$, which may be due to either internal instrument errors or, for example, a platform tilt angle θ causing a bias $g\sin\theta$. The Doppler radar readings are accompanied by a relatively wideband noise $\tilde{v}_m$, which results from the random temporal variations in the backscattered energy due to the roughness of the terrain [6] .

The filter input–output relationship is readily obtained as follows:

$$[(v+\tilde{v}_m-\hat{v})k_1+(v+\tilde{v}_m-\hat{v})\frac{k_2}{s}+\dot{v}+\dot{v}_b+\dot{v}_n]\frac{1}{s}=\hat{v} \tag{9.1.10}$$

Expanding and rearranging, the solution for the velocity estimate $\hat{v}$ is

$$\hat{v}=v+\frac{k_1s+k_2}{s^2+k_1s+k_2}\tilde{v}_m(s)+\frac{s}{s^2+k_1s+k_2}[\dot{v}_b(s)+\dot{v}_n(s)]$$

$$=v+F_1(s)\tilde{v}_m(s)+F_2(s)[\dot{v}_b(s)+\dot{v}_n(s)] \tag{9.1.11}$$

$F_1(s)$, $F_2(s)$ now are second order low-pass and high-pass filters, respectively, with the relative degree of 1 as before. However, if $\dot{v}_b$ is a constant bias, ignoring for the moment $\dot{v}_n$, the second term in (9.1.10) vanishes in the steady-state. Thus, the conflict in choosing the gains k_1, k_2, as indicated for k in the first-order filter of the previous example, is apparently resolved. This is so by virtue of the integral control provided by the additional integrator. However, a choice of too small values for k_1, k_2 would result in very long settling times to reach the steady-state.

EXAMPLE 9.1-1: Error evaluation in a Doppler-inertial ground speed estimator

A numerical example for the foregoing CF is:
$k_1=0.08\ \sec^{-1}$; $k_2=0.005\ \sec^{-2}$; $\dot{v}_b=0.05\ \mathrm{m}\sec^{-2}$; v_m = filtered white noise w with σ_w $\sigma_w=0.2\ \mathrm{m}\sec^{-1}$, and a first order lag, $a/(a+s)$ *where* $a=10$; The performance of the CF is shown in Figure 9.5, which displays the main error contributors. The velocity estimation error is defined as $\tilde{v}=\hat{v}-v$. The plots demonstrate that with the parameter values chosen, the Doppler noise is attenuated by a factor of ~10, and the effect of accelerometer bias essentially vanishes after one minute.

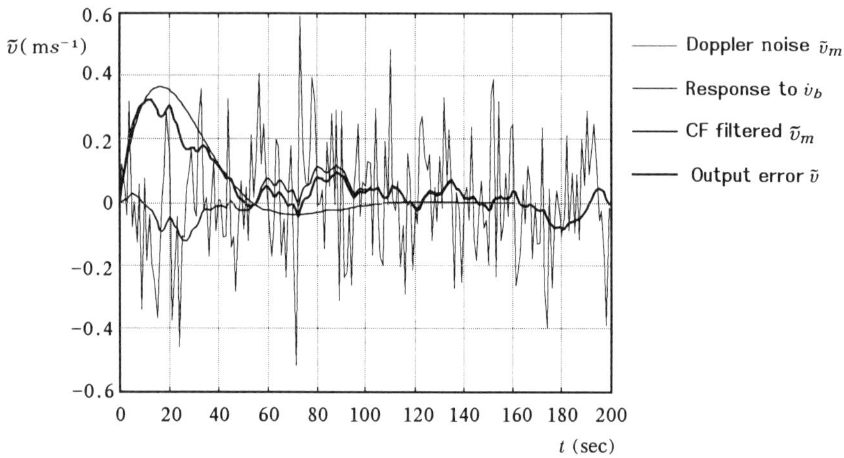

Figure 9.5. Example of Doppler aided-ground speed estimation.

Considering that $\dot{v}_n$ is wideband, s^2 in the denominator of (9.1.11) may be dominant in comparison with $k_1 s+k_2$, especially if k_1, k_2 are small. In this case $F_2(s)$ approaches a pure integrator. Thus, the filter output $\hat{v}$ would diverge and behave as the process illustrated in Chapter 2, (Figure 2.33). This behavior is counterintuitive because the effect of high-frequency accelerometer noise is usually negligible. Fortunately, since the time divergence is proportional to $t^{1/2}$ and the high-frequency noise power in typical accelerometers is on the order of mgs, this divergence is mostly negligible. However, on-board land, marine or airborne vehicles, the accelerometer often outputs noise due to high-frequency vibration or intense acoustic noise, which can be on the order of gs [8] and does not reflect any component of actual vehicle motion. It therefore causes a velocity and, consequently, a position error.

The vanishing of the bias term in (9.1.11) is of particular importance once $\hat{v}$ is integrated to determine position. The zero s in the numerator then cancels the pole $1/s$ and the positional error reaches a constant rms value and does not diverge with time. However, the first term in (9.1.11), whose relative degree is zero, will diverge upon integration to determine position. This divergence, being proportional to $t^{1/2}$, is quite moderate and is not normally a significant contributor to error.

In the design of such a filter, an additional constraint must be considered. In addition the considerations for choosing the magnitude of k_1, k_2, the ratio between them must be chosen to assure good damping. Substituting,

$$k_2 = \omega_0^2 \,;\; k_1 = 2\omega_0\varsigma \tag{9.1.12}$$

Specifying, e. g., $\varsigma = 0.7$, we have $k_1 = 1.4\sqrt{k_2} = 1.4\omega_0$.

The Doppler radar used as the aiding measurement is given here as an example of a long standing-technology. It is by no means the only method. Alternative methods that have been studied are based on estimation of optical flow by means of passive electro-optical imaging sensors [9]. Further progress in aiding inertial measurements has been achieved is by the satellite-based global positioning system (GPS), which has emerged in recent years as the principal technology for precision localization of objects and the foremost method for aiding INS [10].

EXAMPLE 9.1-2: Noise attenuation in a Doppler radar

A Doppler radar has a velocity noise error $\tilde{v}_m$ characterized by the PSD,

$$\Phi_{\tilde{v}_m\tilde{v}_m}(s) = \frac{2\sigma_{\tilde{v}_m}^2 a_v}{a_v^2 - s^2}$$

The rms value of the error is $\sigma_{\tilde{v}_m} = 5$ m/sec; its noise bandwidth is $a_v = 50$ sec^{-1}. It is required to attenuate this noise rms value by a factor of ~ 10. In accordance with (9.1.11) and (9.1.12) and the spectral input-output relationships demonstrated in Example 2.8-1, and denoting the filtered velocity error by $\tilde{v}_f$, we have

$$\Phi_{\tilde{v}_f\tilde{v}_f}(s) = 2\sigma_{\tilde{v}_m}^2 a_v\left|\left(\frac{1.4\omega_0 s + \omega_0^2}{s^2 + 1.4\omega_0 s + \omega_0^2}\right)\left(\frac{1}{a_v + s}\right)\right|^2$$

In order to achieve the required attenuation of 10, ω_0 should be set to about $0.01a_v$, so that $\omega_0 = 0.5$ sec^{-1}. Substituting the above numerical values we have

$$\Phi_{\tilde{v}_f\tilde{v}_f}(s) = 2\times 25\times 50\left|\left(\frac{1.4\times 0.5s + 0.25}{s^2 + 1.4\times 0.5s + 0.25}\right)\left(\frac{1}{50 + s}\right)\right|^2$$

$$= 2500\times\left|\left(\frac{0.7s + 0.25}{s^2 + 0.7s + 0.25}\right)\left(\frac{1}{50 + s}\right)\right|^2$$

$$= 2500\times\left|\left(\frac{0.7s + 0.25}{s^3 + 50.7s^2 + 35.25s + 12.5}\right)\right|^2$$

The spectral integration in accordance with Appendix A2 yields the expression

$$I_3 = P_{\tilde{v}_f} = \frac{1}{2\pi j}\int_{-j\infty}^{j\infty} \Phi_{\tilde{v}_f\tilde{v}_f}(s)ds = 2500\times \frac{c_2^2 d_0 d_1 + (c_1^2 - 2c_0c_2)d_0d_3 + c_0^2 d_2 d_3}{2d_0d_3(-d_0d_3 + d_1d_2)}$$

where

$d_0 = 12.5$; $d_1 = 35.25$; $d_2 = 50.7$; $d_3 = 1$; $c_0 = 0.25$; $c_1 = 0.7$; $c_2 = 0$. The result is

$$P_{\tilde{v}_f} = 0.52 \text{ m}^2/\text{sec}^2; \quad \sigma_{\tilde{v}_f} = 0.723 \text{ m/sec}$$

Thus, the low-pass filtering indeed attenuates the Doppler noise by $0.723/5 = 0.144$, which is a somewhat weaker attenuation than the initially required 0.1.

The wideband accelerometer noise $\dot{v}_n$ is modeled here as a first-order Markov process

$$\Phi_{\dot{v}_n\dot{v}_n}(s) = \frac{2\sigma_{\dot{v}_n}^2 a_n}{a_n^2 - s^2}$$

The large numerical values, including external vibration components, are assumed to be $\sigma_{\dot{v}_n} = 2$ m/sec^2 over the bandwidth of 100 Hz, namely $a_n = 628$ sec^{-1}. The filter error output due to noise therefore is

$$\Phi_{\tilde{v}_n\tilde{v}_n}(s) = 2\sigma_{\dot{v}_n}^2 a_n \left|\left(\frac{s}{s^2 + 0.7s + 0.25}\right)\left(\frac{1}{628 + s}\right)\right|^2$$

$$= 2\times 4\times 628\times \left|\frac{s}{s^3 + 629s^2 + 440s + 157}\right|^2$$

$$I_3 = P_{\tilde{v}_n} = 5024\times \frac{c_2^2 d_0 d_1 + (c_1^2 - 2c_0c_2)d_0d_3 + c_0^2 d_2 d_3}{2d_0d_3(-d_0d_3 + d_1d_2)}$$

where $d_0 = 157$; $d_1 = 440$; $d_2 = 629$; $d_3 = 1$; $c_0 = 0$; $c_1 = 1$ and $c_2 = 0$;

The result is

$$P_{\tilde{v}_n} = 5.78\times 10^{-5} \text{ m}^2/\text{sec}^2; \quad \sigma_{\tilde{v}_n} = 7.6\times 10^{-3} \text{ m/sec};$$

which is quite negligible. An important feature that follows from the foregoing is that $\dot{v}_n$, though being integrated, yields a bounded error $\tilde{v}_n$. For systems that can be represented by this example, the second term in (9.1.10) essentially vanishes both on account of $\dot{v}_b$ and $\dot{v}_n$ so that the filtered output $\hat{v}$ exactly reproduces the real v except for the filtered residual Doppler noise $\sigma_{\tilde{v}_f} = 0.723$ m sec^{-1}. The initial transient due to $\dot{v}_b$ quickly settles in accordance with the effective time constant $\tau = 1/\omega_0\zeta = 1/0.35 = 2.86$ sec.

The baro-inertial altimeter

The autonomous measurement of aircraft altitude naturally calls for the use of a barometric or inertial altimeter. Both of these have serious shortcomings. The barometric altimeter is relatively noisy and suffers from substantial time lags. The inertial measurement of altitude, on the other hand, although essentially noise free, but involves accelerometer drift. Moreover, it is unstable and rapidly diverges with time. These aspects and others are discussed, in [6]. The actual realization is a baro-inertial altimeter in which baro-altimeter readings are fused with inertial readings by means of a complementary filter. This fusion, or *aiding*, as it is often called, eliminates the lags and noise of the barometric readings on one hand and the inertial instability and effects of accelerometer bias on the other hand. Because of these favorable properties, the baro–inertial altimeter has established itself as a widely used technology in conventional aircraft operations.

The following notations and definitions underlie the subsequent analysis:

h — actual altitude;
$\hat{h}$ — estimated altitude;
g_0 — gravity at sea level;
g_h — actual gravity at altitude $h = R + h$, $g_h = g_0 R^2/(R+h)^2$;
$\dot{v}_b$ — accelerometer bias;
$\dot{v}_m$ — accelerometer output $= \ddot{h} + g_h + \dot{v}_b$;
R — radius of Earth;
$g_{\hat{h}}$ — estimated gravity at $\hat{h}$, $g_{\hat{h}} = g_0 R^2/(R+\hat{h})^2$;

We now look at the expression for the estimated altitude $\hat{h}$ based on the double integration of the accelerometer readings:

$$\hat{h} = \int_0^t \int_0^{t'} (\dot{v}_m - \hat{g})\, d\tau\, d\tau' \tag{9.1.13}$$

Denote $\Delta h = h_c - h$

$$\Delta h = \hat{h} - h = \int_0^{t'} \int_0^{t} (\dot{v}_m - g_{\hat{h}})\, d\tau d\tau' - h \tag{9.1.14}$$

$$\Delta \ddot{h} = \dot{v}_m - g_{\hat{h}} - \ddot{h} = \dot{v}_m - g_0 \frac{R^2}{(R+\hat{h})^2} - \ddot{h} \cong \dot{v}_m - g_0 (1 - \frac{2\hat{h}}{R}) - \ddot{h} \tag{9.1.15}$$

Substituting $\dot{v}_m$, we have

$$\Delta \ddot{h} = \ddot{h} + g_0 (1 - \frac{2h}{R}) + \dot{v}_b - g_0 (1 - \frac{2\hat{h}}{R}) - \ddot{h} = \frac{2g_0}{R} (\hat{h} - h) + \dot{v}_b$$

$$= \frac{2g_0}{R} \Delta h + \dot{v}_b \tag{9.1.16}$$

Finally, the following differential equation emerges:

$$\Delta \ddot{h} - \frac{2g_0}{R} \Delta h = \dot{v}_b \tag{9.1.17}$$

Denoting $2g_0/R \triangle \omega^2$, its solution is

$$\Delta h(t) = \Delta h_0 \cosh \omega t + \frac{\Delta \dot{h}_0}{\omega} \sinh \omega t = \frac{\dot{v}_b R}{2g} (\cosh \omega t - 1) \tag{9.1.18}$$

Δh_0 and $\Delta \dot{h}_0$ are the initial conditions of altitude deviation and altitude rate deviaton, resprctively. Note that $\omega = \sqrt{2}\, \omega_s$, where ω_s is the Schuler frequency.

Equation (9.1.17) indicates the instability of the computed altitude deviation. The altitude error grows exponentially with time doubling within an hour. The initial behavior for short periods of time (9.1.17) can be approximated by

$$\Delta h(t) \cong \Delta h_0\left(1+\frac{\omega^2 t^2}{2}\right) + \Delta \dot{h}_0 t\left(1+\frac{\omega^2 t^2}{6}\right) + \frac{\dot{v}_b t^2}{2}\left(1+\frac{\omega^2 t^2}{6}\right) \tag{9.1.19}$$

Clearly, the pure inertial altimeter is useless on account of its fundamental physical properties. Realization of the complementary filter that implements the fusion of the inertial and barometric altimeter is shown in Figure 9.6.

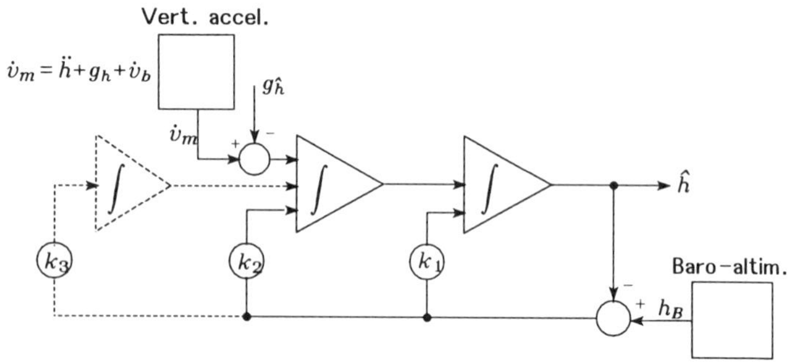

Figure 9.6. Complementary filter implementing a baro-inertial altimeter including accelerometer bias suppression.

The barometric altitude measurement is $h_B \triangleq h + \tilde{h}_B$ where $\tilde{h}_B$ represents the measurement uncertainty and noise. The integration stage shown in dotted lines is optional and, as before, is intended to eliminate accelerometer bias in the steady state.

The analysis of the CF shown in Figure 9.6 proceeds as follows:

$$\left[(h_B - \hat{h})k_1 + (h_B - \hat{h})\frac{k_2}{s} + \frac{\ddot{h} + g_h - g_{\hat{h}} + \dot{v}_b}{s}\right]\frac{1}{s}$$

$$= \left[(h_B - \hat{h})k_1 + (h_B - \hat{h})\frac{k_2}{s} + \frac{\ddot{h} + g_0\left(1 - \frac{2h}{R}\right) - g_0\left(1 - \frac{2\hat{h}}{R}\right) + \dot{v}_b}{s}\right]\frac{1}{s} = \hat{h} \tag{9.1.20}$$

Substituting $h_B = h + \tilde{h}_B$, where $\tilde{h}_B$ is the barometric measurement uncertainty, rearranging, and solving for $\hat{h}$, we have

$$\hat{h}(s) = h(s) + \frac{k_1 s + k_2}{s^2 + k_1 s + k_2 - \omega^2}\tilde{h}_B + \frac{\dot{v}_b}{s^2 + k_1 s + k_2 - \omega^2} \tag{9.1.21}$$

where $\omega^2 \triangleq 2g_0/R$. Equation (9.1.21) discloses the following:

1. The actual altitude, the first term, is perfectly reproduced without any lags.

2. Variations and noise in $\tilde{h}_B$ are low-pass filtered, as given by the second term.

3. The error due to the double integration of $\dot{v}_b$ within the closed loop of the CF is bounded, as seen in the third term.

4. The negative sign of ω^2 indicates the tendency of the basic inertial altimeter to instability. However, if the filter design assures that $k_2 > \omega^2$, the composite altimeter is stable. Since ω^2 is quite a small number, it is not difficult to meet this condition.

5. If it is required to remove the bias in the steady state, the option of the additional filter stage, shown in dotted lines in Figure 9.6, can be implemented.

The foregoing demonstrates that the CF not only bounds errors due to integration. In addition, it has the capability of stabilizing a system that is inherently unstable.

The altimeter is basically barometric and, as such, it measures altitude above sea level. If it is required to measure altitude above ground, the barometric altimeter must be replaced by a radar or a laser altimeter. The analysis, however, basically remains unchanged.

EXAMPLE 9.1–3: Altitude error due to accelerometer bias

Assume the following numerical values: $\dot{v}_b = 1$ mg; $k_1 = 0.1$. For a damping coefficient $\varsigma = 0.7$, $k_2 = k_1^2/2 = 0.005$ and $\omega^2 = 2g/R = 19.6/(6380 \times 1000) = 3.07 \times 10^{-6}$ sec^{-1}.

Clearly, $k_2 \gg \omega^2$, and the system is stable. The steady-state error due to $\dot{v}_b$ is essentially determined by $\dot{v}_b/k_2 = 10 \times 10^{-3}/0.005 = 2$ m. Thus, even a fairly crude accelerometer with the foregoing assumed filter coefficients would cause only a negligible altitude error of 2 m.

Complementary filtering for discrete measurements ⊕

Application of complementary filtering is not necessarily restricted to linear state estimation and additive noise attenuation. It can be applied to remove nonadditive errors caused by sampling and quantization of a process measurement that must retain its analog nature and cannot tolerate errors of this kind. An example of this problem is described in [11] which deals with the measurement of the pilot's head angular orientation with respect to the aircraft body axes. These measurements involve special head-tracking equipment, the output of which requires sampling and quantization of the electromagnetic signals are generated by a Polhemus head motion sensor. The information on head position and orientation is therefore delayed, sampled, and quantized. This information can be used for dynamical shifting the graphic symbols in a head-up display (HUD) or helmet-mounted display (HMD) with the purpose of target tracking and designation [12]. Typical state-of-the-art sampling rates are on the order of 30 Hz. Experience has shown that at these sampling rates, in conjunction with the image generator rate, the dynamic display suffers from aliasing, apparent dead band, and delays. It becomes annoying to the pilot, causing degradation of pointing precision and increased pilot workload. Many of these annoying effects on the head attitude measurement can be alleviated by aiding the measurements with accelerometer pairs mounted on the helmet and a suitably tuned CF. The effect of this kind of aiding is that the twice-integrated angular acceleration output, which would generally rapidly diverge with time, is bounded by the sampled head tracking system. The integrated accelerometer readings eliminate the sampling and quantization errors. Figure 9.7 describes the CF with the sampled measurements of the head tracker. The time delay T is due to the computations followed by the sampling operator.

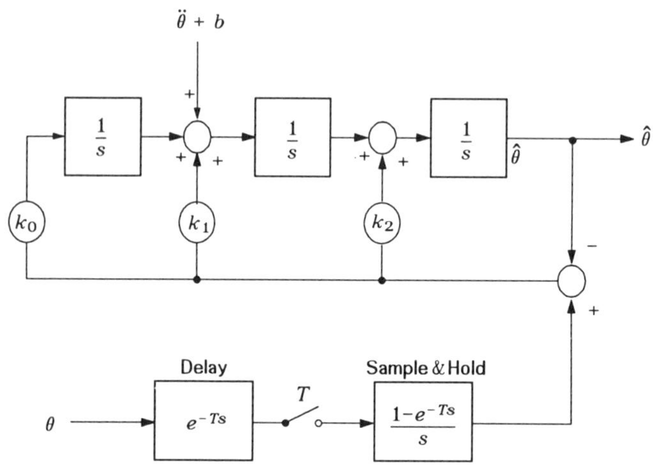

Figure 9.7. Application of CF to the smoothing of a sampled and delayed measurement of a head tracker.

The following relationships apply:

$$\left\{\left[\left(1-e^{-Ts}/Ts\right)e^{-Ts}\theta-\hat{\theta}\right]\frac{k_0}{s}+s^2\theta+b+\left[\left(1-e^{-Ts}/Ts\right)e^{-Ts}\theta-\hat{\theta}\right]k_1\right\}$$

$$+\left[\left(1-e^{-Ts}/Ts\right)e^{-Ts}\theta-\hat{\theta}\right]k_2 = s\hat{\theta} \tag{9.1.22}$$

The solution for $\hat{\theta}$ is

$$\hat{\theta} = \frac{s^3 + k_2\eta s^2 + k_1\eta s + k_0\eta}{s^3+k_2s^3+k_1s+k_0}\theta + \frac{s}{s^3+k_2s^3+k_1s+k_0}b \tag{9.1.23}$$

where

$$\eta = \left[\left(1-e^{-Ts}\right)/Ts\right]e^{-Ts} \tag{9.1.24}$$

From (9.1.23) it is evident that for $s \to \infty$, $\hat{\theta} \to \theta$. For $s \to 0$, $\eta \to 1$ and again $\hat{\theta} \to \theta$. In view of this behavior of $G(s)=\hat{\theta}(s)/\theta(s)$, it is expected that the net phase shift of $G(s)$ will be zero. In order to verify to what extent and over what frequency range this assertion is valid, simulations were carried out. As expected, the second term in (9.1.23) due to the bias b vanishes in the steady state.

An example of simulated sinusoidal head angles θ at the frequency of 1 Hz and amplitude of 15 mrad $\cong$ 1 deg is shown in Figure 9.8. This is a typical angular head vibration amplitude for vertical whole body vibrations of 0.1g. The filter parameters were set at k_2= 4.4, k_1=5.2 and k_0=3. Further details can be found in [11]. The sampled, quantized measurements θ_T and their corresponding smoothed estimates $\hat{\theta}$ are also shown in Figure 9.8. It is interesting to note that the phase shift between $\hat{\theta}$ and θ is indeed negligibly small. The amplitude of $\hat{\theta}$ is somewhat attenuated. This, however, is of lesser consequence in the context of the problem discussed in [11].

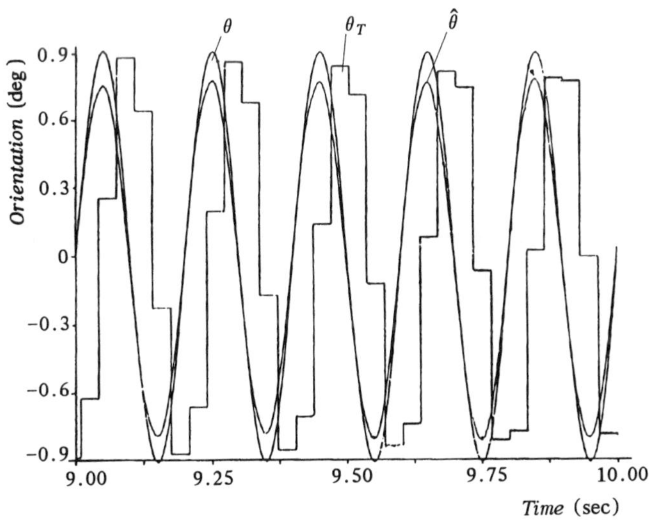

Figure 9.8. Application of CF to the smoothing of sampled and delayed head vibration measurements.

The foregoing example demonstrates that, at least for sufficiently high-frequency sampling rates, the concept of complementary filtering reaches beyond the basic linear assumption of additive wideband noise.

9.2 Equivalence of the CF and the stationary KF

In the development of the CF, the effects of measurement noise in the primary sensor, for instance, the accelerometer noise in the inertial system, and the noise in the aiding sensor, such as the Doppler system, were included in the analysis. However, no attempt was made to determine the values of the gains k_1, k_2 in terms of the sensor noise characteristics with the aim of achieving the smallest estimation error.

Formulation for inertial navigation

We first reformulate the CF in terms of its state-space equation, in particular for the application to inertial navigation, aided by ground based position measurements x_p, which are obtained from systems such as Tacan, Loran, Decca, [6] and more recently from GPS [10]. This formulation presents a special case of Kalman filtering subject to the CF constraint [13].

The corresponding noise-free block diagram is described in Figure 9.9, which describes a highly simplified scheme without regard to Schuler tuning and other terrestrial factors.

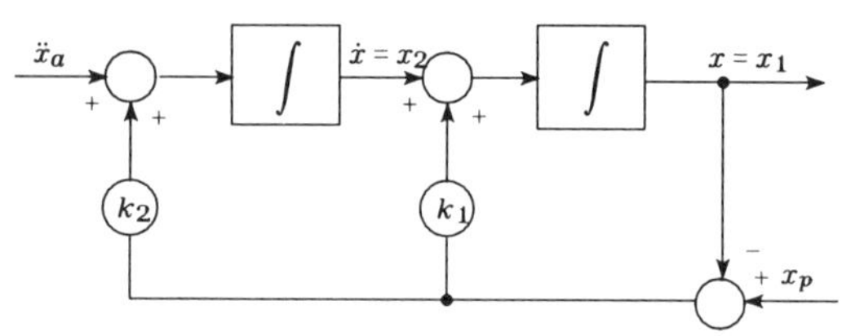

Figure 9.9. State space-representation of the second-order CF.

The noise-free state space equation as directly derived from Figure 9.9 is

$$\begin{bmatrix}\dot{x}_1\\ \dot{x}_2\end{bmatrix}=\begin{bmatrix}0 & 1\\ 0 & 0\end{bmatrix}\begin{bmatrix}x_1\\ x_2\end{bmatrix}+\begin{bmatrix}k_1\\ k_2\end{bmatrix}(x_p-x_1)+\begin{bmatrix}0\\ 1\end{bmatrix}\ddot{x}_a \tag{9.2.1}$$

which can be rewritten in the format

$$\underbrace{\begin{bmatrix}\dot{x}_1\\ \dot{x}_2\end{bmatrix}}_{\dot{x}}=\underbrace{\begin{bmatrix}0 & 1\\ 0 & 0\end{bmatrix}}_{A}\underbrace{\begin{bmatrix}x_1\\ x_2\end{bmatrix}}_{x}-\underbrace{\begin{bmatrix}k_1\\ k_2\end{bmatrix}\begin{bmatrix}1 & 0\end{bmatrix}}_{KH}\underbrace{\begin{bmatrix}x_1\\ x_2\end{bmatrix}}_{x}+\underbrace{\begin{bmatrix}0\\ 1\end{bmatrix}\ddot{x}_a}_{Bu}+\underbrace{\begin{bmatrix}k_1\\ k_2\end{bmatrix}x_p}_{Ky} \tag{9.2.2}$$

namely,

$$\dot{x} = Ax-KHx+Bu+Ky=(A-KH)x+Bu+Ky \tag{9.2.3}$$

The CF as the Luenberger observer ⊕

Equation (9.2.3) is the familiar Lueneberger observer equation described, among others, in Chapter 11 of [14]. It reconstructs the state vector $\breve{\boldsymbol{x}}$ from the *primary* measurement $\ddot{x}_a$, which here plays the role of the control signal u, and from the *aiding* measurement x_p, which here plays the role of the observation $\breve{y}$. It therefore follows that the CF is essentially a linear observer illustrated in its general form in Figure 9.10. $\breve{A}$ and $\breve{B}$ are the representations of the true system A and B. If the observations are completely-noise free, and if $\breve{A} = A$ and $\breve{B} = B$, then $\breve{\boldsymbol{x}} = \boldsymbol{x}$ and $\breve{y} = y$. In the present case of the complementary filter, these equalities are assured since A and B, which by definition represent mathematical operations, consist of ones and zeros only. Generally, A and B represent a physical system, and the equalities $\breve{A} = A$ and $\breve{B} = B$ are not precisely valid. This model mismatch is an additional estimation error not addressed here.

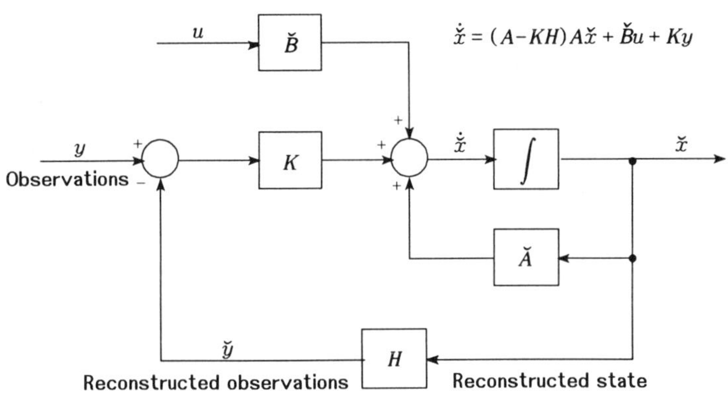

Figure 9.10. Representation of CF as a Luenberger observer.

Inclusion of noise

In reality, however, both u and y are corrupted by noise: w plays the role of process noise while v, associated with the aiding, is the measurement noise. The state reconstruction now takes the role of its corresponding estimates $\hat{\boldsymbol{x}}$, as indicated in Figure 9.11.

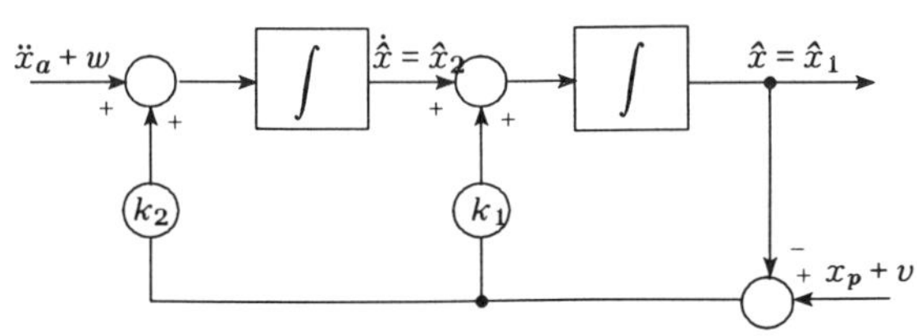

Figure 9.11. Second-order CF with process and measurement noise.

The corresponding state-space equations of the estimator are

$$\begin{bmatrix} \dot{\hat{x}}_1 \\ \dot{\hat{x}}_2 \end{bmatrix} = \begin{bmatrix} 0 & 1 \\ 0 & 0 \end{bmatrix} \begin{bmatrix} \hat{x}_1 \\ \hat{x}_2 \end{bmatrix} + \begin{bmatrix} k_1 \\ k_2 \end{bmatrix} (x_p + v - \hat{x}_1) + \begin{bmatrix} 0 \\ 1 \end{bmatrix} \ddot{x}_a + w \tag{9.2.4}$$

This equation is rewritten in the following format,

$$\begin{bmatrix} \dot{\hat{x}}_1 \\ \dot{\hat{x}}_2 \end{bmatrix} = \begin{bmatrix} 0 & 1 \\ 0 & 0 \end{bmatrix} \begin{bmatrix} \hat{x}_1 \\ \hat{x}_2 \end{bmatrix} - \begin{bmatrix} k_1 \\ k_2 \end{bmatrix} \begin{bmatrix} 1 & 0 \end{bmatrix} \begin{bmatrix} \hat{x}_1 \\ \hat{x}_2 \end{bmatrix} + \begin{bmatrix} 0 \\ 1 \end{bmatrix} (\ddot{x}_a + w) + \begin{bmatrix} k_1 \\ k_2 \end{bmatrix} (x_p + v) \tag{9.2.5}$$

$$\dot{\hat{\boldsymbol{x}}} \qquad A \qquad \hat{\boldsymbol{x}} \qquad KH \qquad \hat{\boldsymbol{x}} \qquad B(u+w) \qquad K(y+v)$$

The general form of this state-space representation, therefore, is

$$\dot{\hat{\boldsymbol{x}}} = A\hat{\boldsymbol{x}} - KH\hat{x} + B(u+w) + K(y+v)$$

$$= (A - KH)\hat{\boldsymbol{x}} + B(u+w) + K(y+v) \tag{9.2.6}$$

where $\hat{x}_2$ obviously represents the velocity estimate $\dot{\hat{x}}$, while $\hat{x}_1$ represents the position estimate $\hat{x}$. Clearly, the observer provides the estimate $\hat{x}$ based on the measurements and given the gain matrix K. The estimation error $\boldsymbol{e} \triangleq \boldsymbol{x} - \hat{\boldsymbol{x}}$ is the measure that determines the performance of the estimator. Using the foregoing formulations, we have

$$\dot{\boldsymbol{e}} = \dot{\boldsymbol{x}} - \dot{\hat{\boldsymbol{x}}} = A\boldsymbol{x} + Bu + Bw - A\hat{\boldsymbol{x}} - Bu - Ky + KH\hat{\boldsymbol{x}}$$

$$= A(\boldsymbol{x} - \hat{\boldsymbol{x}}) - KH(\boldsymbol{x} - \hat{\boldsymbol{x}}) + Bw - Kv \tag{9.2.7}$$

Thus,

$$\dot{\boldsymbol{e}} = (A - KH)\boldsymbol{e} + Bw - Kv \tag{9.2.8}$$

Equation (9.2.8) represents the dynamics of the estimation error $\boldsymbol{e}$, where w and v are the forcing functions and the eigenvalues of $A-KH$ determine the nature of the convergence process of the estimate given the initial condition $\boldsymbol{e}(0) = \boldsymbol{e}_0$.

The optimum observer

The central issue in the preceding observer formulation is the gain matrix K, which minimizes the expectation of e^2, namely, the covariance of e, which will be denoted by P. The value of this minimizing gain K will be denoted by K^*. We now make the not necessarily realistic assumption that v and w are white noise processes. Therefore the weighted sum of w and v is also a white noise process:

$$\xi = Bw - Kv \tag{9.2.9}$$

Thus,

$$\dot{\boldsymbol{e}} = (A - KH)\boldsymbol{e} + \xi \tag{9.2.10}$$

is a linear system excited by white noise. Clearly, if all the eigenvalues of the matrix $A-KH$ have negative real parts, the observer is stable and, in the absence of noise, i.e., $\xi=0$, $\boldsymbol{e}$ converges from some initial value $\boldsymbol{e}_0$ to zero.

In the analysis that follows, it is essential to formulate the covariance matrix Q_ξ:

$$\begin{aligned} Q_\xi = E\{\xi(t)\xi^T(\tau)\} &= B(t)E\{w(t)w^T(\tau)\}B^T(\tau) \\ &-K(t)E\{v(t)w^T(\tau)\}B^T(\tau) - B(t)E\{w(t)v^T(\tau)\}K^T(\tau) \\ &+K(t)E\{v(t)v^T(\tau)\}K^T(\tau) \end{aligned} \tag{9.2.11}$$

For the sake of generality, the matrices B and K are formulated as time-dependent. All the expected values on the right-hand side of (9.2.11) are assumed to be white noise processes. Thus,

$$E\{w(t)w^T(\tau)\} = W(t)\delta(t-\tau) \tag{9.2.12}$$

$$E\{w(t)v^T(\tau)\} = X(t)\delta(t-\tau) \tag{9.2.13}$$

$$E\{v(t)v^T(\tau)\} = V(t)\delta(t-\tau) \tag{9.2.14}$$

So that,

$$E\{\xi(t)\xi^T(\tau)\} = Q_\xi(t)\delta(t-\tau) \tag{9.2.15}$$

where $Q_\xi(t)$, which is the weight of the Dirac function $\delta(t-\tau)$, is the spectral density of $\xi(t)$. In view of (9.2.11), $Q_\xi(t)$ it is given by

$$\begin{aligned} Q_\xi(t) &= B(t)W(t)B^T(t) - K(t)X^T(t)B^T(t) - B(t)X(t)K^T(t) \\ &+K(t)V(t)K^T(t) \end{aligned} \tag{9.2.16}$$

We now define the covariance of the error $\boldsymbol{e}$,

$$P \triangleq E\{\boldsymbol{e}\boldsymbol{e}^T\} \tag{9.2.17}$$

In view of (9.2.8), we have

$$\begin{aligned} \dot{P} &= (A-KH)P + P(A^TH^TK^T + Q_\xi \\ &= (A-KH)P + P(A-H^TK^T) + BWB^T - KX^TB^T - BXK^T + KVK^T \end{aligned} \tag{9.2.18}$$

The cross covariance between the observation noise v and the process noise w is in most cases absent, i.e., $X=0$. With this assumption it can be shown by a variational approach, e.g., [14] Chapter 9, that in order to assure that the cost functional is non-negative definite, the linear term in the variation must vanish. The coefficient of this linear term is therefore, set to zero, namely,

$$K^*V - P^*H^T = 0 \tag{9.2.19}$$

Thus,

$$K^* = P^*H^TV^{-1} \tag{9.2.20}$$

Substituting this solution into (9.2.18) while dropping X results in the well known Riccati matrix differential equation

$$\dot{P}^* = AP^* + P^*A^T - P^*H^TV^{-1}HP^* + BWB^T \tag{9.2.21}$$

the solution of which yields the minimal value of the error covariance P^*. The foregoing procedure is known as the *continuous linear Kalman filter*. In principle, it can be solved analytically, but for systems of order higher than three, the computation becomes prohibitively complex.

The steady-state Kalman filter

The substitution of K^* in the observer renders it as the *optimal observer*, better known as the *time-varying Kalman filter*. The interpretation is that initially K^* may be quite large, which is advantageous because it imposes a rapid reduction of $\boldsymbol{e}$. Assuming that $(A-KH)$ in (9.2.18) is asymptotically stable, then, as time progresses, $\dot{P} \to 0$ and the solution converges to its steady state. Under these conditions, (9.2.21) reduces to the algebraic Riccati equation:

$$0 = AP^* + P^*A^T - P^*H^TV^{-1}HP^* + BWB^T \tag{9.2.22}$$

which is a quadratic in P^*. It is clear that the solution for P^*, and therefore for K^*, is determined by the noise parameters V and W. Recalling (9.2.2), we have

$$A = \begin{bmatrix} 0 & 1 \\ 0 & 0 \end{bmatrix}; \; B = \begin{bmatrix} 0 \\ 1 \end{bmatrix}; \; C = [1 \;\; 0]; \; P = \begin{bmatrix} p_{11} & p_{12} \\ p_{21} & p_{22} \end{bmatrix}; \; V = \sigma_v^2; \; W = \sigma_w^2$$

Since P is symmetric, $p_{12} = p_{21}$. Substituting these expressions into (9.2.21) and carrying out the required matrix operations, (9.2.21) takes the form

$$\begin{bmatrix} 2p_{12} & p_{22} \\ p_{22} & 0 \end{bmatrix} = \begin{bmatrix} p_{11}/\sigma_v^2 & p_{11}p_{12}/\sigma_v^2 \\ p_{11}p_{12}/\sigma_v^2 & p_{12}^2/\sigma_v^2 - \sigma_w^2 \end{bmatrix} \tag{9.2.23}$$

Equating element by element on both sides of (9.2.22), the solutions are

$$p_{11} = \sqrt{2\sigma_w\sigma_v^3}; \quad p_{12} = \sigma_w\sigma_v; \quad p_{22} = \sigma_w\sqrt{2\sigma_v}$$

From (9.2.20), the solution of the optimal gain matrix K^* is determined by

$$K^* = P^*\begin{bmatrix} 1 \\ 0 \end{bmatrix}\frac{1}{\sigma_v^2} = \begin{bmatrix} \sqrt{2\sigma_w/\sigma_v} \\ \sigma_w/\sigma_v \end{bmatrix} = \begin{bmatrix} k_1^* \\ k_2^* \end{bmatrix} \tag{9.2.24}$$

Thus, if a valid assumption can be made for W and V, or rather for the rms values of the process and measurement noises σ_w and σ_v, the Kalman gain matrix K can be a priori determined, and it remains constant. This kind of filter is referred to as the *stationary, or constant-gain Kalman filter.* As stated earlier, it is an observer whose gain matrix has been optimized on the basis of statistical models of the process and measurement noises. Clearly, if the actual noise statistics varies with time, the optimality of the filter is not retained. Equation (9.2.24) discloses that K^* is determined by the ratio σ_w/σ_v. The larger it is, the higher the filter gain that can be afforded and the more rapid its convergence. On the other hand, we conclude that a relatively large measurement noise v would allow only a small K^*. In view of (9.2.8), this would result in a correspondingly slow convergence of the estimate. The physical interpretation of these results is that if σ_w/σ_v is zero, the accelerometer is ideal and k_1, k_2 vanish, the loop is open and pure open-loop double integration takes place. On the other hand, if σ_v is large, the resulting vanishing k_1, k_2 indicate that these noisy measurements should be ignored. The adjustment of the assumed σ_w and σ_v to what are thought to be their true values is known as the *tuning* of the Kalman filter. Thus, if the actual noise statistics change, the filter gets detuned, and it will not render its best performance.

The foregoing computations become very tedious, especially for dimensions higher than two. In practice the Riccati equation is solved numerically. In view of the great importance of optimal filter design, the computational procedure has been developed as an easily applicable linear quadratic estimation (lqe) MATLAB algorithm, the synopsis of which is:

$$[\mathrm{L,P}] = \mathrm{lqe(A,G,C,Q,R)} \tag{9.2.25}$$

L stands for K^*, P for P, A for A, G for G, C for C, Q for W, and R for V. G is the process noise matrix, defined in the standard general state-space equation rather than the specialized form of (9.2.4), namely,

$$\dot{x} = Ax + Bu + Gw \tag{9.2.26}$$

Thus, once the parameters of A,G,C,Q,R are selected, the algorithm returns the required numerical values of $\mathrm{L} = K^*$ and the minimal mean square estimation error $\mathrm{P} = P^*$.

The foregoing example for the KF design, which is based on white noise models, is in many cases far from realistic especially in the context of inertial navigation. Although it serves the purpose of illustrating the computational procedure, the dominant errors in accelerometer measurements $\ddot{x}_a$ are unknown biases or very slow Markov processes that cannot be represented by white noise. However, this lack of realism can be overcome by formulating augmenting state-space equations that represent such Markov processes. These augmented equations are driven by white noise. Thus, the realism of a narrow-band Markov process is represented, while the white noise model whith its computational simplifications and advantages is retained.

The augmented system model

In order to design the KF for colored noise, we assume that in the actual system w, is white noise ν filtered by a first-order lag, as presented in (2.6.13), namely,

$$\dot{w} = -aw + a\nu \tag{9.2.27}$$

and its model in the observer would be

$$\dot{\hat{w}} = -a\hat{w} + a\nu \tag{9.2.28}$$

We now denote $\hat{w} \triangleq \hat{x}_3$ and use it to augment the state space-equation (9.2.5) repeated here

$$\begin{bmatrix} \dot{\hat{x}}_1 \\ \dot{\hat{x}}_2 \end{bmatrix} = \begin{bmatrix} 0 & 1 \\ 0 & 0 \end{bmatrix} \begin{bmatrix} \hat{x}_1 \\ \hat{x}_2 \end{bmatrix} - \begin{bmatrix} k_1 \\ k_2 \end{bmatrix} \begin{bmatrix} 1 & 0 \end{bmatrix} \begin{bmatrix} \hat{x}_1 \\ \hat{x}_2 \end{bmatrix} + \begin{bmatrix} 0 \\ 1 \end{bmatrix} (\ddot{x}_a + w) + \begin{bmatrix} k_1 \\ k_2 \end{bmatrix} (x_p + v)$$

to yield

$$\begin{aligned} \dot{\hat{x}}_1 &= 0.\hat{x}_1 + 1.\hat{x}_2 - k_1\hat{x}_1 + 0.\ddot{x}_a + 0.\hat{x}_3 + k_1(x_p + v) \\ \hat{x}_2 &= 0.\hat{x}_1 + 0.\hat{x}_2 - k_2\hat{x}_2 + 1.\ddot{x}_a + 1.\hat{x}_3 + k_2(x_p + v) \\ \dot{\hat{x}}_3 &= 0.\hat{x}_1 + 0.\hat{x}_2 - a\hat{x}_3 + a\nu \end{aligned} \tag{9.2.29}$$

Rearranging, we have

$$\begin{bmatrix} \dot{\hat{x}}_1 \\ \dot{\hat{x}}_2 \\ \dot{\hat{x}}_3 \end{bmatrix} = \begin{bmatrix} 0 & 1 & 0 \\ 0 & 0 & 1 \\ 0 & 0 & -a \end{bmatrix} \begin{bmatrix} \hat{x}_1 \\ \hat{x}_2 \\ \hat{x}_3 \end{bmatrix} - \begin{bmatrix} k_1 \\ k_2 \\ 0 \end{bmatrix} \begin{bmatrix} 1 & 0 & 0 \end{bmatrix} \begin{bmatrix} \hat{x}_1 \\ \hat{x}_2 \\ \hat{x}_3 \end{bmatrix} + \begin{bmatrix} 0 & 0 \\ 1 & 0 \\ 0 & a \end{bmatrix} \begin{bmatrix} \ddot{x}_a \\ \nu \end{bmatrix} \tag{9.2.30}$$

$\hat{x}$ $\quad A \quad$ $\hat{x}$ $\quad K \quad H \quad x \quad B \quad u$

In view of (9.2.7) and (9.2.8), u, namely $\ddot{x}_a$, drops out in formulating the error equation; v is retained here as white noise, but it could be modeled as colored noise. In this case, the dimension of A would increase to 4×4. These augmented matrices are now substituted into the basic Riccati equation (9.2.21) in which $W(t)\delta(t-\tau)$ is replaced by $N(t)\delta(t-\tau) = E\{\nu(t)\nu'(\tau)\}$. In conjunction with (9.2.20), the optimal gain matrix $[k_1, k_2, 0]'$ is determined. For given numerical values, it is solved by (9.2.23). Clearly, if $a \to 0$, (9.2.26) reduces to $\dot{\hat{w}} = 0$, which corresponds to the case of a random (unknown in magnitude and sign) bias in the accelerometer. Note that this relationship is equivalent to the notion of integral control in the example of the baro-inertial altimeter shown in Figure 9.6. Recall that the output of the third integrator fed by $(h_B - \hat{h})k_3$ must be zero so that its output reaches a constant value that cancels the accelerometer bias $\dot{v}_b$.

of the inertial components, may even be higher, but it is compact, potentially reliable, and its use may be advantageous in a suitable technical-operational context.

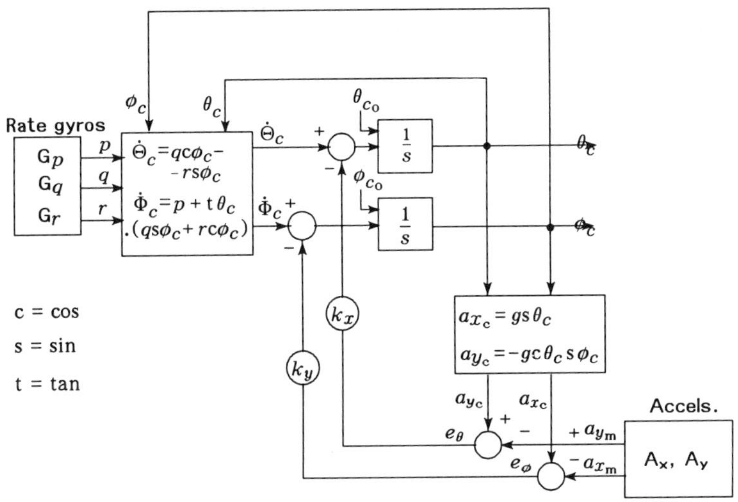

Figure 9.12. Gravity-aided angular rate measurements constituting an attitude reference system.

The computed quantities a_{x_c} and a_{y_c} do not incorporate acceleration components, while the measured quantities a_{x_m}, a_{y_m} do. Here these components play the role of measurement noise in terms of the linear observer introduced in (9.2.5). Since $\theta_c \approx \theta$ and $\phi_c \approx \phi$, we have, to a very good approximation,

$$e_\theta = \theta_c - \theta_m - \dot{U} - QW + RV = \theta_c - \theta_m + n_\theta \tag{9.3.5}$$

$$e_\phi = \phi_c - \phi_m - \dot{V} - RU + PW = \phi_c - \phi_m + n_\phi \tag{9.3.6}$$

where n_θ and n_ϕ are the horizontal acceleration measurement noise terms, which are relatively wideband. Therefore, if k_x and k_y are made sufficiently small, the effect of n_θ and n_ϕ can be rendered acceptably small. If not, additional filtering can be introduced in series with e_θ and e_ϕ as described in [16]. If the vehicle undergoes a sustained acceleration, for example, due to a prolonged turn, a cutout mechanism can be incorporated in a manner similar to that in the vertical gyro. The foregoing is an example of an instrumentation concept in which the environmental conditions may not be sufficiently isolated to render satisfactory performance under all operational conditions.

It appears that the small rotation angles can be readily solved from the vector increment $\Delta \boldsymbol{H}$ and the measured field intensities H_x, H_y, H_z. In each step the previous $[H_x' \; H_y' \; H_z']^T$ serves as the new $[H_x \; H_y \; H_z]^T$ in the present step.

Magnetic aiding

Pietila and Dunn [17] studied the derivation of attitude information from Earth's magnetic field by using as a basis a triad of orthogonal body fixed magnetometers. Each of these measures a component of the local magnetic field intensity vector $\boldsymbol{H}$. The magnitude of each component clearly depends on the aircraft attitude angles defined as the Euler angles Φ, Θ, Ψ with respect to the local-level local north. The relationships of attitude variations as a function of the magnetic vector component variations are determined by matrix representations of the orthogonal Euler transformation $[E]$ given in Appendix A9. Let H_x, H_y, H_z be the magnetic components measured at a given air frame attitude and let H_x', H_y', H_z' be the components after a body rotation. The vector $\boldsymbol{H}' \triangleq [H_x' \; H_y' \; H_z']^T$ is related to the vector $\boldsymbol{H} \triangleq [H_x \; H_y \; H_z]^T$ by the Euler transformation matrix $[E]$. Thus, $\boldsymbol{H}' = [E]^T \boldsymbol{H}$, and explicitly

$$\begin{bmatrix} H_x' \\ H_y' \\ H_z' \end{bmatrix} = \begin{bmatrix} c\Psi c\Theta & c\Psi s\Theta s\Phi - s\Psi c\Phi & c\Psi s\Theta c\Phi + s\Psi s\Phi \\ s\Psi c\Theta & s\Psi s\Theta s\Phi + c\Psi c\Phi & s\Psi s\Theta c\Phi - c\Psi s\Phi \\ -s\Theta & c\Theta \sin\Phi & c\Theta c\Phi \end{bmatrix} \begin{bmatrix} H_x \\ H_y \\ H_z \end{bmatrix} \tag{9.3.7}$$

Assume now that the angular rotations over a given time interval are small in terms of radians and are denoted by ϕ, θ, ψ; then, $\sin\phi \simeq \phi$; $\sin\theta \simeq \theta$; $\sin\psi \simeq \psi$; $\cos\phi \simeq \cos\theta \simeq \cos\Psi \simeq 1$, and products of small terms are negligible. The transformation (9.3.7) then reduces to the following linear skew-symmetric transformation matrix:

$$\begin{bmatrix} H_x' \\ H_y' \\ H_z' \end{bmatrix} = \begin{bmatrix} 1 & -\psi & \theta \\ \psi & 1 & -\phi \\ -\theta & \phi & 1 \end{bmatrix} \begin{bmatrix} H_x \\ H_y \\ H_z \end{bmatrix} = \begin{bmatrix} H_x \\ H_y \\ H_z \end{bmatrix} + \begin{bmatrix} H_x 0 - H_y \psi + H_z \theta \\ H_x \psi + H_y 0 - H_z \phi \\ -H_x \theta + H_y \phi + H_z 0 \end{bmatrix}$$

$$= \begin{bmatrix} H_x \\ H_y \\ H_z \end{bmatrix} + \begin{bmatrix} 0 & H_z & -H_y \\ -H_z & 0 & H_x \\ -H_y & -H_x & 0 \end{bmatrix} \begin{bmatrix} \phi \\ \theta \\ \psi \end{bmatrix} \tag{9.3.8}$$

The increment $\Delta \boldsymbol{H} = \boldsymbol{H}' - \boldsymbol{H}$ is explicitly given by

$$\begin{bmatrix} \Delta H_x \\ \Delta H_y \\ \Delta H_z \end{bmatrix} = \begin{bmatrix} 0 & H_z & -H_y \\ -H_z & 0 & H_x \\ H_y & -H_x & 0 \end{bmatrix} \begin{bmatrix} \phi \\ \theta \\ \psi \end{bmatrix} \tag{9.3.9}$$

However, it is easily verified that the skew-symmetric matrix in (9.3.9) is singular, implying that there is no unique solution for $[\phi \;\; \theta \;\; \psi]^T$ and that only two out of the three angles can be determined. To determine all three, one angle must be measured or calculated independently. From (9.3.9), we have

$$\Delta H_x = H_z\theta - H_y\psi \tag{9.3.10}$$

$$\Delta H_y = -H_z\phi + H_x\psi \tag{9.3.11}$$

$$\Delta H_z = H_y\phi - H_x\theta \tag{9.3.12}$$

Assume now that the heading angle ψ is independently measured, for example, by a directional gyro, and denoted by ψ_m. We then have the solutions for roll and pitch:

$$\phi = \frac{-\Delta H_y}{H_z} + \frac{H_x}{H_z}\psi_m \tag{9.3.13}$$

$$\theta = \frac{\Delta H_x}{H_z} + \frac{H_y}{H_z}\psi_m \tag{9.3.14}$$

Thus, a three-axis orthogonal magnetometer can determine the pitch and roll angles, provided that heading is independently measured. This, however, may not be easily implementable. An alternative choice would be to independently measure or compute θ from indirect air data measurements and to determine ϕ and ψ. Denoting these independent measurements by θ_c, since they are actually computed from direct measurements from (9.3.12), we have

$$\phi_c = \frac{\Delta H_z}{H_y} - \frac{H_x}{H_y}\theta_c \tag{9.3.15}$$

$$\psi_c = \frac{\Delta H_y}{H_x} - \frac{H_z}{H_x}\phi = \frac{\Delta H_y}{H_x} - \frac{H_z}{H_x}\frac{\Delta H_z}{H_y} - \frac{H_z H_x}{H_x H_y}\theta_c = \frac{\Delta H_y}{H_x} - \frac{H_z}{H_x}\frac{\Delta H_z}{H_y} - \frac{H_z}{H_y}\theta_c$$

$$= \frac{\Delta H_y}{H_x} - \frac{H_z}{H_y}\left(\frac{\Delta H_z}{H_x} - \theta_c\right) \tag{9.3.16}$$

It is clear that the three component projection measurements H_x, H_y, H_z of the single magnetic vector $\boldsymbol{H}$ can determine only two out of three angles. In any situation designed to derive directional information from sensed projections of that vector, such as Euler angles, the number of independent measurements that is needed is greater by one than the number of angles to be determined. Thus, for example, to determine the roll angle from the components of the gravitation vector $\boldsymbol{g}$ by means of body-mounted accelerometers, the two measurements $a_y = -g\cos\theta\sin\phi$ and $a_z = -\cos\theta\cos\phi$ (see (3.1.11)) are needed.

The body-mounted magnetometer, supported by anadditional measurement, as shown earlier, can provide long-term, stable, yet noisy attitude information. The blending of the magnetically derived attitude measurements with rate gyro or angular acceleration measurements can yield much better results typical of the complementary filter. This topic is covered in detail in [15]. Here we show the basic scheme in which the Euler differential equations (9.3.1) to (9.3.3) again provide the required kinematic relationships. These, however, render the CFs as nonlinear devices. The basic set of filters, along with the coupling between the estimated attitude angles, is shown in Figure 9.13.

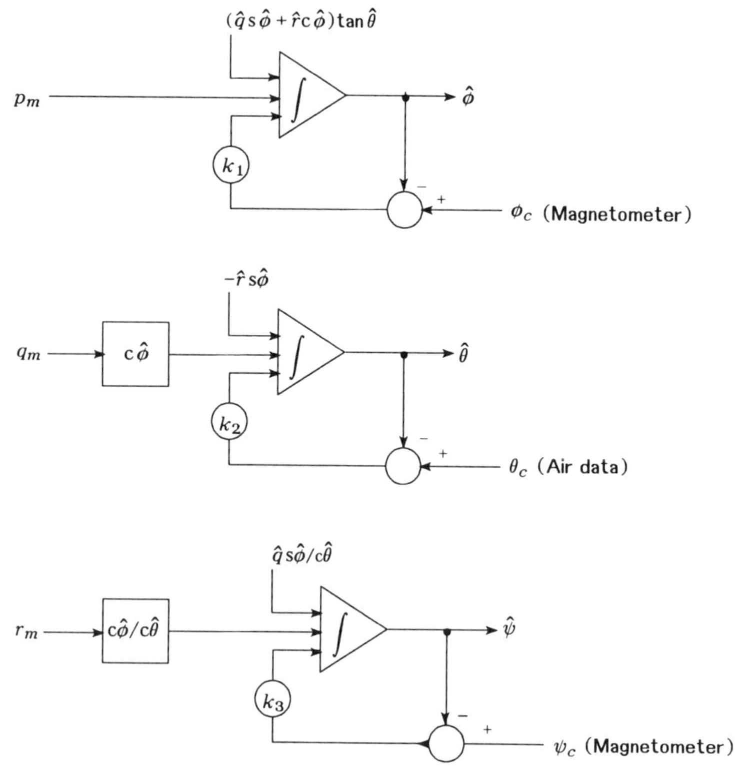

$$\gamma_c = \sin^{-1}(\dot{h}_m/V_{a_m})$$

$$\theta_c = \alpha_m \cos\hat{\phi} + \gamma_c$$

$$\dot{\Psi} = Q\frac{\sin\Phi}{\cos\Theta} + R\frac{\cos\Phi}{\cos\Theta}$$

Figure 9.13. Magnetic and air-data aiding rate gyro measurements constituting an attitude reference system.

Since it is without the integral control stage, the CF filter diagram in Figure 9.13 only bounds the rate gyro drift rates but does not eliminate them. Additional integration stages can, however be incorporated to remove the effect of drift rate, as previously shown, for instance, in Figure 9.6.

As stated, θ_c is determined from air data measurements. To accomplish this, the following auxiliary measurements are needed:

1. the measured altitude rate $\dot{h}_m$;

2. the measured angle of attack α_m;

3. the measured longitudinal air speed V_m.

With this additional information, we derive the computed flight path angle γ_c

$$\gamma_c = \sin^{-1}(\dot{h}_m / V_{am}) \tag{9.3.17}$$

and with the aid of α_m, the computed pitch angle is

$$\theta_c = \alpha_m \cos\hat{\phi} + \gamma_c \tag{9.3.18}$$

This method, named Method 1 in [15], yielded accuracies in ϕ, θ, ψ, on the order of 0.5 to 4 deg in a series of actual flight tests with a Convair CV990.

In conclusion, sole magnetic field information is not sufficient to provide the necessary aiding. In this example, two additional air data measurements are required.

Aiding by aircraft kinematics

A different version of aiding instrumentation for attitude determination is described in [18]. The concept presented there is based on elementary kinematic angular relationships in the vertical and horizontal planes as will be shown below. It is intended for the Boeing *Brave 200*, which is conceived as a low-cost expendable RPV. At the outset, heading is separately provided by a magnetometer, and the primary measurements for pitch and roll are provided by two rate gyros. An additional yaw rate gyro is required for the estimation of the roll angle ϕ_c, as shown below. The principle of operation is shown in Figure 9.13.

The underlying equations are

$$\theta_c = \alpha_m c\hat{\phi} + \beta_m s\hat{\phi} + \gamma_c \tag{9.3.19}$$

This is the same as (9.3.17), except that the-side slip angle β is also incorporated and γ_c is derived in accordance with (9.3.17).

From the second equation of (6.1.10), we have for the lateral specific force in body axes the linearized equation

$$a_y = \dot{v} + Ur - Wp - g\,c\theta\,s\phi \tag{9.3.20}$$

Assuming moderate angles of attack α and that p is only occasionally nonzero during transients, Wp can be disregarded under steady-state conditions. Also, $\dot{v}$ is nonzero only during transients or gusts and can be regarded as a high-frequency force component. With these assumptions, the steady-state output a_y of a lateral accelerometer can be considered zero. The remaining terms constitute the relationship

$$Ur - gc\theta\,s\phi = 0 \tag{9.3.21}$$

or

$$\sin\phi = \frac{Ur}{gc\theta} \tag{9.3.22}$$

where $U \simeq V_a$ especially if α is small, and r can be measured by a yaw rate gyro to deliver r_m. Consequently, we have the computed roll angle:

$$\phi_c = \sin^{-1}\left[\frac{V_{am} r_m}{gc\theta_c}\right] \tag{9.3.23}$$

The composite block diagram for obtaining the attitude estimates $\hat{\phi}$ and $\hat{\theta}$ is shown in Figure 9.14.

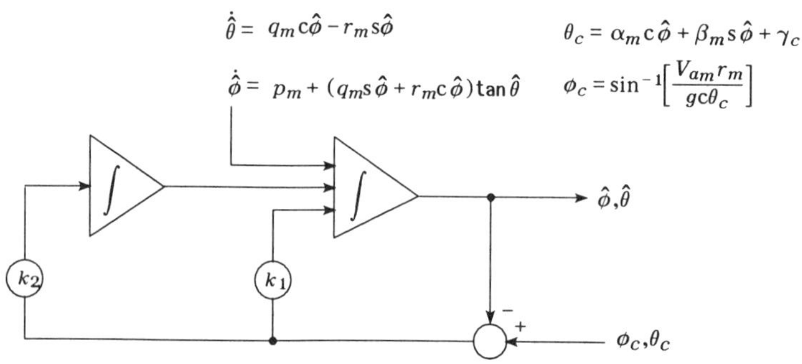

Figure 9.14. Aiding by aircraft kinematics for attitude angle estimation.

This CF configuration employs p_m, q_m, as the primary measurements; r_m is auxiliary, as manifested in (9.3.22). The drifts in p_m, q_m are eliminated by the extra integration stage. The computed aiding quantities ϕ_c, θ_c are entirely derived from air data and r_m. A bias in r_m only slightly changes the scale factor of ϕ_c. Since it does not involve integration, it does not diverge with time.

According to [18], the heading angle ψ is derived from a gimbaled three-axis magnetometer. The accuracies achieved in pitch and roll by six-degrees-of freedom simulation are quoted to be substantially less than 1 deg.

Aiding by the aircraft dynamical model ⊕

In the context of low-cost technologies for attitude estimation, in particular for expendable unmanned vehicles, the extensive search for low-cost sensor-estimation technologies by means of constant-gain complementary filters has been of great interest since the advent of RPVs in the early 1970s. In [15] and [18], for example, the primary measurements are inertial angular rates, or angular accelerations. The aiding measurements typically are angle of attack, barometric altitude, altitude rate, air speed, acceleration and magnetic flux. Such measurements, besides providing direct readings of for example, heading, are the basis of the steady-state estimates of variables such as θ_{ss} or ϕ_{ss}, which aid the short-term attitude estimates derived by the rate gyros. In those studies, the computed aiding variables are based on linearized dynamical models of the aircraft that permit only relatively small perturbations of variables around nominal values. The complementary filter approach in those studies disregards the coupling between external gust loads and angular perturbations and it treats the latter as independent noise that should be attenuated by the CF. It also disregards the complex dynamical coupling between variables and the effect of nonlinearities that become important if substantial variations in attitude take place. Furthermore, in the early stages of research, it was considered to be of paramount importance to provide simple analog type computational algorithms because of the modesty of computer memories and speed at the time. However, with the advent of modern high-speed large-memory miniature and low-cost microcomputers, the barriers imposed by computational load have practically been removed. It is by now feasible to provide a full six-degrees-of-freedom digital model of aircraft motion as detailed by the twelve nonlinear coupled equations given in Appendix A9, along with an EKF algorithm for real time operation, as detailed in Appendix B9. This approach, which is here termed *aiding by the dynamical model*, is described in [19] and briefly reviewed here.

Atmospheric random gust disturbances are described by a vector velocity field $[u_g, v_g, w_g]^T$ known as *turbulence*, modeled after Dryden [20] as an air speed, altitude-dependent first-order Gauss-Markov process with an unknown mean value.

The measurement system consists of:

1. a triad of orthogonal body-mounted rate gyros to measure p, q, r;

2. a magnetic heading sensor to measure ψ;

3 an air speed sensor to measure V_a;

4. a barometric altimeter to measure h_B.

All measurements are contaminated by random noise and unknown biases. The state vector of the aircraft is defined by equations (A9.1)–(A9.9), which are augmented by the variables describing the atmospheric turbulence and the gyro biases. Thus, in the nonlinear state-space equation, we have

$$\dot{X}(t) = f[X(t),\ U(t),t] + W(t) \tag{9.3.24}$$

where $X(t) \in R^{16}$ is the augmented state vector

$$X(t) \triangleq [u,\ v,\ w,\ p,\ q,\ r,\ \phi,\ \theta,\ \psi,\ h,\ u_g,\ v_g,\ w_g,\ \mu_p,\ \mu_q,\ \mu_r]^T \tag{9.3.25}$$

The first ten terms in this vector are aircraft state variables, the next three are the gust components and the last three are the constant rate gyro biases. Initially, it is assumed that $\dot{\mu}_p, = \dot{\mu}_q, = \dot{\mu}_r = 0$. The situation where the biases are time-varying is addressed later. The aircraft control vector consists of elevator, aileron, rudder, and throttle commands and is defined as

$$U(t) \triangleq [\delta_a,\ \delta_e,\ \delta_r,\ \delta_T]^T \tag{9.3.26}$$

where $U(t) \in R^4$. The process noise vector $W(t)$ is $W(t) \in R^{16}$ and is defined as

$$W(t) \triangleq [0,\ 0,\ 0,\ 0,\ 0,\ 0,\ 0,\ 0,\ 0,\ 0,\ w_1,\ w_2,\ w_3,\ 0,\ 0,\ 0]^T \tag{9.3.27}$$

The linear measurement equation is defined by

$$Z_k = HX_k + V_k \tag{9.3.28}$$

where

$$Z_k \triangleq [\ V_{m_k},\ p_{m_k},\ q_{m_k},\ r_{m_k},\ \psi_{m_k},\ h_{m_k}]^T \tag{9.3.29}$$

The measurement matrix H is defined by

$$H \triangleq \begin{bmatrix} 1 & 0 & 0 & 0 & 0 & 0 & 0 & 0 & 0 & 0 & -1 & 0 & 0 & 0 & 0 & 0 \\ 0 & 0 & 0 & 1 & 0 & 0 & 0 & 0 & 0 & 0 & 0 & 0 & 0 & 1 & 0 & 0 \\ 0 & 0 & 0 & 0 & 1 & 0 & 0 & 0 & 0 & 0 & 0 & 0 & 0 & 0 & 1 & 0 \\ 0 & 0 & 0 & 0 & 0 & 1 & 0 & 0 & 0 & 0 & 0 & 0 & 0 & 0 & 0 & 1 \\ 0 & 0 & 0 & 0 & 0 & 0 & 0 & 0 & 1 & 0 & 0 & 0 & 0 & 0 & 0 & 0 \\ 0 & 0 & 0 & 0 & 0 & 0 & 0 & 0 & 0 & 1 & 0 & 0 & 0 & 0 & 0 & 0 \end{bmatrix} \tag{9.3.30}$$

$V_k \in R^6$ is a vector of discrete white noise components defined by

$$V_k \triangleq [\ V_{v_k},\ V_{p_k},\ V_{q_k},\ V_{r_k},\ V_{\psi_k},\ V_{h_k}]^T \tag{9.3.31}$$

Equations (9.3.24) and (9.3.27) permit the formulation of the EKF algorithm in accordance with Appendix B9. For preliminary testing of the algorithm, a series of computer simulations was performed. The aircraft chosen as a model was a slow, subsonic, conventionally configured RPV. Its aerodynamical data and aeromechanical parameters are given in Appendix C5. The simulation represents a 300 sec run of uncontrolled flight at an initial velocity of 40 m/sec, an initial altitude of 1000 m, and a turbulence of 1m/sec rms. The particular model chosen had unstable lateral dynamics. It therefore executed a spiral dive as indicated in the simulation plot shown in Figure 9.15.

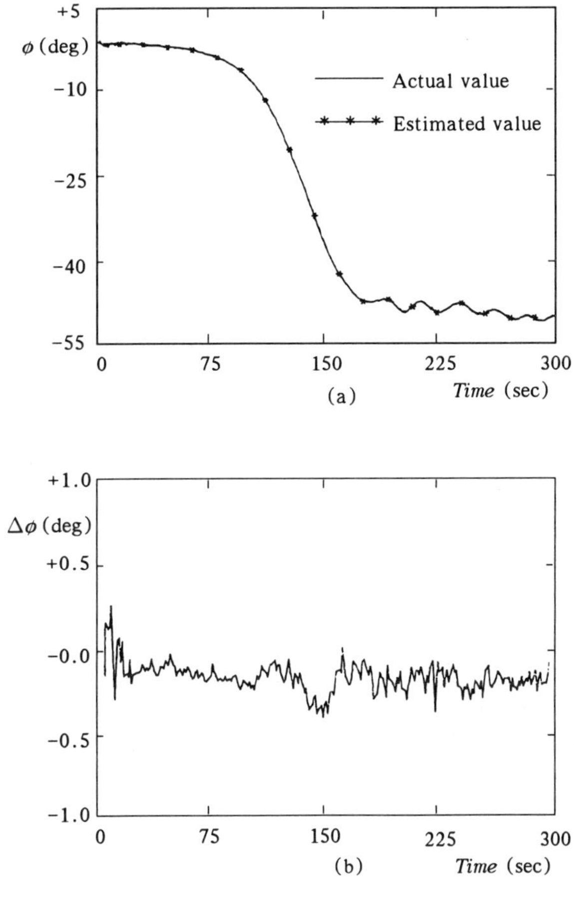

Figure 9.15. Simulation of aircraft attitude angles in ϕ, θ in an uncontrolled spiral dive.

Under these conditions, all the aircraft parameters undergo large variations. However, as seen in Figure 9.15, the estimation errors of ϕ and θ converge to less than 0.1 deg. Additional simulation runs were performed for a variety of flight trajectories and turbulence levels, yielding consistently similar results. The foregoing assumes that the aircraft parameters are exactly known, which is clearly not the case in reality. On the basis of state-of-the-art experience in the determination of aerodynamic coefficients, these uncertainties were assumed to be on the order of 5%. The effects of these uncertainties were investigated by flight simulations, both for straight and level flight and for intense "slalom" maneuvers at 500 m altitude and an air speed of 40 m/sec. For each run, a different parameter was varied. The results disclosed that in all cases, 5% variations could be tolerated without exceeding attitude estimation errors of 1 deg. The largest sensitivity turned out to be to variations in the lift slope $C_{L\alpha}$, which is essentially a constant for moderate values of α. However, this coefficient is relatively well known, and its uncertainty is small. Effects of uncertainty in control derivatives can be interpreted as deviations in stability derivatives. Trim offsets can be canceled by the flight control system, and they would not affect the EKF.

Clearly, the algorithm can operate only after the aircraft becomes airborne because the aircraft model is not valid on the ground . Therefore, initially, and for a short time before takeoff, the attitude angles are determined by the integration of Euler equations (9.3.1) to (9.3.3). Initial errors, which may accumulate during this period, vanish after awhile because the estimator converges, within the steady-state errors $\Delta\phi$, $\Delta\theta$, to the real aircraft attitude angles.

The foregoing, carried out with the full-state EKF $X \in R^{16}$ as described in [19], was not feasible at the time for real-time operation. Examination of the computational cycle time disclosed that it required 1 sec CPU time on the IBM VAX 750, the computer then available, whereas in reality the propagation of the estimated state vector must be carried out with a step not greater than 10 m sec to assure a sufficiently high rate for display and control. In [19], it is shown how the state vector can be reduced to a lower order by giving up some of the measurements and taking advantage of the sparseness of the measurement matrix H to simplify the computations. The result of these reductions and simplifications was that the algorithm provided the attitude angle estimates at the rate of 16 Hz. However, presently available state of the art microprocessors in the class of 806486 or higher perform the entire computational process detailed in Appendices A9 and B9 within 10 m sec, that is, in real time without resorting to simplifications and approximations. Enhanced computational economy, as described in [19], has its merits if the computational task is viewed in the broader sense of an entire avionic system that performs additional real-time functions such as navigation, reconnaissance, and system management. The implementation of the EKF algorithm is based on the formulations given in [21]. Specifically, the reduction of dim(X) is achieved by the following steps:

1. The altitude h is removed from X, and its measurement h_m from Z; h_m, however, is used to determine the air density for updating aerodynamic coefficients.

2. The azimuth ψ is removed from X. There is no point in trying to derive its

estimate because in this context, the estimate cannot be better than the measurement.

3. Instead of estimating u, v, w, u_g, v_g, and w_g, the relative air speeds, $u_r = u - u_g$, $v_r = v - v_g$, $w_r = w - w_g$, are estimated. These are of V_a and are the relevant quantities that generate the aerodynamic forces that operate on the aircraft. Consequently, the reduced order state and measurement vectors are

$$X \triangleq \left[\mu_p, \mu_q, \mu_r, p, q, r, u_r, v_r, w_r, \phi, \theta\right]^T \tag{9.3.32}$$

$$Z \triangleq \left[p_m, q_m, r_m, V_{am}\right]^T \tag{9.3.33}$$

Thus, $\dim(X)$ and $\dim(Z)$ have been reduced to 11 and 4, respectively. Since the computational cycle time for the EKF is approximately proportional to the cube of $\dim(X)$, the cycle time is reduced approximately by a factor of 3. This reduction, however, was not sufficient to achieve on-line operation.

First, note that the propagation of P requires the solution of n^2 differential equations. But, recalling that P is symmetric, only $(n+1)n/2$ are needed. In the present case, with $n = 16$, 136 solutions would be needed; in the reduced state, with $n = 11$, only 66 solutions are required.

Next, one is tempted to use a relatively large integration step and the simple Euler method. This, however, easily causes numerical instability. To prevent this, a small integration step is indicated, but the penalty would be a longer computation time. The alternative could be a more precise integration algorithm such as Runge-Kutta, but again, at the expense of increased computation time.

To circumvent this conflict, the discrete formulation of (B9.2) which is also its solution is chosen. Thus, we have,

$$P_{k+1|k} = \Phi\left(\hat{X}, U, \Delta t\right) P_{k|k} \Phi^T\left(\hat{X}, U, \Delta t\right) + \tilde{Q}(\Delta t) \tag{9.3.34}$$

where Φ is the transition matrix describing the system dynamics that satisfies the matrix differential equation

$$\frac{\partial \Phi}{\partial t} = F\Phi \tag{9.3.35}$$

and $F\left(\hat{X}, U, \Delta t\right)$ is defined by (B9.6). Equation (9.3.34) is determined by the relationship

$$\Phi\left(\hat{X}, U, \Delta t\right) = \exp\left\{F\left(\hat{X}, U, \Delta t\right)\right\} = \sum_{n=0}^{\infty} \frac{F\left(\hat{X}, U, \Delta t\right)^n}{n!} \tag{9.3.36}$$

The discrete noise covariance matrix $\tilde{Q}(\Delta t)$ is determined by

$$\tilde{Q}(\Delta t) = \int_0^{\Delta t} \Phi(\hat{X}, U, \tau) Q(\tau) \Phi(\hat{X}, U, \tau)^T d\tau \tag{9.3.37}$$

By neglecting terms of order $(\Delta t)^2$ and higher, (9.3.36) can be simplified to

$$\tilde{Q}(\Delta t) \cong Q\Delta t \tag{9.3.38}$$

In accordance with (9.3.35), the transition matrix Φ is given by an infinite series. However, inspection discloses that it is quite adequate to retain only the first three terms to obtain satisfactory precision.

Toward the recursive computation of the gain matrix K_{k+1} in accordance with (B9.9), further simplifications can be made by utilizing the properties of Φ and of the measurement matrix H. Φ can be partitioned into submatrices as follows:

$$\Phi(\hat{X}, U, \Delta t) = \begin{bmatrix} I & 0 & 0 \\ 0 & \Phi_{11} & \Phi_{12} \\ 0 & \Phi_{21} & \Phi_{22} \end{bmatrix} \tag{9.3.39}$$

I is a (3×3) identity matrix that originates in augmentation of X by the three constant unknown gyro drift rates μ_p, μ_q, μ_r. Φ_{11}, Φ_{12}, Φ_{21}, Φ_{22} are (4×4) submatrices that, together with I, constitute the (11×11) transition matrix Φ. The covariance matrix P is symmetrical and can be partitioned as follows:

$$P \triangleq \begin{bmatrix} P_{11} & P_{12} & P_{13} \\ P_{12}^T & P_{22} & P_{23} \\ P_{13}^T & P_{23}^T & P_{33} \end{bmatrix} \tag{9.3.40}$$

P_{11} is (3×3); P_{12} and P_{13} are (3×4); P_{22}, P_{23}, P_{33} are (4×4). Expressing (9.3.33) with the aid of (9.3.38) and (9.3.39), we have the covariance matrix propagation in explicit form:

$$\begin{bmatrix} P_{11} & P_{12} & P_{13} \\ P_{12}^T & P_{22} & P_{23} \\ P_{13}^T & P_{23}^T & P_{33} \end{bmatrix}_{k+1|k} = \begin{bmatrix} I & 0 & 0 \\ 0 & \Phi_{11} & \Phi_{12} \\ 0 & \Phi_{21} & \Phi_{22} \end{bmatrix} \begin{bmatrix} P_{11} & P_{12} & P_{13} \\ P_{12}^T & P_{22} & P_{23} \\ P_{13}^T & P_{23}^T & P_{33} \end{bmatrix}_{k|k} \begin{bmatrix} I & 0 & 0 \\ 0 & \Phi_{11}^T & \Phi_{21}^T \\ 0 & \Phi_{12}^T & \Phi_{22} \end{bmatrix}$$

$$+ \begin{bmatrix} 0 & 0 & 0 \\ 0 & \tilde{Q}_1 & 0 \\ 0 & 0 & \tilde{Q}_2 \end{bmatrix} \tag{9.3.41}$$

From (9.3.40), the explicit covariance propagation equations are

$$
\begin{aligned}
\left[P_{11}\right]_{k+1|k} &= \left[P_{11}\right]_{k|k} \\
\left[P_{12}\right]_{k+1|k} &= \left[P_{12}\right]_{k|k}\Phi_{11}^{T} + \left[P_{13}\right]_{k|k}\Phi_{12}^{T} \\
\left[P_{13}\right]_{k+1|k} &= \left[P_{12}\right]_{k|k}\Phi_{21}^{T} + \left[P_{13}\right]_{k|k}\Phi_{22}^{T} \\
\left[P_{22}\right]_{k+1|k} &= \Phi_{11}^{T}\left[P_{22}\right]_{k|k}\Phi_{11}^{T} + \Phi_{11}^{T}\left[P_{23}\right]_{k|k}\Phi_{12}^{T} + \left\{\Phi_{11}\left[P_{23}\right]_{k|k}\Phi_{12}^{T}\right\}^{T} \\
&\quad + \Phi_{12}\left[P_{33}\right]_{k|k}\Phi_{12}^{T} + \tilde{Q}_{1} \\
\left[P_{23}\right]_{k+1|k} &= \Phi_{11}^{T}\left[P_{22}\right]_{k|k}\Phi_{21}^{T} + \Phi_{11}^{T}\left[P_{23}\right]_{k|k}\Phi_{22}^{T} + \left\{\Phi_{21}\left[P_{23}\right]_{k|k}\Phi_{12}^{T}\right\}^{T} \\
&\quad + \Phi_{12}\left[P_{33}\right]_{k|k}\Phi_{12}^{T} \\
\left[P_{33}\right]_{k+1|k} &= \Phi_{21}^{T}\left[P_{22}\right]_{k|k}\Phi_{21}^{T} + \Phi_{21}\left[P_{23}\right]_{k|k}\Phi_{22}^{T} + \left\{\Phi_{21}\left[P_{23}\right]_{k|k}\Phi_{22}^{T}\right\}^{T} \\
&\quad + \Phi_{22}\left[P_{33}\right]_{k|k}\Phi_{22}^{T} + \tilde{Q}_{2}
\end{aligned}
\tag{9.3.42}
$$

The number of arithmetic operations involved in (9.3.34) is 2662 products and 2541 summations, while using (9.3.42) involves 1280 products and 1128 summations.

Further reduction in computation time is achieved by rearranging the reduced matrix to an (11×4) measurement matrix H as follows:

$$
H=\left[\begin{array}{ccc|cccc|cccc}
1 & 0 & 0 & 1 & 0 & 0 & 0 & 0 & 0 & 0 & 0 \\
0 & 1 & 0 & 0 & 1 & 0 & 0 & 0 & 0 & 0 & 0 \\
0 & 0 & 1 & 0 & 0 & 1 & 0 & 0 & 0 & 0 & 0 \\
0 & 0 & 0 & 0 & 0 & 0 & 1 & 0 & 0 & 0 & 0
\end{array}\right]
\tag{9.3.43}
$$

which can be rewritten as

$$
H=[D \mid I \mid 0]
\tag{9.3.44}
$$

Exploiting the sparseness of H and substituting (9.3.43) and (9.3.40) into (B9.9), we obtain the following equations for computing the gains:

$$
\left[K_1\right]_{k+1} = G_1 M^{-1}
\tag{9.3.45}
$$

$$
\left[K_2\right]_{k+1} = G_2 M^{-1}
\tag{9.3.46}
$$

$$
\left[K_3\right]_{k+1} = G_3 M^{-1}
\tag{9.3.47}
$$

where K_1 is (3×4); K_2 and K_3 are (4×4) and are submatrices of the (11×4) overall gain matrix K. G_1, G_2, and G_3 are given by the relationships

$$G_1=[P_{11}]_{k+1|k}H^T+[P_{12}]_{k+1|k} \tag{9.3.48}$$

$$G_2=[P_{12}]_{k+1|k}H^T+[P_{22}]_{k+1|k} \tag{9.3.49}$$

$$G_3=[P_{13}]_{k+1|k}H^T+[P_{23}]_{k+1|k} \tag{9.3.50}$$

and M, with the aid of Appendix B9, is given by

$$M=H[P_{11}]_{k+1|k}H^T+H[P_{12}]_{k+1|k}+\{H[P_{12}]_{k+1|k}\}^T+R_k \tag{9.3.51}$$

where R_k is the covariance matrix of the measurement noise V_k.

Taking advantage of the sparseness and the special form of H further reduces the computational load in comparison with the standard matrix operations which compute every element of a $(n\times n)$ matrix regardless of the numerous zero elements. Thus, for updating the covariance matrix P, the substitution of (9.3.44) to (9.3.47) into (B9.12), yields the following simplified updating equations of the elements of P:

$$\begin{aligned}
[P_{11}]_{k+1|k+1}&=[P_{11}]_{k+1|k}-G_1[K_1]^T_{k+1}-\{G_1[K_1]^T_{k+1}\}^T+[K_1]_{k+1}M[K_1]^T_{k+1}\\
[P_{12}]_{k+1|k+1}&=[P_{12}]_{k+1|k}-G_1[K_2]^T_{k+1}-[K_1]^T_{k+1}G_2^T+[K_1]_{k+1}M[K_2]^T_{k+1}\\
[P_{13}]_{k+1|k+1}&=[P_{13}]_{k+1|k}-G_1[K_3]^T_{k+1}-[K_1]^T_{k+1}G_3^T+[K_1]_{k+1}M[K_3]^T_{k+1}\\
[P_{22}]_{k+1|k+1}&=[P_{22}]_{k+1|k}-G_2[K_1]^T_{k+1}-\{G_2[K_2]^T_{k+1}\}^T+[K_2]_{k+1}M[K_2]^T_{k+1}\\
[P_{23}]_{k+1|k+1}&=[P_{23}]_{k+1|k}-G_2[K_3]^T_{k+1}-[K_2]^T_{k+1}G_3^T+[K_2]_{k+1}M[K_3]^T_{k+1}\\
[P_{33}]_{k+1|k+1}&=[P_{33}]_{k+1|k}-G_3[K_3]^T_{k+1}-\{G_3[K_3]^T_{k+1}\}^T+[K_3]_{k+1}M[K_3]^T_{k+1}
\end{aligned} \tag{9.3.52}$$

The propagation of P in time requires the linearization of the equations of the aircraft motion in order to derive the Jacobian F and the computation of the transition matrix Φ. Attempting to use the original EKF algorithm requires the computation of F and Φ in each computational cycle, thus causing a very substantial computational load. However, the dynamics of the aircraft model indicate that even the fastest modes—roll, dutch roll, and short period pitch—are on the order of 1sec time constants. Since the linearization of F is done in the neighborhood of $\hat{x}_{k+1|k}$, and assuming that $\hat{x}_{k+1|k} \rightarrow x_{k+1}$, the computation of Φ at a rate greater than a few times per second is not needed. That is, a rate much slower than the 10 m sec called for in the original full-state EKF is sufficient. A high rate of computation practically does not contribute to the precision because, Φ_{k+1} does not differ much from Φ_k.

The reduced-order EKF, with the foregoing computational simplifications, was implemented on a VME Motorola system 1131 minicomputer. Tests of this algorithm disclosed that the cycle time of the EKF was reduced to 0.111 sec, almost ten times faster than for the original full-state EKF, but not quite satisfactory for real-time operation. Furthermore, this cycle time, if used for the propagation of X, could easily cause numerical divergence. To provide a workable compromise between the required filter output rate on the order of 20 Hz and the computational requirements, the filter was implemented by a multi-rate scheme using two separate computation rates, so that the time-consuming computation of Φ was not executed in every cycle. This saving in the computation rate of Φ permitted the increase of the update rate of the filter, which eventually resulted in the compromise rate of 16 Hz. The computational cycle consisted of the following steps:

I. The first cycle:
1) Propagation of the state vector X in accordance with (B9.4).
2) Computation of the Jacobian in accordance with (B9.6).
3) Computation of Φ in accordance with (9.3.365) retaining only the first three terms in the series expansion.
4) Propagation of P in accordance with equations (9.3.42).

II. The second cycle:
1) Propagation of the state vector X in accordance with (B9.4).
2) Propagation of P in accordance with equations (9.3.41) in which Φ is computed in the first cycle..
3) Computation of the gain matrix K in accordance with equations (9.3.45) to (9.3.49).
4) Updating the state vector X in accordance with (B9.10).
5) Updating P in accordance with equations (9.3.52).

The second cycle is executed five times, after which the computation in the first cycle is resumed, and so on. This 5:1 ratio was determined by tuning the filter and by simulation trials. It represents a fair compromise between the need to reduce the update interval of Φ and the need to increase the output rate of the EKF. A flow chart that describes this multirate EKF algorithm is shown in Figure 2 of [19].

A simulation run for the roll angle estimate $\hat{\phi}$ and the estimation error $\Delta\hat{\phi}$ for a duration of 600 sec is shown in Figure 9.16. The flight conditions are the same as for the simulation test of the full state EKF described in Figure 9.15. The rms values $\sigma_{\Delta\hat{\phi}}$ (and $\sigma_{\Delta\hat{\theta}}$, not shown here) are 0.21 and 0.26 deg, respectively, which is $\sim$2.5 larger than for the full-state EKF shown earlier. However, these values are compatible with flight control and display needs.

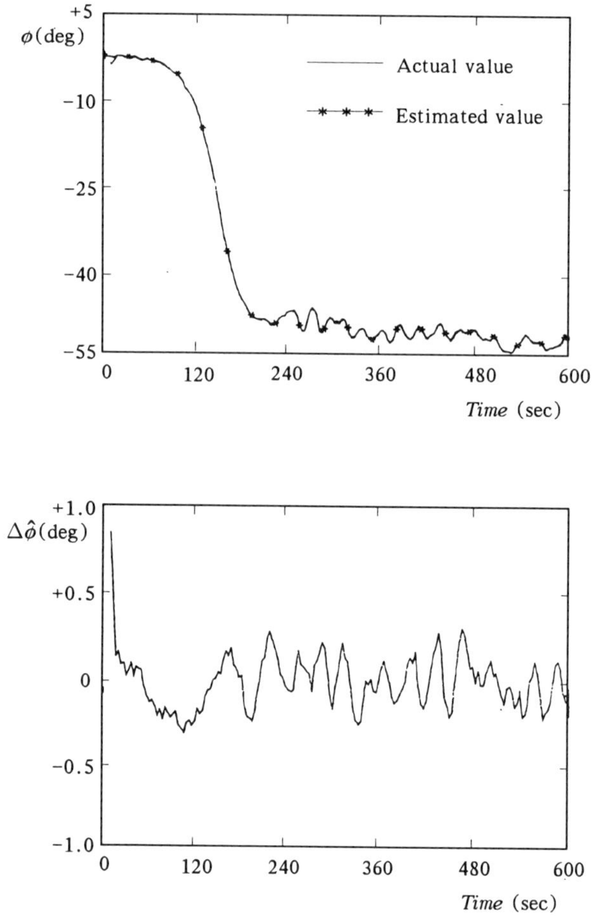

Figure 9.16. Simulation of reduced order EKF attitude angle estimation.

Comparing Figure 9.16 to Figure 9.15 discloses that the convergence time of the reduced-order filter is longer than of the full-state filter. On the other hand, the relative sensitivity of the estimates to possible parameter variations was found to be smaller than for the full-state filter. This reduced sensitivity is characteristic of suboptimal solutions. The validity of the dual-rate filter technique was tested by simulation, under conditions identical to the above, of a single-rate filter running at 16 Hz in which both F and Φ were computed in each cycle. The results showed that there was practically no difference compared to the performance displayed in Figure 9.15. This result supports the choice of the dual-rate filter.

To further validate the performance of the reduced-order dual-rate filter, simulation tests were carried out for a controlled flight path of a slalom trajectory, in which ϕ and θ varied with a period of 40 sec and amplitudes of 10 and 15 deg respectively. A sample of these tests is displayed in Figure 9.17. This sample demonstrates the validity of the algorithm under conditions of substantial maneuvering in which the nonlinearities of the dynamical model come to bear. It is interesting to note that after the settling of the initial transient, the steady-state estimation errors $\Delta\hat{\phi}$ and $\Delta\hat{\theta}$ are smaller than those shown in the uncontrolled flight simulation shown in Figure 9.16. This is apparently the result of the enhanced observability of the system due to the consistent excitation by the slalom maneuver.

In the present example, the rate gyros were assumed to have drift rates of 0.1 deg/sec. These resulted in offsets in the attitude estimates of ~ 1 deg. Therefore, the gyro biases were included in the state vector in accordance with (9.3.31). However, further examination by simulation tests disclosed that a somewhat better gyro with a drift rate of 0.03 deg/sec would result in estimation errors of only ~ 0.25 deg. This is comparable to the random error and it does not significantly increase to total rms estimation error. Therefore, the estimation of the drift rates is not necessarily essential and it could be discarded. This implies that dim(X) would be reduced from 11 to 8, thus achieving further substantial saving in the cycle time and computational load.

The assumption that $\dot{\mu}_p, \dot{\mu}_q, \dot{\mu}_r = 0$ is not quite valid in reality. In the case of low grade gyros, the drift rate is determined by numerous factors such as temperature variations, dead space, hysteresis and sudden jumps due to mechanical uncertainties which are sensitive to shocks and vibration. The drift history, therefore may be a combination of random variations and sudden jumps.

The EKF, as described above, is not designed to cope with such variations. If no special provisions are made, sudden jumps in the drift rates will not be abruptly estimated. The reason, of course, is that in principle, $P \to 0$ and consequently, $K \to 0$. In the professional slang, this is sometimes referred to as the estimator "going to sleep". In reality, in the presence of noise, P and K do not quite converge to zero. Thus, several minutes may elapse before the estimate reconverges to the new value. This situation is demonstrated by the simulation shown in Figure 9.18, which shows the response of the estimator to sudden jumps in the roll, pitch and yaw rate gyros.

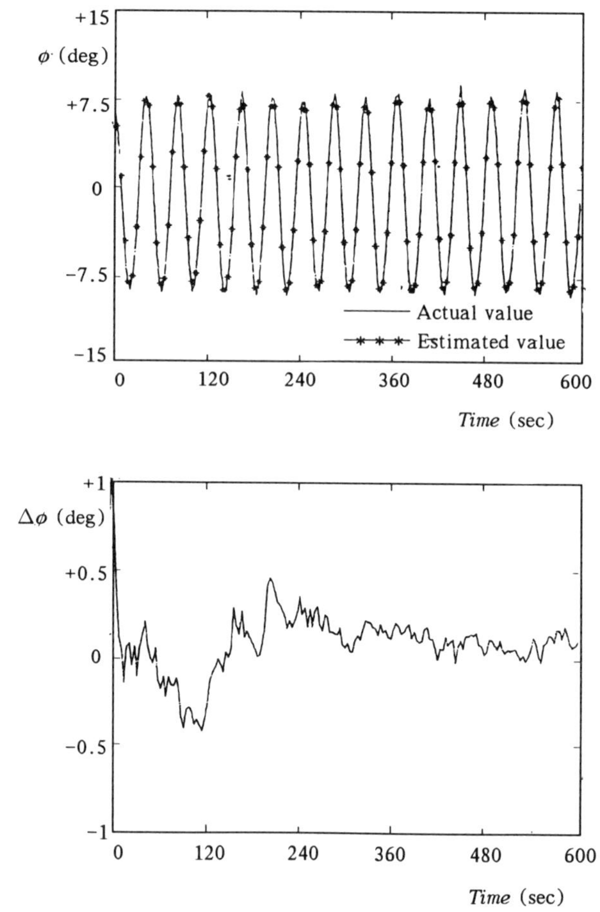

Figure 9.17. Simulation of roll angle estimation and error in an intensive slalom maneuver.

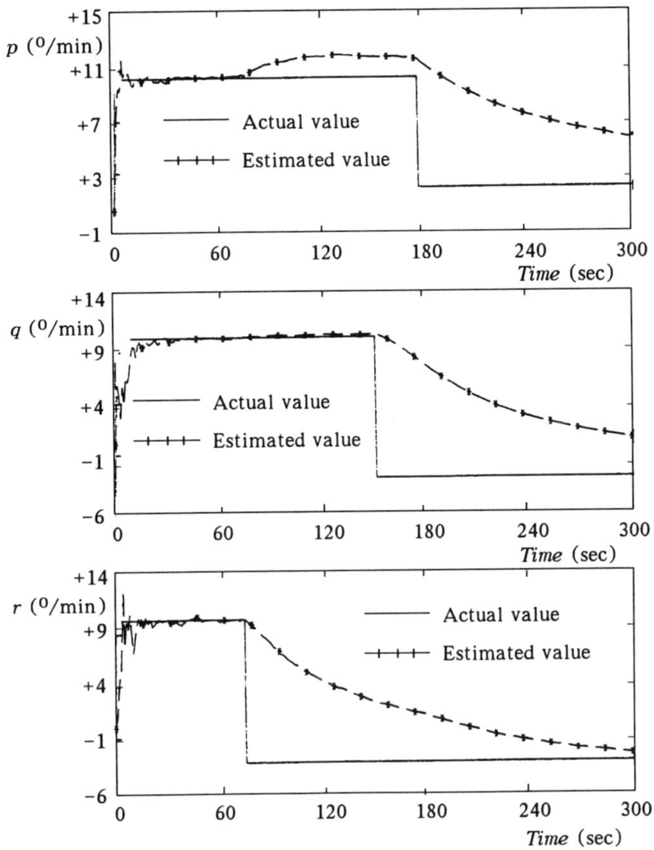

Figure 9.18. Response of the angular rate estimate to sudden jumps in the actual values.

So far, the gyro drift rates have been modeled as constants. In reality, this assumption may be valid only for limited periods of time. It is not feasible to mathematically model biases for low-grade rate gyros because of their complex and erratic behavior. Consequently, their drift rate has been modeled as piecewise constant with jumps at unpredictable instants in time. The EKF model for constant biases estimates the drift rates following such a jump, only after substantial time has elapsed, as shown in Figure 9.19. To implement an EKF that maintains the alertness to quickly estimate sudden changes in drift rates, the following approach was adopted: The biases continue to be modeled as constants, but to prevent the corresponding gain elements in K from converging toward zero, the diagonal elements in the error covariance P, which are due to μ_p, μ_q, μ_r, are reinitialized from time to time whenever a significant change in the biasvector occurs. This implies that a detector for bias change is required. A suitable scheme was proposed by Hinkley [22] and Basseville [23]. For its implementation, a new variable Y is defined as

$$Y_k \triangleq \omega_{m_k} - \hat{\omega}_k - \hat{\mu}_k \tag{9.3.53}$$

ω_{m_k} is the actual measured gyro output in the kth interval, $\hat{\omega}_k$ is the estimated angular rate and $\hat{\mu}_k$ is the estimated drift rate. It is assumed that in the steady state, i.e., after the decay of transients, $\hat{\omega}_k \to \omega_k$ and $\hat{\mu}_k \to \mu_k$ so that $Y_k = V_k$. Since the measurement noise V_k is zero mean, we have

$$E(Y_k) = E(V_k) = 0 \tag{9.3.54}$$

Assume now that at a given instant t_k a change in a bias occurs. This can be formulated as follows:

$$\begin{aligned} \mu_k &= \mu_0, & t &< t_k \\ \mu_k &= \mu_0 + \Delta\mu & t &\geq t_k \end{aligned} \tag{9.3.55}$$

Therefore, since the estimator is unable to update its current estimate after the event of the change, we have

$$E(Y_k) = \Delta\mu \tag{9.3.56}$$

This change in $E(Y_k)$ is the indication of the change that has occurred in the bias. For the actual implementation of the detection algorithm, a minimal detectable threshold value $\Delta\mu_m$ is chosen and two algorithms are operated in parallel, one to detect an increase and the other to detect a decrease in $E(Y_k)$. The following constitutes the recursive algorithm

$$\left.\begin{aligned} T_0 &= 0 \\ T_n &= T_{n-1} + Y_n + \Delta\mu_m/2 \\ M_n &= \max_{1 \leq k \leq n} T_k \\ \text{Alarm when: } & M_n - T_n \geq \lambda \end{aligned}\right\} \tag{9.3.57}$$

$$\left.\begin{aligned} U_0 &= 0 \\ U_n &= U_{n-1} + Y_n - \Delta\mu_m/2 \\ m_n &= \min_{1 \leq k \leq n} U_k \\ \text{Alarm when: } & U_n - m_n \geq \lambda \end{aligned}\right\} \tag{9.3.58}$$

where λ is the threshold value, the choice of which is a compromise between too many false alarms if λ is set too small and too many undetected changes if λ is set too large. and λ is determined by the following relation:

$$\lambda = \gamma \frac{2\sigma^2}{\Delta \mu_m} \tag{9.3.59}$$

where σ^2 is the variance of the measurement noise V_k and γ is a constant suitably chosen between 3 and 5.

Simulations were performed with jumps identical to those shown in Figure 9.18, but in which the Hinkley detector was activated. Sample results are shown in Figure 9.19 that demonstrate the essentially abrupt detection of the jump in the drift biases but at the expense of more noise. This additional noise is the result of false alarms on one hand and undetected jumps on the other. This additional noise, however, which is relatively high-frequency, has only a small effect on the estimated attitude angles, as can be seen in Figure 9.20.

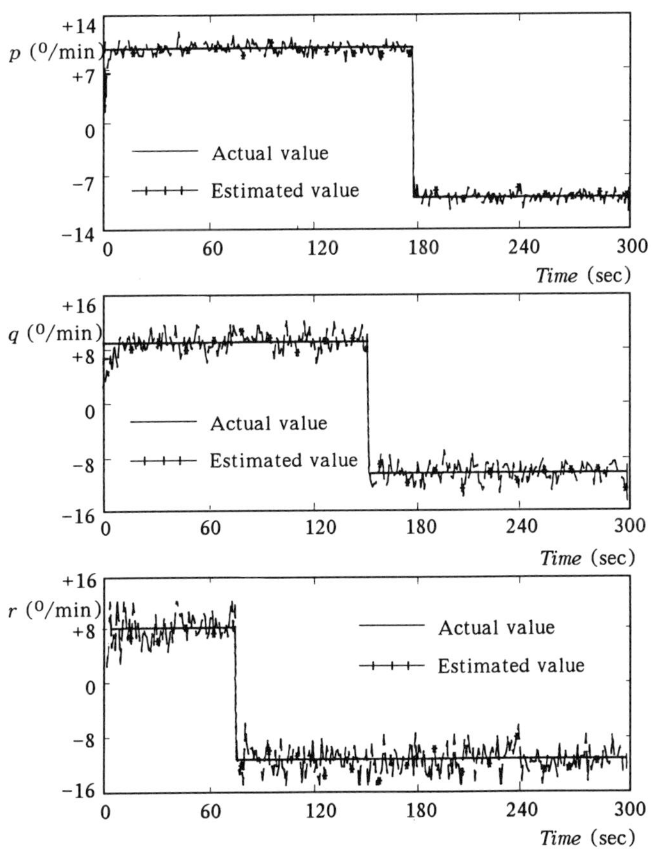

Figure 9.19. Abrupt jumps in the drift rate biases and their corresponding estimates.

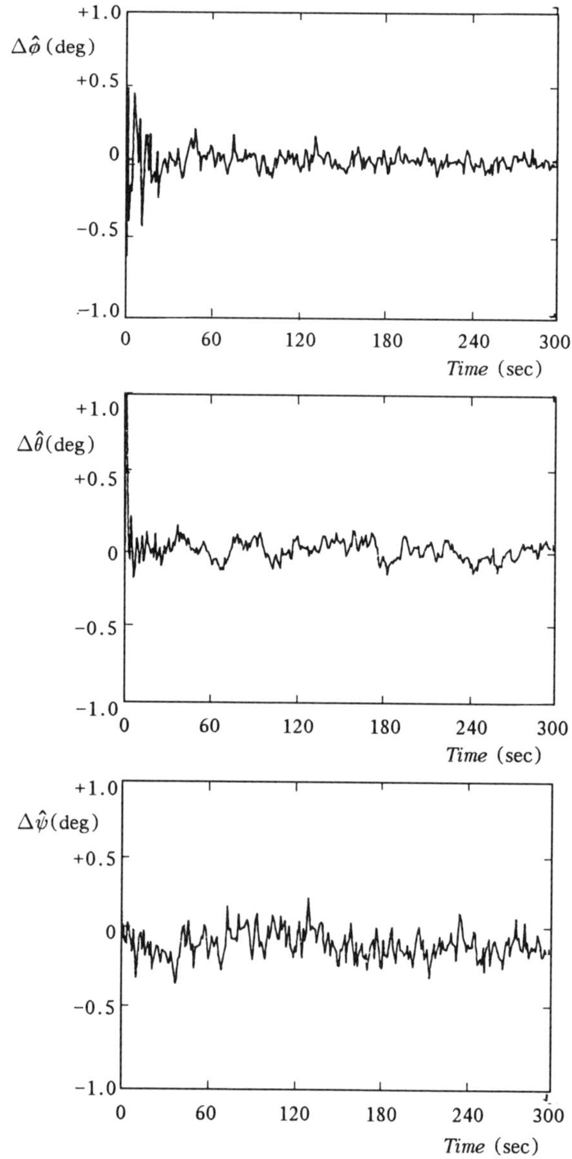

Figure 9.20. Time histories of attitude angle estimation errors for the drift rate jumps of Figure 9.18 and with an activated Hinkley detector.

The foregoing computer simulation results were experimentally validated in laboratory tests using real rate gyros mounted on a high-precision three-axis flight table that executed typical aircraft attitude variations. The details of the components, equipment and filter tuning via Q and R are given in [15]. The results showed excellent correspondence with the computer simulation described earlier.

The precision and robustness of the estimator and the feasibility of its implementation in real time render it a candidate method in conjunction with modern miniature low-cost, low-grade RLGs and FOGs, or vibratory micromechanical gyros which are highly reliable but of quite poor precision when operating on their own without aiding.

Problems

Problem 9.1– Aided distance estimation: Longitudinal accelerometer and TACAN

An aircraft is equipped with an accelerometer aligned with its longitudinal axis. The idea is to integrate the output of the accelerometer to obtain aircraft distance from a known starting point. However, the accelerometer has unknown bias and also reads $g\sin\theta$, where θ is the trim pitch angle, which provides the required angle of attack $\alpha=\theta-\gamma$ that is to sustain the required lift force. For horizontal flight, $\gamma=0$, so that $\alpha=\theta$ which is not quite constant because it varies with variations in airspeed. Thus, $g\sin\theta$ can be the source of a very large distance error. The accelerometer-based distance estimate is aided by zero-mean but noisy TACAN (Tactical Air Navigation System) [7] range measurements.

a) Lay out the scheme of a second-order well-damped complementary filter that, in principle, removes the $g\sin\theta$ error in the steady state.

The position noise of the TACAN is characterized by the autocorrelation function $\phi_{xx}(\tau)=\sigma_x^2 e^{-a|\tau|}$, where σ_x= 15 m and a= 10 sec^{-1}.

The temporal variation in θ is characterized by the autocorrelation function $\phi_{\theta\theta}(\tau) =\sigma_\theta^2 e^{-b|\tau|}$, where σ_θ = 1^0 and b = 0.015 sec^{-1}.

b) Choose the gain constants of the filter so that it provides substantial attenuation of the TACAN noise while assuring a damping factor $\varsigma\geq 0.5$.

c) For the choice in b), calculate the mean square errors due to the accelerometer θ-induced bias, the TACAN noise, and the combined noise.

Problem 9.2– Vertical gyro: A special case of an aided free gyro

Describe the vertical gyro in one of its axes, say ξ, in terms of a complementary filter.

a) Draw the equivalent schematic, and define the measurement and reference variables and the errors due to measurement and process noise as defined for a complementary filter.

b) Indicate the position of the cutout switch in the schematic.

c) Show the function of optional integral control to remove steady-state errors.

Hint: Describe the angular deviation ς as the time integral of $\dot{\varsigma}_d$ and δ as the reference angle.

Problem 9.3– Schuler oscillations: Damping by external measurement of velocity

Describe a single axis INS systems in the format of a complementary filter including Schuler tuning.

a) Show the effect of constant accelerometer bias and of a constant gyro drift rate bias.

b) Choose typical values for these biases and derive the explicit expressions of $\Delta X(t)$ as a result of these biases.

c) Add an external ground speed measurement and show that it can damp the Schuler oscillation. Choose the gain for the velocity measurement so that the damping factor ς is 0.6.

Problem 9.4— Autonomously aided roll angle estimation: By yaw rate aiding

Yaw angular rate $\dot{\psi}$ and roll angle ϕ are related through $g\tan\phi = \dot{\psi}V$, where V is the aircraft ground speed. The aircraft is equipped with a strap-down yaw rate gyro. Thus, the computed roll angle is $\phi_c = \tan^{-1}[(V/g)\dot{\psi}]$. The aircraft is also equipped with a roll rate gyro that outputs $\dot{\phi}_m \triangleq \dot{\phi} + \dot{\phi}_d$.

a) Develop a complementary filter scheme that shows how ϕ_c can be used to aid ϕ_m, which is obtained by the integration of $\dot{\phi}_m$.

b) What is the expression for the steady-state roll angle estimation error $\Delta\hat{\phi}$ as a result of $\dot{\phi}_d$?

c) Assume $\dot{\phi}_d$ = 8 ^{0}/min. Choose the filter gains so that $\Delta\hat{\phi} \leq 1^0$.

Problem 9.5—Aided roll angle estimation by yaw rate aiding: By accelerometer pair

Basically, the problem is similar to 9.4. However, ϕ measurement is derived from a pair of accelerometers mounted on the aircraft wing tips, as described in Chapter 6.

a) Devise the required scheme of the CF to implement the concept. What is the order of the CF with and without integral control?

b) Assume a mismatch bias error Δa = 0.1 mg. Designing for a good damping factor of the CF, compute the gain vector of the CF so that $\Delta\phi$ does not exceed 1 deg.

c) Discuss the feasibility of the concept.

Problem 9.6— Signal-to-noise ratio in a CF or KF: Interpretation

Interpret the CF in terms of a Luenberger observer and represent the accelerometer noise and Doppler noise of a Doppler-inertial system in terms of measurement and process noise, respectively. Define precisely the meaning of signal-to-noise ratio.

Problem 9.7— Kalman filtering: Interpretation of low gain values

In a complementary filter, it is deemed desirable to reduce the gain matrix as much as possible in order to attenuate the process noise.

a) What is the meaning of "possible" in this context. Is an effect of diminishing returns involved?

b) What would be the effect at the filter output of wideband measurement noise if the filter gain is very low?

c) Does the effect indicated in b) appear in a Kalman filter whose gain is optimized in accordance with the signal-to-noise ratio? Explain.

Appendix A9: Equations of aircraft dynamics

Notations of aircraft parameters:

S — Reference wing area
b — Wing span
c — Wing cord
T — Engine thrust

1) Dynamical translatory equations of motion in body axes:

$$\dot{u} = -\frac{1}{2}\rho V_a^2 (S/m)[C_D \cos\alpha \cos\beta + C_Y \cos\alpha \sin\beta - C_L \sin\alpha] + T/m - g\sin\theta - qw + rv \tag{A9.1}$$

$$\dot{v} = -\frac{1}{2}\rho V_a^2 (S/m)[C_D \sin\beta - C_Y \cos\beta] + g\sin\phi \cos\theta - ru + pw \tag{A9.2}$$

$$\dot{w} = -\frac{1}{2}\rho V_a^2 (S/m)[C_D \sin\alpha \cos\beta + C_Y \sin\alpha \sin\beta + C_L \cos\alpha] - g\cos\theta \sin\phi - pv + qu \tag{A9.3}$$

2) Angular dynamical equations of motion:

$$\dot{p} = \frac{1}{I_x I_z - I_{xz}^2}\left\{I_z\left[\frac{1}{2}\rho V_a^2 SbC_l + (I_y - I_z)qr\right] + I_{xz}\left[\frac{1}{2}\rho V_a^2 SbC_n + (I_x - I_y + I_z)pq - I_{xz}qr\right]\right\} \tag{A9.4}$$

$$\dot{q} = \frac{1}{I_y}\left[\frac{1}{2}\rho V_a^2 ScC_m + I_{xz}(r^2 - p^2) + (I_z - I_x)rp\right] \tag{A9.5}$$

$$\dot{r} = \frac{1}{I_x I_z - I_{xz}^2}\left\{I_z\left[\frac{1}{2}\rho V_a^2 SbC_n + (I_z - I_y)qr\right] + I_{xz}\left[\frac{1}{2}\rho V_a^2 SbC_l + (I_y - I_x - I_z)qr + I_{xz}pq\right]\right\} \tag{A9.6}$$

3) The kinematic Euler equations:

$$\dot{\phi} = p + \tan\theta(q\sin\phi + r\cos\phi) \tag{A9.7}$$

$$\dot{\theta} = q\cos\phi - r\sin\phi \tag{A9.8}$$

$$\dot{\psi} = q\frac{\sin\phi}{\cos\theta} + r\frac{\cos\phi}{\cos\theta} \tag{A9.9}$$

4) Transformation of ground speed from body to Earth coordinates:.

$$\dot{Z}_E = -\dot{h} = -u\sin\theta + v\sin\phi\cos\theta + w\cos\phi\cos\theta \tag{A9.10}$$

$$\dot{X}_E = u\cos\theta\cos\psi + v(\sin\phi\sin\theta\cos\psi - \cos\phi\sin\psi) + w(\cos\phi\sin\theta\cos\psi + \sin\phi\sin\psi) \tag{A9.11}$$

$$\dot{Y}_E = u\cos\theta\sin\psi + v(\sin\phi\sin\theta\sin\psi - \cos\phi\cos\psi) + w(\cos\phi\sin\theta\sin\psi + \sin\phi\cos\psi) \tag{A9.12}$$

5) Aerodynamic coefficients:

Longitudinal:

$$C_L = C_{L0} + C_{L\alpha}\alpha + C_{L\beta} + \frac{c}{2V_a}C_{Lq}q + C_{L\delta e}\delta_e \tag{A9.13}$$

$$C_m = C_{m0} + C_{m\alpha}\alpha + C_{m\beta}\beta + \frac{c}{2V_a}(C_{m\dot{\alpha}}\dot{\alpha} + C_{mq}q) + C_{m\delta_e}\delta_e \tag{A9.14}$$

$$C_D = C_{D0} + C_k C_L^2 \tag{A9.15}$$

Lateral:

$$C_Y = C_{Y\alpha}\alpha + C_{Y\beta}\beta + \frac{b}{2V_a}(C_{Yp}p + C_{Yr}r) + C_{Y\delta a}\delta_a + C_{Y\delta r}\delta_r + C_{Y\delta e}\delta_e \tag{A9.16}$$

$$C_l = C_{l\alpha}\alpha + C_{l\beta}\beta + \frac{b}{2V_a}(C_{lp}p + C_{lr}r) + C_{l\delta a}\delta_a + C_{l\delta r}\delta_r + C_{l\delta e}\delta_e \tag{A9.17}$$

$$C_n = C_{n\alpha}\alpha + C_{n\beta}\beta + \frac{b}{2V_a}(C_{np}p + C_{nr}r) + C_{n\delta a}\delta_a + C_{n\delta r}\delta_r + C_{n\delta e}\delta_e \tag{A9.18}$$

The absolute value of air speed V_a, angle of attack α, and angle of side slip β are given by,

$$V_a = \left[(u-u_g)^2 + (v-v_g)^2 + (w-w_g)^2\right]^{1/2} \tag{A9.19}$$

$$\alpha = \tan^{-1}\left[\frac{w-w_g}{u-u_g}\right] \tag{A9.20}$$

$$\beta = \sin^{-1}\left[\frac{v-v_g}{V_a}\right] \tag{A9.21}$$

where u_g, v_g, w_g are wind gust components along the body axes x_b, y_b, z_b. The relationship between air density ρ and altitude h is given by

$$\rho = 82.90 \times \frac{\{1 - [h/(4.4\times10^4)]\}^{5.2067}}{289.15 - 6.5\times10^{-3}h} \tag{A9.22}$$

The foregoing set of nonlinear equations constitute the mathematical model that underlies the formal expression $f[X(t),\ u(t), t] + w(t)$ in Appendix B9.

Appendix B9: Formulation of the extended Kalman filter

In the general case, the system model may be nonlinear and time-arying so that the closed-form solutions of the linear time-invariant solution of the KF are not valid. This obstacle is surmounted by the extended Kalman filter, which allows system nonlinearities and temporal variations. In general, this is the case in modeling aircraft dynamics. All the linear models are approximations that must be carefully scrutinized for their validity in a given application. The solution of the Riccati equation and the propagation of the state vector and the error covariance matrix can only be accomplished by a numerical recursive algorithm, as will be shown below. The most common approach is to model the aircraft in continuous format and the KF in discrete format. This known as the *continuous-discrete formulation.*

A general form of the continuous time nonlinear state-space equation, which is representative of an aircraft, can be expressed [17] as

$$\dot{X} = f[X(t),\ u(t), t] + w(t) \tag{B9.1}$$

where X is the $(n \times 1)$ state vector, u is the $(r \times 1)$ control vector, dim $u \leq$ dim X, and w, dim $w \leq$ dim X, is modeled as a white process noise vector, which is essential in the development of the EKF algorithm.

The sampled noisy measurement equation is

$$z_k = HX_k + v_k \tag{B9.2}$$

The role of the EKF is to provide estimates $\hat{X}_k$ of X_k from the noisy measurements z_k. The estimation error is defined as

$$x \triangleq X - \hat{X} \tag{B9.3}$$

The computational process proceeds as follows:

1) Define the system model

$$\dot{\hat{X}} = f(\hat{X}, u) \tag{B9.4}$$

and solve it (numerically) from the instant t_k to t_{k+1}. Denote $\hat{X}(t_k) = \hat{X}_{k|k}$ and, accordingly, the solution at t_{k+1}, by $\hat{X}(t_{k+1}) = \hat{X}_{k+1|k}$.

2) Propagate the estimation covariance matrix $P \triangleq E(xx')$ by

$$\dot{P} = F(\hat{X}, u)P + PF^T(\hat{X}, u) + Q \tag{B9.5}$$

where

$$F(\hat{X}, u) \triangleq \left.\frac{\partial f(X, u)}{\partial X}\right|_{X = \hat{X}_{k+1|k}} \tag{B9.6}$$

is the Jacobian of $f(X,u)$ defined for the end of the sampling interval

t_{k+1} and $Q(t)$ is the PSD matrix of white process noise $W(t)$,

$$E[W(t)W(t-\tau)] = Q(t)\delta(t-\tau) \tag{B9.7}$$

Equation (B9.5) must be solved for P_{k+1} from the instant t_k to t_{k+1} under the initial condition

$$P(t_k) = P_{k|k} \tag{B9.8}$$

3) Solve for the optimum gain matrix K_{k+1} by means of

$$K_{k+1} = P_{k+1|k}H^T[HP_{k+1|k}H^T + V_k]^{-1} \tag{B9.9}$$

where V_k is the covariance matrix of the white measurement noise v_k. With this updated gain matrix, compute the updated estimation error

$$x_{k+1|k+1} = K_{k+1}[Z_{k+1} - H\hat{X}_{k+1|k}] \tag{B9.10}$$

and update the estimate $X_{k+1|k+1}$,

$$\hat{X}_{k+1|k+1} = \hat{X}_{k+1|k} + \hat{x}_{k+1|k+1} \tag{B9.11}$$

Finally, update the covariance matrix P

$$P_{k+1|k+1} = [I-K_{k+1}H]P_{k+1|k}[I-K_{k+1}H]^T + K_{k+1}V_kK_{k+1}^T \tag{B9.12}$$

The detailed derivation of (B9.9)–(B9.12), which constitutes the actual continuous-discrete (EKF), can be found in Chapter 6 of Gelb [21].

Appendix C9: Aircraft aerodynamic coefficients

1) Dimensions, mass and moments of inertia:

$c = 0.54$ m
$b = 4.2$ m
$S = 2.27$ m
$m = 110$ kg
$I_x = 185$ kg m sec^2
$I_y = 24.4$ kg m sec^2
$I_z = 37.9$ kg m sec^2
$I_{xz} = 0$

2) Aerodynamic coefficients:

$C_{l_{\delta_r}} = 0.015$
$C_k = 0.035$
$C_{L_0} = -0.22$
$C_{L_\alpha} = -5.31$
$C_{L_\beta} = 0$
$C_{L_q} = 0$
$C_{L_{\delta_e}} = -0.7$
$C_{m_0} = 0.09$
$C_{m_\beta} = 0$
$C_{m_q} = -17$
$C_{m_{\delta_e}} = -0.8$
$C_{m_{\dot{\alpha}}} = -3$
$C_{Y_\alpha} = 0$
$C_{Y_\beta} = -0.58$
$C_{Y_p} = 0$

$C_{l_{\delta_a}} = 0.1$
$C_{n_\alpha} = 0$
$C_{n_\beta} = 0.19$
$C_{n_p} = -0.125$
$C_{n_r} = -0.125$
$C_{n_{\delta_e}} = 0$
$C_{n_{\delta_r}} = -0.16$
$C_{n_{\delta_a}} = -0.065$
$C_{Y_r} = 0$
$C_{Y_{\delta_a}} = 0$
$C_{l_\alpha} = 0$
$C_{l_\beta} = -0.08$
$C_{l_r} = 0.26$
$C_{l_p} = -0.65$
$C_{l_{\delta_e}} = 0$

References

1. Brown, R.G. "Integrated Navigation Systems and Kalman Filtering: A perspective," *Navigation, J. Institute of Navigation*, Vol. 19, No. 4, Winter 1972, pp. 355–362.

2. Kalman, R.E. "A New Approach to Linear Filtering and Prediction Problems," *Transactions ASME, J. Basic Engineering*, Vol. 82D, March 1960, pp. 34-45.

3. Papoulis, A., *Probability, Random Variables, and Stochastic Processes*, third edition, McGraw-Hill, New York, 1991.

4. Widrow, B., et al, "Adaptive Noise Cancelling: Principles and Applications," *Proceedings, IEEE*, Vol. 63, Dec. 1975, pp. 1692–1716.

5. Higgins, W.T. "A Comparison of Complementary and Kalman Filtering," *IEEE Transactions on Aerospace and Electronic Systems*, Vol. AES-11, No. 3, May 1975, pp. 321–325.

6. Kayton, M. and Fried, W.R. , *Avionics Navigation Systems*, John Wiley & Sons, New York, 1969.

7. Garg, S.C., Morrow, L.D., and Mamen R., "Strapdown Navigation Technology: A Literature Survey," *AIAA Journal of Guidance and Control*, Vol. 1, No. 3, May–June 1978, pp. 161–172.

8. Harris, C.M., *Shock and Vibration Handbook*, McGraw-Hill, New York, 1989.

9. Merhav, S.J., and Bresler, Y., "On-line Vehicle Motion Estimation from Visual Terrain Information, Pt. I: Recursive Image Registration; Pt. II: Ground Velocity and Position Estimation," IEEE Transactions on Aerospace annd Electronic Systems, Vol. AES-22, No. 5, September 1986, pp. 583–604.

10. Parkinson, W.P., "History and Operation of NAVSTAR, the Global Positioning System," *IEEE Transactions on Aerospace and Electronic Systems*, Vol. 30, No. 4, October 1994, pp. 1145–1161.

11. Merhav, S., and Velger, M., "Compensating Sampling Errors in Stabilized Helmet Mounted Displays Using Auxiliary Acceleration Measurements," *AIAA Journal of Guidance, Control and Dynamics*, Vol. 14, No. 5, Sept.-Oct. 1991, pp. 1067–1069.

12. Lifshitz S., and Merhav, S., "Adaptive Suppression of Biodynamic Interference in Helmet-Mounted Displays and Head Teleoperation," *AIAA Journal of Guidance Control and Navigation*, Vol. 14, No. 6, Nov.–Dec. 1991, pp 1173–1180.

13. Meditch, J.S., *Stochastic Optimal Linear Estimation and Control*, McGraw-Hill, New York, 1969.,

14. Friedland, B., *Control System Design, an Introduction to State-Space Methods*, McGraw-Hill, New York, 1986.

15. Sorensen, J.A. "Laboratory Demonstration of Aircraft Estimation Using Low-Cost Sensors," NASA CR-152049, prepared for Ames Research Center, Moffett Field, CA, 1975.

16. Bar Itzhack, I.Y., and Ziv, I., "Frequency and Time Domain Designs of a Strapdown Vertical Determination System," *Proceedings AIAA Guidance, Navigation and Control Conference*, Williamsburg, August 1986, paper 86-2149.

17. Pietila, R. and Dunn, W.R., "A Vector Autopilot System," *IEEE Trans. on Aerospace and Electronic Systems*, Vol. AES 12, No. 3, 1976, pp. 341–347.

18. Daniel, J.A., Knox, J.R., and Raney, L.H., "Advances in Flight Control Navigation and Terminal Guidance Development for a Low-Cost Multifunction Unmanned Air Vehicle System," *Unmanned Systems*, Winter, 1989/90, pp. 17–25.

19. Koifman, M., and Merhav S., "Autonomously aided Strapdown Attitude Reference System," *AIAA Journal of Guidance, Dynamics and Control*, Vol. 14, No. 6, Nov.–Dec. 1991, pp. 1164–1172.

20. Etkin, B., *Dynamics of Atmospheric Flight*, John Wiley & Sons, New York, 1972.

21. Gelb, A., *Applied Optimal Estimation*, Chapter 6, MIT Press, Cambridge MA, 1974.

22. Hinkley, D.V., "Inference about the Change Point from Cumulative Sum Tests," *Biometrica*, Vol. 58, No. 3, 1971, pp. 509–523.

23. Basseville, M., et al., Pt. I, "Edge Detection using Sequential Methods for Change in Level,"; Pt. II, "Sequential Detection of Change in Mean," *IEEE Transactions on Acoustics, Speech and Signal Processing*, Vol. ASSP-29, No. 1, 1981, pp. 32–50.

Index